Amphibian Ecology and Conservation

Techniques in Ecology and Conservation Series

Series Editor: William J. Sutherland

Bird Ecology and Conservation: A Handbook of Techniques
William J. Sutherland, Ian Newton, and Rhys E. Green

Conservation Education and Outreach Techniques
Susan K. Jacobson, Mallory D. McDuff, and Martha C. Monroe

Forest Ecology and Conservation: A Handbook of Techniques
Adrian C. Newton

Habitat Management for Conservation: A Handbook of Techniques
Malcolm Ausden

Conservation and Sustainable Use: A Handbook of Techniques
E.J. Milner-Gulland and J. Marcus Rowcliffe

Invasive Species Management: A Handbook of Techniques
Mick N. Clout and Peter A. Williams

Amphibian Ecology and Conservation: A Handbook of Techniques
C. Kenneth Dodd, Jr

Amphibian Ecology and Conservation

A Handbook of Techniques

Edited by

C. Kenneth Dodd, Jr

OXFORD
UNIVERSITY PRESS

Great Clarendon Street, Oxford, OX2 6DP,
United Kingdom

Oxford University Press is a department of the University of Oxford.
It furthers the University's objective of excellence in research, scholarship,
and education by publishing worldwide. Oxford is a registered trade mark of
Oxford University Press in the UK and in certain other countries

© Oxford University Press, 2010

The moral rights of the authors have been asserted

First published 2010

Published in the United States of America by Oxford University Press
198 Madison Avenue, New York, NY 10016, United States of America

British Library Cataloguing in Publication Data

Data available

Library of Congress Cataloging in Publication Data

Data available

ISBN 978–0–19–954119–5

Preface

As this volume is completed, more than 6400 amphibian species have been recognized, with new taxa being described nearly every day. The last few decades have seen an explosion in systematic research, particularly in the tropics. Long-recognized centers of diversity have been explored using increasingly sophisticated sampling techniques, yielding many new taxa. At the same time, new centers of speciation, such as Sri Lanka and the Western Ghats of India, have been discovered, while molecular techniques have yielded previously unsuspected diversity within some well-known taxa, such as the plethodontid salamanders of southeastern North America and the green toad (*Bufo viridis*) complex of Eurasia. For amphibian systematists, these are exciting times.

Unfortunately, amphibians are now at greater peril than at any time in recent geologic history, a situation chronicled in two recent data-rich books (Lannoo 2005; Stuart *et al.* 2008). Habitats are being lost at alarming rates because of expanding human populations and generally favorable economic conditions fostering development; emerging infectious diseases, particularly amphibian chytrid fungus (*Batrachochytrium dendrobatidis*), threaten worldwide impacts; non-indigenous species proliferate, affecting amphibians and their habitats; and amphibians, with their permeable skins, diverse life histories, and often biphasic life cycles requiring both terrestrial and aquatic habitats, are being saturated by a host of lethal and sublethal toxic substances. New threats, such as the effects of global climate change, further imperil amphibians, especially those with limited distributions and dispersal capabilities. Fully one-third of all amphibians are now considered threatened (Stuart *et al.* 2004), and 168 species have become extinct within the last two decades. Clearly, these are treacherous times for many frogs, salamanders, and caecilians.

Amphibians are, quite frankly, engaging animals. Despite Linnaeus' early characterization of amphibians in the context of "Terrible are thy works, O God", biologists have come to appreciate that their diverse life histories and shear numbers offer a wealth of material for research on basic ecological principles, such as trophic interactions, phenotypic plasticity, predator–prey interactions, community structure, mate choice and recognition, water balance, and many others. In response to threats, conservation biologists have probed these and other questions in hopes of understanding amphibian biology in order to prevent declines and extinctions. The basic and applied themes of biology merge in these disciplines: understanding ecology leads to conservation options (see Gascon *et al.* 2007), and conservation-based research leads to a better appreciation of ecological principles.

To say that there are a great many techniques available in ecological and conservation-based research on amphibians is an understatement. The pages of journals such as *Herpetological Review* and *Applied Herpetology* contain techniques papers with every issue. Specialized books, such Heyer *et al.* (1994), Henle and Veith (1997), and Gent and Gibson (1998), offer additional summaries that are as applicable today as when they were published. No one volume can include all techniques. The current volume is meant not to supplant these earlier works, but to supplement them and add new areas not previously summarized, such as occupancy modeling, landscape ecology, genetics, telemetry, and disease biosecurity. Our objectives have been to delineate important new developments, to give an idea as to what the techniques tell or do not tell a researcher, to focus attention on biases and data inference, and to get readers to appreciate sampling as an integral part of their science, rather than just a means of capturing animals. The techniques used will set the boundaries within which results can or should be interpreted.

As noted earlier, amphibian systematics is a flourishing field, with many new opportunities made available by combining large datasets using molecular and morphological data with powerful computer analysis. The phylogeny of amphibians is undergoing increasingly sophisticated analysis. Some analyses, such as those of Frost *et al.* (2006), suggest relationships that differ substantially from "traditional" concepts. If accepted, extensive nomenclatural changes will be warranted. Although Frost *et al.* (2006) have advocated substantive changes in nomenclature, many of which will likely be accepted with further study, other amphibian biologists disagree with automatically accepting every change proposed by these authors. In this book, I have decided to retain the older nomenclature rather than make a taxonomic decision each time a name is mentioned. The intended audience of this volume (biologists starting their careers in ecology and conservation) likely will be more familiar with the older generic names of *Bufo*, *Rana*, and *Hyla*, and initially may be confused by the unfamiliar replacement names. Readers should be aware, however, that names such as *Lithobates* (= *Rana*, in part), *Anaxyrus* (= *Bufo*, in part), and others currently unfamiliar may soon be more commonplace.

I wish to thank the following for taking their valuable time to review manuscripts and offer suggestions for improving this volume: James Austin, Larissa Bailey, Bruce Bury, Dan Cogălniceanu, Sarah Converse, Steve Corn, Rafael Ernst, Alisa Gallant, Marian Griffey, Kerry Griffis-Kyle, Richard Griffiths,

Margaret Gunzburger, Tibor Hartel, Robert Jehle, Steve Johnson, Y.-C. Kam, Sarah Kupferberg, Frank Lemckert, Harvey Lillywhite, John Maerz, Joseph Mitchell, Clinton Moore, Erin Muths, James Petranka, Benedikt Schmidt, Ulrich Sinsch, Kevin Smith, Lora Smith, Joseph Travis, Susan Walls, and Matthew Whiles. I greatly appreciate the support from Ian Sherman and Helen Eaton at Oxford University Press and editorial help from freelance copy-editor Nik Prowse, and thank series editor, Bill Sutherland, for inviting me to edit the amphibian volume. This volume is dedicated to all the biologists who take up the challenge of amphibian ecology and conservation.

C. Kenneth Dodd, Jr

References

Frost, D. R., Grant, T., Faivovich, J., Bain, R. H., Haas, A., Haddad, C. F. B., de Sá, R. O., Channing, A.,Wilkinson, M., Donnellan, S. C. *et al.* (2006). The amphibian tree of life. *Bulletin of the American Museum of Natural History*, **297**, 1–370.

Gascon, C., Collins, J. P., Moore, R. D., Church, D. R., McKay, J. E., and Mendelson, III, J. R. (eds) (2007). *Amphibian Conservation Action Plan.* IUCN/SSC Amphibian Specialist Group, Gland and Cambridge.

Gent, T. and Gibson, S. (eds) (1998). *Herpetofauna Worker's Manual.* Joint Nature Conservation Committee, Peterborough.

Henle, K. and Veith, M. (eds) (1997). *Naturschutzrelevante Methoden der Feldherpetologie.* Mertensiella 7.

Heyer, W. R., Donnelly, M. A., McDiarmid, R. W., Hayek, L.-A., and Foster, M. S. (eds). (1994). *Measuring and Monitoring Biological Diversity. Standard Methods for Amphibians.* Smithsonian Institution Press, Washington DC.

Lannoo, M. J. (ed) (2005). *Amphibian Decline. The Conservation Status of United States Species.* University of California Press, Berkeley, CA.

Stuart, S., Chanson, J. S., Cox, N. A., Young, B. E., Rodrigues, A. S. L., Fishman, D. L., and Waller, R. W. (2004). Status and trends of amphibian declines and extinctions worldwide. *Science*, **306**, 1783–6.

Stuart, S., Hoffman, M., Chanson, J. S., Cox, N. A., Berridge, R. J., Ramani, P., and Young, B. E. (eds) (2008). *Threatened Amphibians of the World.* Lynx Edicions, Barcelona; IUCN, Gland; Conservation International, Arlington, VA.

Contents

5 Dietary assessments of larval amphibians   **71**

Matt R. Whiles and Ronald Altig

14 Area-based surveys 247

David M. Marsh and Lillian M.B. Haywood

15 Rapid assessments of amphibian diversity 263

James R. Vonesh, Joseph C. Mitchell, Kim Howell, and
Andrew J. Crawford

16 Auditory monitoring of anuran populations   281

Michael E. Dorcas, Steven J. Price, Susan C. Walls, and William J. Barichivich

17 Measuring habitat   299

Kimberly J. Babbitt, Jessica S. Veysey, and George W. Tanner

Part 5 Amphibian communities   319

18 Diversity and similarity   321
C. Kenneth Dodd, Jr

19 Landscape ecology and GIS methods   339
Viorel D. Popescu and James P. Gibbs

Part 6 Physiological ecology and genetics 361

22 Genetics in field ecology and conservation   407

Trevor J. C. Beebee

26 Disease monitoring and biosecurity   481

D. Earl Green, Matthew J. Gray, and Debra L. Miller

27 Conservation and management   507

C. Kenneth Dodd, Jr

Contributors

Ross A. Alford School of Marine and Tropical Biology, James Cook University, Townsville, Queensland 4811, Australia. E-mail: ross.alford@jcu.edu.au

Ronald Altig Department of Biological Sciences, Mississippi State University, Mississippi State, Mississippi 39762-5759, USA. E-mail: raltig@biology.msstate.edu

Kimberly J. Babbitt Department of Natural Resources and the Environment, University of New Hampshire, Durham, NH 03824, USA. E-mail: kbabbitt@cisunix.unh.edu

Larissa L. Bailey Department of Fish, Wildlife and Conservation Biology, Colorado State University, Fort Collins, CO 80523, USA. E-mail: Larissa.Bailey@colostate.edu

William J. Barichivich Florida Integrated Science Center, US Geological Survey, 7920 NW 71st Street, Gainesville, FL 32653, USA. E-mail: wbarichivich@usgs.gov

Trevor J. C. Beebee School of Life Sciences, University of Sussex, Falmer, Brighton BN1 9QG, UK. E-mail: t.j.c.beebee@sussex.ac.uk

Michelle D. Boone Department of Zoology, 212 Pearson Hall, Miami University, Oxford, Ohio 45056, USA. E-mail: BooneMD@muohio.edu

Andrew J. Crawford Smithsonian Tropical Research Institute, Apartado Postal 0843-03092, Balboa, Ancón, Republic of Panama. E-mail: andrew@dna.ac

Dan Cogălniceanu University Ovidius Constanţa, Faculty of Natural Sciences, Bvd. Mamaia 124, Constanţa, Romania. E-mail: dcogalniceanu@univ-ovidius.ro

Paul Stephen Corn U.S. Geological Survey, Aldo Leopold Wilderness Research Institute, 790 E. Beckwith Ave., Missoula, MT 59801, USA. E-mail: scorn@usgs.gov

Martha L. Crump Department of Biological Sciences, Northern Arizona University, Flagstaff, Arizona 86011, USA. E-mail: marty.crump@ nau.edu

C. Kenneth Dodd, Jr Department of Wildlife Ecology and Conservation, University of Florida, Gainesville, FL 32611, USA. E-mail: Terrapene600@gmail.com

Michael E. Dorcas Department of Biology, Davidson College, Davidson, NC 28035, USA. E-mail: midorcas@davidson.edu

John W. Ferner Department of Biology, Thomas More College, Crestview Hills, Kentucky 41017, USA. E-mail: john.ferner@thomasmore.edu

J. Whitfield Gibbons Savannah River Ecology Laboratory, P.O. Drawer E, Aiken, South Carolina 29802, USA. E-mail: wgibbons@uga.edu

James P. Gibbs Department of Environmental and Forest Biology, State University of New York, Syracuse, NY 13210, USA. E-mail: jpgibbs@esf.edu

Matthew J. Gray Center for Wildlife Health, Department of Forestry, Wildlife and Fisheries, University of Tennessee, 274 Ellington Plant Sciences Building, Knoxville, TN 37996, USA. E-mail: mgray11@utk.edu

D. Earl Green U.S. Geological Survey, National Wildlife Health Center, 6006 Schroeder Drive, Madison, WI 53711, USA. E-mail: david_green@usgs.gov

Elizabeth B. Harper State University of New York, College of Environmental Sciences and Forestry, 1 Forestry Drive, Syracuse, New York 13210, USA. E-mail: eharper@esf.edu

Reid N. Harris Department of Biology, James Madison University, Harrisonburg, Virginia 22807, USA. E-mail: harrisRN@jmu.edu

Lillian M. B. Haywood Department of Biology, Washington and Lee University, Lexington, VA 24450, USA. E-mail: haywoodl@wlu.edu

Kim Howell Department of Zoology and Wildlife Conservation, PO Box 35064, University of Dar es Salaam, Dar es Salaam, Tanzania. E-mail: kmhowell@udsm.ac.tz

Victor S. Lamoureux Binghamton University, PO Box 6000, Binghamton, New York 13902, USA. E-mail: vlamoureux@stny.rr.com

Harvey B. Lillywhite Department of Zoology, University of Florida, Gainesville, FL 32611, USA. E-mail: hbl@zoo.ufl.edu

Dale M. Madison Binghamton University, PO Box 6000, Binghamton, New York 13902, USA. E-mail: dmadison@binghamton.edu

David M. Marsh Department of Biology, Washington and Lee University, Lexington, VA 24450, USA. E-mail: marshd@wlu.edu

Roy W. McDiarmid Patuxent Wildlife Research Center, US Geological Survey, National Museum of Natural History, Washington DC 20036, USA. E-mail: mcdiarmidr@si.edu

Claude Miaud Laboratoire d' Alpine UMR CNRS 5553, Université de Savoie, 73376 Le Bourget-du-Lac, France. E-mail: claude.miaud@univ-savoie.fr

Debra L. Miller Veterinary Diagnostic and Investigational Laboratory, The University of Georgia, College of Veterinary Medicine, 43 Brighton Road, Tifton, GA 31793, USA. E-mail: millerdl@uga.edu

Joseph C. Mitchell Mitchell Ecological Research Service, LLC, PO Box 5638, Gainesville, FL 32627–5638, USA. E-mail: dr.joe.mitchell@gmail.com

James D. Nichols U.S. Geological Survey, Patuxent Wildlife Research Center, 12100 Beech Forest Road, Laurel, MD 20708, USA. E-mail: jnichols@usgs.gov

Peter W. C. Paton Department of Natural Resources, University of Rhode Island, Kingston, Rhode Island 02881, USA. E-mail: ppaton@uri.edu

Joseph H. K. Pechmann Department of Biology, 132 Natural Sciences Building, Western Carolina University, Cullowhee, North Carolina 28723, USA. E-mail: jpechmann@email.wcu.edu

Jérôme Pellet A. Mailbach Sàrl, CP 99, Ch. de la Poya 10, CH-1610 Oron-la-Ville, Switzerland. E-mail : jerome.pellet@amailbach.ch

James W. Petranka Department of Biology, University of North Carolina at Asheville, Asheville, North Carolina 28804, USA. E-mail: petranka@unca.edu

Viorel D. Popescu Department of Wildlife Ecology, University of Maine, Orono, ME 04469, USA. E-mail: dan.popescu@umit.maine.edu

Steven J. Price Department of Biology, Wake Forest University, Winston-Salem, NC 27109, USA. E-mail: sjprice@davidson.edu

Jonathan L. Richardson School of Forestry and Environmental Studies, Yale University, 370 Prospect Street, New Haven, Connecticut 06511, USA. E-mail: jonathan.richardson@yale.edu

Dennis Rödder Zoological Research Museum Alexander Koenig, Adenauerallee 160, D-53113, Bonn, Germany and Faculty of Geography/Geosciences, Trier University, Wissenschaftspark Trier-Petrisberg, Am Wissenschaftspark 25–27, D-54286 Trier, Germany. E-mail: d.roedder.zfmk@uni-bonn.de

Jodi J. L. Rowley School of Marine and Tropical Biology, James Cook University, Townsville, Queensland 4811, Australia. E-mail: Jodi.Rowley@austmus.gov.au

Benedikt R. Schmidt Zoologisches Institut, Universität Zürich, Winterthurerstrasse 190, CH-8057 Zürich, Switzerland. E-mail: bschmidt@zool.uzh.ch

Raymond D. Semlitsch Division of Biological Sciences, 105 Tucker Hall, University of Missouri, Columbia, Missouri 65211, USA. E-mail: SemlitschR@missouri.edu

David K. Skelly School of Forestry and Environmental Studies, Yale University, 370 Prospect Street, New Haven, Connecticut 06511, USA. E-mail: david.skelly@yale.edu

Mirco Solé Department of Biology, Universidade Estadual de Santa Cruz, Rodovia Ilhéus – Itabuna, km 16, Ilhéus, Bahia, Brazil. E-mail: mksole@uesc.br

Donald W. Sparling Cooperative Wildlife Research Laboratory, Life Science II, MS6504, Southern Illinois University, Carbondale, Illinois 62901, USA. E-mail: dsparl@siu.edu

George W. Tanner Department of Wildlife Ecology and Conservation, University of Florida, Gainesville, FL 32611, USA. E-mail: tannerg@ufl.edu

Valorie R. Titus Binghamton University, PO Box 6000, Binghamton, New York 13902, USA. E-mail: vtitus1@binghamton.edu

Jessica S. Veysey Department of Natural Resources and the Environment, University of New Hampshire, Durham, NH 03824, USA. E-mail: jss4@unh.edu

James R. Vonesh Department of Biology, Virginia Commonwealth University, 1000 West Cary Street, Richmond, VA 23284, USA. E-mail: jrvonesh@vcu.edu

Susan C. Walls Florida Integrated Science Center, US Geological Survey, 7920 NW 71st Street, Gainesville, FL 32653, USA. E-mail: swalls@usgs.gov

Matt R. Whiles Department of Zoology and Center for Ecology, Southern Illinois University, Carbondale, Illinois 62901-6501, USA. E-mail: mwhiles@zoology.siu.edu

John D. Willson Savannah River Ecology Laboratory, P.O. Drawer E, Aiken, South Carolina 29802, USA. E-mail: willson@uga.edu

Part 1
Introduction

Amphibian diversity and life history

Martha L. Crump

When the first crossopterygian crawled out of the rich Devonian waters and cast the first envious vertebrate gaze at the terrestrial world, a boundless empire awaited colonization. Although the change from an ungainly lobe-finned locomotion to a terrestrial walking gait . . . was agonizingly slow, generations succeeded generations, archotypes [*sic*] gave way to new evolutionary experiments, and the land became the home for the first quadrupeds—the amphibians.

William E. Duellman (1970)

1.1 Introduction

Over the past 350 million years, amphibian descendents of lobe-finned fishes have radiated into most habitats on Earth. In doing so, they have acquired spectacular and sometimes bizarre physiological, morphological, behavioral, and ecological attributes that mold their innovative life histories. Amphibians have highly permeable skin, which makes them both vulnerable to losing water and able to absorb water. Their eggs, covered with jelly capsules rather than hard shells, lose water rapidly. For these reasons, amphibians require relatively moist environments.

Many sampling techniques have been developed in North America or Europe where most amphibians exhibit the complex life cycle of aquatic eggs, aquatic larvae, and metamorphosis into terrestrial adults that return to water to breed. Not all amphibians fit this stereotype. The spectacular diversity of amphibian life histories provides a focus for studying their natural history, as well as presents a challenge since researchers must ensure that field methods are appropriate for the target species.

1.2 Amphibian species richness and distribution

Scientists have named approximately 1.75 million species of living organisms (Groom *et al.* 2006). About 72% of all named animals are insects; about 5% are vertebrates. Approximately 0.5% of all animal species—6347 species (see AmphibiaWeb; http://amphibiaweb.org)—belong to the class Amphibia. The class gets it name from the Greek words *amphi* meaning "two" and *bios* meaning "mode of life" because many species have a diphasic life history: they spend part of their lives in water and part on land. Biologists divide the class into three orders: Gymnophiona (caecilians), Urodela (salamanders), and Anura (frogs) (Figure 1.1).

Gymnophiona, from the Greek words *gymnos* and *ophis* meaning "naked serpent," encompasses 174 species. Caecilians, which resemble large earthworms, are long, skinny animals with no legs and reduced eyes. Annuli (grooves) encircle their bodies. Their tails are either greatly reduced or absent, and a sensory tentacle sits between each eye and nostril. Some caecilians have small dermal scales beneath the surface of their mucus-covered skin. These scales, composed mainly of collagen fibers and minerals, are not found in salamanders or anurans. Adult caecilians range in length from a little more than 10 cm to about 1.5 m. Most are highly specialized for burrowing. Others live on the ground but are cryptic and secretive. Some are aquatic. Caecilians occur in tropical habitats around the world except for Madagascar and the Papuan–Australian region (Pough *et al.* 2004).

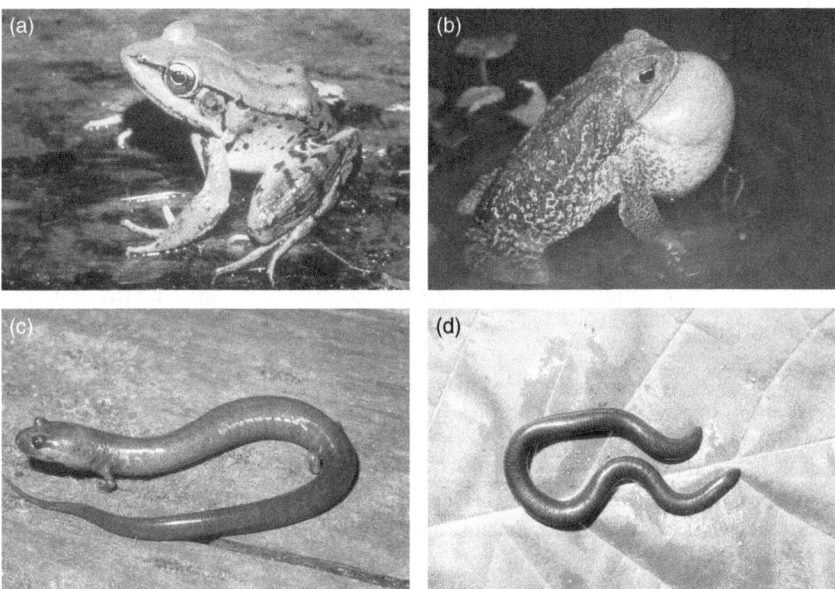

Fig. 1.1 Representatives of the three orders of amphibians. (a) Anura: *Rana palmipes*, from Ecuador, (b) Anura: *Bufo arenarum*, from Argentina, (c) Urodela: *Phaeognathus hubrichti*, from Alabama, USA, (d) Gymnophiona: *Hypogeophis rostratus*, Seychelles. Photographs (a) and (b) by Martha L. Crump; photographs (c) and (d) by C. Kenneth Dodd, Jr.

Five hundred and seventy-one species belong to Urodela, from the Greek words *uro* and *delos* meaning "tail evident." All salamanders have tails, and adults have elongate bodies. Most have front and back legs of about the same length; the limbs of a few aquatic species are greatly reduced or absent. Salamanders are completely aquatic, terrestrial, combined aquatic and terrestrial, fossorial, or arboreal. Adults range in size from about 30 mm to nearly 2 m. Most salamander species occur in eastern and western North America and temperate Eurasia, although plethodontids have radiated extensively in Central and South America. There are no salamanders in sub-Saharan Africa, Australasia, Australia, or much of tropical Asia, and they are missing from most islands (Pough *et al.* 2004).

Frogs, which include toads, make up the order Anura from the Greek words *an* and *oura*, meaning "without tail." Although tadpoles have tails, adults do not. Anurans live in the water, on the ground, underground, and in the trees. Most have long, strong back legs well suited for jumping. Males of most species call to attract females for mating. Adults of the 5602 recognized species range in size from about 13 mm to 30 cm. Anurans live almost everywhere except where restricted by cold temperatures or extremely dry conditions, and except for many oceanic islands (Pough *et al.* 2004).

Duellman identified 43 areas worldwide with exceptionally high numbers of amphibian species, endemic species (those found nowhere else), or both (Duellman 1999). Nineteen of these high diversity areas are in the western hemisphere; the others are in Eurasia, Africa, and the Papuan–Australian region. The neotropical region houses 54% of the world's amphibian species.

Amphibians live in nearly every habitat except for open oceans, most oceanic islands, polar regions, and some extremely dry deserts (Wells 2007). These restrictions are imposed on them because of their highly permeable skin that loses water, and because they are ectothermic: the energy needed to raise their body temperatures comes from the sun. Thus, amphibians become inactive at low temperatures. These characteristics, however, work to their advantage as well. In dry areas and those with seasonal rainfall, amphibians absorb water through their skin by contacting moist soil. Their low metabolic rates translate into low energy requirements and allow them to estivate, often underground, during unfavorable conditions.

1.3 Amphibian lifestyles and life history diversity

Textbooks, management guides, and monitoring manuals, especially those originating in Europe and North America, often give an oversimplified impression of amphibian life histories. In fact, amphibian life histories are often complex, and many are still poorly understood. Not all species have an aquatic larval stage and a terrestrial adult stage.

Reproductive mode, a central aspect of amphibian life history, refers to the site of egg deposition, egg and clutch characteristics, type and duration of embryonic and larval development, and type of parental care if any (Duellman and Trueb 1986). Many amphibians are not tied to aquatic habitats for reproduction. Instead, they reproduce on land, underground, or in trees, even in the temperate zones. The following brief discussion of selected lifestyles and life histories reveals that similar behaviors have evolved in diverse taxonomic groups and geographical areas: amphibian "experiments" toward greater independence from standing or flowing water and perhaps from lower predation pressure as well.

Readers wishing more information concerning amphibian life histories should consult Duellman (2007) and Wells (2007). For reviews of reproductive modes, see Salthe (1969), Salthe and Duellman (1973), Wake (1982, 1992), Haddad and Prado (2005), and Duellman (2007). For reviews of parental care see Crump (1995, 1996).

1.3.1 Caecilians

Caecilian lifestyles include aquatic, combined aquatic and terrestrial, terrestrial, and fossorial. Fully aquatic caecilians generally have compressed bodies with well-developed dorsal fins on the posterior portion. Fossorial species generally have blunt heads, used for pushing and compacting the soil while burrowing.

Although details of reproductive biology are unknown for many species, caecilians display two basic modes: oviparous (egg-laying) and viviparous (bearing live young) (Wake 1977, 1992). Oviparous caecilians lay eggs on land. In some species the eggs hatch into larvae that wriggle to water. Caecilian larvae exhibit less dramatic metamorphosis than do salamanders or frogs. They hatch almost fully developed, and the larval period is short. The eggs of some oviparous species undergo direct development. That is, development occurs within the egg capsule; there is no free-living larval stage. Some oviparous females stay with their eggs, which probably protects them from predators and from drying out. Viviparity evolved independently several times in caecilians. Females retain eggs in their oviducts until development is complete (Wake 1993; Wilkinson and Nussbaum 1998). After the developing young exhaust their yolk reserves, they scrape lipid-rich secretions from their mothers' oviductal epithelium with their fetal teeth. Gestation lasts for many months, and the newborn are large relative to their mothers.

1.3.1.1 Aquatic

All species in the South American family Typhlonectidae are either fully aquatic or semi-aquatic. Some in the latter group spend the day in burrows they construct next to water, then emerge at night to feed in shallow water. All typhlonectids are assumed to be viviparous. Soon after the young are born, they shed their gills and quickly acquire adult morphology.

1.3.1.2 Combination aquatic and terrestrial

Caecilians of the two most primitive families—Ichthyophiidae and Rhinotrematidae—all appear to have complex life cycles encompassing both water and land. As far as is known, ichthyophiids from southeastern Asia lay their eggs in burrows or under vegetation at the edge of water. The female coils around her eggs. The hatchling larvae wriggle to water. Rhinotrematids from South America likewise appear to be oviparous, and free-living aquatic larvae have been found for some species. Although egg-laying sites are unknown, females presumably oviposit on land near water and the larvae make their way to water. The developmental mode is not known for uraeotyphlids, from southern India, but presumably they also lay eggs on land and have aquatic larvae. Some species of the large, widespread family Caeciliidae also oviposit on land and have free-living aquatic larvae.

1.3.1.3 Terrestrial and/or fossorial

Many caecilians burrow underground, although some forage at night on the ground surface. Some terrestrial and fossorial caeciliids from South America, Africa, India, and the Seychelles lay direct-developing eggs. In some of these, females have been found coiled around their eggs. Some terrestrial and fossorial caeciliids are viviparous. All members of the family Scolecomorphidae from Africa are terrestrial. One is viviparous; in the other five species the mode is unknown.

1.3.2 Salamanders

Salamander lifestyles include fully aquatic, combined aquatic and terrestrial, terrestrial, arboreal, and fossorial. Body shapes of aquatic salamanders range from the slender and eel-like sirenids and amphiumids to the flattened and robust cryptobranchids. Many permanently aquatic species retain external gills as adults. Most arboreal salamanders are small and have extensively webbed feet; some have prehensile tails. Burrowing salamanders generally have long slender bodies and tails, reduced limbs and feet, and small body size.

As a group salamanders exhibit diverse reproductive modes. Some salamanders go through a complex life cycle with aquatic larvae and metamorphosis. Others are paedomorphic: they become sexually mature and reproduce while retaining juvenile characteristics. Still others have direct-developing eggs or give birth to live young.

1.3.2.1 Aquatic

Some aquatic salamanders lay eggs in still or slowly flowing water. *Siren* and *Pseudobranchus* from southeastern USA and northeastern Mexico live in swamps, lakes, marshes, and sluggish streams, where they attach their eggs to vegetation. The larvae develop into eel-like paedomorphic adults: they lack eyelids, have external gills, and their skin resembles larval skin. When their habitat dries out, sirenids secrete a mucous cocoon and burrow into the mud where they estivate until conditions improve. The Mexican axolotl (*Ambystoma mexicanum*) is permanently aquatic and paedomorphic. Its larvae fail to metamorphose fully, and the gonads mature in the larval body form.

Proteus anguinus, from southeastern Europe, lives in subterranean lakes and streams of limestone caves where it lays aquatic eggs that hatch into aquatic larvae. Females attend their eggs. In North America, *Typhlomolge* and *Haideotriton* also live in cave waters and lay aquatic eggs that hatch into aquatic larvae. All these species are paedomorphic.

Some aquatic salamanders oviposit in flowing water of cold streams. Male hellbenders (*Cryptobranchus alleganiensis*) from eastern North America construct nests under rocks. More than one female might lay her strings of eggs in the nest. The male guards the eggs through their early stages. Likewise, male *Andrias japonicus* from Japan and male *Andrias davidianus* from central China guard their nests. In all three cryptobranchid species, aquatic larvae undergo incomplete metamorphosis. The adults retain certain larval features such as lidless eyes and the absence of a tongue pad.

1.3.2.2 Combination of aquatic and terrestrial

In Eurasia, many salamandrids (e.g. *Triturus*) live on land but lay their eggs in ponds, lakes, or streams. Likewise, most Asian hynobiids are terrestrial as adults but migrate to aquatic sites to lay aquatic eggs that hatch into aquatic larvae. The same is true for many North American terrestrial salamanders. Tiger salamanders (*Ambystoma tigrinum*) and spotted salamanders (*Ambystoma maculatum*) migrate to breeding ponds in the spring. Once the aquatic larvae metamorphose and leave the water, they return to their aqueous beginnings only during breeding season.

In contrast, female marbled salamanders (*Ambystoma opacum*) lay their eggs under leaf litter or logs in depressions on dry ground. They stay until their eggs hatch when the nests flood during winter rains. Following aquatic larval development and metamorphosis, marbled salamanders are completely terrestrial.

Female *Amphiuma* from the southeastern USA lay large yolky eggs in dried-out swamps or in cavities under logs near ponds: places that will flood during spring and summer rains. Females stay with their eggs until then. Once they hatch, the well-developed larvae metamorphose within a few weeks. Adults live in swamps or slow-moving streams, but move overland during rains. They survive droughts by burrowing underground for 2 years or more.

Some plethodontid salamanders lay eggs on land, and after hatching the larvae make their way to water. Female *Hemidactylium scutatum* oviposit just above or at the water line in bogs and swamps. After hatching, the larvae wriggle or flip into the water. Some *Desmognathus* lay their eggs under rocks, logs, or leaf litter at water's edge. The newly hatched larvae stay with their mother in the nest site for several days, then wriggle to water.

The complex life cycle of red-spotted newts (*Notophthalmus viridescens*) from eastern North America involves several stages. Aquatic females attach their eggs to underwater vegetation. The eggs hatch into aquatic larvae that eventually metamorphose into an immature terrestrial stage called an eft. The orange-red efts stay on land for 1–14 years. Eventually they return to ponds where they turn dull green with a row of small red dots along each side of their body and transform into the aquatic adult body form. In some populations, however, paedomorphic adults reproduce in the larval body form.

1.3.2.3 Terrestrial

Many species of salamander live under leaf litter or logs, retreating into crevices or holes during dry conditions. Terrestrial salamanders, including many in the temperate zone, have various ways of reproducing independent of water bodies. The seepage salamander (*Desmognathus aeneus*) oviposits in a nest near water. The larvae hatch at an advanced stage and metamorphose within a few days in the nest without feeding. Other plethodontid salamanders (e.g. *Desmognathus wrighti*, *Bolitoglossa*, and *Plethodon*) lay large, direct-developing eggs in cavities or inside hollow logs. In many species, the female remains with the eggs; in a few species the male remains instead. Montane populations of European fire salamanders (*Salamandra salamandra*) often retain eggs in their oviducts. The young absorb nutrients from their yolk and are born live, although lowland populations usually have aquatic eggs and larvae. *Salamandra atra* are viviparous; after the developing young exhaust their yolk reserves, they obtain nutrients from the female.

1.3.2.4 Fossorial

Most *Oedipina*, fossorial or semi-fossorial plethodontid salamanders ranging from Mexico to northern South America, lay direct-developing eggs underground. Some fossorial plethodontids (e.g. *Lineatriton lineola*) attend their direct-developing eggs.

1.3.2.5 Arboreal

Neotropical arboreal plethodontids (e.g. *Bolitoglossa* and *Nototriton*) lay their eggs under mats of mosses and liverworts on tree branches and in bromeliads. The eggs undergo direct development, and some female *Bolitoglossa* attend their eggs. *Aneides lugubris* from western North America lays direct-developing eggs as high as 10 m above the ground.

1.3.3 Anurans

Anuran lifestyles include purely aquatic, aquatic and terrestrial, terrestrial, arboreal, and fossorial. A frog with relatively short hind legs is most likely a terrestrial hopper or fossorial species. One with long hind legs is likely to be aquatic, arboreal, or a jumping terrestrial species. Aquatic anurans tend to have their eyes on the top of their heads rather than at the sides, and they often have fully webbed feet. Some have flattened bodies. Arboreal frogs generally have expanded pads on the ends of their toes. Many fossorial species have small heads with pointed snouts and depressed bodies. Some have spadelike tubercles on their hind feet used for burrowing.

Anurans have evolved remarkably diverse life histories, from aquatic eggs to viviparity. In between those extremes, frogs from numerous families on many continents lay their eggs out of water yet have aquatic larvae. In the section headings below, the first designation refers to post-metamorphic stages, the second to egg and/or larval stages.

1.3.3.1 Aquatic/aquatic

Many aquatic frogs lay eggs that hatch into tadpoles. African clawed frogs (*Xenopus*) attach their eggs to submerged vegetation in standing water. The South American paradox frog (*Pseudis paradoxa*) oviposits among vegetation in shallow water of ponds and lakes. Aquatic larvae grow to 25 cm—the largest of any frog—then they metamorphose into relatively small juveniles. Tailed frogs (*Ascaphus truei*) from northwestern USA and adjacent Canada live in cold, torrential streams where they lay eggs under rocks. In some areas larvae require several years to metamorphose.

Some fully aquatic anurans brood their young. Female South American *Pipa* carry eggs embedded in their backs. In some species of *Pipa* the eggs hatch into tadpoles. In others they undergo direct development. Female Australian gastric-brooding frogs (*Rheobatrachus*) swallowed their late-stage eggs or early-stage larvae, and the tadpoles absorbed yolk reserves while developing in their mothers' stomachs. No *Rheobatrachus* have been seen since the 1980s; both species are assumed extinct.

1.3.3.2 Terrestrial/aquatic

Terrestrial anurans from many families lay aquatic eggs, and the aquatic larvae metamorphose into terrestrial juveniles. Species that oviposit in standing water produce clutches in compact masses (some ranids), floating rafts on the water surface (some ranids), strings (most *Bufo* and some pelobatids), scattered individually or in small packets on the bottom substrate (*Bombina* and *Discoglossus*), attached individually or in small groups onto submerged plants (some species of *Pseudacris*, *Acris*, *Hyla*, and *Spea*), or as a film on the water surface (some microhylids).

Some anurans breed in moderately fast streams or mountain torrents. Female *Atelopus* from Central and South America lay their eggs in strings attached to rocks. The larvae have ventral sucker-like discs that allow them to adhere to rocks while feeding. Tadpoles of bufonid stream-breeding *Ansonia* from Asia and *Werneria* from Africa have similar suckers. Other stream-adapted frogs, such as many ranids, lay large eggs in compact masses attached to rocks in areas where the current is slow and the eggs are less likely to be swept away.

Some terrestrial anurans produce foam nests in which their eggs are suspended. Male *Leptodactylus*, *Physalaemus*, and *Pleurodema* from Central and South America kick their hind legs during amplexus, whipping the females' eggs and mucus and their sperm and mucus into foamy masses (Figure 1.2a). The outermost layer of foam dries quickly and provides some protection against desiccation and predation. Some foam-nesters produce their nests on the water surface, others in cavities or holes next to ponds. *Leptodactylus bufonius* constructs mud nests at the margin of temporary ponds and deposits its foam nests inside (Figure 1.2b). The eggs hatch into tadpoles that remain in the nests until rains dissolve the nests and flood the area. Some Australian myobatrachids also construct foam nests on the water surface.

Some terrestrial frogs oviposit in very small bodies of water on land. Brazilian *Bufo castaneoticus* lay their eggs in water-filled fruit capsules of the Brazil nut tree, and the tadpoles feed on detritus. *Eupsophus* from Chile and

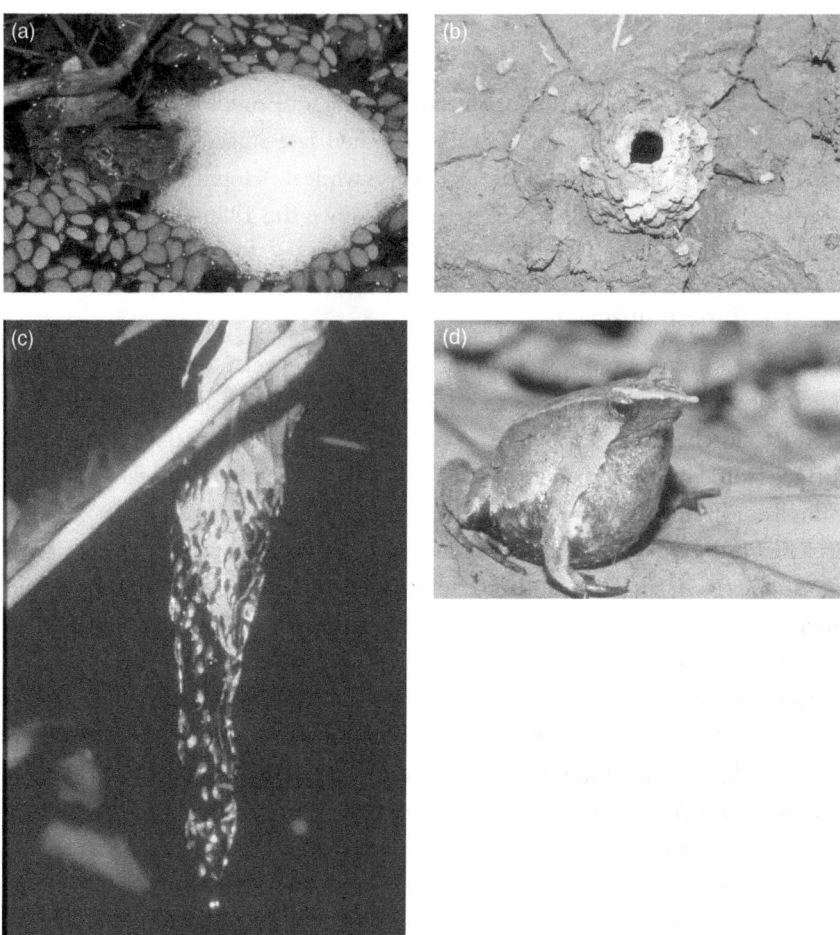

Fig. 1.2 Representatives of four anuran modes of reproduction. (a) *Pleurodema borelli* pair constructing a foam nest on the surface of water, from Argentina, (b) *Leptodactylus bufonius* mud nest by the edge of a depression, from Argentina, (c) *Hyla bokermanni* eggs on a leaf above water, from Ecuador, (d) male *Rhinoderma darwinii* brooding tadpoles in its vocal sac, from Chile. Photographs by Martha L. Crump.

Crinia georgiana from Australia oviposit in small water-filled depressions, seeps, or crevices on the ground where non-feeding larvae absorb their large yolks before metamorphosing.

Temperate and tropical frogs with terrestrial eggs and aquatic larvae accomplish this feat in diverse ways. Dendrobatids lay their eggs on land, but then a parent transports the larvae to water. In a few species of *Dendrobates*, the female also provides her young with unfertilized eggs as food. *Rhinoderma rufum*, from Chile, lays its eggs on land. The male takes late-stage eggs into his vocal sac where they hatch, then transports the larvae to water. In Europe, male midwife toads (*Alytes*) carry their eggs wrapped around their hind legs. Eventually they hop to ponds where the eggs hatch into aquatic larvae.

1.3.3.3 Arboreal/aquatic

Taxonomically diverse arboreal frogs lay their eggs on vegetation overhanging water. After the eggs hatch, the larvae fall into the water below where they continue to develop. *Agalychnis*, *Phyllomedusa*, and some *Hyla* from the New World tropics lay their eggs over standing water (Figure 1.2c), and neotropical centrolenids oviposit over flowing water. In some centrolenids, a parent protects the eggs from predators and keeps them moist by resting on them. Female *Afrixalus* from sub-Saharan Africa oviposit on leaves above water, then fold the leaf edges together and glue them in place with oviductal secretions. Some arboreal Old World rhacophorids and hyperoliids construct foam nests on vegetation overhanging temporary pools or slow-moving streams.

Water-filled basins offer oviposition sites that presumably lessen the risk of eggs getting swept away and reduce predation. Males of several neotropical gladiator frogs (e.g. *Hyla boans* and *Hyla rosenbergi*) construct basins beside streams or rivers. Water seeps in and fills the nests, and the frogs lay eggs as a surface film. After developing in the basin, the tadpoles metamorphose into froglets that take to the trees.

Some arboreal anurans oviposit in water-filled tree holes and axils of aerial plants. The eggs of many of these frogs (e.g. *Anodonthyla*, *Platypelis*, and *Plethodonthyla* from Madagascar) have large amounts of yolk. The tadpoles typically lack mouthparts, and they are non-feeding. In contrast, female *Osteopilus brunneus* from Jamaica lay eggs in water-filled leaf axils of bromeliads and continue to deposit about 250 more eggs in the bromeliad every few days throughout the tadpoles' development. The tadpoles—up to about 170 in a clutch—feed on the later-arriving eggs until they metamorphose into arboreal froglets.

Some arboreal frogs attach their eggs to the walls of water-filled cavities in trees. After hatching, the tadpoles drop into the water. In *Chirixalus eiffingeri*, an Asian rhacophorid, the female returns periodically and deposits fresh eggs for her tadpoles to eat. In others of this reproductive mode, the aquatic larvae feed on algae and debris.

In the New World tropics, female *Flectonotus* carry their eggs in dorsal pouches. After the eggs hatch as advanced tadpoles, the females transport them to water-filled bromeliads or bamboo where they complete development. Females of some neotropical *Gastrotheca* also brood their eggs in dorsal pouches and transport the tadpoles to aquatic sites.

1.3.3.4 Fossorial/aquatic

Scaphiopus and *Spea*, North American spadefoot toads, spend much of their lives underground but emerge following heavy rains and lay their eggs in newly formed ponds. The tadpoles develop quickly, which increases the probability of metamorphosing before the ponds dry. *Rhinophrynus dorsalis*, from southern Texas to Costa Rica, likewise lives underground and emerges after the first heavy rains to breed in temporary ponds. Female *Hemisis marmoratum* from Africa lay their eggs in subterranean chambers and stay with their eggs until after they hatch. At that point, the female digs a tunnel into adjacent water for the tadpoles.

1.3.3.5 Terrestrial/non-aquatic

Terrestrial anurans exhibit diverse life histories that free them from aquatic breeding sites. *Adenomera* from South America deposits foam nests under logs or in terrestrial cavities, and non-feeding tadpoles develop in the nests until they metamorphose. Neotropical *Eleutherodactylus*, *Oreophrynella* from Guyana and southern Venezuela, and New Guinean microhylids lay direct-developing eggs under logs or leaf litter. Many attend their eggs.

Some completely terrestrial anurans brood their young. Female *Assa darlingtoni* from Australia attend their terrestrial eggs. When the eggs are about 12 days old, the father climbs into the egg mass, rupturing the capsules. The newly hatched tadpoles wriggle into brood pouches, one along each side of the male's body. He broods his non-feeding larvae until they metamorphose into froglets. Female Darwin's frogs (*Rhinoderma darwinii*) from Chile and Argentina lay their eggs on moist ground. Just before the eggs hatch the males gobble them into their mouths and into the vocal sacs (Figure 1.2d). The young ingest secretions from the vocal-sac lining and emerge from their fathers' mouths as froglets.

Several *Nectophrynoides*, African bufonids, retain eggs in their oviducts and give birth to live young. In *Nectophrynoides occidentalis*, after depleting their yolk reserves the developing embryos feed on "uterine milk" secretion produced by glands in the mother's oviduct walls. These frogs live at high elevations, exposed to long periods of cold and drought. The females have a 9-month gestation period during which they estivate underground.

1.3.3.6 Arboreal/non-aquatic

Some neotropical *Eleutherodactylus* lay their direct-developing eggs in tree holes, bromeliads, moss, or on leaves. Some attend their eggs, others do not. *Eleutherodactylus jasperi* from Puerto Rico lived in arboreal bromeliads and gave birth to live young. The direct-developing eggs were retained in the oviducts, and nutrition came entirely from the embryo's yolk reserves. This species has not been seen since 1981 and is assumed extinct.

Female *Cryptobatrachus*, *Stefania*, and *Hemiphractus*, neotropical hylids, carry their direct-developing eggs exposed on their backs, secured by mucous gland secretions. Females of some *Gastrotheca* brood direct-developing eggs in dorsal pouches that protect the developing embryos from predators and desiccation and also function in gaseous exchange between the females and their embryos.

1.3.3.7 Fossorial/non-aquatic

The burrowing microhylid *Synapturanus salseri* from Colombia lays its eggs in burrows just below the root mat on the forest floor. Non-feeding tadpoles hatch at an advanced stage and absorb their yolk reserves. The Brazilian burrowing leptodactylid *Cycloramphus stejnegeri* likewise oviposits in underground nests and has non-feeding tadpoles. Other fossorial anurans, such as *Geocrinia* and *Arenophryne* (Australian myobatrachids), *Callulops* (New Guinean microhylids), and *Breviceps* (African microhylids), lay direct-developing eggs in underground burrows. Female *Breviceps* stay with their eggs and presumably keep them moist.

1.4 Amphibian declines and why they matter

The world is experiencing a "biodiversity crisis": rapid and accelerating loss of species and habitat (Ehrlich and Ehrlich 1981; Myers 1990; Raven 1990; Wilson 1992). Amphibians are part of this overall loss. Populations of amphibians are declining and disappearing worldwide at an increasing rate as compared to pre-1980 decades, even from protected areas (Blaustein and Wake 1990; Phillips 1994; Stuart *et al.* 2004). During the late 1980s and early 1990s, many declines seemed mysterious because there was no obvious cause. Skeptics argued that declines might be simply natural population fluctuations. Since the late 1980s, scientists worldwide have focused on determining the extent of declines and identifying the causes. We now know these declines are real.

The International Union for the Conservation of Nature (IUCN) assesses the status of species on a global scale and maintains and updates a catalog of taxa that face a high risk of global extinction: the IUCN *Red List of Threatened Species*. The 2008 update lists 30% of described amphibians as threatened with extinction (IUCN 2008). Since 1500, at least 39 species of amphibians have become extinct.

Scientists have hypothesized six major threats to amphibians: habitat modification and destruction, commercial over-exploitation, introduced species, environmental contaminants, global climate change, and emerging infectious diseases, especially the chytrid fungus *Batrachochytrium dendrobatidis* (Collins and Storfer 2003). Most agree the primary threat is habitat modification and destruction. For the past 100 years, human population growth has been exponential and has occurred largely in areas with the highest amphibian species richness: the tropics and subtropics (Gallant *et al.* 2007). As a result, these landscapes are being heavily modified to support agriculture and other human activities. The chytrid fungus also is exerting a major impact in many areas and on many species (Smith *et al.* 2006). Thus far the chytrid has caused the decline or extinction of about 200 species of frog (Skerratt *et al.* 2007). Many factors make the chytrid a significant concern, including its wide distribution in both the New and Old Worlds, its rapid spread and high virulence, and the fact that it infects a broad diversity of host species (Daszak *et al.* 1999).

Why should we care if we lose amphibians? It is for the same basic reasons we should care if other animals and plants disappear: economics, ecosystem function, esthetics, and ethics (Noss and Cooperrider 1994; Groom *et al.* 2006).

1.4.1 Economics

Selfishly, we should care if we lose amphibians because we use them for our own benefit, including for food and as pets. We use literally tonnes of frogs each year in medical research and teaching. We have isolated novel chemical compounds from granular glands of anuran skin and have used these compounds to develop new drugs.

1.4.2 Ecosystem function

Amphibians play a key role in energy flow and nutrient cycling because they serve as both predator and prey. By eating huge quantities of algae, tadpoles reduce the rate of natural eutrophication, the over-enrichment of water with nutrients, which leads to excessive algal growth and oxygen depletion. Most adult amphibians eat insects and other arthropods. As ectotherms, amphibians are efficient at converting food into growth and reproduction. Unlike endothermic birds and mammals that generate heat metabolically, amphibians expend relatively little energy to maintain themselves. Birds and mammals use up to 98% of their ingested energy to maintain their body temperatures, leaving as little as 2% to be converted to new animal tissue: food for predators. In contrast, amphibians convert about 50% of their energy gained from food into new tissue, which is transferred to the next level in the food chain

(Pough *et al.* 2004). If amphibians disappeared, would the world be overrun with houseflies, mosquitoes, and crop-eating insect pests? Would their predators go extinct?

1.4.3 Esthetics

Imagine the silence of rainy spring evenings without the lively croaking of male frogs. The monotonous roads without spring migrations of salamanders. People worldwide consider frogs to be good luck because of their association with rain. Amphibians provide inspiration for our artistic endeavors, from literature to music and the visual arts.

1.4.4 Ethics

Every species is a unique product of evolution. In 1982 the United Nations General Assembly adopted the World Charter for Nature, which states: "Every form of life is unique, warranting respect regardless of its worth to man, and, to accord other organisms such recognition, man must be guided by a moral code of action" (Noss and Cooperrider 1994). More than 100 nations signed the charter. Like all other living species, amphibians have intrinsic value and a right to exist.

Amphibians, amazing descendants of terrestrial pioneers, are fighting for their lives in a world greatly modified by humans.

1.5 References

Blaustein, A. R. and Wake, D. B. (1990). Declining amphibian populations: a global phenomenon? *Trends in Ecology and Evolution*, **5**, 203–4.

Collins, J. P. and Storfer, A. (2003). Global amphibian declines: sorting the hypotheses. *Diversity and Distributions*, **9**, 89–98.

Crump, M. L. (1995). Parental care. In H. Heatwole and B. K. Sullivan (eds), *Amphibian Biology*, vol. 2: *Social Behaviour*, pp. 518–67. Surrey Beatty and Sons, Chipping Norton, NSW.

Crump, M. L. (1996). Parental care among the Amphibia. In J. S. Rosenblatt and C. T. Snowdon (eds), *Advances in the Study of Behavior*, vol. 25: *Parental Care: Evolution, Mechanisms, and Adaptive Significance*, pp. 109–44. Academic Press, New York.

Daszak, P., Berger, L., Cunningham, A. A., Hyatt, A. D., Green, D. E., and Speare, R. (1999). Emerging infectious diseases and amphibian population declines. *Emerging Infectious Diseases*, **5**, 735–48.

Duellman, W. E. (1970). *The Hylid Frogs of Middle America*, vol. 1. Monograph of the Museum of Natural History, Number 1. University of Kansas, Lawrence, KA.

Duellman, W. E. (1999). Global distribution of amphibians: patterns, conservation, and future challenges. In W. E. Duellman (ed.), *Patterns of Distribution of Amphibians: a Global Perspective*, pp. 1–30. John Hopkins University Press, Baltimore, MD.

Duellman, W. E. (2007). Amphibian life histories: their utilization in phylogeny and classification. In H. Heatwole and M. J. Tyler (eds), *Amphibian Biology*, vol. 7. *Systematics*, pp. 2843–92. Surrey Beatty and Sons, Chipping Norton, NSW.

Duellman, W. E. and Trueb, L. (1986). *Biology of Amphibians*. McGraw-Hill, NewYork.

Ehrlich, P. R. and Ehrlich, A. H. (1981). *Extinction: The Causes and Consequences of the Disappearance of Species*. Random House, New York.

Gallant, A. L., Klaver, R. W., Casper, G. S., and Lannoo, M. J. (2007). Global rates of habitat loss and implications for amphibian conservation. *Copeia*, **2007**, 967–79.

Groom, M., Meffe, G. K., and Carroll, C. R. (2006). *Principles of Conservation Biology*, 3rd edn. Sinauer Associates, Sunderland, MA.

Haddad, C. F. B. and Prado, C. P. A. (2005). Reproductive modes in frogs and their unexpected diversity in the Atlantic Forest of Brazil. *BioScience*, **55**, 207–17.

IUCN (International Union for the Conservation of Nature) (2008). *Red List of Threatened Species*. IUCN, Gland. http://www.iucn.org/themes/ssc/redlist.htm.

Myers, N. (1990). Mass extinctions: what can the past tell us about the present and the future? *Global and Planetary Change*, **82**, 175–85.

Noss, R. F. and Cooperrider, A. Y. (1994). *Saving Nature's Legacy: Protecting and Restoring Biodiversity*. Island Press, Washington DC.

Phillips, K. (1994). *Tracking the Vanishing Frogs: an Ecological Mystery*. St. Martin's Press, New York.

Pough, F. H., Andrews, R. M., Cadle, J. E., Crump, M. L., Savitzky, A. H., and Wells, K. D. (2004). *Herpetology*, 3rd edn. Prentice Hall, Upper Saddle River, NJ.

Raven, P. H. (1990). The politics of preserving biodiversity. *BioScience*, **40**, 769–74.

Salthe, S. N. (1969). Reproductive modes and the number and sizes of ova in the urodeles. *American Midland Naturalist*, **81**. 467–90.

Salthe, S. N. and Duellman, W. E. (1973). Quantitative constraints associated with reproductive mode in anurans. In J. L. Vial (ed.), *Evolutionary Biology of the Anurans*, pp. 229–49. University of Missouri Press, Columbia, MO.

Skerratt, L. F., Berger, L., Speare, R., Cashins, S., McDonald, K. R., Phillott, A. D., Hines, H. B., and Kenyon, N. (2007). Spread of chytridiomycosis has caused the rapid global decline and extinction of frogs. *EcoHealth*, **4**, 125–34.

Smith, K. F., Sax, D. F., and Lafferty, K. D. (2006). Evidence for the role of infectious disease in species extinction and endangerment. *Conservation Biology*, **20**, 1349–57.

Stuart, S. N., Chanson, J. S., Cox, N. A., Young, B. E., Rodrigues, A. S.L., Fischman, D. L., and Waller, W. (2004). Status and trends of amphibian declines and extinctions worldwide. *Science*, **302**, 1783–6.

Wake, M. H. (1977). The reproductive biology of caecilians: an evolutionary perspective. In D. H. Taylor and S. I. Guttman (eds), *The Reproductive Biology of Amphibians*, pp. 73–101. Plenum, New York.

Wake, M. H. (1982). Diversity within a framework of constraints. Amphibian reproductive modes. In D. Mossakowski and G. Roth (eds), *Environmental Adaptation and Evolution*, pp. 87–106. Gustav Fischer, New York.

Wake, M. H. (1992). Reproduction in caecilians. In W. C. Hamlett (ed.), *Reproductive Biology of South American Vertebrates*, pp. 112–20. Springer-Verlag, New York.

Wake, M. H. (1993). Evolution of oviductal gestation in amphibians. *Journal of Experimental Zoology*, **266**, 394–413.

Wells, K. D. (2007). *The Ecology and Behavior of Amphibians*. University of Chicago Press, Chicago, IL.

Wilkinson, M. and Nussbaum, R. A. (1998). Caecilian viviparity and amniote origins. *Journal of Natural History*, **32**, 1403–9.

Wilson, E. O. (1992). *The Diversity of Life*. The Belknap Press of Harvard University Press, Cambridge, MA.

2

Setting objectives in field studies

Dan Cogălniceanu and Claude Miaud

2.1 Basic concepts for a good start

Considerable financial and human resources may be wasted due to poor research design and implementation. In addition, poor planning can result in taking non-repeatable or unreliable data that are of no or limited use, and leads to the paradox of "data-rich, information-poor." Rigorous study planning and the definition of clear and realistic goals can help avoid wasted effort. Before undertaking field studies, biologists first must define their objectives and then assess the availability of resources necessary to accomplish them. This is a key stage, and any uncertainty or vagueness at this point could prove detrimental to the success and usefulness of the study.

There are two different approaches in scientific investigation: inductive and deductive methods. The inductive or exploratory method relies on gathering data without first identifying hypotheses to be tested; results may then be explained based on the data gathered. However, science involves more than accumulating data and then searching for patterns. A useful study must start with an observation that identifies a problem or question to be resolved. The observation should have biological significance and be based on current evolutionary theory (Wolff and Krebs 2008). For example, amphibian populations are declining: what are the causes? Two related species hybridize on just a narrow contact zone: why? Habitats are fragmented by human activity: what effects does this have on population persistence, and why (e.g. dispersal abilities, food resources, population demography, genetics, susceptibility to predators, and human activities)?

As science is an activity focused on understanding natural processes and, especially in modern times, in devising solutions to problems, the hypothesis-based deductive method was developed (James and McCulloch 1985). An *hypothesis* is a statement related to an observation which may be true, but for which proof has not yet been found. The function of the hypothesis is to direct the search for order among facts. Fieldwork then serves to test one's hypothesis, thereby scientifically demonstrating correlation, causation, or the absence thereof. The hypothesis must of necessity regard some facts as more significant than others, based on the researcher's previous experience, familiarity with the literature, and individual interpretation.

As scientific knowledge is rarely complete, any list of potential alternative hypotheses is also unlikely to be complete. Therefore, alternative hypotheses for the particular problem or question being considered must also be formulated. Thus, it is necessary to conduct research with both the hypothesis and alternative hypothesis in mind, as more than one cause may be contributing to any single effect (Wolff and Krebs 2008). For example, Jaeger (1972) used enclosures to test the mechanism of interspecific competition between two species of plethodontid salamanders, hypothesizing that it resulted from either differential exploitation of food resources or through interference. Both primary and alternative hypotheses are formulated and tested so that researchers can determine which explanations best fit the results obtained. Unsupported hypotheses can be rejected, but even the most parsimonious hypothesis may not be fully accepted because there still may be underlying explanations as yet untested (Senar 2004). Repeated experimentation involving alternative hypotheses eventually allows researchers to gain a measure of confidence in the validity of empirically supported hypotheses (Jaeger and Halliday 1998). Both hypotheses and data are essential for credible science since, as Krebs (1999) concluded, hypotheses without data are not very useful, and data without hypotheses are useful only for an inductive approach.

Ecology is an empirical science that requires data from the "real" world, and a field study is a tool which can ultimately lead to the acceptance or rejection of an hypothesis. After repeated experimentation and validation, data and information from many field studies are synthesized and organized into concepts of how nature functions. These concepts are the result of the integration between what scientists think they know (based on previous research and observation) and newly acquired data (Ford 2000). Asking the right question is important, because the type of question asked and the particular techniques and methods used in hypothesis-driven research strongly influence what is discovered and the direction of future research. During the course of this feedback process, scientists come to a better understanding of the phenomena they study (Figure 2.1).

Scientists present what they want to accomplish during a field study in terms of goals and objectives. A *goal* is a statement that explains what the study is designed to accomplish. It is usually a broad and general statement, inclusive of

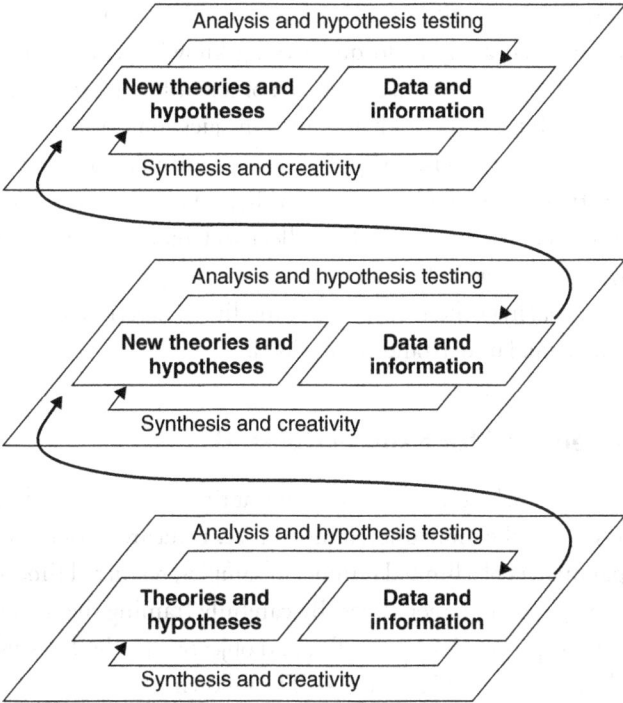

Fig. 2.1 The cycle of scientific investigation and the shift towards the spiral of knowledge.

a long-term direction. The goal is then split into specific, measurable *objectives* that indicate how the goal can be achieved within a specified time frame, and the expected results.

Defining a study's objectives clearly is the first and probably the most important single step in research planning, and is a key element in a successful project. The goals of research can be *non-applied* (i.e. aimed at increasing or changing existing knowledge) or *applied* (i.e. focused on solving practical questions), or both. Rather than addressing an effect, a study should focus on the cause of the phenomenon in question. Separating effects from causes is a major challenge when setting objectives. For example, acid rain has well-known causes, but much research today focuses on the effects or seeks solutions to limit its impact. In another example, many recent studies have reported on amphibian declines (Stuart *et al.* 2004), but many fewer have identified specific causes (e.g. Becker *et al.* 2007). It is important that amphibian biologists do not limit their focus to specific taxa; otherwise, they lose sight of the fact that similar effects may be occurring in other taxa. Amphibian biologists need to establish closer links with researchers studying similar problems in other taxa (Halliday 2005).

Ideally, the cycle of scientific investigation should proceed from data and information gathering and analysis, to quantifying knowledge, and to conceptual understanding. A logical, stepwise approach of scientific inquiry (Lehner 1996) should be followed: (1) perceive that a problem/question exists; (2) formulate a possible explanation (i.e. devise an hypothesis); (3) formulate alternative hypotheses; (4) identify the best approach to test the hypothesis (i.e. theoretical models, experiments, or field observations); (5) collect and analyze data; (6) support or reject the hypothesis; and (7) understand the meaning and implications of the results. The original hypothesis can then be modified, experiments repeated and, with time, conceptual understanding attained.

2.2 Steps required for a successful study

Ecological systems are large and complex and, at times, unpredictable. The more complex the system, the more uncertainties arise. Data are most often obtained within the parameters of a limited number of samples, restricted time, or specific area, and then applied to larger scales. By carefully framing the hypothesis and deciding the most practical, meaningful, and objective method of study, degree of error can be minimized (Hayek 1994). The following section presents some of the most important issues that must be considered when planning a field study.

2.2.1 Temporal and spatial scales

The issue of scale has three components (Schneider 2001): (1) pressing problems in ecology often exist within timescales of decades or centuries, and cover large areas; (2) most data can be gathered directly for only short periods of time and over small areas; (3) patterns measured at small scales do not necessarily hold true at larger scales; nor do processes observed at smaller scales necessarily exist at larger ones. Thus, an inappropriate scale limits the inference of the results.

Spatial and temporal scales can be classified as gradient, interval, discrete, and continuous (Bernstein and Goldfarb 1995). Setting temporal and spatial scales involves defining their boundaries and helps to answer three practical questions: where, for how long, and how often? Selecting the appropriate scale is difficult because of environmental heterogeneity in both time and space. Amphibian populations and communities also vary widely throughout time and space, since they are influenced by a multitude of environmental variables (Meyer *et al.* 1998; Pechmann *et al.* 1991; Pellet *et al.* 2006).

Temporal scales are rarely discussed explicitly, but they are often assumed to span years rather than generations. Many time-dependent phenomena such as extinction, predator–prey interactions, competition, and succession reveal important insights if considered on a generational scale (Frankham and Brook 2004). The current distribution and abundance of animal species inhabiting an area is the result, in part, of the impact of geological processes (e.g. plate tectonics, orogenesis, sea-level changes, major catastrophes), ecological processes (e.g. competition, predation, climate), life-history traits (e.g. dispersal), and recent human impacts (e.g. changes in vegetation and habitat structure, overexploitation, introduction of nonindigenous species). For example, in the Northern Hemisphere the Pleistocene–Holocene post-glacial recolonizations of species and human activities (e.g. habitat destruction and alteration) are the two major factors shaping the current distribution of amphibians. Objectives of a time-scale component of a study plan should include such factors as short-term (less than one generation), intermediate (more than one to a few generations), and long-term (many generations, and inclusive of a large spatial scale and monitoring activities).

Spatial scales and objectives are strongly interconnected. Selection of a study site includes such factors as the complexity of local macro- and microhabitats and, at large spatial scales, biogeographical provinces and landscapes (Morrison 2002). Of great importance to the overall study plan are the criteria for describing vegetation, whether gross (e.g. foliage height, diversity), physiognomy (physical structure), or floristic (plant taxonomic description). The relative usefulness of structural or floristic measures depends on the spatial scale of the analysis. For example, whereas most studies do not require a detailed description of plant taxa, species lists might be essential for characterizing microhabitat and trophic (resource) availability.

The importance of choosing the correct scale can be exemplified by a study assessing population fluctuations within a particular species: the first step would be to understand the distribution of the species in order to thoroughly sample all areas, rather than by biasing results by sampling in only one small portion of the range. In another example, a study of the movement patterns of a species may focus on individual distributional patterns, daily or weekly movements, seasonal migrations related to reproduction or dispersal, or on region-wide range shifts through years or decades.

2.2.2 Choosing the model species

Many amphibian species have restricted distributions and complex habitat requirements. Some amphibian populations, especially among temperate pond-breeding species, are maintained by episodic reproduction that occurs in a sporadic and unpredictable manner (Alford and Richards 1999). If a number of different species is available for study, selecting an experimental species should be based on an understanding of its life history (i.e. feeding, habitat use, dispersal abilities, reproduction, behavior, predators, phylogeography, population genetics) and its suitability in resolving the particular question being asked (Wolff and Krebs 2008). For example, a species with low detectability is not a good choice for a mark–recapture study since it will require a considerable sampling effort and entail potential capture bias (Weir *et al.* 2005). The role of keystone species (e.g. *Eleutherodactylus bransfordii* in Costa Rican forests), umbrella species (e.g. *Rana sylvatica* in the Milwaukee river basin), or flagship species (e.g. *Salamandra lanzai* in the Western Alps) can also be considered. Rare or threatened species are often selected because of conservation applications, but research on them could incur potential risks due to ethical considerations. Researchers should avoid choosing a rare or threatened species if common or less vulnerable surrogate species can be selected.

2.2.3 Pilot/desk study

A preliminary pilot study usually helps to develop and test realistic and achievable objectives, and avoids later shortcomings and failures. Usually a pilot study is carried out on short temporal and small spatial scales, and allows the testing of conceptual models and methods. What might seem simple during office planning might prove completely different or unworkable in the field. Even a simple review of the literature can prove helpful in avoiding mistakes, however.

2.2.4 Elaborate a conceptual model

A conceptual model (e.g. a simple box diagram showing components and linkages) is a simplified model of the system to be studied. There are no ideal methods to employ; instead, a multitude of models are available to choose from with various degrees of complexity. Conceptual models are helpful in that they can be used to select the variables to be measured that might be considered important to the study. Since professionals with diverse backgrounds have different philosophies or approaches to using models, it is recommended that each team member contribute knowledge of and/or expertise with various models, then develop an integrated study model from which to work (Maher *et al.* 1994).

After model development, many of the design questions become more obvious. A model need not embrace all components of the system; it needs only to be adequate for the scope of the investigation. There is always a possible risk that the conceptual model is inappropriate or over-simplified; thus, a slightly more complex model may need to be developed. Adopting a more complex model focuses researchers on collecting additional data that might prove important, despite the extra costs involved with sampling and measurements. In other words, it might be better to collect more data than appears necessary at first, than to discover later that some important parameter was omitted. A pilot study helps to avoid under- or over-collecting data. Although conceptual models help researchers identify what to measure, the timing of studies is determined by the natural history of the species of interest.

2.2.5 The SMART approach

Specific objectives allow for greater chances of conducting a successful research project. SMART stands for specific, measurable, attainable, relevant, and time (Piotrow *et al.* 1997), an approach used in project writing that also helps in formulating specific and measurable objectives within study plans. SMART helps devise objectives that are clear and concise, indicates what is to be achieved, addresses only attainable results, indicates when each stage will be completed, and is not encumbered by idealistic aspirations.

- S: The *specific* part of an objective defines what will be done and where it will occur. When setting objectives, ask simple questions that can be answered, and avoid ambiguities, the use of buzzwords or jargon (e.g. "cutting-edge"), and pompous phrasing.
- M: *Measurable* is an attribute of an activity or its results. The source of and mechanism for collecting measurable data are identified, and collection of these data is determined feasible. If the objective of the study is to document trends (i.e. an increase or decrease of one or more measurable variables), then a baseline is required to act as a reference point (e.g. habitat availability, characteristics and use; amphibian community structure; population size). If a baseline is not yet available then it will be useful to first have it established.
- A: *Attainable* refers to the probability of conducting the proposed activities within the established time frame with the available resources and support. It also includes the external factors critical to success. Doing research today, for example, can become difficult due, in part, to increasing administrative restrictions (Prathapan *et al.* 2008). Other external factors include unforeseen costs, shifts in exchange rates, obtaining collecting and access permits, changes in legislation, and political, security, and health (both human and animal) issues. In coping with external factors, a risk analysis is useful because it allows for planning alternative strategies.
- R: Some useful measures for the *relevance* of the study's objectives are the utility and value of the results for practical purpose (e.g. management of protected areas, better conservation measures, and ecological restoration). It

may be difficult to determine how relevant the objective may be, especially since the true relevance may not be apparent until the study is completed (e.g. if a drought were to affect a long-term study of amphibian breeding). Perhaps an easy way to evaluate the relevance is to answer the "so what" question, thus avoiding undertaking studies of limited or no interest.

- T: Finally, the *time* required to complete the project will depend on the parameters discussed above.

2.2.6 Applying the SMART approach to plan an amphibian inventory

In the following example, the SMART approach is used as the basis for setting up an inventory of amphibian species in a national park or reserve.

- *Specific*: determine whether to inventory all amphibian species, or only those of special interest, such as rare, threatened, or endemic species. Set the spatial limits of the study, such as the administrative boundary of the park or reserve, or specific habitats of interest (e.g. temporary ponds; roads in areas which bisect amphibian dispersal corridors; high elevations containing unique habitats or species richness, such as cloud forests).

- *Measurable*: select parameters to be measured (e.g. presence/not detected, relative abundance, density, sex or life stage, percentage of area occupied), select methods appropriate for each species (especially taking detection probabilities into account), and decide habitat parameters (e.g. temperature, humidity, vegetative cover, characteristics of aquatic habitats).

- *Attainable*: identify resources and available support (e.g. funding, work force) and administrative considerations (e.g. collecting and access permits, training staff for field data collection, animal care and use requirements, security issues).

- *Relevance*: provide concise statements as to the importance of the research, and whether it has practical applications. The relevance of the proposed research is particularly important in field studies, especially if it will aid in the management of protected areas and species conservation. Utility also assists in determining research approaches. For example, incorporating measures of abundance with detection probabilities, in addition to occupancy models, is much more informative in determining status than occupancy models alone.

- *Time*: the research schedule (planning, fieldwork, data analysis, report and publication preparation) should be clearly stated based on the previously outlined parameters and limiting factors.

2.2.7 Experimental versus field studies

Biological communities, apart from their high internal complexity, are subject to random, naturally occurring fluctuations involving both physical and biotic parameters (e.g. heat, cold, drought, effects of disease, and spread of invasive species). Stochastic fluctuations such as these are a major source of statistical variance in nature, resulting in a shift by some researchers towards less complex and more controlled field and laboratory experimental designs (see Chapter 6). Such a reductionist approach can be framed within a simpler conceptual model where the number of variables of interest is reduced in return for minimizing uncontrolled environmental fluctuation.

Still, there is an experimental continuum between laboratory and field studies (Figure 2.2). Perhaps some of the most powerful experimental tools which can provide the statistical rigor required for hypothesis testing are outdoor experiments (Fauth 1998; Rowe and Dunson 1994; Chapter 6). Controlled experiments maximize a researcher's ability to detect a response to variables of interest (e.g. food, density of individuals, hydroperiod). The best approach is to use observations in nature coupled with experimental analysis. For example, Wilbur (1997) used both natural and artificial ponds to investigate complex food webs in temporary ponds, and their effects on the larval amphibian community.

2.2.8 Methods for sampling, data storage, and analysis

Selecting an appropriate research technique is not an easy task, especially since "trendy" methods may not be the best ones to use. Methods selected must be adequate for the proposed objectives, and it is best to test them first to determine

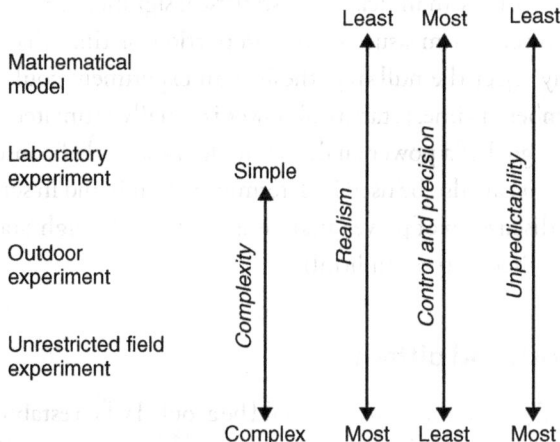

Fig. 2.2 Trade-offs between complex and simple experiments.

efficiency and effectiveness. Researchers should adopt complimentary models, applicable and relevant methods, and an altogether SMART use of study plans, resources, and results. The most important elements in any study plan are accurate and precise methods that meet the objectives of the study. A cost-benefit analysis can prove helpful in the final selection of the methods (Arntzen *et al.* 1995, 2004).

The sampling design should be both *efficient* (a better use of sampling effort in obtaining results) and *simple* (easily comprehended and easily implemented) (Scott and Köhl 1993). Statistical analyses often have biases or limitations in ecological studies (Krebs 1999; Pollock *et al.* 2002). Statistical significance may not be the equivalent of biological significance. Focusing too much on statistical issues may prove deceptive or lead to inaccurate conclusions. One way to minimize the potential of error is to incorporate power analysis into study designs (Yoccoz 1991). Statistical power refers to the ability of a test to correctly reject a null hypothesis, and is frequently evaluated in the context of the sample size required to detect an effect of a given magnitude (Michener 2000).

When using probability statistics, it is possible to make two kinds of error: a researcher may claim there is a difference when one does not exist, or can fail to detect a difference when one does exist (Underwood 1997). The first type of error is estimated by the traditional probability value ($\propto = 0.05$) associated with statistical tests. If a researcher rejects the null hypothesis at this level, then there is still a 5% chance that he/she is in error.

The second type of error is more difficult to estimate because it depends on the sample size, the magnitude of effect, and sample variability. Researchers are likely to correctly reject a null hypothesis with larger sample sizes, larger effect sizes, or less variable samples. Alternatively, a higher critical value may be selected (e.g. $\propto = 0.1$) as an indication of statistical significance.

Statistical power is a measure of the proportion of time that a researcher would correctly reject the null hypothesis if an experiment could be repeated an infinite number of times; statistical power is usually estimated by computer simulation. The goal of a power analysis is to define a level of confidence in the research results; it can also be used in determining trends and in setting optimal sample size. A discussion of power analysis is available through StatSoft (www.statsoft.com/textbook/stpowan.html).

2.3 Trade-offs and pitfalls

There are several pitfalls that can and should be avoided when establishing objectives in field studies (Bardwell 1991). The most common ones are: (1) addressing the wrong problem; (2) stating the problem in a way that no solution is possible;

(3) prematurely accepting a solution as the only possible answer; and (4) using data and information that are either incorrect (e.g. inaccurate information in the scientific literature) or irrelevant. There are several additional pitfalls that may be avoided by careful planning and a thorough understanding of the questions to be asked (Tucker *et al.* 2005). Four of the most frequent are listed below.

1) The statistical framework might be inadequate, since many techniques developed in the context of controlled experimentation are sometimes incorrectly applied to field data, resulting in an inappropriate use of the null hypothesis (Johnson 1999).
2) Researchers and technicians might differ in their skills, use non-comparable methods, or have different personal goals. Training prior to the start of field collection of data and a comparison of each person's abilities helps to minimize these problems.
3) Methods may be changed during a study. This could lead to an incompatibility of data sets and limit the interpretation of results.
4) The locations of permanent sample sites are not properly recorded so that different areas are subsequently revisited or sampled.

When designing fieldwork, researchers need to be aware of potential options and trade-offs, and try to balance them (Hairston 1989). Examples include: (1) complexity versus simplicity (e.g. choices ranging from theoretical models to field experiments), (2) confidence in results versus general application (e.g. high confidence can be achieved at short temporal and small spatial scales and with relatively simple goals and conceptual models, but the results will be of limited value), and (3) replication versus sophistication of experimental design, recognizing that it is impossible to simultaneously maximize precision, realism, and generality (Levins 1968).

2.4 Ethical issues

While ecological field studies have generated a wealth of useful and important data, the environmental impacts of these studies have rarely been quantified. A basic assumption is that the relative benefits of research outweigh the potential short-term costs to the study animals or habitats (Farnsworth and Rosovsky 1993). Even the simple act of observation, however, may affect the behavior of the study organisms. Repeatedly visiting different sites during fieldwork can have negative impacts, either by spreading or introducing nonindigenous species (Whinam *et al.* 2005) or diseases, such as amphibian chytrid fungus (Garner *et al.* 2005), or by microhabitat alteration (e.g. turning logs). Preventive measures, such as equipment disinfection (Chapter 26) and routine checks for unwanted "passengers" (e.g. seeds) are now widely used (ARG-UK 2008).

Controversies continue regarding the negative effects of some common techniques. For example, toe-clipping, historically one of the most widely used marking techniques, has recently received much criticism (Chapter 8; May 2004; McCarthy and Parris 2004). Critics, however, have not provided much evaluation of the impacts of alternative procedures. Apart from dorsal or ventral pattern mapping by photography or computer imaging, which are non-invasive, all other marking techniques have disadvantages (Phillott *et al.* 2007). The effects of toe-clipping vary among species, and therefore must be assessed accordingly. Other marking methods for amphibians are available, although they are not as economic or as easy to use, but which have fewer risks (Chapter 8; Ferner 2007). Toe-clipping also may be prohibited by regulatory constraints; information on regulations is best obtained and evaluated during the planning stage. Thus, there may be a trade-off between the risks associated with methodology and the knowledge to be gained, even when the species may be benefited (Funk *et al.* 2005).

Field studies often involve years of hard work. In the end, the results may provide few insights compared to the amount of effort to acquire them, unless careful planning precedes the initiation of research activities. Careful planning optimizes researcher effort and helps ensure that the data recorded will be statistically accurate, with beneficial results in advancing knowledge of amphibian ecology and conservation biology.

2.5 Acknowledgments

Cristina Vâlcu, Tibor Hartel, Marian Griffey, and Ken Dodd provided helpful comments on previous versions of this chapter.

2.6 References

Alford, R.A. and Richards, S.J. (1999). Global amphibian declines: a problem in applied ecology. *Annual Review of Ecology and Systematics*, **30**, 133–65.

ARG-UK (Amphibian and Reptile Groups of the UK) (2008). *Amphibian Disease Precautions: a Guide for UK Fieldworkers*. ARG-UK Advice Note 4, pp. 1–5. www.arg-uk.org.uk/Downloads/ARGUKAdviceNote4.pdf.

Arntzen, J. W., Oldham, R. S., and Latham, D. M. (1995). Cost effective drift fences for toads and newts. *Amphibia-Reptilia*, **16**, 137–45.

Arntzen, J. W., Goudie, I. B., Halley, J.J., and Jehle, R. (2004). Cost comparison of marking techniques in long-term population studies: PIT-tags versus pattern maps. *Amphibia-Reptilia*, **25**, 305–15.

Bardwell, L. V. (1991). Problem-framing: a perspective of environmental problem solving. *Environmental Management*, **15**, 603–12.

Becker, C. G., Fonseca, C. R., Haddad, C. F. B., Batista, R. F., and Prado, P. I. (2007). Habitat split and the global decline of amphibians. *Science*, **318**, 1775–7.

Bernstein, B. B., and Goldfarb, L. (1995). A conceptual tool for generating and evaluating ecological hypotheses. *BioScience*, **45**, 32–9.

Farnsworth, E. J., and Rosovsky, J. (1993). The ethics of ecological field experimentation. *Conservation Biology*, **7**, 463–72.

Fauth, J. E. (1998). Investigating geographic variation in interspecific interactions using common garden experiments. In W. J. Resetaridis Jr and J. Bernardo (eds), *Experimental Ecology, Issues and Perspectives*, pp. 394–415. Oxford University Press, New York.

Ferner, J. W. (2007). *A Review of Marking and Individual Recognition Techniques for Amphibians and Reptiles*. Herpetological circular no. 35. Society for the Study of Amphibians and Reptiles, Salt Lake City, UT.

Ford, E. D. (2000). *Scientific Method for Ecological Research*. Cambridge University Press, Cambridge.

Frankham, R. and Brook, B. W. (2004). The importance of time scale in conservation biology and ecology. *Annales Zoologici Fennici*, **41**, 459–63.

Funk, W. C., Donnelly, M. A., and Lips, K. R. (2005). Alternative views of amphibian toe-clipping. *Nature*, **433**, 193.

Garner T. W. J., Walker, S., Bosch, J., Hyatt, A.D., Cunningham, A.A., and Fisher, M.C. (2005). Chytrid fungus in Europe. *Emerging Infectious Diseases*, **11**, 1639–41.

Hairston, N. G. (1989). Hard choices in ecological experimentation. *Herpetologica*, **45**, 119–22.

Halliday, T. (2005). Diverse phenomena influencing amphibian population declines. In M. Lannoo (ed.), *Amphibian Declines: the Conservation Status of United States Species*, pp. 3–6. University of California Press, Berkeley, CA.

Hayek, L. A. (1994). Research design for quantitative amphibian studies. In W. R. Heyer, M. A. Donnelly, R. W. McDiarmid, L. A. Hayek, and M. Foster (eds), *Measuring and Monitoring Biological Diversity. Standard Methods for Amphibians*, pp. 21–39. Smithsonian Institution Press, Washington.

Jaeger, R. G. (1972). Food as a limited resource in competition between two species of terrestrial salamanders. *Ecology*, **53**, 535–46.

Jaeger, R. G. and Halliday, T. R. (1998). On confirmatory versus exploratory research. *Herpetologica*, **54** (supplement), 64–6.

James, F. C. and McCulloch, C. E. (1985). Data analysis and the design of experiments in ornithology. In R. F. Johnston (ed.), *Current Ornithology*, vol. 2, pp. 1–63. Plenum Press, New York.

Johnson, D. H. (1999). The insignificance of statistical significance testing. *Journal of Wildlife Management*, **63**, 763–72.

Krebs, C. J. (1999). *Ecological Methodology*, 2nd edn. Benjamin/Cummings, Menlo Park, CA.

Lehner, P. N. (1996). *Handbook of Ethological Methods*, 2nd edn. Cambridge University Press, Cambridge.

Levins, R. (1968). *Evolution in Changing Environments*. Princeton University Press, Princeton, NJ.

Maher, W. A., Cullen, P. W., and Norris, R. H. (1994). Framework for designing sampling programs. *Environmental Monitoring and Assessment*, **30**, 139–62.

May, R. M. (2004). Ethics and amphibians. *Nature*, **431**, 403.

McCarthy, M. A. and Parris, K. M. (2004). Clarifying the effect of toe clipping on frogs with Bayesian statistics. *Journal of Applied Ecology*, **41**, 780–6.

Meyer A. H., Schmidt, B.R., and Grossenbacher, K. (1998). Analysis of three amphibian populations with quarter-century long time-series. *Proceedings of the Royal Society of London Series B Biological Sciences* **265**, 523–8.

Michener, W. K. (2000). Research design: translating ideas to data. In W. K. Michener and J. W. Brunt (eds), *Ecological Data: Design, Management and Processing*, pp. 1–24. Blackwell Science, Oxford.

Morrison, M. L. (2002). *Wildlife Restoration. Techniques for Habitat Analysis and Animal Monitoring*. Society for Ecological Restoration. Island Press, Washington DC.

Pechmann, J.H.K., Scott, D.E., Semlitsch, R.D., Caldwell, J.P., Vitt, L.J., and Gibbons, J.W. (1991). Declining amphibian populations: the problem of separating human impacts from natural fluctuations. *Science*, **253**, 892–5.

Pellet J., Schmidt, B.R., Fivaz, F., Perrin, N., and Grossenbacher, K. (2006). Density, climate and varying return points: an analysis of long-term population fluctuations in the threatened European tree frog. *Oecologia*, **149**, 65–71.

Phillott, AD, Skerratt, L. F., McDonald, K. R., Lemckert, F. L., Hines, H. B., Clarke, J. M., Alford, R. A., and Speare, R. (2007). Toe-clipping as an acceptable method of identifying individual anurans in mark recapture studies. *Herpetological Review*, **38**, 305–8.

Piotrow, P. T., Kincaid, D. L., Rimon, J. G., and Rinehart W. (1997). *Health Communication: Lessons from Family Planning and Reproductive Health*. Johns Hopkins School of Public Health, Center for Communication Programs. Praeger Publishers, Westport, CT.

Pollock, K. H., Nichols, J. D., Simons, T. R., Farnsworth, G. L., Bailey, L. L., and Sauer J. R. (2002). Large scale wildlife monitoring studies: statistical methods for design and analysis. *Environmetrics*, **13**, 105–19.

Prathapan, K. D., Rajan, P. D., Narendran, T. C., Viraktamath, A. C., Aravind, N. A., and Poorani, J. (2008). Death sentence on taxonomy in India. *Current Science*, **94**, 170–1.

Rowe, C. L. and Dunson, W. A. (1994). The value of simulated pond communities in mesocosms for studies of amphibian ecology and ecotoxicology. *Journal of Herpetology*, **28**, 346–56.

Schneider, D. C. (2001). The rise of the concept of scale in ecology. *BioScience*, **51**, 545–53.

Scott, C. T. and Köhl, M. (1993). A method of comparing sampling design alternatives for extensive inventories. *Mitteilungen der Eidgengessischen Anstalt fuer Wald, Schnee und Landschaft, Birmensdorf*, **69**, 1–62.

Senar, J. C. (2004). *Mucho mas que plumas*. Monografies del Museu de Ciències Naturals 2. Museu de Ciències Naturals i l'Institut Botànic de Barcelona, Barcelona.

Stuart, S. N., Chanson, J. S., Cox, N. A., Young, B. E., Rodrigues, A. S. L., Fischman, D. L., and Waller, W. (2004). Status and trends of amphibian declines and extinctions worldwide. *Science*, **306**, 1783–5.

Tucker, G., Bubb, P., de Heer, M., Miles, L., Lawrence, A., Bajracharya, S.B., Nepal, R.C., Sherchan, R., and Chapagain, N. R. (2005). *Guidelines for Biodiversity Assessment and Monitoring for Protected Areas.* KMTNC, Kathmandu, Nepal.

Underwood, A. J. (1997). *Experiments in Ecology. Their Logical Design and Interpretation Using Analysis of Variance.* Cambridge University Press, Cambridge.

Weir, L. A., Royle, A., Nanjappa, P., and Jung, R. E. (2005). Modeling anuran detection and site occupancy on North American Amphibian Monitoring Program (NAAMP) routes in Maryland. *Journal of Herpetology*, **39**, 627–39.

Whinam, J., Chilcott, N., and Bergstrom, D. M. (2005). Subantarctic hitchhikers: expeditioners as vectors for the introduction of alien organisms. *Biological Conservation*, **121**, 207–19.

Wilbur, H. M. (1997). Experimental ecology of food webs: complex systems in temporary ponds. *Ecology*, **78**, 2279–2302.

Wolff, J. O. and Krebs, C. J. (2008). Hypothesis testing and the scientific method revisited. *Acta Zoologica Sinica*, **54**, 383–6.

Yoccoz, N. G. (1991). Use, overuse and misuse of significance tests in evolutionary biology and ecology. *Bulletin of the Ecological Society of America*, **72**, 106–11.

Part 2
Larvae

3

Morphology of amphibian larvae

Roy W. McDiarmid and Ronald Altig

3.1 Background

The larvae of amphibians are non-reproductive and usually aquatic. Most undergo metamorphosis prior to attaining an adult morphology and sexual maturity. Species within each amphibian order that develop by other modes (e.g. direct development; Altig and Johnston 1989; Thibaudeau and Altig 1999) have non-feeding larvae or embryos, and we do not discuss them in this chapter.

Amphibian larvae have some generalized morphological features that are useful for identification. In contrast to fish, they lack bony supports in the tail fins, and the vent is a longitudinal slit (not obvious in tadpoles). Amphibian larvae also lack eyelids, and most have external gills that are visible at some stage in their ontogeny.

Most caecilians (Gymnophiona) whether terrestrial or aquatic as adults, have aquatic larvae that look grossly like the legless, elongate adults. The larvae of salamanders (Caudata) look much like the adults. Unlike the condition in caecilians and frogs, salamanders may occur in larval (i.e. larval morphology, non-reproductive, and metamorphic) or larviform (i.e. larval morphology, reproductive, may or may not metamorphose) states. Larviform salamanders may exist as pedotypes (i.e. larval relative to the normal developmental trajectory of the taxon, reproductive, will metamorphose if the environmental conditions change to the detriment of being in the larval environment; some ambystomatids and salamandrids; terminology of Reilly *et al.* 1997) or pedomorphs (i.e. larval relative to the developmental pattern of their ancestors, do not metamorphose; all amphiumids, cryptobranchids, proteids, sirenids, and some plethodontids). Larval and larviform salamanders grossly resemble metamorphosed individuals in general body form but retain a number of larval features. Pond-adapted forms have a more bulky body and larger gills and tail fins than the more streamlined, stream-adapted forms.

The larvae of frogs and toads (Order Anura), called tadpoles, are grossly different from adults and have many developmental (Altig and Johnston 1989) and morphological (e.g. Altig and McDiarmid 1999a; also various morphologies documented in staging tables, see Duellman and Trueb 1986, pp. 128–9) features not seen in other amphibian larvae. Tadpoles live in many kinds of habitats; the most common types of tadpoles are found in lentic or lotic water, spend most of their time on the bottom, and feed by rasping material from submerged surfaces.

Because amphibians are ectotherms, their inherent developmental rates are modified by temperature and other environmental variables; size is thus an inaccurate estimator of chronological age. Consequently, biologists describe tadpole ontogeny using a staging table that divides their development into recognizable stages based on the attainment of specific morphological landmarks. With such a table the degree of development of morphological features of tadpoles can be compared among populations and across taxa occurring in the same or different habitats regardless of chronological age or attainable size.

Larval amphibians are exceptionally variable within and among species, although the degree and patterns of that variation are poorly documented and their sources rarely investigated. The many papers on induced morphological changes published in recent years (e.g. Relyea and Auld 2005, among many others) have made it abundantly clear that every tadpole of a given taxon collected at any site is a variant within the broad phenotypic range of its taxon. Although results from mesocosm experiments with controlled combinations and densities of species provide some insight into understanding phenotypic variation, predicting morphological variation from random samples of ponds is highly unlikely. The presence of different sets of predators in natural situations complicates the picture even more, and one has to keep these factors in mind when evaluating the morphology of tadpoles.

We mention in passing that less is known about amphibian eggs (e.g. Altig and McDiarmid 2007), hatchlings (Gosner stages 21–24; Altig 1972), and metamorphs than is known about tadpoles (stages 25–41) and other amphibian larvae. We urge workers to preserve and describe positively identified samples of these stages. Here we summarize data on the morphology, ontogeny, and diversity of larvae in each amphibian order.

3.2 Larval caecilians

3.2.1 Morphology and ontogeny

The vermiform body (Figure 3.1) has primary annuli homologous to the costal folds of salamanders that form during early development; secondary and tertiary annuli may form later. External gills with branched rami (Figures 3.1a and c) are

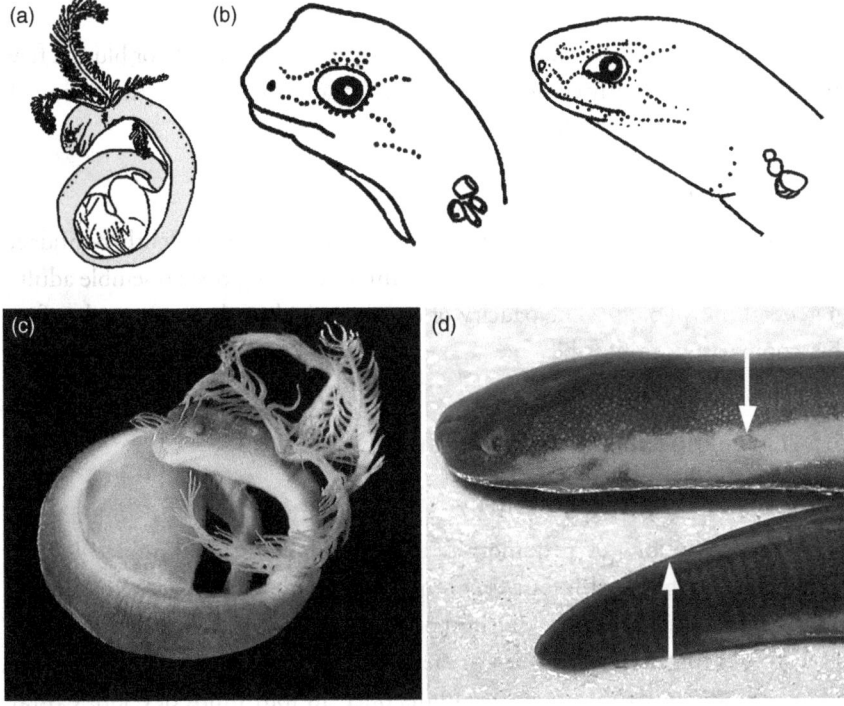

Fig. 3.1 Morphology of a larval caecilian. (a) Embryo of *Ichthyophis glutinosus*, modified from figure in Sarasin and Sarasin (1887–1890); (b) heads of embryos of *Ichthyophis kohtaoensis*, stages 27 (left) and 33 (right), modified from drawings in Dünker *et al.* (2000); (c) hatchling of *I. kohtaoensis*, modified from photograph by A. Summers on front cover, *Journal of Morphology* 2000, **243**(1); and (d) larva of *Ichthyophis bananicus* showing the gill slit (upper arrow) and tail fin (lower arrow), photograph by R. Nussbaum.

present during embryological development and lost at hatching (Figure 3.1b). The single gill slit (Figure 3.1d) closes at metamorphosis, and a small fin is present on a short tail (Figure 3.1d). The sensory tentacle, situated adjacent or anterior to the eye of caecilians, develops at metamorphosis. Some caecilians have small bony scales (i.e. osteoderms) embedded in the skin in the grooves between the annuli. These scales that are quite small when they first appear, get progressively larger as they develop, beginning in the posterior annuli and moving anteriorly. The definitive pattern of occurrence varies among species. Neuromasts, or lateral line organs, are present throughout larval life; labial folds, which modify the shape and size of the mouth opening during suction feeding, are present as they are in larval salamanders. The most complete information on early ontogeny can be found in Dünker *et al.* (2000) (also see Sarasin and Sarasin 1887–1890; Brauer 1897, 1899; Sammouri *et al.* 1990).

3.2.2 Coloration

Most caecilians are somewhat drab shades of gray, brown, black, or blue. A few species are more brightly colored and slightly banded or striped (terminology of Altig and Channing 1993).

3.2.3 Diversity

Larvae of species of caecilians in the families Rhinatrematidae, Ichthyophiidae, and some Caeciilidae that are known are similar to and grossly resemble adults in general morphology. The paucity of ontogenetic data, however, makes further comparisons impossible.

3.3 Larval and larviform salamanders

3.3.1 Morphology and ontogeny

With the exceptions of pedomorphs in the families Amphiumidae (cylindrical, elongate body with four tiny limbs, each with one to three tiny digits) and Sirenidae (cylindrical, elongate bodies with front limbs only, each with three or four fingers), salamander larvae have a typical quadruped morphology (Figure 3.2a) similar to that of the adults once all four limbs develop. Costal grooves divide the myotomic muscle bundles of the trunk into costal folds. Gills of various shapes are often prominent (Figures 3.2a, b, and d), one to four gill slits open during larval life eventually close in metamorphic taxa, gill rakers are prominent to absent, and a gular fold that typically is free from adjacent throat tissue is usually present. Fleshy labial folds modify mouth size and shape during suction feeding, and neuromasts are present, although sometimes difficult to see without special techniques (Lannoo 1985). Major metamorphic modifications include the loss of tail fins, gills, gill slits, gular folds, and neuromasts, the development of eyelids, and many other integumentary, osteological, and physiological changes.

Pond-inhabiting larvae (e.g. most species of Ambystomatidae and Hynobiidae) have robust bodies and heads, tall dorsal fins that can originate as far anterior as the back of the head, and long gill rami with plumose fimbriae. Stream-inhabiting larvae (e.g. certain species of Hynobiidae, Plethodontidae, and some Salamandridae) are more streamlined; they have low fins that usually originate near the tail/body junction, and shorter, less plumose gills. In species within the Rhyacotritonidae and desmognathine Plethodontidae, the gill rami are exceptionally short with few fimbriae. Gill rami are branched in species in the families Amphiumidae (gills lost soon after hatching) and Sirenidae (gills persist

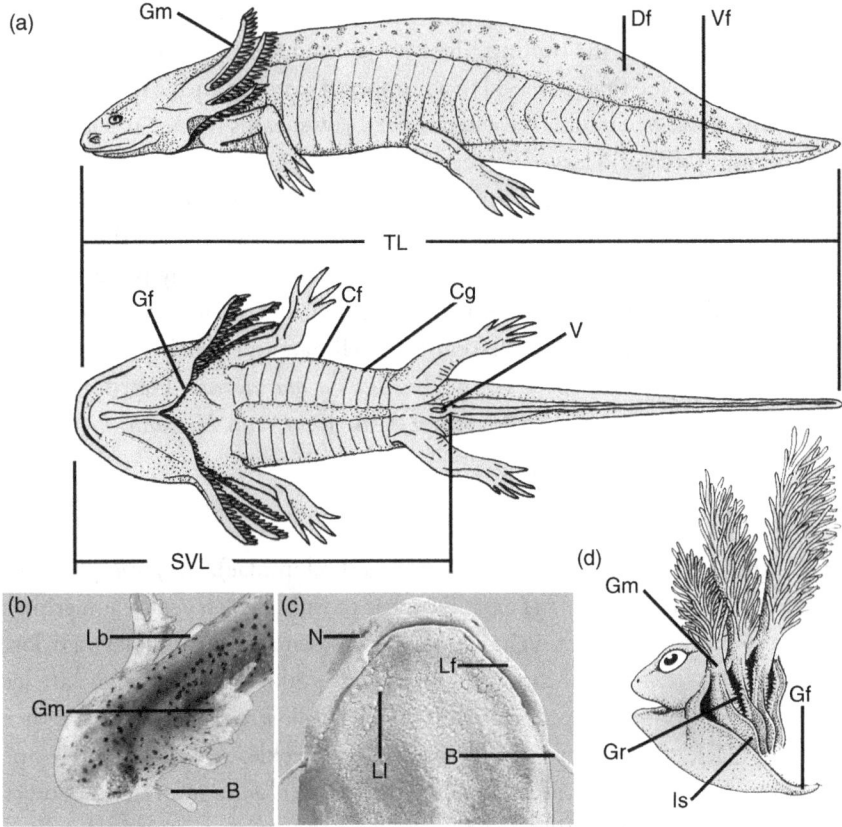

Fig. 3.2 Measurements and body parts of a larval salamander. (a) Lateral (upper) and ventral (lower) views of a typical *Ambystoma* salamander larvae, drawing by D. Karges; (b) dorsal view of the head and anterior body region of a hatchling *Ambystoma maculatum*, photograph by A.M. Richmond; (c) ventral view of head of *A. maculatum*; and (d) stylized drawing of gill structure of a larval salamander, modified from drawing in Pfingsten and Downs (1989). B, balancer; Cf, costal fold; Cg, costal groove; Df, dorsal fin; Gf, gular fold; Gm, gill ramus with fimbriae attached to posterior surface; Gr, gill rakers; Is, interbranchial septum; Lb, limb bud; Lf, labial fold; Ll, lateral line organs (neuromasts); N, naris; SVL, snout–vent length; TL, total length; V, vent; Vf, ventral fin.

throughout life although they atrophy during aestivation); rami in larvae of all other families are non-branched. Larvae of sirenid salamanders have low, fleshy, pigmented fins restricted to the tail, as in adults, and hatchlings have a transparent dorsal fin that originates well forward on the body. Amphiumid larvae have a very short, low caudal fin that is lost soon after hatching. All other salamander larvae have tail fins throughout their ontogeny. Regression of the dorsal fin usually starts long before other metamorphic changes are noticeable. Fleshy flaps occur on the trailing edges of the hind legs of *Onychodactylus* larvae. Keratinized toe tips are found in a number of taxa but are most common in stream inhabitants. Sirenids have keratinized jaw sheaths (upper and lower) as do some ambystomatid larvae (lower).

Hatching occurs before the limbs are fully developed, and the front limbs usually develop faster than the hind ones. Some hatchlings (e.g. in the families Ambystomatidae and Salamandridae) have a balancer, a fleshy projection on each side of the lower part of the head (Figures 3.2b and c), that is lost soon after hatching. Staging tables (compilation in Duellman and Trueb 1986, p. 128) are available for several species (e.g. *Ambystoma maculatum*, Harrison 1969; *Ambystoma mexicanum*, Cano-Martinez *et al.* 1994; and *Hynobius nigrescens*, Iwasawa and Yamashita 1991).

3.3.2 Coloration

In contrast to *Stereochilus marginatus* (Plethodontidae), *Rhyacotriton* spp. (Rhyacotritonidae), and most pedomorphs, most of which retain something similar to the larval coloration as adults, larval salamanders often have a coloration distinct from that of the metamorph or the adult. Larval sirenids are jet black with contrasting stripes and bands of red or yellow, while adults have either a gray or black ground color usually overlain by speckles and small blotches of gold to greenish iridophores. Color and pattern (i.e. coloration) in larvae of most species can vary considerably during ontogeny, throughout a day, and among sites in response to substrate color, temperature, and water clarity. Colors are typically muted grays, browns, and blacks, and patterns range from none (unicolored), blotched, and mottled through striped (longitudinal or diagonal contrasting markings) and banded (transverse contrasting markings). The dorsum of the tail muscle of small *Ambystoma* is often banded, and the pattern may be retained throughout ontogeny (e.g. *Ambystoma talpoideum*) or change to a totally different pattern sometime after hatching and then again after metamorphosis. In some species and populations, larval *Desmognathus* (Plethodontidae) have a distinct pattern that is kept throughout life. Although colors are usually more muted, the adult patterns in other salamanders (e.g. *Ambystoma tigrinum* group, many plethodontids) may appear at metamorphosis or a different pattern may appear (e.g. most *Ambystoma*) after metamorphosis and slowly develop into the adult pattern, which is achieved long before sexual maturity.

3.3.3 Diversity

Larval salamanders show much less ecomorphological diversity than tadpoles. By definition, pedotypes and pedomorphs retain a larval morphology even though they become reproductive, and species that metamorphose and are adapted for either pond or flowing water are the most easily recognized groups. Cannibal morphotypes with enlarged heads and altered dentition occur in some species (Ambystomatidae, Hynobiidae). Pond inhabitants often do not over-winter, whereas some stream inhabitants may grow as larvae for several years before undergoing metamorphosis. In some parts of their range *Notophthalmus viridescens* (Salamandridae) larvae metamorphose into a brilliantly colored eft that lives on the forest floor for several years before returning to the ponds for an aquatic existence as a reproductive adult.

3.4 Anuran tadpoles

3.4.1 Morphology and ontogeny

The transition between the body and tail across all stages and taxa of tadpoles is difficult to define. The demarcation between the two is most consistently and accurately described as the juncture of the axis of the tail myotomes with the posterior body surface (Figure 3.3a, bottom). The tails of most tadpoles lack vertebrae and are composed of dorsal and ventral fins and a long series of pro-gressively smaller myotomic muscle bundles surrounding the notochord. The tadpoles of some species of Megophryidae do have tail vertebrae (e.g. Haas *et al.* 2006; Handrigan *et al.* 2007). The shapes and extents of the fins vary among taxa and habitats (i.e. tallest in pond dwellers, lowest in fast-water and semi-terrestrial forms). The eyes are either dorsal (i.e. lie totally within the dorsal sil-houette; Figure 3.3b, left) or lateral (i.e. included as part of the dorsal silhouette; Figure 3.3b, right). All free-living tadpoles have a spiracle(s) through which water that has been pumped in through the mouth by the buccopharyngeal muscula-ture and passed over the gills and food filtering system passes out of the body. In the vast majority of species, the spiracle is single and situated somewhere on the left side of the body (Figure 3.3a). Tadpoles of *Ascaphus* (Leiopelmatidae) have a single, ventral spiracle on the chest, while those of *Bombina* (Bombinatoridae) and *Alytes* and *Discoglossus* (Alytidae) have a single spiracle located almost midventrally on the abdomen. In microhylid larvae, the midventral spiracle is located at the posterior part of the abdomen or near the vent. The tadpoles of pipids, rhinophrynids and the leptodactylid genus *Lepidobatrachus* have dual lateral spiracles. The spiracles of *Lepidobatrachus* develop differently from the other two taxa (Ruibal and Thomas 1988).

Scent-laden water passes through the nares, over the olfactory epithelium of the nasal sacs, and into the buccal cavity via the internal nares. The shape of the external apertures varies from round to elliptical and may have a variety of papillae

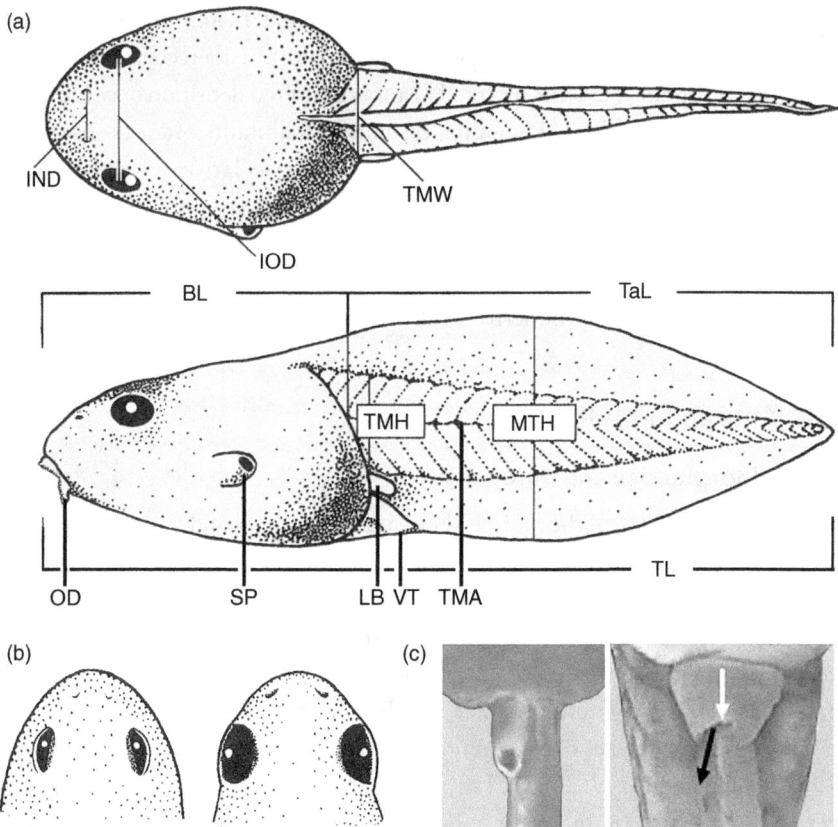

Fig. 3.3 Measurements and body parts of a tadpole. (a) Dorsal (upper) and lateral (lower) views of a typical tadpole, drawing of *Rana* sp. by D. Karges; (b) dorsal (left) and lateral (right) eye positions of tadpoles, stylized drawings by D. Karges; and (c) examples of medial (left, *Bufo boreas*) and dextral (right, *Rana catesbeiana*) vent tubes; white arrow, plane of ventral fin; black arrow, outflow of vent tube. BL, body length; IND, internarial distance; IOD, interorbital distance; LB, hind limb bud; MTH, maximum tail height; OD, oral disc; SP, spiracle; TMA, tail muscle axis; TMH, tail muscle height; TMW, tail muscle width; TL, total length; TaL, tail length; VT, vent tube.

associated with the margins. A vent tube extends posteriorly from the midventral abdomen. Two major types are recognized: dextral, where the aperture lies to the right of the sagittal plane of the tail fin (e.g. hylids and ranids), and medial, where the aperture lines parallel with the plane of the tail fin (e.g. bufonids and scaphi-opodids; Figure 3.3c). As with the spiracular tube configurations, there are many subtle variations in the shape and position of the vent tube.

The lateral line system (i.e. neuromasts; Hall *et al.* 2002; Lannoo 1985) is composed of many depressions in the skin with sensory cells in the center that signal the patterns of water flow over various parts of the body and tail. The distribution and arrangement of neuromasts may be useful in distinguishing between closely related species. In darkly pigmented tadpoles these sites are often pale and obvious, but evaluation of stitch patterns in most tadpoles requires separating the epidermis from the underlying dermis, clearing in glycerin, and viewing the skin with dark-field illumination (see Lannoo 1985).

The oral apparatus, the composite of upper and lower labia and all keratinized mouthparts, is highly variable across taxa and ecological types. The most common oral apparatus (Figure 3.4a and c) occurs in many taxa in lentic and lotic sites. An assembly of the two infralabial and two Meckel's cartilages with three joints forms the lower jaw that is surmounted by a serrated, keratinized jaw sheath. The supralabial cartilage of the upper jaw is surmounted by a similar keratinized jaw sheath, and during a bite, the lower jaw passes totally behind the upper (Figure 3.4d). The interactions of the serrated margins of the sheaths serve as cutting/gouging surfaces when a tadpole feeds. The highly variable shapes of the jaw sheaths suggest different feeding abilities.

The face of the oral disc has fleshy transverse tooth ridges (Figure 3.4a, c, and d) surmounted by a row(s) of keratinized labial teeth. In most cases, several replacement teeth are interdigitated below a presently erupted tooth (Figure 3.4b, lower right), and they successively move into position as the erupted tooth wears out. The tooth rows are numbered from the anterior edge of the disc toward the mouth on the upper labium and from the mouth to the posterior edge of the disc on the lower labium. A fractional designation indicates the number of tooth rows on each labium; some rows have naturally occurring medial gaps denoted parenthetically. For example, a Labial Tooth Row Formula (LTRF) of 2(2)/3(1) indicates two upper rows with a gap in the second one and three lower rows with a gap in the first one (Figures 3.4a and c). Some tadpoles lack tooth rows (i.e. 0/0), and the maximum LTRF known is 17/21 in a tadpole of an undescribed hylid frog from the Guayana Highlands of southern Venezuela.

The papillate margins of the oral disc may be complete and encircle the disc (e.g. tadpoles of Scaphiopodidae, Pelobatidae, and many stream-inhabiting tadpoles of several families), have a medial dorsal gap (most common; Figure 3.4a), or have both dorsal and ventral gaps (e.g. Bufonidae and those of some Hylidae, Mantellidae, Ranidae, and Rhacophoridae; Figure 3.4c). Although the number of rows of papillae on different parts of the disc margin may vary, the lengths of the papillae are typically somewhat uniform; several species of *Phrynobatrachus* (Ranidae) have exceptionally elongate papillae along the posterior margin of the disc. Submarginal papillae occur on the face of the disc away from the margin

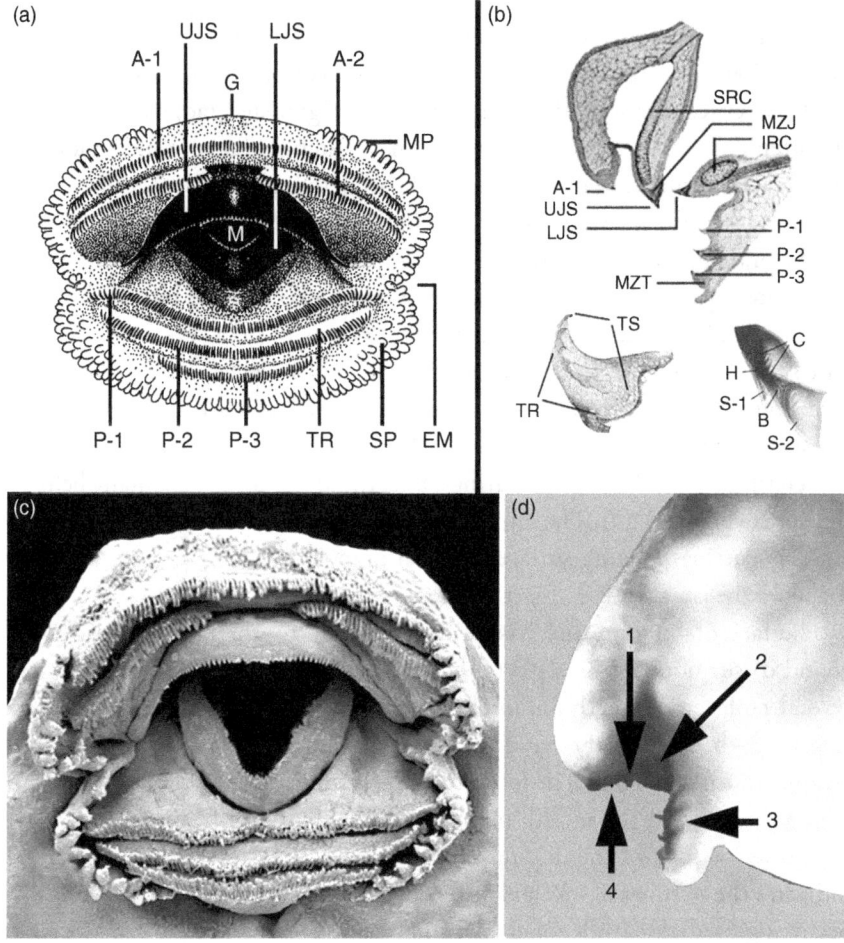

Fig. 3.4 Components of the oral disc of a tadpole. (a) Oral disc of a typical tadpole, schematic drawing by D. Karges; (b) sagittal sections of the oral apparatus of a common benthic tadpole (upper) and a tooth ridge (lower left), schematic drawings modified from ones in Heron-Royer and Van Bambeke (1889); two labial teeth of *Hyla chrysoscelis* (lower right) removed from a tooth series and in natural position; (c) scanning electron photomicrograph of the oral apparatus of a tadpole of *Bufo fowleri*, actual oral disc width, about 2.3 mm, photograph by M. Penuel-Matthews; and (d) lateral view of the mouthparts of *H. chrysoscelis* (1, upper jaw sheath; 2, lower jaw sheath; 3, lower tooth rows; 4, upper tooth row). A-1, A-2, anterior tooth rows 1 and 2; P-1, P-2, P-3, posterior tooth rows 1 to 3; S-1, S-2, sheaths of presently erupted (1) and first replacement (2) teeth; B, body of first replacement tooth; C, cusps on first replacement tooth; EM, lateral emargination in oral disc; G, dorsal gap in marginal papillae; H, head of presently erupted tooth; IRC, infrarostral cartilage; LJS, lower jaw sheath; M, mouth; MP, marginal papillae; MZJ, mitotic zone for production of jaw sheath; MZT, mitotic zone for production of labial teeth; SP, submarginal papillae; SRC, suprarostral cartilage; TR, tooth ridge; TS, tooth series; UJS, upper jaw sheath.

and form various patterns (Figure 3.4a). The margin of the disc may be emarginate (i.e. indented; Figure 3.4a) or not.

Hatchlings (Gosner 1960; stages 21–24) usually have external gills but lack eyes and limb buds. The forelimb buds develop beneath the operculum after it closes, and the hind-limb buds grow from the posteroventral intersection of the body and tail muscle (Figure 3.3a, bottom). Staging tables have been made for a number of taxa (Duellman and Trueb 1986, pp. 128–129), but using a common table allows for meaningful comparisons among taxa. Gosner (1960; general) and Nieuwkoop and Faber (1956; *Xenopus*) are the two most commonly cited. Recent summaries of tadpole morphology and terminology are included in Altig (2007b) and Altig and McDiarmid (1999a, 1999b).

3.4.2 Coloration

Except for notations in descriptions, surprisingly little has been written about tadpole coloration. As in other larval amphibians, three basic populations of pigment-containing cells interact to produce both color and pattern. Melanophores contain melanins that produce browns and blacks, iridophores contain reflective guanine crystals that produce whites and silvers, and xanthophores contain carotenoids that produce yellows and reds. The pigments are retained inside the cells and can be dispersed in various patterns under the influence of temperature and light. Altig and Channing (1993) summarized the diversity of colorations in tadpoles, and Caldwell (1982) tested the functions of coloration in tadpoles experimentally.

3.4.3 Diversity

Most morphological characters of tadpoles reflect their ecology. Suctorial tadpoles in a number of families have streamlined bodies and mouthparts modified to maintain position in fast-flowing water as they feed. A typical increase in the number of tooth rows, to a maximum of 17/21, is usually accompanied by a larger oral disc with complete marginal papillae. Other unusual morphological structures found in stream-inhabiting tadpoles include a belly modified as a sucker (a few bufonid and ranid species), and lateral sacs (or lymphatic sacs) on the ventrolateral parts of the body of other stream-dwelling tadpoles (Arthroleptidae). Tadpoles of *Mertensophryne* (Bufonidae) have a hollow crown on the head that encircles the eyes and nares, and tadpoles of *Schismaderma carens* (Bufonidae) have a semicircular, transverse flap of skin behind the eyes.

Suspension-feeding tadpoles in the families Microhylidae, Pipidae, and Rhinophrynidae have reduced, soft mouthparts that lack keratinized structures. They usually hang in midwater and capture suspended particles as water is pumped in through the mouth and out the spiracles. Even so, not all tadpoles that lack keratinized mouthparts are suspension feeders, and tadpoles with keratinized mouthparts that are infected with the amphibian chytrid fungus (*Batrachochytrium dendrobatidis*) often lose most or all such structures.

Carnivores and other macrophagous feeders have a diversity of mouthparts related to how they feed. For example, tadpoles of the leptodactylid frog *Lepidobatrachus* have huge mouths but almost no soft or keratinized mouthparts; they engulf entire organisms, including other tadpoles. Tadpoles of other leptodactylid frogs, *Ceratophrys* spp., have huge jaws and many tooth rows and efficiently tear their victims to pieces. Carnivorous tadpoles of *Spea* (Scaphiopodidae) feed similarly. A number of other tadpoles (e.g. Hylidae: *Hyla leucophyllata* group; Ranidae: *Occidozyga*) lack all or most tooth rows but have huge jaw sheaths. Tadpoles that occupy tree holes and bromeliad tanks (e.g. some species of Dendrobatidae, Hylidae, and Rhacophoridae) are of several morphological types (Lannoo *et al.* 1987; Lehtinen *et al.* 2004) and have different diets; some are non-feeding, several are detritovores or macrophagous carnivores, many eat frog eggs (fertilized or not) of their own (cannibals) or other species, and some eat only trophic eggs supplied by their mother. Surface-film feeders have large oral discs, but keratinized structures are reduced or absent; the disc is turned upward (i.e. umbelliform) and captures material carried in the surface film. The oral discs of these surface-feeding tadpoles may be formed primarily from the lower labium (e.g. Microhylidae) or from parts of both labia (e.g. Megophryidae). This convergent morphology occurs in six families, and most tadpoles of this sort occur only in the slow reaches of streams.

Attempts have been made to define ecomorphological guilds or groups of taxa with suites of common morphological characters that are presumed to reflect a common ecology (e.g. Altig and Johnston 1989). Because of the lack of ecological data for many species and an incomplete understanding of how some of their morphologies actually function, we advise caution in assigning species to specific guilds without some knowledge of their natural history. For example, the morphologies of *Mantidactylus lugubris* (Altig and McDiarmid 2006) and of some taxa that occur in phytotelms suggest that one might find them in fast-flowing water. In fact, tadpoles of *M. lugubris* live in leaf packs in slow-flowing water.

3.5 Summary

Amphibian larvae show considerable morphological diversity from the relatively conserved forms of caecilians and salamanders to the unusual and often novel structures found in tadpoles of frogs and toads. The extreme variability of tadpoles is almost certainly a product of ontogenetic challenges, the recently discovered influences of predators and competitors (e.g. Relyea and Auld 2005), and the selective effects of different habitats. The morphological variability manifest under these different conditions makes identifications and ecological evaluations especially difficult. This situation, combined with our lack of understanding of geographic variation in larval amphibians, especially tadpoles, emphasizes the relative poor state of our knowledge of larval biology. Much remains to be learned. The discovery of various anomalies (e.g. Drake *et al.* 2007), often with a weak understanding of their causes (Altig 2007a; see also Lannoo 2008), adds yet another impediment to our total understanding of larval morphology.

3.6 References

Altig, R. (1972). Notes on the larvae and premetamorphic tadpoles of four *Hyla* and three *Rana* with notes on tadpole color patterns. *Journal of the Elisha Mitchell Scientific Society*, **88**, 113–19.

Altig, R. (2007a). Comments on the descriptions and evaluations of tadpole mouthpart anomalies. *Herpetological Conservation and Biology*, **2**, 1–4.

Altig, R. (2007b). A primer for the morphology of anuran tadpoles. *Herpetological Conservation and Biology*, **2**, 73–6.

Altig, R. and Channing, A. (1993). Hypothesis: functional significance of colour and pattern of anuran tadpoles. *Herpetological Journal*, **3**, 73–5.

Altig, R. and Johnston, G. F. (1989). Guilds of anuran larvae: relationships among developmental modes, morphologies, and habitats. *Herpetological Monographs*, **3**, 81–109.

Altig, R. and McDiarmid, R. W. (1999a). Body plan: development and morphology. In R. W. McDiarmid and R. Altig (eds), *Tadpoles, the Biology of Anuran Larvae*, pp. 24–51. University of Chicago Press, Chicago, IL.

Altig, R. and McDiarmid, R. W. (1999b). Diversity: familial and generic characterizations. In R. W. McDiarmid and R. Altig (eds), *Tadpoles, the Biology of Anuran Larvae*, pp. 295–337. University of Chicago Press, Chicago, IL.

Altig, R. and McDiarmid, R. W. (2006). Descriptions and biological notes on three unusual mantellid tadpoles (Amphibia: Anura: Mantellidae) from southeastern Madagascar. *Proceedings of the Biological Society of Washington*, **119**, 418–25.

Altig, R. and McDiarmid, R. W. (2007). Morphological diversity and evolution of egg and clutch structure in amphibians. *Herpetological Monographs*, **21**, 1–33.

Brauer, A. (1897). Beiträge zur Kenntnis der Entwicklungsgeschichte und Anatomie der Gymnophionen. *Zoologische Jahrbucher Anatomie*, **10**, 277–472.

Brauer, A. (1899). Beiträge zur Kenntnis der Entwicklung und Anatomie Gymnophionen. *Zoologische Jahrbucher Anatomie*, **12**, 477–508.

Caldwell, J. P. (1982). Disruptive selection: a tail color polymorphism in *Acris* tadpoles in response to differential predation. *Canadian Journal of Zoology*, **60**, 2818–27.

Cano-Martínez, A., Vargas-González, A., and Asai, M. (1994). Metamorphic stages in *Ambystoma mexicanum*. *Axolotl Newsletter*, **23**, 64–71.

Drake, D. L., Altig, R., Grace, J. R., and Walls, S. C. (2007). Occurrence of oral defects in larval anurans from protected sites. *Copeia*, **2007**, 449–58.

Duellman, W. E. and Trueb, L. (1986). *Biology of Amphibians*. McGraw-Hill, New York.

Dünker, N., Wake, M. H., and Olson, W. M. (2000). Embryonic and larval development in the caecilian *Ichthyophis kohtaoensis* (Amphibia, Gymnophiona): a staging table. *Journal of Morphology*, **243**, 3–34.

Gosner, K. L. (1960). A simplified table for staging anuran embryos and larvae with notes on identification. *Herpetologica*, **16**, 183–90.

Haas, A., Hertwig, S., and Das, I. (2006). Extreme tadpoles: the morphology of the fossorial megophryid larva, *Leptobrachella mjobergi*. *Zoology*, **109**, 26–42.

Hall, J. A., Larsen, Jr, J. H., and Fitzner, R. E. (2002). Morphology of the prometamorphic larva of the spadefoot toad, *Scaphiopus intermontanus* (Anura: Pelobatidae), with an emphasis on the lateral line system and mouthparts. *Journal of Morphology*, **252**, 114–30.

Handrigan, G. R., Haas, A., and Wassersug, R. J. (2007). Bony-tailed tadpoles: the development of supernumerary caudal vertebrae in larval megophryids (Anura). *Evolution and Development*, **9**, 190–202.

Harrison, R. G. (1969). Harrison stages and description of the normal development of the spotted salamander, *Ambystoma punctatum* (Linn.). In R. G. Harrison (ed.), *Organization and Development of the Embryo*, pp. 44–6. Yale University Press, New Haven, CT.

Héron-Royer, L. F. and Van Bambeke, C. (1889). Le vestibule de la bouche chez les têtards des batraciens anoures d'Europe; sa structure, ses caractères chez les diverses espèces. *Archives de Biologie*, **9**, 185–309.

Iwasawa, H. I. and Yamashita, K. (1991). Normal stages of development of a hynobiid salamander, *Hynobius nigrescens* Stejneger. *Japanese Journal of Herpetology*, **14**, 39–62.

Lannoo, M. J. (1985). Neuromast topography in *Ambystoma* larvae. *Copeia*, **1985**, 535–9.

Lannoo, M. J. (2008). *Malformed Frogs. The Collapse of Aquatic Ecosystems*. University of California Press, Berkeley, CA.

Lannoo, M. J., Townsend, D. S., and Wassersug, R. J. (1987). Larval life in the leaves: arboreal tadpole types, with special attention to the morphology, ecology, and behavior of the oophagous *Osteopilus brunneus* (Hylidae) larva. *Fieldiana Zoology*, **38**, 1–31.

Lehtinen, R. M., Lannoo, M. J., and Wassersug, R. J. (2004). Phytotelm-breeding anurans: past, present and future research. In R. M. Lehtinen (ed.), *Ecology and Evolution of Phytotelm-Breeding Anurans*, pp. 1–9. Miscellaneous Publication 193. Museum of Zoology, University of Michigan, Ann Arbor, MI.

Nieuwkoop, P. D. and Faber, J. (1956). *Normal tables of Xenopus laevis (Daudin)*. North-Holland Publishing Company, Amsterdam.

Pfingsten, R. A. and Downs, F. L. (1989). Salamanders of Ohio. *Bulletin of the Ohio Biological Survey*, **7**, 1–350.

Reilly, S. M., Wiley, E. O., and Meinhardt, D. J. (1997). An integrative approach to heterochrony: the distinction between interspecific and intraspecific phenomena. *Biological Journal of the Linnean Society*, **60**, 119–43.

Relyea, R. A. and Auld, J. R. (2005). Predator- and competitor-induced plasticity: how changes in foraging morphology affect phenotypic trade-offs. *Ecology*, **86**, 1723–9.

Ruibal, R. and Thomas, E. (1988). The obligate carnivorous larvae of the frog, *Lepidobatrachus laevis* (Leptodactylidae). *Copeia*, **1988**, 591–604.

Sammouri, R., Renous, S., Exbrayat, J. M., and Lescure, J. (1990). Développement embryonnaire de *Typhlonectes compressicaudus* (Amphibia, Gymnophiona). *Annales de Sciences Naturelles-Zoologie et Biologie Animale, Paris*, **11**, 135–63.

Sarasin, P. and Sarasin, S. (1887–1890). Zur Entwicklungsgeschichte und Anatomie der Ceylonesischen Blindwühle *Ichthyophis glutinosus*. In *Ergebnisse Naturwissenschaftlicher Forschungen auf Ceylon in den Jahren 1884–1886*. Band II. C. W. Kreidel's Verlag, Wiesbaden.

Thibaudeau, D. G. and Altig, R. (1999). Endotrophic anurans: development and evolution. In R. W. McDiarmid and R. Altig (eds), *Tadpoles, the Biology of Anuran Larvae*, pp. 170–88. University of Chicago Press, Chicago, IL.

4

Larval sampling

David K. Skelly and Jonathan L. Richardson

4.1 Introduction

Most amphibian species are metamorphic, and among those, the majority have larvae that are fully aquatic (Duellman and Trueb 1986). These larvae are found in an enormous variety of contexts ranging from bromeliads and tree holes, to brackish pools and the largest rivers and lakes (Duellman and Trueb 1986). Add to this mix of environments the fact that amphibian larvae differ dramatically in their microhabitat use, from the water surface to benthic mud, and researchers sampling larval amphibians are confronted with a non-trivial challenge of matching techniques with study goals and logistic limits. Fortunately, there is a wealth of experience that can be tapped when making decisions about where, when, and how to sample (Shaffer *et al.* 1994; Olson *et al.* 1997).

4.1.1 Why sample larvae?

The reasons to sample amphibian larvae are many, but most emphasize either strategic or logistic considerations. From the strategic perspective, larvae represent a critical life history stage. Many of the reasons that motivate scientists to study amphibians are related to the dynamics and fate of their populations. Monitoring of larval cohorts can provide critical information regarding the trajectory of a population and the factors that may affect abundance and distribution. Larvae also represent concrete evidence of breeding. While many studies of anurans use male calling as an index of breeding distribution, males can call from wetlands where no breeding takes place. Larval surveys are less likely to overestimate breeding distribution.

For many species larvae are also convenient. Many adult amphibians are found at low density spread across large areas of terrestrial habitat. In addition they are often cryptic, arboreal, or fossorial, making it extremely difficult to sample their populations systematically. For amphibians with fully metamorphic life histories and aquatic larvae, breeding represents the point in the life cycle where a species can be indexed in a relatively small, defined area. The techniques described below offer a variety of systematic approaches to estimating population attributes that would be extremely challenging to determine for the adult stages of many amphibian species.

Because of their strategic and logistic advantages, amphibian larvae have been used in a wide range of research aimed at addressing fundamental questions in fields ranging from ecology and evolution to physiology and developmental biology. Well before amphibian larvae were widely studied by scientists concerned about declines and species extinctions, larval amphibians were held up as a model system by biologists.

4.1.2 Target responses

Once the motivation for sampling larvae is known, the target response variables may be selected (Table 4.1). Information on species occupancy is frequently sought in both basic and applied settings. Presence and absence data can be important in determining species richness within a wetland as well as providing evidence for changes in species range. Any of the techniques described in this chapter can be used to estimate species presence and absence. However, if presence/absence is the sole motivation for sampling, some techniques such as dip-netting (described below) may be much more efficient than others.

Table 4.1 Larval amphibian sampling methods and resulting response variables

Response	Metrics	Sampling methods	Inferences
Species occupancy	Presence/absence	All methods	Species distribution, richness
Relative density	Catch per unit effort (CPUE)	Time- or area-constrained sampling, trapping, litterbags	Comparing among populations and species
Density	Individuals per unit area or volume	Area- or volume-constrained sampling including box and pipe sampling	Comparing among populations and species cohort survival
Population size	Total number of individuals in population	Extrapolation from area- or volume-constrained samples, mark–recapture	Population dynamics extinction risk

Beyond estimating presence and absence, sampling can be directed at estimating density. Any technique which can be scaled by effort (sometimes expressed as catch per unit effort, or CPUE) can offer estimates of relative density; that is, the density of a given species relative to another. As one example, the number of larvae from one species recovered during 30 person minutes of dip-netting may be compared with a second species as an index of their relative density (both expressed as individuals recovered per person minute). A smaller number of techniques, identified below, allow direct estimates of the number of individuals per area, or per volume. In general, these techniques operate on the principle that a defined area or volume is sampled effectively. As one example, each use of a standard box sampler covers 0.5 m². After completing a set of 10 box samples within a wetland, the number of larvae of some species recovered from 5 m² can be expressed as the number of individuals per square meter. Finally, using mark–recapture techniques, estimates of the total number of individuals within a larval cohort may be made.

4.1.3 Timing

Amphibian larvae range from practically immobile, yolk-laden hatchlings, to 30-cm-long salamander larvae capable of moving extremely rapidly. Deciding when to sample during the larval period is best done in conjunction with decisions about how to sample. Small larvae move slowly and are often found in shallow areas. These attributes can make sampling relatively straightforward using a variety of techniques. This is something to consider if your study aims allow flexibility in the developmental stage sampled. It also suggests that effective larval sampling depends on a close knowledge of the life histories of the species to be sampled and their developmental progress within a given year. Late snowpack melt, droughts, and a particularly cold or warm spring can all shift the timing of larval development substantially. In tropical climates the time of year can be far less important than the onset of wet season rains in triggering breeding and the timing of larval sampling.

4.1.4 Sampling effort

How much is enough? In any sampling context, researchers are confronted with the task of deciding how to allocate effort. As expected, a satisfactory answer depends on study goal and tolerance of uncertainty. Any of the techniques described below can be scaled in effort. The effort adopted, however, will require direct estimates of how species detection, larval density estimates, or whatever the target variable is, changes with increased effort. In our own research, we have intentionally varied per wetland sampling effort (samples per visit, number of visits, timing of visits) to determine the sensitivity of target responses (e.g. Werner *et al.* 2007). There is no shortcut here; rules of thumb can be misleading resulting in wasted effort and more collection than necessary or, more likely, incomplete and inaccurate information (Skelly *et al.* 2003).

If you are new to a system, plan to learn during your initial sampling. Intentionally trying multiple techniques and different degrees of sampling effort will provide information that will ultimately save you time and produce more reliable information. Before you step in the water, be prepared to spend more time in each habitat if it is needed and to consider using multiple techniques to capture the range of species and larval stages you intend to study.

4.2 Sampling techniques

4.2.1 Box/pipe sampler

4.2.1.1 Description

These area-based samplers are dropped rapidly through the water column in areas of less than 1 m in depth. The earliest versions were sheet metal or plywood boxes, 0.5 × 1.0 m, used in conjunction with a metal frame net designed to fit snugly within the width of the box (Harris *et al.* 1988). Repeated sweeps of the trapped volume of water are used to clear and count the amphibian larvae within. A second form, the pipe sampler (Skelly 1996), ranges in size but is typically constructed of an approximately 1 m length of aluminum pipe, 30 cm or so diameter. Polyvinyl-coated aluminum pipe manufactured for air handling in commercial heating and cooling applications is particularly effective. Pipe samplers are used with dip nets constructed to measure half of the diameter of the pipe. Entrapped larvae are cleared using repeated circular sweeps of the water volume (Figure 4.1a).

Researchers have developed algorithms to ensure complete, or at least consistent, clearance of animals. As one example, Werner *et al.* (2007) report sweeping the water column of a pipe sampler a minimum of 10 times, and for at least 10 null sweeps after the last collected animal has been removed from the net. For both types of sampler, multiple samples are collected in a single wetland. Researchers can use grids to lay out sampling points which can be distributed among habitat types within a wetland. Alternatively, a minimum distance between sampling points can be specified and placement of individual samples set haphazardly within that constraint. If desired, data can be kept individually for each box or pipe providing information on spatial distribution, variation in local density, patterns of species coincidence, or association with particular microenvironments (Freidenburg 2003).

Fig 4.1 Representative sampling techniques targeting larval amphibians in aquatic habitats: (a) Pipe sampling in which a pipe is used to trap larvae that are then cleared using a dip net (section 4.2.1); (b) leaf litterbags are used for sampling salamander larvae in stream habitats (from Waldron *et al.* 2003; section 4.2.4); (c) a collapsible funnel trap deployed along a pond edge (section 4.2.5); (d) a larval *Ambystoma talpoideum* salamander marked with a lateral orange visible implant elastomer (VIE) tag (photograph courtesy of Kristen Landolt of Murray State University; section 4.2.6).

4.2.1.2 Application

Box samplers are most effective within vernal ponds with an open water column and simple bottom substrates such as decaying leaves. Pipe samplers were later developed to capture the advantages of a box sampler while enabling the sampling of a wider variety of environments including those where water is interspersed with emergent vegetation (Skelly 1996). We are unaware of the use of area-based larval amphibian samplers within stream environments, although the use of comparable samplers for benthic macroinvertebrates suggests the potential for such an application.

4.2.1.3 Considerations

The major advantage of box and pipe samplers is that sampling of a known area of wetland bottom provides direct estimates of larval density (individuals/m^2). If depth within each pipe is recorded, estimates of density per unit volume can be determined. In either case, if wetland bottom area (or volume) is known, researchers have extrapolated pipe- and box-sample-based density estimates into estimates of the size of entire larval cohorts (e.g. Werner *et al.* 2007).

A major disadvantage of these area-based samplers is their time-intensiveness. Repeatedly placing and clearing a box or pipe is relatively slow and methodical work even for experienced users. In addition, when sampling takes place in remote areas that require hiking into and out of, box and pipe samplers can be inconvenient because of their size.

4.2.2 Dip net

4.2.2.1 Description

Dip nets are probably the most common sampling tools used to collect amphibian larvae. Dip-netting can be relatively unstructured if the goal is simply to capture representatives of a particular species. Alternatively, dip-net surveys in which the elapsed time (Werner *et al.* 2007) or number of sweeps (Gunzburger 2007) is counted can be used to provide estimates of relative abundance of species. With calibration from other methods such as pipe sampling, time-constrained dip-net surveys have been used to provide per-unit-area density estimates (Werner *et al.* 2007). Effective dip-netting depends strongly on the species targeted and the environment type sampled. However, most neophytes tend to collect a great deal of substrate along with each sweep. With experience, users can learn how far into the water column and substrate the net needs to pass to capture the larvae without burying them in an overabundance of mud and vegetation.

Dip nets used in larval amphibian sampling range in size and shape from small aquarium nets used in confined volumes of water (Thoms *et al.* 1997) to very large nets approaching the size of some small seines (S. Cortwright, personal communication). Regardless of net size, it is common for amphibian biologists to construct their own nets or to customize store-bought nets. Dip nets used for larval amphibian sampling can have much shallower net bags and tend to require finer mesh than those used for fish and other larger organisms. A shallow net bag facilitates rapid processing of each sweep and can speed sampling significantly. Researchers can also construct net dimensions that fit the structure of the environments they sample (e.g. narrower nets may be appropriate for dipping out of marshes with emergent vegetation, small stream pools, or tire ruts).

4.2.2.2 Application

Dip nets are used in most of the places where amphibian larvae are found. Nets can be of lighter construction in vernal ponds (e.g. using mosquito mesh for the net bag) compared with those used in streams and other places with abrasive substrates.

4.2.2.3 Considerations

Dip-netting is fast, requires a minimum of equipment, and can be performed in a wide range of environments. Nets can be constructed inexpensively to perform well in particular contexts. On the downside, as much as any technique, dip-netting effectively relies on experience. An experienced user can vastly outperform a beginner working side by side. A second disadvantage relates to population density estimates. While dip-netting can be used by itself to estimate relative density in terms of catch per unit effort (where effort is often measured as time spent sampling), estimates of area- or volume-based density must rely on other techniques or through calibration from samples collected using other methods in the same or comparable environments (Werner *et al.* 2007).

4.2.3 Seine

4.2.3.1 Description

Seines have long been used to collect amphibian larvae (Routman 1984). They are particularly effective in sampling open, deeper areas that cannot easily be sampled using box and pipe samplers or dip nets. Seining typically requires at least two people. One person at each end sets the seine in a line and then begins moving in an agreed direction until a position is reached where they move together and begin gathering the ends of the seine up and out of the water, forcing the entrapped larvae down into the middle. This is done most easily if the seine is being gathered onto the shore. Sometimes this is not possible in which case, some sort of floating platform such as a foam bucket float can be used. When the captured sample is concentrated in the center of the seine, it is picked up and moved onto the shore where the contents are sorted and the larvae are processed as desired. Most seines used for amphibian larvae are relatively small, on the order of 3–5 m long and 1 m or so deep (Shaffer *et al.* 1994). Larger seines often have a bag sewn in the middle. The bag can greatly increase sampling effectiveness, but may also require a third person to keep it from getting caught or rolled up in vegetation.

4.2.3.2 Application

Seines are typically used in open-water areas of ponds and lakes, although large stream pools may also be sampled using seines in conditions where flow is not too great.

4.2.3.3 Considerations

Often seines are the only means of sampling larger amphibian larvae that live in open-water regions of ponds (e.g. large Ambystomatid salamander larvae). A major disadvantage of seines is that they are hard to handle in vegetation-choked ponds and lakes often frequented by larval amphibians. In such an environment, it can take over half an hour to sort through the vegetation and muck gathered in a 5-min seine haul. During this sorting process, smaller amphibian larvae can be hard to detect (and seine mesh is often coarse enough to enable their escape) meaning that seines are most often used for large larvae. As with dip-netting, seines are typically used to estimate catch per unit effort, although it is possible to estimate number captured per unit area in some conditions (Shaffer *et al.* 1994).

4.2.4 Leaf litterbags

4.2.4.1 Description

Leaf litterbag sampling is a relatively new method for sampling stream habitats for amphibians, particularly salamander larvae and adults (Pauley and Little 1998). Litterbags create an artificial habitat refugium reproducing leaf packs commonly found within streams, yet enclosed within mesh bags that can be easily removed and sampled for salamanders. Leaf litterbags are constructed using plastic mesh netting. Mesh gauge varies, but the 1.9 cm mesh (commonly found in deer-exclusion netting and similar products) has proven useful. The mesh is cut into squares, with 50×50, 70×70, and 90×90 cm mesh sections representing small, medium, and large bags, respectively. Several rocks are then placed on the netting to anchor the bag, followed by leaves, needles, and other material likely to be found within a particular stream environment. Once filled with litter, the corners of the mesh are pulled together and cinched at the top using a cable tie (Figure 4.1b). Colored flagging may be added to facilitate retrieval; this can be critical in contexts where the bag may be displaced downstream by a high-flow event. Litterbags tend to work best when deployed in locations where debris packs are likely to naturally accumulate (e.g. pools, channel bends). In higher-flow environments, bags can be secured by partially covering them with larger rocks or by tethering bags to roots or pinning them using a stake driven into the substrate (Waldron *et al.* 2003; Talley and Crisman 2007). Bags are sampled by placing a dip net underneath the bag in the water column, quickly removing the bag from the water, and placing it over a light-colored dishpan (Pauley and Little 1998; Jung *et al.* 2000). While gently shaking the bag, salamander larvae and adults will often fall into the dishpan where species identification and enumeration can take place. The litter pile should also be examined to ensure that all captured individuals are included. The litterbag can then be reassembled and deployed again, although the leaf litter may need to be replaced to compensate for decomposition over time.

4.2.4.2 Application

Litterbags have been used in stream environments where debris collects, and this technique targets species known to utilize leaf packs. They are easy to deploy, and quick and inexpensive to construct. They also manage to exploit an attribute of species that makes them otherwise hard to sample. In the absence of litterbag sampling, larvae of many stream-dwelling species are difficult to collect. They are also able to capture more secretive or uncommon species that might be missed in a dipnet survey. While litterbag sampling can produce species presence and relative density data for a stream reach, it does not likely provide accurate estimates of absolute population size or density, since it is unclear what exact area of the stream is being sampled with each bag (Chalmers and Droege 2002; Waldron *et al.* 2003).

4.2.4.3 Considerations

Litterbags are a highly specialized sampling tool. They are useful only for species that dwell in streams and use leaf packs. But if such a species is being targeted, the advantages of litterbags are substantial. Litterbag size is an important consideration, as medium and large bags can capture more individuals and species; however, the size of the focal stream may only accommodate a smaller bag (Waldron *et al.* 2003; Talley and Crisman 2007). Additionally, samples can be biased if potential competitors or predatory individuals colonize the bag. Researchers can discourage predatory (usually larger) adults and species by using finer mesh or submerging bags in deeper water, leading to a preferential capture of larvae (Waldron *et al.* 2003). Finally, the utility of litterbags may vary seasonally, as the abundance of natural leaf pack habitats can vary throughout the year, depending on the surrounding habitat cover.

4.2.5 Trapping

4.2.5.1 Description

Traps can be an effective means to capture amphibian larvae, requiring the researcher to simply deploy the traps and check them after a time period sufficient to have captured resident larvae. Most traps used by amphibian researchers are of a funnel design, which channels larvae into a large holding section that can be accessed by the researcher to recover captured animals. Commercially available wire minnow traps have been used in many amphibian studies (e.g. Fronzuto and Verrell 2000; Ghioca and Smith 2007). Home-made funnel traps using plastic bottles (e.g. 2-L plastic drinks bottles) have also been used successfully (Calef 1973; Richter 1995). Collapsible traps made of fine nylon mesh and available commercially (Promar, Gardena, CA, USA; Figure 4.1c), have capture rates equal to or better than traditional wire minnow traps (Adams *et al.* 1997; C. Pearl, personal communication). The finer mesh of these collapsible traps allows for the retention of much smaller larvae, and the compact size and weight make them suitable for backcountry work. Pyramid-shaped crayfish traps (Johnson and Barichivich 2004) are an alternative to minnow and collapsible mesh traps that can be particularly effective when it is important for part of the trap to extend above the water surface. In all cases, traps are deployed by dropping each in a predetermined area of the pond with a line attached and tied to a tree or float to make locating and retrieving it easier.

4.2.5.2 Application

Traps can be effective in capturing amphibian larvae present within many aquatic habitats with, perhaps, the exception of fast-moving water. Trapping is particularly suited to detection of species presence, and to estimate catch per unit effort (a metric of relative abundance). Trapping is sometimes the only suitable method in deep water, steep-sided pond basins, or frozen ponds where wading in to conduct sampling is not feasible. Additionally, it may be easier to sample habitats with structurally complex bottoms or vegetation-choked areas using traps, where seining and dip-netting may be difficult. Lastly, funnel trapping will often capture rare, secretive, or more nocturnal species not detected using other methods conducted in a short time period and during the daytime.

4.2.5.3 Considerations

Whereas traps can be left overnight, this practice requires a sampling location to be revisited within a short time period (usually 12–24 h) to avoid trap mortality, especially of non-target species or life stages. It is especially important, when traps are to be left for extended periods, to keep part of the trap above water to allow access to water surface for trapped animals. Secondly, wire mesh size in commercial minnow traps (around 6 × 6 mm) may be too large to effectively trap smaller larvae, in which case sealing the trap with window screening composed of finer mesh or the use of nylon collapsible traps can address this issue. Also, any trapping methods and resulting data are based on an assumption of equal capture probability among individuals and populations. This can be violated if the presence of conspecifics, competitors, or predators in the cage alters capture probabilities. There could also be community-level biases if populations being sampled differ in species composition. For instance, predators present in one habitat but not the other may alter the behaviour of a target species and subsequent capture probabilities. There can be other specific sampling biases in trapping certain species and in some habitat types (e.g. playa wetlands; Ghioca and Smith 2007). Weather, larval size and developmental stage, and resource availability can all affect capture rates using traps (Adams *et al.* 1997). Some researchers use baits when trapping amphibian larvae. However, for at least some species, it appears that baiting traps does not increase trap effectiveness (Adams *et al.* 1997).

4.2.6 Mark–recapture

4.2.6.1 Description

Mark–recapture techniques are commonly used in studies of amphibian adult populations, but can also be useful for the larval stage as well. However, rapid growth often accompanied by a dramatic shift in body design can render marking methods (often developed for juvenile and adults; see Chapter 8) unsuitable for larval amphibians. Successful marking techniques for larvae include temporary injectible organic dyes (Seale and Borass 1974) and externally staining dyes, which typically stain amphibian larvae for less than 24 h (Pfennig 1999; Jung *et al.* 2002; Harris *et al.* 2003), but can also slow growth rates (Travis 1981). More permanent, yet onerous, marking techniques using paint sprays, stains, dimethyl sulfoxide, and Super Glue (Ireland 1989) have largely been replaced by more convenient and robust visible implant elastomer (VIE) tagging (Figure 4.1d). Passive integrated transponder (PIT) tags may have limited use in all but the largest larval species due to tag size (down to about 8 mm) and surgery required to implant, although the development of smaller "injectable" PIT tags, applied using a hypodermic needle, may expand the potential for this technique (Biomark, Boise, ID, USA).

VIE tagging appears to hold the most promise for amphibians, balancing the ease of marking and longevity of the actual mark. Elastomer marks consist of a silicone-based polymer material that is injected subcutaneously and cures into a pliable and biologically inert solid (Northwest Marine Technology, Shaw Island, WA, USA). Whereas some elastomer colors are visible to the naked eye when present under translucent skin, fluorescent colors are often used and easily detected using an ultraviolet light source (portable lights are available for field purposes). Lowe (2003) used VIE to mark larvae of the spring salamander (*Gyrinophilus porphyriticus*) and has indicated that marks can still be seen 10 years after marking (W. Lowe, personal communication). Fading of the elastomer does not appear to be a common problem, although marks can migrate from the point of injection or be lost altogether. Grant (2008) found that wood frog (*Rana sylvatica*) tadpoles can retain marks through metamorphosis and that larger marks (> 2 mm) were more likely to migrate in two larval stream salamander species. Additionally, it was indicated that stream salamanders could be marked without anesthesia, while wood frog tadpoles required anesthesia and also had poorer mark retention (E. Grant, personal communication).

4.2.6.2 Application

Mark–recapture studies can be conducted in just about any habitat type, assuming that the same method of capture is used for each sampling period. Assuming that a sufficient proportion of the population is originally marked, mark–recapture techniques can provide robust estimates of absolute population size, especially when combined with robust capture–recapture estimation models (Chapter 24). Jung *et al.* (2002) found that mark–recapture methods provided the most accurate estimates of population size for two species of tadpoles in desert pool habitats.

4.2.6.3 Considerations

Regardless of technique, small amphibian larvae are relatively difficult to mark. The VIE technique will be more useful for species with larger larvae, or at least with individuals farther along in development. *Gyrinophilus porphyriticus* larvae as small as 2 cm in total length have been marked successfully, as well as *R. sylvatica* individuals down to 2.5 cm in total length. Additionally, any substantial loss of tags within a marked cohort can seriously bias population size estimates. Consider this when deciding which technique to use and for what exact purpose.

4.3 Other techniques

A number of additional techniques have been used in sampling amphibian larvae. Because they have limited application or because they have been used relatively infrequently, we mention them only briefly here and point readers to sources where more detailed descriptions are available.

4.3.1 Bottom net

A bottom net is placed on the bottom in a set position such that, when triggered, floats on an upper frame carry the net to the surface, entrapping larvae in open-water areas of small ponds (Shaffer *et al.* 1994). This area-based sampling technique has been used relatively little for amphibians, but could be effective in water where other techniques are not feasible.

4.3.2 Electroshocking

Electroshocking, developed initially to sample fish, is used primarily within streams and rivers. However, it can stun and facilitate capture of amphibian larvae as well (Brown and May 2007). It appears to be most effective for lentic species found in slow-moving parts of rivers (Shaffer *et al.* 1994). Many stream-dwelling amphibian species use retreats or bury themselves in the substrate in ways that prevent their detection and capture even if stunned during electroshocking.

4.3.3 Visual encounter survey

Visual encounter surveys, discussed more thoroughly in Chapter 15, may also be used, with some considerations, to sample larvae. Visual encounter surveys alone will likely be insufficient to detect a significant proportion of larvae present in dark or turbid water (such as a tannin-rich vernal pool, an algae-/vegetation-choked pond, or a stream with turbulent waters). Larval behavior and microhabitat preference (e.g. aggregation, crypsis, water-column basking, hiding in the substrate, preference for deep water or thick vegetation) can lead to species-specific and highly variable rates of detection. However, where water conditions allow, and the potential for confusing species is low, visual encounter surveys can be an efficient technique.

4.4 Conclusions

As in many aspects of field biology, the techniques used for sampling amphibian larvae are often passed without criticism or comment from one generation of researchers to the next. In many cases, there has been little effort to ask why one technique should be used as opposed to its alternatives or how a given technique may be most effectively applied. The many techniques outlined in this chapter are connected through the references listed below to an enormous cumulative effort to understand the most effective and efficient means to estimate the presence and density of larvae. Most of the techniques require little equipment, and that equipment is typically relatively inexpensive. Neither are the techniques difficult to master. Collectively, this means that there is little reason not to try multiple techniques and to calibrate and understand the consequences of altering the timing and intensity of sampling. The modest effort to do so will greatly increase the reliability of the information gathered and, in all probability, lead to unforeseen insights into the biology of the species being studied.

4.5 Acknowledgments

We thank S. Cortwright, M. McPeek, E. Werner, and H. Wilbur for teaching us about larval amphibian sampling. M. Adams, E. Grant, B. Hossack, W. Lowe, and C. Pearl provided helpful insight and details into the techniques they use.

4.6 References

Adams, M.J., Richter, K.O., and Leonard, W.P. (1997). Surveying and monitoring amphibians using aquatic funnel traps. In D.H. Olson, W.P. Leonard, and R. Bury (eds), *Sampling Amphibians in Lentic Habitats: Methods and Approaches for the Pacific Northwest: Northwest Fauna Number 4*, pp. 47–54. Society for Northwestern Vertebrate Biology, Olympia, WA.

Brown, L.R. and May, J.T. (2007). Aquatic vertebrate assemblages of the Upper Clear Creek Watershed, California. *Western North American Naturalist*, **67**, 439–51.

Calef, G.W. (1973). Natural mortality of tadpoles in a population of *Rana aurora*. *Ecology*, **54**, 741–58.

Chalmers, R.J. and Droege. S. (2002). Leaf litter bags as an index to populations of northern two-lined salamanders (*Eurycea bislineata*). *Wildlife Society Bulletin*, **30**, 71–4.

Duellman, W.E. and Trueb, L. (1986). *Biology of Amphibians*. John Hopkins University Press, Baltimore, MD.

Freidenburg, L.K. (2003). *Spatial Ecology of the Wood Frog (Rana sylvatica)*. PhD Dissertation, University of Connecticut, Storrs, CT.

Fronzuto, J. and Verrell, P. (2000) Sampling aquatic salamanders: tests of the efficiency of two funnel traps. *Journal of Herpetology*, **34**, 146–7.

Ghioca, D.M. and Smith, L.M. (2007). Biases in trapping larval amphibians in playa wetlands. *Journal of Wildlife Management*, **71**, 991–5.

Grant, E.H.C. (2008). Visual implant elastomer mark retention through metamorphosis in amphibian larvae. *Journal of Wildlife Management*, **72**, 1247–52.

Gunzburger, M.S. (2007). Evaluation of seven aquatic sampling methods for amphibians and other aquatic fauna. *Applied Herpetology*, **4**, 47–63.

Harris, R.N, Alford, R.A., and Wilbur, H.M. (1988). Density and phenology of *Notophthalmus viridescens dorsalis* in a natural pond. *Herpetologica*, **44**, 234–42.

Harris, R.N., Vess, T.J., Hammond, J.I., and Lindermuth, C.J. (2003). Context-dependent kin discrimination in larval four-toed salamanders *Hemidactylium scutatum* (Caudata: Plethodontidae). *Herpetologica*, **59**, 164–77.

Ireland, P.H. (1989). Larval survivorship in two populations of *Ambystoma maculatum*. *Journal of Herpetology*, **23**, 209–15.

Johnson, S.A. and Barichivich, W.J. (2004). A simple technique for trapping *Siren lacertina*, *Amphiuma means*, and other aquatic vertebrates. *Journal of Freshwater Ecology*, **19**, 263–9.

Jung, R.E., Droege, S., Sauer, J.R., and Landy R.B. (2000). Evaluation of terrestrial and streamside salamander monitoring techniques at Shenandoah National Park. *Environmental Monitoring and Assessment*, **63**, 65–79.

Jung, R.E., Dayton, G.H., Williamson, S.J., Sauer, J.R., and Droege, S. (2002). An evaluation of population index and estimation techniques for tadpoles in desert pools. *Journal of Herpetology*, **36**, 465–72.

Lowe, W.H. (2003). Linking dispersal to local population dynamics: a case study using a headwater salamander system. *Ecology*, **84**, 2145–54.

Olson, D.H., Leonard, W.P., and Bury, R.B. (eds) (1997). *Sampling Amphibians in Lentic Habitats: Methods and Approaches for the Pacific Northwest: Northwest Fauna Number 4*. Society for Northwestern Vertebrate Biology, Olympia, WA.

Pauley, T.K. and Little, M. (1998). A new technique to monitor larval and juvenile salamanders in stream habitats. *Banisteria*, **12**, 32–6.

Pfennig, D.W. (1999). Cannibalistic tadpoles that pose the greatest threat to kin are most likely to discriminate kin. *Proceedings of the Royal Society of London Series B Biological Sciences* **266**, 57–61.

Richter, K.O. (1995). A simple aquatic funnel trap and its application to wetland amphibian monitoring. *Herpetological Review*, **26**, 90–1.

Routman, E.J. (1984). A modified seining technique for single person sampling of deep or cold water. *Herpetological Review*, **15**, 72–3.

Seale, D. and Borass, M. (1974). A permanent mark for amphibian larvae. *Herpetologica*, **30**, 160–2.

Shaffer, H.B., Alford, R.A., Woodward, B.D., Richards, S.J., Altig, R.G., and Gascon, C. (1994). Quantitative sampling of amphibian larvae. In W.R. Heyer, M.A. Donnelly, R.W. McDiarmid, L.C. Hayek, and M.S. Foster (eds), *Measuring and Monitoring Biological Diversity: Standard Methods for Amphibians*, pp. 130–41. Smithsonian Institution Press, Washington DC.

Skelly, D.K. (1996). Pond drying, predators, and the distribution of *Pseudacris* tadpoles. *Copeia*, **1996**, 599–605.

Skelly, D.K., Yurewicz, K.L., Werner, E.E., and Relyea, R.A. (2003). Estimating decline and distributional change in amphibians. *Conservation Biology*, **17**, 744–51.

Talley, B.L. and Crisman, T.L. (2007). Leaf litterbag sampling for larval plethodontid salamander populations in Georgia. *Environmental Monitoring and Assessment*, **132**, 509–15.

Thoms, C., Corkran, C.C., and Olson, D.H. (1997). Basic amphibian survey for inventory and monitoring in lentic habitats. In D.H. Olson, W.P. Leonard, and R. Bury (eds), *Sampling Amphibians in Lentic Habitats: Methods and Approaches for the Pacific Northwest: Northwest Fauna Number 4*, pp. 35–46. Society for Northwestern Vertebrate Biology, Olympia, WA.

Travis, J. (1981). The effect of staining on the growth of *Hyla gratiosa* tadpoles. *Copeia*, **1981**, 193–6.

Waldron, J.L., Dodd, Jr., C.K., and Corser, J.D. (2003). Leaf litterbags: factors affecting capture of stream-dwelling salamanders. *Applied Herpetology*, **1**, 23–36.

Werner, E.E., Skelly, D.K., Relyea, R.A., and Yurewicz, K.L. (2007). Amphibian species richness across environmental gradients. *Oikos*, **116**, 1697–1721.

5

Dietary assessments of larval amphibians

Matt R. Whiles and Ronald Altig

5.1 Background

Diet analyses of larval amphibians provide information on foraging patterns, nutritional requirements, and trophic interactions in aquatic food webs, information that is critical for successful conservation and management. Numerous amphibian species around the globe are declining or presumed extinct (Stuart *et al.* 2004). Thus, there is an urgent need for detailed information on larval diets so that we can understand whether food-related factors are limiting wild populations and facilitate successful rearing of species in captivity.

Some species of all three orders of living amphibians have free-living, feeding, larval forms. Larval caecilians and salamanders, both of which occur in a variety of lentic and lotic habitats, are all carnivores that eat small aquatic organisms by suction-feeding, and their diets and trophic status are thus relatively easily assessed. Anuran larvae (tadpoles) occur in almost every conceivable type of freshwater habitat and show great diversity in feeding modes. As such, morphological diversity of the oral structures, which glean materials from substrata, and the buccopharyngeal apparatus, which selectively captures food particles, of tadpoles is large.

Whereas a few tadpoles are macrocarnivores (e.g. *Lepidobatrachus*, Leptodactylidae), most either rasp materials associated with substrata such as biofilms, periphyton, and deposited organic particles (e.g. ranids and many hylids), harvest naturally suspended particles in the water column (e.g. microhylids, pipids, and rhinophrynids), capture particles in the surface film (i.e. neustonic tadpoles with umbelliform oral structures), or consume conspecific or heterospecific eggs. Most of these feeding modes need further study (Altig *et al.* 2007). Although many tadpoles are classified as general herbivores or detritivores, they likely obtain a fair amount of energy and nutrients from more easily assimilated heterotrophic components of their diets (e.g. animal parts and microbes; Altig *et al.* 2007).

Herein, we focus on descriptive and observational methods for assessing diet and trophic status of larval amphibians. However, other approaches that are beyond the scope of this review can also provide valuable insight into the trophic ecology of larval amphibians. In particular, manipulative approaches such as cafeteria experiments can be used to examine amphibian food preferences (e.g. Kupferberg 1997) and field enclosure/exclosure experiments can illuminate functional roles and trophic interactions (e.g. Solomon *et al.* 2004). Combined approaches, such as experimental manipulations performed in conjunction with gut content analyses (e.g. Ranvestel *et al.* 2004; Whiles *et al.* 2006) have been used to examine amphibian ecology at multiple scales, from individual foraging to roles of amphibians in ecosystem processes. Manipulative approaches for studying various aspects of amphibian ecology are discussed in Chapters 6 and 12 in this volume.

5.2 Larval caecilians and salamanders

With some modifications (e.g. different foraging strategies; benthic versus nektonic and day versus night feeding), general habitats (e.g. lentic versus lotic), and morphologies (e.g. gape size and size and number of gill rakers), the method of food capture is essentially the same in all larval caecilians and salamanders. These groups form a negative pressure by rapidly depressing the throat as they open the mouth so as to suck food items into the buccal cavity. Plant materials found in the guts of these groups are generally considered as incidental. As such, gut content analyses can be effective for assessing diet in these groups, although robust sampling strategies that account for spatiotemporal foraging patterns and feeding behaviors should be used.

5.3 Anuran tadpoles

Most tadpoles appear to be omnivorous, although this contention is debated and the true trophic status of many tadpoles has not been accurately assessed (Altig *et al.* 2007). When tadpoles consume a range of food types that vary in digestibility and nutritional quality, simply quantifying the constituents of their diet may provide limited information on the relative importance of specific foods to growth and development. Accurate assessments of the trophic status of omnivorous tadpoles require information on assimilation which can be obtained through isotopic analyses, fatty acid analyses, or by applying ecological efficiencies (e.g. assimilation and/or production efficiencies) to gut content data. Consummatory diet information *can* be useful for studying feeding behaviors and the types of materials that the oral apparatus and buccopharyngeal structures can harvest and capture.

5.4 Assessing food sources and diets

5.4.1 Category I: preparatory studies

Along with analyses of larval amphibians and/or their gut contents, many studies also include some assessment of food availability. In most freshwater habitats, food resources are dynamic. For example, algal blooms can occur over short time periods, and thus frequent sampling of potential food items is necessary. Coupled with studies of foraging behavior, information on the seasonal phenology of resources can guide the timing of sampling and aid in data interpretation. On an even shorter diurnal time scale, samples of larvae and their prey collected during daylight may not accurately reflect trophic interactions of salamanders active in the dark.

Sampling protocols, including number of samples collected at a given time, apportioning samples among available micro- or mesohabitats (e.g. pools versus riffles in a stream, macrophyte beds versus non-vegetated areas of a pond), will vary with the specific system examined and associated questions. Preliminary sampling can be used to obtain estimates of variability for *a priori* power analyses to assist with decisions regarding sampling intensities (e.g. Steidl *et al.* 1997).

Diets of many larvae change through development, and thus size and stage should be accounted for when sampling. Larval size is important, but does not always reflect age. For this reason, a staging table based on the sequence of acquisition of discrete morphological markers should be used. Dünker and Wake (2000) provide a staging table for caecilians, and those for salamander larvae and tadpoles are summarized in Duellman and Trueb (1986).

Collection and preservation techniques for amphibians are often taxonomically specific (see Heyer *et al.* 1994). For collection of potential food items, Hauer and Lamberti (2006) provide a wealth of information on proper collection techniques of materials ranging from detritus to algae. Smith (2001), Thorpe and Covich (2001), and Merritt *et al.* (2008) provide methods and taxonomic keys for invertebrate prey. Prey items should be identified to the lowest taxonomic level possible. For both amphibian larvae and prey, specimens should be placed in a readily accessible, cataloged collection. This allows others to verify identifications and use materials for future studies.

5.4.2 Category II: gut contents

Most studies of amphibian diets have relied on examinations of gut contents. However, gut-content data should be interpreted cautiously because the types and amounts of materials present are a function of the time and place of collection (time of day, season, age of the individual, etc.) and euthanization and preservation methods. While gut contents acquired with robust sampling designs that account for spatial and temporal variations can be reliable indicators of foraging patterns and functional roles of larval amphibians, they may not be good indicators of true trophic status because of differential assimilation of materials (Altig *et al.* 2007). As noted above, we assert that larval caecilians and salamanders are the only groups for which true trophic position can be assessed via examination of gut contents alone. All methods have limitations, so the best approach is to assess diets and trophic relations using multiple approaches (e.g. examination of stomach contents combined with stable-isotopic analyses of tissues). Further, approaches should be iterative to account for time lags between ingestion of available foods and actual changes in isotopic signatures or fatty acid profiles, which are a function of tissue turnover rates.

Gut contents can be removed from larger larval salamanders and caecilians without killing the individual. Although gut contents may be removed with forceps from large individuals that are anesthetized, the safest and most effective techniques involve forced regurgitation. This is generally accomplished with stomach flushing or gastric lavage, a technique that involves displacing materials from the stomach and back out the mouth with a low-pressure stream of water (Solé *et al.* 2005). Approaches vary and should be adjusted for an individual's size and mouth and gut morphologies, but generally involve gently forcing water from a syringe through a lubricated (water or a water-based lubricant) catheter tube that is inserted into the esophagus. During this procedure, the animal is held with the head facing down and over a vessel where displaced materials and fluid are collected. Contents can then be rinsed through an appropriately sized sieve and preserved in ethanol or buffered formalin. Excess water should be removed before preservation. If performed correctly, this technique is harmless and very effective because fresh materials are collected from the stomach.

5.4.3 Procedures: anurans and small predators

For other amphibians (e.g. tadpoles, small salamanders), specimens are generally sacrificed, in which case individuals should be collected as gently as possible, anesthetized with MS-222®, Orajel®, or similar materials, and quickly euthanized in a humane manner to avoid vomiting and/or continued digestion of food. The gut and its contents can be removed in the field or later from properly preserved specimens. In cases where fatty acid or isotope analyses are also planned, specimens may need to be frozen rather than placed in fluid preservatives. Freezing leads to distortion of some materials, and removal of guts and processing or preservation should take place immediately after specimens are thawed.

5.4.4 Processing and analysis of gut contents

Processing methods for gut contents must account for differences in gut tracts among taxa. Most tadpoles have long, coiled guts that are typically several times their body length, whereas salamanders and caecilians have shorter guts with a well-developed stomach that is typical of carnivores. These differences warrant different approaches.

For salamanders and caecilians, prey items are removed from the gut, identified to the desired taxonomic level, and counted and measured. In cases where guts contain many small prey items any number of subsampling techniques can be used, so long as they are unbiased toward different types of prey. A simple technique involves spreading the gut contents evenly in a Petri dish with numbered grids and then using a table of random numbers to select grids in which contents are identified and measured until a representative subsample has been examined. Protocols for degree of subsampling range from fixed counts (e.g. 100 individuals) to proportion of total examined (e.g. one-eighth of the total grids or area examined). Methods should be calibrated by progressively examining larger portions of a subset of samples to assess the accuracy of various subsample sizes. For most purposes, expressing prey in terms of mass or volume is best, although simple numerical analyses are sufficient for some assessments of foraging behaviors and prey availability. Mass analyses are more appropriate for assessments of trophic status and nutrition because they can be converted to organic content or caloric equivalents. Mass is generally estimated by actual weighing (generally as dry mass (DM) or ash-free dry mass (AFDM), which is a proxy for organic content; see methods below) or use of length/mass relationships, which are generally in the form: $M = aL^b$ where M is the mass of the organism, L is a linear dimension such as length, and a and b are regression constants (reviewed by Benke *et al.* 1999). Length/mass relationships for a variety of potential prey items can also be found in Rogers *et al.* (1976), Bottrell *et al.* (1976), Dumont *et al.* (1975), and Culver *et al.* (1985). In many cases, prey items are fragmented, and thus care must be taken to reassemble them or estimate total lengths. Body length and head width, which are the most common measures used, are generally performed with a dissecting microscope equipped with an ocular micrometer, graduated stage, or image-analysis capabilities. Mass is usually expressed as DM or AFDM. For more in-depth analyses, caloric content is very informative. Estimates of caloric contents for many types of prey are available (e.g. Cummins and Wuychek 1971; Rodgers and Qadri 1977), but these estimates tend to be quite variable. Depending on the level of precision desired, analyses such as bomb calorimetry should be performed on actual prey items or other materials from the guts or specimens collected from the same habitats at the same time. As with DM and AFDM, size/caloric-content relationships can be developed.

While collection and identification of gut contents from salamanders and caecilians is relatively straightforward, tadpoles present a challenge because their long guts contain mixtures of materials of different digestibility that are at different stages of digestion. In this case, the most recently ingested materials from the foregut should be the foci of analyses. This is generally accomplished by removing a section of the foregut that is a consistent proportion of the total gut length (e.g. the anterior quarter of the gut) and preparing it for examination under a microscope (for identification of individual food particles) or for analyses of caloric content, fatty acid profiles, or isotopic composition (see methods for fatty acids and isotopes below). Obviously, method of collection and preservation can limit the types of analyses that can be performed on gut contents.

For identification of foregut contents, materials are generally rinsed from the gut, slide-mounted, and examined under a microscope. One rather simple and effective method is to place materials in glycerin and gently homogenize them. If needed, water can be added and the materials sonicated to further break up contents. Depending on the abundance of particles, either the entire sample or a subsample is filtered with low vacuum on to a 0.45 μm gridded membrane filter to obtain a reasonable dispersion of particles across the filter (e.g. 10–20 particles/grid). Filters should be allowed to dry at ambient temperature for 2–3 h to remove excess water, and then two or three drops of Type A immersion oil, which clears the filter, are added. After approximately 24 h the filters will clear and can then be placed on a slide and covered with a coverslip, which is sealed with enamel nail polish. Slide-mounted materials are generally subsampled (e.g. some fixed number of fields of view examined, a fixed number of grids examined, or all particles along transects are identified and measured). Measurements of individual particles can be made using methods ranging from an ocular micrometer to sophisticated image analysis systems. Data on identified food types are generally expressed as area or converted to mass, or, in the case of algae, biovolume (Lowe and LaLiberte 2006; Steinman et al. 2006). There are many variations for mounting and analyzing gut contents (for recent examples using various taxonomic groups, see Evans-White et al. 2003; Ranvestel et al. 2004; Rosi-Marshall and Wallace 2002). Regardless of modifications, the key points are to avoid damaging materials beyond recognition and to obtain an evenly dispersed, representative sample.

Masses of materials in guts can be estimated with methods ranging from simply weighing materials to using size/mass relationships (e.g. body length/mass relationships for prey; see above section on salamander and caecilian guts) either developed by the investigator or gleaned from the literature. For direct weighing, DM or AFDM estimates are best. Dry mass estimates are generally developed by drying materials at 55–60°C to constant mass. For AFDM, dried materials are weighed, then ashed in a furnace ($\approx$500°C) for 1–2 h, and weighed again; mass of remaining ash is subtracted from the original dry mass to obtain AFDM. In either case, samples should be allowed to cool in a dessicator before weighing so that moisture is not absorbed in the cooling process. Actual drying and ashing times will vary with sample size. Small amounts of materials can be processed in this manner on small pieces of heavy duty aluminum foil or glass fiber filters. Development of wet mass/DM or AFDM relationships can streamline procedures and decrease the number of samples that are destroyed. Likewise, if caloric equivalents are derived through calorimetry, similar relationships can be developed.

Gut content data expressed as mass or area (or in some cases biovolume for algae (see Lowe and LaLiberte 2006) can be combined with estimates of assimilation efficiencies for more accurate assessments of trophic status. Procedures vary, but a simple approach is to express food types as percent of total gut contents and multiply each by its respective assimilation efficiency. Resultant values for the food types are then summed and the percentage contribution of each to assimilated materials can be calculated. If caloric content is estimated, indices of prey or food-type importance can be calculated using basically the same basic procedure; masses of each food type in the gut are multiplied by their corresponding caloric density and percentage caloric contribution of each food type is the total number of calories attributable to a given food type divided by the total caloric content in the gut. Regester *et al.* (2008) provide examples of these types of analyses and related procedures applied to pond-breeding ambystomatid communities. Examples using various freshwater omnivores can be found in Bowen *et al.* (1995) and Evans-White *et al.* (2003). Regester *et al.* (2008) and Skelly and Golon (2003) provide examples of methods for estimating assimilation efficiencies of amphibian larvae.

In cases where algae are obviously consumed, analyses of photosynthetic pigments may be performed on gut contents to estimate algal biomass. For this procedure, specimens and/or their gut contents must be frozen immediately upon collection and kept dark and frozen prior to analyses. Chlorophyll *a*, which can be used to estimate algal biomass, can be extracted from samples of gut contents with an organic solvent and concentration is then estimated with spectrophotometry, fluorometry, or high-performance liquid chromatography. Steinman *et al.* (2006) provide a review of methods to estimate algal biomass via photosynthetic pigment analyses. Although informative to a degree, pigment analyses alone cannot account for the digestibility of algal materials, and some tadpoles will re-ingest feces containing undigested algal materials. The ability of tadpoles to digest algal materials can be assessed through microscopic examination of cells in feces (e.g. whether full chloroplasts are present) or culturing algae from feces can identify the quantity and types of materials that are still viable after passing through guts (Peterson and Boulton 1999). Simple counts of prey items or food types in guts may not provide reliable information on trophic status, but can be valuable for assessing feeding preferences and interactions with other consumers. A variety of indices, including Horn's index and Morisita's index, can be applied to gut-content data to quantify diet overlap among species (see Brower *et al.* 1998; Krebs 2001; Chapter 18). Likewise, if data on the abundance of prey and/or food type in the environment are also available, feeding selectivity can be assessed and quantified using a variety of available indices. Relatively simple and commonly used indices include Ivlev's index of electivity (Ivlev 1961) and the selectivity index (Chesson 1983).

5.5 Category III: assimilatory diet

Analyses of isotopic composition and fatty acid profiles can provide reliable information on diets and trophic status because they reflect assimilation over long periods of time. However, these approaches can be more costly and complicated than simple gut analyses because they require expensive analytical equipment, and they also have their own limitations, some of which are discussed below. In most cases, samples are simply collected and preserved in a manner that does not alter isotopic or fatty acid composition, depending on the analyses to be performed, and then sent to an analytical laboratory for analyses.

5.5.1 Stable-isotope analysis

Isotopes of carbon (^{13}C) and nitrogen (^{15}N) are commonly used in studies of freshwater food webs because they allow for tracking transfers of organic carbon and nitrogen from living plant and detrital sources to primary consumers and predators (Peterson and Fry 1987; Fry 2006). Other isotopes, including deuterium (^2H or D) and ^{18}O, are gaining increasing attention from ecologists for tracing movements of organisms. We focus on carbon and nitrogen, as they can provide valuable information on assimilatory diets and methods for their analyses are standardized. Stable-isotope analyses focus on ratios of heavier (e.g. ^{15}N and ^{14}C) and lighter isotopes (^{14}N and ^{13}C). Ratios are expressed as δ, which is parts per thousand (‰) deviation from a standard, calculated as $\delta^{13}C$ or $\delta^{15}N = [(R_{sample} - R_{standard})/R_{standard}] \times 1000$ where $R = {}^{13}C/{}^{12}C$ or $^{15}N/{}^{14}N$. Common standards are Pee Dee belemnite limestone for C and atmospheric N (‰ value set to 0). A positive (N) or less negative (C) value indicates the sample is enriched with the heavy isotope.

Carbon isotope signatures of consumers are generally similar to those of their food source or sources (i.e. you are what you eat), which can have $\delta^{13}C$ values ranging from slightly higher than atmospheric CO_2 (–8‰) to –40‰ for a variety of reasons that influence degree of fractionation (see Peterson and Fry 1987; Fry 2006; Hershey *et al.* 2006). In particular, in streams, water velocity can influence carbon isotopic compositions of foods such as algae (Finlay *et al.* 1999). Differences in C_3 and C_4 photosynthetic pathways result in differences among some plants and their detritus. Likewise, there are often measurable differences in carbon isotope ratios between terrestrial and aquatic autotrophs and their detritus.

Nitrogen is more often used to assess trophic position, as $\delta^{15}N$ shows a fairly consistent change with each trophic step ($\approx$3.4‰ increase in $\delta^{15}N$ with each trophic step up from primary consumer to predator; Vander Zanden and Rasmussen 1999; Post 2002). This fairly predictable fractionation results from the tendency for organisms to excrete more of the lighter nitrogen isotope than the heavy one. However, fractionation can vary and there is evidence that the increase in $\delta^{15}N$ with each trophic step is considerably less than 3.4‰ in some tropical stream food webs (Kilham and Pringle 2000). In general, basal resources such as periphyton and detritus in freshwater habitats will have $\delta^{15}N$ values ranging from near 0 to 3‰, primary consumers range from 3 to 6‰, and predators from 6 to 12‰, but values can vary considerably for a variety of reasons (Fry 2006). Whiles *et al.* (2006) and Verburg *et al.* (2007) provide examples of isotopic analyses of stream food webs with abundant and diverse tadpole assemblages.

Most studies examine naturally occurring isotopic ratios (natural abundance studies) in consumer tissues and food resources, but additions of enriched materials (e.g. tracer studies using ^{15}N enriched ammonium or nitrate or ^{13}C enriched acetate) are also used, particularly when natural isotopic ratios among food types are similar. Tracer studies can be informative for assessing diets and trophic interactions, as well as examining roles of consumers in biogeochemical cycling, because materials (e.g. N or C) are followed from basal resources through top consumers. Stable-isotope tracer additions have been used to examine carbon cycling through consumers in streams (Hall 1995), nitrogen cycling in stream food webs (Hall *et al.* 1998), and movement of nutrients from streams to riparian habitats via consumers (Sanzone *et al.* 2003). Tracer studies could be particularly informative for assessing roles of amphibians in freshwater ecosystem processes, as well as material and energy exchanges between aquatic and terrestrial systems.

5.5.2 Procedures

Stable-isotope analyses can be performed on entire organisms or, in the case of larger individuals, muscle tissue (generally 1–2 g wet weight). While muscle tissue is the usual focus for larger animals, blood and other tissues such as liver or skin are sometimes examined because they have different incorporation and turnover rates, and can thus be used to assess temporal patterns of food availability and assimilation (Dalerum and Angerbjorn 2005). For example, while isotopic analyses of muscle tissues may reflect assimilation over a period of months, liver or blood plasma, which turnover much more rapidly, can reflect a period of days or weeks. This isotopic memory is an important consideration, as diets of some amphibians can change considerably through development and with changing resource availability.

Muscle tissues from large individuals can be sampled non-lethally using sterile biopsy punches, which are available from many surgical supply companies. If biopsies are collected, wounds should be treated with an appropriate antibiotic before releasing the specimen. Live amphibians and prey items that are to be processed in their entirety should be allowed to clear their guts as much as possible by placing them in filtered water from the same habitat for approximately 12 h. Alternatively, gut tracts can be removed before processing, but this can be a tedious process for small specimens. The best methods for preservation of samples for isotopic analyses are freezing or drying at low temperature (55–60°C to constant mass). In the field, samples should be placed on ice or in liquid nitrogen and then transferred to a freezer (−80°C) or drying oven when available. Dried samples should be stored in a desiccant such as drierite or silica. Cross-contamination must be avoided throughout sampling and processing of all materials.

Comprehensive analyses of food webs require sampling of all potential food items that are available for larval amphibians or their prey. Sampling of prey, algae, biofilms, detritus, and other potential resources simply requires collection of representative samples. These samples can be collected using standard procedures (e.g. Hauer and Lamberti 2006), but care must be taken to avoid cross-contamination. Sampling of fine particles from substrates or the water column can be collected and processed on glass-fiber filters. As with any other sampling, replicates should be collected to account for natural variability. In flowing waters, areas of different velocities should be sampled, as current velocity can alter the carbon isotope compositions of periphyton (Finlay *et al.* 1999).

Prior to isotopic analyses, all samples are dried to constant weight at 55–60°C, ground into a fine powder, weighed, and packed into tin capsules. Samples of particulates on glass-fiber filter can either be carefully scraped from the filters, or a portion of the filter, with sampled material on it, can be packed into the tin capsule. While muscle tissue is the usual focus for larger animals, for most prey items (e.g. aquatic invertebrates) the entire organism is processed, or, in the case of very small prey, individuals are combined. Samples for N isotope analysis should contain at least approximately 100 µg of DM of N; samples for C isotope analysis should contain approximately 2 mg of DM of C. Sample sizes for analysis of both C and N isotopes vary with C:N ratio of the sample, with larger sample sizes needed (e.g. 10–60 mg) for materials with high C:N ratio (e.g. sediments with low %N). Samples of basal resources and sediments from calcareous regions should be processed to remove carbonates before analysis. Carbonates can easily be removed by acidification via exposure to HCl vapors (Yamamuro and Kayanne 1995). Isotopic analyses are performed using a continuous-flow isotope-ratio mass spectrometer.

Analyses and interpretation of stable-isotope data range from simply comparing values of resources and consumers to assess relative contributions, to quantifying trophic status (Vander Zanden *et al.* 2000; Post 2002) or niche breadth (Layman et al. 2007). Depending on the nature of the system, stable-isotope analyses will sometimes reveal clear relationships among consumers and resources. In other cases, environmental complexities and high degrees of omnivory can confound results. In cases where multiple food sources are likely contributing to the isotopic composition of a consumer, as would be the case with most tadpoles, mixing models can be used to assess relative contributions of different foods. Simple two-source mixing models can be used to estimate relative contributions of two resources (e.g. periphyton and detritus; see Hershey *et al.* 2006), whereas more complex models are necessary when more than two food sources are likely.

5.5.3 Fatty acid analyses

Fatty acids are the essential foundations of energy storage lipids in consumers, but they are not readily synthesized by animals. As such, they must be obtained through their food. Differences in fatty acid content among food types can contribute to many of the qualitative differences among larval food resources in terms of their importance for growth and development. Given their nutritional value, that they are conserved, and that various resources and consumers have unique combinations of fatty acids, fatty acid profiles can be good indicators of diet and trophic status. Analyses of fatty acid profiles have been successfully used to assess food web structure in marine, lake, and stream systems (Muller-Navarra *et al.* 2000; Stübing *et al.* 2003; Sushchik *et al.* 2003; as well as aquatic–terrestrial food web linkages, Koussoroplis *et al.* 2008). In some cases, fatty acid analyses can provide detailed taxonomic information, including specific types of algae assimilated (e.g. Napolitano *et al.* 1997).

Samples of muscle tissues (large amphibians), whole organisms (small amphibians, prey items), and basal resources (algae, detritus) can be collected in the same manner as described above for isotopic analyses, except that all materials should be frozen at −80°C as quickly as possible. Moisture content of materials is quantified with freeze drying so that lipids are not altered or volatilized. Lipids are extracted from homogenized materials using chloroform/methanol solvent extraction (Bligh and Dyer 1957). Analyses involve separation of polar and non-polar lipids with a solid-phase extraction system, conversion into fatty acid methyl esters, and separation using a gas chromatograph with a flame ionization or mass spectrometer detector.

Discriminant function analysis and similar approaches can be used to assess trophic interactions among sampled resources and consumers (Budge *et al.* 2002). As with isotopic analyses, resolution will vary among systems, and in situations where trophic interactions are complex multiple approaches (e.g. fatty acid analyses combined with isotopic analyses and/or gut content analyses) will obviously produce more robust results.

5.6 Summary

Understanding the food habits of larval amphibians is actually the first step in understanding the interactions of amphibians and the communities in which they live and ultimately their ecological significance (Altig *et al.* 2007). Given the current status of many amphibian species and populations, and the likelihood that past and ongoing declines have ecosystem-level consequences (Ranvestel *et al.* 2004; Whiles *et al.* 2006), there is urgent need for detailed, quantitative information on the myriad ecological roles of larvae. The methods reviewed herein provide the basic tools for assessing diet, nutritional ecology, and trophic interactions. Choosing the proper technique ultimately depends on the research question and resource limitations. Because each approach has inherent limitations, we recommend combined approaches.

5.7 References

Altig, R., Whiles, M. R., and Taylor, C. L. (2007). What do tadpoles really eat? Assessing the trophic status of an understudied and imperiled group of consumers in freshwater habitats. *Freshwater Biology*, **52**, 386–95.

Benke, A. C., Huryn, A. D., Smock, L. A., and Wallace, J. B. (1999). Length-mass relationships for freshwater macroinvertebrates in North America with particular reference to the southeastern United States. *Journal of the North American Benthological Society* **18**, 308–43.

Bligh, E.G. and Dyer, W. J. (1959). A rapid method of total lipid extraction and purification. *Canadian Journal of Biochemistry and Physiology* **37**: 911–917.

Bottrell, H. H., Duncan A., Gliwicz, Z. M., Grygierek, E., Herzig A., Hillbricht-Ilkowska A., Kurasawa H., Larsson P., and Weglenska, T. (1976). A review of some problems in zooplankton production studies. *Norwegian Journal of Zoology*, **24**, 419–56.

Bowen, S. H., Lutz, E. V., and Ahlgren, M. O. (1995). Dietary-protein and energy as determinants of food quality—trophic strategies compared. *Ecology*, **76**, 899–907.

Brower, J. E., Zar, J. H., and von Ende, C. N. (1998). *Field and Laboratory Methods for General Ecology*. McGraw-Hill, Boston, MA.

Budge, S. M., Iverson, S. J., Bowen, W. D., and Ackman, R. G. (2002). Among- and within-species variability in fatty acid signatures of marine fish and invertebrates on the Scotian Shelf, Georges Bank, and southern Gulf of St. Lawrence. *Canadian Journal of Fisheries and Aquatic Sciences*, **59**, 886–98.

Chesson, J. (1983). The estimation and analysis of preference and its relationship to foraging models. *Ecology*, **64**, 1297–1304.

Culver, D. A., Boucherle, M. M., Bean, D. J., and Fletcher, J. W. (1985). Biomass of freshwater crustacean zooplankton from length-weight regressions. *Canadian Journal of Fisheries and Aquatic Sciences*, **42**, 1380–90.

Cummins, K. W. and Wuycheck, J. C. (1971). Caloric equivalents for investigations in ecological energetics. *Mitteilungen Internationale Vereinigung für Theoretische und angewandte Limnologie*, **18**, 1–158.

Dalerum, F. and Angerbjorn, A. (2005). Resolving temporal variation in vertebrate diets using naturally occurring stable isotopes. *Oecologia*, **144**, 647–58.

Duellman, W. E. and Trueb, L. (1986). *Biology of Amphibians*. McGraw-Hill, New York.

Dumont, H. J., Van de Velde, I., and Dumont, S. (1975). The dry weight estimate of biomass in a selection of Cladocera, Copepoda, and Rotifera from the plankton, periphyton and benthos of continental waters. *Oecologia*, **19**, 75–97.

Dünker, N., Wake, M. H., and Olson, W. M. (2000). Embryonic and larval development in the caecilian *Icthyophis kohtaoensis*: a staging table. *Journal of Morphology*, **243**, 3–34.

Evans-White, M. A., Dodds, W. K., and Whiles, M. R. (2003). Ecosystem significance of crayfishes and central stonerollers in a tallgrass prairie stream: functional differences between co-occurring omnivores. *Journal of the North American Benthological Society*, **22**, 423–41.

Finlay, J. C., Power, M. E., and Cabana, G. (1999). Effects of water velocity on algal carbon isotope ratios: implications for river food web studies. *Limnology and Oceanography*, **44**, 1198–1203.

Fry, B. (2006). *Stable Isotope Ecology*. Springer, Berlin.

Hall, Jr, R. O. (1995). Use of a stable carbon-isotope addition to trace bacterial carbon through a stream food-web. *Journal of the North American Benthological Society*, **14**, 269–77.

Hall, Jr, R. O., Peterson, B. J., and Meyer, J. L. (1998). Testing a nitrogen-cycling model of a forest stream by using a nitrogen-15 tracer addition. *Ecosystems*, **1**, 283–98.

Hauer, F. R. and Lamberti, G. A. (eds) (2006). *Methods in Stream Ecology*. Academic Press, San Diego, CA.

Hershey, A. E., Fortino, K., Peterson, B. J., and Ulseth, A. J. (2006). Stream food webs. In F. R. Hauer and G. A. Lamberti (eds), *Methods in Stream Ecology*, pp. 637–9. Academic Press, San Diego, CA.

Heyer, W. R., Donnelly, M. A., McDiarmid, R. W., Hayek, L.-A. C., and Foster, M. S. (eds) (1994). *Measuring and Monitoring Biological Diversity. Standard Methods for Amphibians*. Biological Diversity Handbook Series, Smithsonian Institution Press, Washington DC.

Ivlev, V. S. (1961). *Experimental Ecology of the Feeding of Fishes*. Yale University Press, New Haven, CT.

Kilham, S. S. and Pringle, C. M. (2000). Food webs in two neotropical systems as revealed by stable isotope ratios. *Verhandlungen Internationale Vereinigung für theoretische und angewandte Limnologie*, **27**, 1768–75.

Koussoroplis, A. M., Lemarchand, C., Bec, A., Desvilettes, C., Amblard, C., Fournier, C., Berny P., and Bourdier, G. (2008). From aquatic to terrestrial food webs: decrease of the docosahexaenoic acid/linoleic acid ratio. *Lipids*, **43**, 461–6.

Krebs, C. J. (2001). *Ecological Methodology*, 2nd edn. Harper Collins Publishers, New York.

Kupferberg, S. (1997). Facilitation of periphyton production by tadpole grazing: Functional differences between species. *Freshwater Biology*, **37**, 427–39.

Layman, C. A., Quattrochi, J. P., Peyer, C. M., and Allgeier, J. E. (2007). Niche width collapse in a resilient top predator following ecosystem fragmentation. *Ecology Letters*, **10**, 937–44.

Lowe, R. L. and LaLiberte, G. D. (2006). Benthic stream algae: distribution and structure. In F. R. Hauer and G. A. Lamberti (eds), *Methods in Stream Ecology*, pp. 327–56. Academic Press, San Diego, CA.

Merritt, R. W., Cummins, K. W., and Berg, M. B. (2008). *An Introduction to the Aquatic Insects of North America*, 4th edn. Kendall/Hunt Publishing Company, Dubuque, IO.

Muller-Navarra, D. C., Brett, M. T., Liston, A. M., and Goldman, C. R. (2000). A highly unsaturated fatty acid predicts carbon transfer between primary producers and consumers. *Nature*, **403**, 74–7.

Napolitano, G. E., Pollero, R. J., Gayoso, A. M., MacDonald, B. A., and Thompson, R. J. (1997). Fatty acids as trophic markers of phytoplankton blooms in the Bahia Blanca Estuary (Buenos Aires, Argentina) and in Trinity Bay (Newfoundland, Canada). *Biochemistry and Systematic Ecology*, **25**, 739–55.

Peterson, B. J. and Fry, B. (1987). Stable isotopes in ecosystem studies. *Annual Review of Ecology and Systematics*, **18**, 293–320.

Peterson, C. G. and Boulton, A. J. (1999). Stream permanence influences microalgal food availability to grazing tadpoles in arid-zone springs. *Oecologia*, **118**, 340–52.

Post, D. M. (2002). Using stable isotopes to estimate trophic position: models, methods, and assumptions. *Ecology*, **83**, 703–18.

Ranvestel, A. W., Lips, K. R., Pringle, C. M., Whiles, M. R., and Bixby, R. J. (2004). Neotropical tadpoles influence stream benthos: evidence for the ecological consequences of decline in amphibian populations. *Freshwater Biology*, **49**, 274–85.

Regester, K. J., Whiles, M. R., and Lips, K. R. (2008). Variation in the trophic basis of production and energy flow associated with emergence of larval salamander assemblages (Ambystomatidae) among forest ponds. *Freshwater Biology* **53**, 1754–67.

Rodgers, D. W. and Qadri, S. U. (1977). Seasonal variations in calorific values of some littoral benthic invertebrates of the Ottawa River, Ontario. *Canadian Journal of Zoology*, **55**, 881–4.

Rogers, L. E., Hinds W. T., and Buschom, R. L. (1976). A general weight vs. length relationship for insects. *Annals of the Entomological Society of America*, **69**, 387–9.

Rosi-Marshall, E. J. and Wallace, J. B. (2002). Invertebrate food webs along a stream resource gradient. *Freshwater Biology*, **47**, 129–41.

Sanzone, D. M., Meyer, J. L., Marti, E., Gardiner, E. P., Tank, J. L., and Grimm, N. B. (2003). Carbon and nitrogen transfer from a desert stream to riparian predators. *Oecologia*, **134**, 238–50.

Skelly, D. K. and Golon, J. (2003). Assimilation of natural benthic substrates by two species of tadpoles. *Herpetologica* **59**, 37–42.

Smith, D. G. (2001). *Pennak's Freshwater Invertebrates of the United States, Porifera to Crustacea*. John Wiley and Sons, New York.

Solé, M., Beckmann, O., Pelz, B., Kwet, A., and Engels, W. (2005). Stomach-flushing for diet analysis in anurans: an improved protocol evaluated in a case study in *Araucaria* forests, southern Brazil. *Studies on Neotropical Fauna and Environment*, **40**, 23–8.

Solomon, C. T., Flecker, A. S., and Taylor, B. W. (2004). Testing the role of sediment-mediated interactions between tadpoles and armored catfish in a neotropical stream. *Copeia*, **2004**, 610–16.

Steidl, R. J., Hayes, J. P., and Schauber, E. (1997). Statistical power analysis in wildlife research. *Journal of Wildlife Management*, **61**, 270–9.

Steinman, A. D., Lamberti, G. A., and Leavitt, P. R. (2006). Biomass and pigments of benthic algae. In F. R. Hauer and G. A. Lamberti (eds), *Methods in Stream Ecology*, pp 352–79. Academic Press, San Diego, CA.

Stuart, S. N., Chanson, J. S., Cox, N. A., Young, B. E., Rodrigues, A. S. L., Fischman, D. L., and Waller, R. W. (2004). Status and trends of amphibian declines and extinctions worldwide. *Science*, **306**, 1783–6.

Stübing, D., Hagen, W., and Schmidt, K. (2003). On the use of lipid biomarkers in marine food web analyses: an experimental case study on the Antarctic krill, *Euphausia superba*. *Limnology and Oceanography*, **48**, 1685–1700.

Sushchik, N. N., Gladyshev, M. I., Moskvichova, A. V., Makhutova, O. N., and Kalachova, G. S. (2003). Comparison of fatty acid composition in major lipid classes of the dominant benthic invertebrates of the Yenisei River. *Comparative Biochemistry and Physiology*, **134B**, 111–22.

Thorpe, J. H. and Covich, A. P. (2001). *Ecology and Classification of North American Freshwater Invertebrates*. Academic Press, San Diego, CA.

Vander Zanden, M. J. and Rasmussen, J. B. (1999). Primary consumer $\delta^{15}N$ and $\delta^{13}C$ and the trophic position of aquatic consumers. *Ecology*, **80**, 1395–1404.

Vander Zanden, M. J., Shuter, B. J., Lester, N. P., and Rasmussen, J. B. (2000). Within- and among-population variation in the trophic position of a pelagic predator, lake trout *(Salvelinus namaycush)*. *Canadian Journal of Fisheries and Aquatic Sciences*, **57**, 725–31.

Verburg, P., Kilham, S. S., Pringle, C. M., Lips, K. R., and Drake, D. L. (2007). A stable isotope study of a neotropical stream food web prior to the extirpation of its large amphibian community. *Journal of Tropical Ecology*, **23**, 643–51.

Whiles, M. R., Lips, K. R., Pringle, C. M., Kilham, S. S., Bixby, R. J., Brenes, R., Connelly, S., Colon-Gaud, J. C., Hunte-Brown, M., Huryn, A. D. *et al.* (2006). The effects of amphibian population declines on the structure and function of Neotropical stream ecosystems. *Frontiers in Ecology and Environment*, **4**, 27–34.

Yamamuro, M. and Kayanne, H. (1995). Rapid direct determination of organic-carbon and nitrogen in carbonate-bearing sediments with a Yanaco Mt-5 Chn Analyzer. *Limnology and Oceanography*, **40**, 1001–5.

6

Aquatic mesocosms

Raymond D. Semlitsch and Michelle D. Boone

6.1 Introduction

We have often told our students that any experiment can be easily criticized for lack of realism because all experiments, whether in the laboratory or field, by design are contrived by humans and are not natural. However, in spite of such criticism the more challenging issue is coming up with new venues or innovative designs to answer important ecological or conservation questions and to balance trade-offs. As amphibian ecology has become more experimental, we have been confronted with numerous trade-offs concerning the choice of venue, design, realism, and replication. Each choice is often coupled with benefits as well as limitations. As in many rapidly developing fields, criticism is an expected and healthy process that pushes investigators to think about their choices and develop solutions. The use of mesocosms in studies of aquatic ecology across an array of taxonomic groups has been central in that debate (e.g. Kimball and Levin, 1985; Carpenter 1996; Schindler 1998). Within the field of amphibian biology, a number of reviews also have stimulated a healthy discussion on the use of experiments (Hairston 1989a, 1989b), and in particular, use of mesocosms (e.g. Jaeger and Walls 1989; Rowe and Dunson 1994; Skelly and Kiesecker 2001; Skelly 2002).

The purpose of our chapter is to provide background on the use of aquatic mesocosms in ecological and conservation studies of amphibians. We describe the history, various types, set-ups, and designs and present selected case studies to emphasize their versatility. We also provide examples and comments on limitations of the technique and strengths of inference that come with the results. Our goal is to provide readers with an effective technique to answer a range of ecological and conservation questions.

6.2 Historical background

The use of aquatic mesocosms in amphibian ecology was born out of the need to move away from purely descriptive field studies and correlative analyses and towards more manipulative studies that could test hypotheses to establish cause and effect (see Wilbur 1987). Studies using aquatic mesocosms are often viewed as the best compromise between the variability of natural wetlands, confounding factors, and the lack of control that is common in field studies and the ability to manipulate single variables, replicate treatments, and randomly assign treatments to experimental units in laboratory experiments. They have sometimes been referred to as "hybrid" or "semi-field" experiments (Rowe and Dunson 1994). Aquatic mesocosms differ from "microcosms" in that they are self-sustaining once established and they do not need input of additional nutrients or resources after initial set-up, thus allowing study organisms to complete critical phases of their life cycle. To be effective, mesocosms must be capable of rearing populations of amphibian larvae, simulate important components of the aquatic ecosystem, and be reproducible and affordable. For purposes of this discussion, we refer primarily to two common types of aquatic mesocosm: cages or pens (Figure 6.1) that are placed in a natural pond or stream to capture location effects such as habitat variability and containers including plastic tubs, wading pools, or cattle watering tanks (Figure 6.2) that are filled with water and have components of aquatic ecosystems added to simulate natural variation. Artificial streams fit in the category of container mesocosms but have a flow-through or recirculating water source or system. Any of these types can be large or small and contain few or many components of a natural system; in this way they form a continuum between laboratory microcosms and natural field studies.

To the best of our knowledge, Warren Brockelman was the first to use pens placed in natural ponds for experimental amphibian studies in 1966–7 in Michigan, USA (Brockelman 1969). He used wooden framed pens (0.61 m wide $\times$ 2.44 m long $\times$ 0.6 m high) covered with fiberglass window screening and with lids, half of which had screen bottoms and half had bottoms open to sediments. His study addressed the effects of intraspecific competition on *Bufo americanus* tadpole growth, development, and survival.

The first use of containers for experimental amphibian studies was by Henry Wilbur and Joe Travis in 1977–9 in North Carolina (Wilbur and Travis 1984). They used cattle watering tanks (galvanized steel, round, 3.58 m² surface area, 0.61 cm deep) buried in the ground flush to the surface and filled with natural pond water. Replicate tanks were positioned in different habitat types to test differences in the colonization of aquatic habitats by invertebrates and amphibians,

Fig. 6.1 Examples of different cages used to rear amphibian larvae.

High-Shade Treatment

Full-Sun Treatment

Low-Shade Treatment

Fig. 6.2 Examples of different cattle tanks and pools used to rear amphibian larvae.

and the formation of community structure. In 1980, one of Henry Wilbur's graduate students, Peter Morin, subsequently developed a more efficient and novel venue for experiments using cattle tanks to manipulate factors important to community regulation (Morin 1981). He placed cattle tanks (epoxy-coated galvanized steel, round, 1.52 m diameter, 1000 L) in a field, above ground, so that they could be filled, manipulated, and easily drained to initiate new experiments each year. Each tank was filled with tap water and then each received an equal amount of dried litter collected from the edge of a natural pond (550 g) and Purina Trout Chow (50 g) for substrate and a source of nutrients. Equal amounts of inoculum of plankton derived from pooled plankton collections from eight natural ponds were added to each tank to initiate a diverse food web and prey base for larvae. This latter approach of using cattle tanks as aquatic mesocosms that could be renewed for each question and study has been critical in the advancement experimental amphibian ecology.

6.3 Why use mesocosms?

Studies in nature and in the laboratory allow researchers to answer important questions, but both have limitations (Diamond 1986). While laboratory studies allow us to maximize control and manipulate factors so that cause–effect relationships can be established, the laboratory lacks realism and complexity, reducing its application in the real world. In contrast, field studies offer realism and can encompass the complexity of natural food webs, but it can be difficult to distinguish cause–effect relationships and there can be high levels of variation among replicates. Mesocosms are the middle ground. Proponents argue that mesocosms offer the best of both worlds: control, the ability to determine cause–effect relationships, natural food webs, and realism (Wilbur 1987; Rowe and Dunson 1994; Drake *et al.* 1996; Boone and James 2005). Critics, however, contend that mesocosms offer little gain over laboratory studies in that they are still artificial and may have limited application to the real world, yet they require more work and time, and they have greater variation among replicates than laboratory studies (Carpenter 1996; Skelly 2002). Certainly, mesocosms would not be mistaken for natural ponds, but they incorporate many natural elements that allow them to mimic some but not all of the processes occurring in larger natural ponds or lakes (Schindler 1998).

There are a number of reasons why mesocosms can be a useful technique in your research program. First, mesocosms allow you to incorporate some components of natural systems into your experiments—the changes in season, photoperiod, temperature, and habitat that animals experience in nature—something that you typically cannot alter in the laboratory. Secondly, if you have evaluated a mechanism within or between organism(s), or developed a theoretical model, then a logical next step is to see if your predictions hold in a more complex and natural environment; mesocosm studies are a logical next step. Thirdly, if you want to evaluate single factors in isolation, a mesocosm may be ideal. Fourthly, manipulations in mesocosms can help you eliminate less important factors and find system drivers, so that you can design a better, more efficient large-scale field study.

If you need a short-term answer to a basic experimental question, then a laboratory experiment can offer an expedient answer: can trematodes infect limb buds? What concentration of an insecticide is lethal to tadpoles or zooplankton? Do overwintered bullfrog tadpoles prey upon recently hatched tadpoles? If your question is site- or habitat-specific or requires a vast area to encompass the entire ecosystem, then field experiments may be the best way to go: can ponds in city parks and golf courses support native fauna? What effects do forest-management practices have on amphibian communities? Studies in mesocosms provide a useful surrogate for the field and allow us to evaluate single and multiple factors that may be important at a larger scale, which allows us to better design larger field studies and provides insight into the system: do changes in food webs from eutrophication cause increases in number of trematodes and in the number of infected amphibians? Does a sublethal pesticide used on golf courses cause changes in aquatic community dynamics? Although mesocosms are often referred to as a "black box," suggesting that the mechanisms underlying the outcome are not clear, additional studies can be done to discern mechanisms by behavioral observations, periodic sampling of covariates (e.g. temperature, primary production, dissolved carbon), or with a paired laboratory study (e.g. prey palatability).

Mesocosms are useful in addressing a variety of questions in ecology, evolution, behavior, and conservation (see case studies below). We surveyed the literature from 2004 to 2007 through the Ecological Society of America (including the journals *Ecology, Ecological Applications, Frontiers in Ecology and Evolution,* and *Ecological Monographs*), *Oecologia, Evolution, Animal Behavior, and Behavior* by searching the terms "amphibian" or "frog" or "salamander" and then selecting all studies that used mesocosms. A total of 30 articles using mesocosms or enclosures were found only in the journals *Oecologia* or of the Ecological Society of America. The studies addressed a range of questions in ecology (e.g. species interactions, population biology, community structure), evolution (e.g. phenotypic plasticity), behavior (e.g. oviposition selection), and conservation (e.g. effects of pathogens, habitat fragmentation, pesticides). Whereas mesocosms have been historically used for basic questions of ecology and evolution, they are becoming more frequently used in the field of conservation and have enormous potential for use in regulatory fields such as ecotoxicology (reviewed by Boone and James 2005).

The appropriateness of a mesocosm as a representative of a natural system may depend on many things, including the size of the mesocosm and the species or community of interest. For studies with amphibian, invertebrate, and plankton communities, aquatic mesocosms ranging from 200 to more than 1500 L (see Figure 6.3a) were self-sustaining and maintained a similar temperature regime

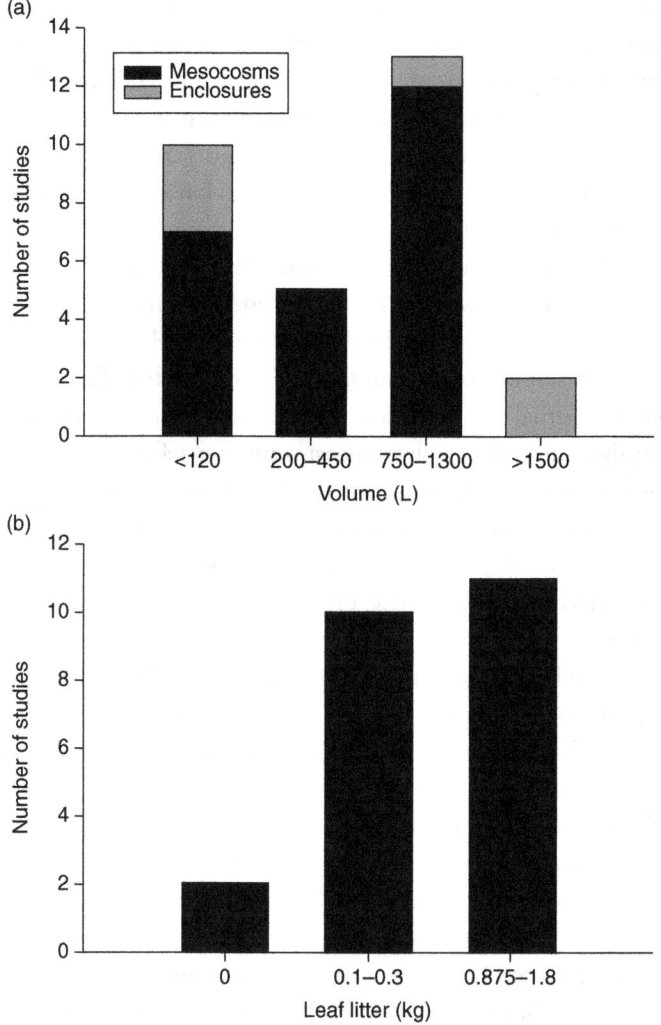

Fig. 6.3 Abundance of mesocosm studies conducted: (a) using different volumes and (b) with different amounts of leaf litter (see text for methods and results).

to natural communities. This is an important limitation because mesocosms of these sizes may not be appropriate for studies with larger species like wide-ranging predatory fish or larger-scale ecosystem processes like nutrient cycling. What is important for effective use is that the amphibians of interest, invertebrate prey, and plankton resources can complete major portions of their life cycle and be self-sustaining in the mesocosm. Isolated, temporary pond communities range in size from tire ruts to large shallow wetlands, and have characteristics that differ from larger lake systems that may be deeper and experience turnover and temperature inversions. Therefore, effective use of mesocosms is likely limited to species and processes that occur in shallow, temporary wetlands.

Additionally, there is evidence indicating that results generated from mesocosm experiments have high predictability to the field. Mesocosm studies evaluating the role of competition, predation, and pond drying (e.g. Wilbur 1972, 1987; Morin 1981, 1983; Semlitsch 1987) have proven useful in understanding the structure and regulation of amphibian communities. Long-term monitoring at a natural wetland (Rainbow Bay) in South Carolina, USA, has shown that community structure is determined by competitive and predatory interactions occurring along a gradient of pond drying (Semlitsch et al. 1996), which was predicted by mesocosm research. Manipulations with sublethal pesticide exposure in mesocosms show zooplankton populations are reduced and periphyton abundance increases for short periods of time, which can result in positive effects on tadpoles despite the pesticide having sublethal but negative effects on tadpoles directly (e.g. Boone and James 2003; Boone and Bridges-Britton 2006). These studies suggest that changes in the food web may be more important than direct negative effects of the contaminant alone (Mills and Semlitsch 2004). When these studies were replicated under more natural field conditions in ponds, similar outcomes were found despite the fact that predators were not excluded and that zooplankton populations could have recovered more quickly due to ephippia present in the soil sediments (Boone et al. 2004). Studies of dragonfly predation on tadpoles have documented qualitatively similar results between natural pools and mesocosms (Gascon 1992) and between different sizes of mesocosms (Gascon and Travis 1992). Such examples lend strong support to the usefulness of mesocosms, particularly given that mesocosm manipulations require less time, money, and effort to conduct, relative to the large field studies.

6.4 Types of mesocosm

There are a number of types of mesocosm, including cattle watering tanks or wading pools (Figure 6.2) and field enclosures (Figure 6.1). Cattle tank ponds can be purchased in galvanized steel (which must be sealed with an epoxy paint or lined) and polyethylene plastic (Behlen or Rubbermaid produce tanks in blue or black). The polyethylene ponds are most commonly used today because they are long-lasting (10 years), are not prone to rusting, and do not require sealing as metal tanks do. Wading pools are useful and cheap but short-lived (1–2 years) and usually made of PVC/vinyl, which can leach phthalates, chemicals that are associated with endo-crine disruption. Water tests found that Behlen polyethylene tanks did not leach estrogenic compounds (S.I. Storrs, personal communication). However, wading pools can be especially useful to address short-term behavioral questions related to factors that may influence oviposition (Vonesh and Buck 2007) or plasticity in tad-pole morphological development associated with predators and competition which may occur in a matter of weeks (Relyea 2004; Relyea and Auld 2005).

Field caging studies enable us to study a discrete location or habitat while moving toward greater realism. Field cages can range in size from small to large. In our survey, most of the cages were relatively small (2.4–120 L; Garcia *et al.* 2004; Griffis-Kyle and Ritchie 2007; Ireland *et al.* 2007) or large (pond size or sections; Scott 1990; Loman 2004; Boone *et al.* 2004). Aquatic field cages are often made of fine screen mesh, such as fiberglass window screen, mosquito net-ting, or plastic meshes, which are exposed to the larger matrix allowing for water exchange and influx of aquatic life smaller than the mesh size. Alternatively, field cages can also be made of polyethylene plastics that will prevent the flow of materials into the enclosure after initial set-up and also prevent things like pesticides or predators from moving into or out of the enclosure. These field cages whether made of permeable or impermeable material can be attached to untreated lumber into a box or rectangle to be placed in a pond (Figure 6.2). If the screen, mesh, or fabric has rigidity, it is also possible to use it to construct circular cages to hold animals (Figure 6.2). Alternatively, "bags" can be sewn or fabricated that can be supported with an external structure like multiple stakes, drinking water pipe, and/or floatation devices (Figure 6.2).

The advantage of field cages is that you can study sites of particular interest and ask questions in the relevant environment giving your study greater realism and the ability to manipulate factors of interest. In contrast, field cages often do not represent independent observations (unlike wading pools or tanks) given that there is the potential for exchange among enclosures. For example, if a researcher was examining how a fish predator may influence growth and devel-opment of tadpoles, the presence of fish in the pond or the presence/absence of fish as a treatment could influence all the enclosures if there was water exchange. However, these potential problems can be eliminated or minimized with choice of mesh and some design considerations for the cages.

6.5 Setting up mesocosms

There are a number of ways to set-up mesocosms for larval amphibian studies, with cattle tank ponds and wading pools generally established with a standardized volume of water (tap or well water), a nutrient base (leaves or grass and/or rabbit chow), and plankton from natural ponds. Larger volumes of water (900–1500 L) have the advantage of being more stable environments, especially in terms of temperature extremes in summer, and less prone to predator invasion. Smaller volumes of water can be shaded and raised above the ground to reduce likelihood of predators moving through. In our survey mentioned above, most mesocosm studies were conducted in tanks holding 750–1300 L of water (Figure 6.1). Water is typically allowed to sit between 2 days and 2 weeks so that chlorine from the tap water will dissipate and so that algal and zooplankton inoculum can establish.

The addition of leaf litter, grass, rabbit chow, and/or some other nutrient source is a standard practice in mesocosm experiments. Natural litter substrates serve as a slow-release nutrient source. Rabbit chow serves as a fast-release nutrient source. Both of these will support the aquatic food web throughout the field season and can influence other relevant factors like the amount of ultraviolet radiation penetration of the water (by influencing abundance of dissolved organic compounds) and dissolved oxygen. Researchers have used different types and amounts of nutrients, which can impact the aquatic system. In our survey, we found that roughly an equal number of studies used relatively low or high amounts of leaf litter (Figure 6.3). Most enclosure studies did not add nutrients unless it was being explicitly tested, presumably because there were adequate flow of nutrients and food into the enclosures. However, leaf litter or vegetation can be added to serve as refugia for amphibian prey within the enclosures.

The amount and type of nutrient addition can have an effect on the aquatic community because it sets the energy base for the system. A recent study by Williams et al. (2008) using dead grass or deciduous leaf litter found higher productivity in systems with grass, which resulted in greater tadpole growth. In studies manipulating the amount of leaf litter, Rubbo et al. (2008) and Boone (unpublished data) showed that increasing the amount of leaf litter increased the productivity of the system, and hence the survival and growth of tadpoles. Additionally, Rubbo and Kiesecker (2004) found the species from which the leaf litter originated had an effect on amphibian biomass and survival, with oak leaves having positive effects on amphibians compared to maple leaves. The type of leaf litter or grass can influence the nutrient base of the community and influence the algal and bacterial communities that flourish there. Consideration of the type of natural system you are attempting to represent (e.g. forested pond, grassy pond, or marsh) is the best guide to selecting the appropriate nutrient base to add.

It is important to establish the primary components of an aquatic food web in mesocosm tanks in order for them to be self-sustaining. A complex community of both phyto- and zooplankton is critical to help balance nutrient exchange, bacterial growth, and nitrogenous waste production as well as serve as a food source for amphibian larvae. Temporary wetlands are often an ideal source of inoculum for mesocosms because there will be fewer insect predators and no fish. All inoculum added to mesocosms should be strained to eliminate detrimental insect or fish larvae, and eggs or larvae of non-target amphibian species. Several additions over time and from varying natural ponds will help ensure adequate diversity and complexity.

6.6 Common experimental designs

Establishing cause–effect relationships is the hallmark of mesocosm approaches and allows complex multifactorial designs. Designs using pools or tanks allow clear tests of each factor independently. Mesocosm studies allow researchers to address how incremental changes in a treatment level impact the study organism and how different treatments may interact with one another. This level of manipulation would be difficult to achieve under most field conditions, but is a critical tool for understanding how amphibian populations and communities function.

Replicate tanks or pools are physically separate pond communities that can be established with standardized procedures but diverge if treatment differences occur. Each tank represents an independent replicate and experimental unit. Cages located within ponds are typically not independent from one another; therefore, the question of interest will determine how these field experiments are replicated. If you are interested in how a particular type of habitat influences an amphibian community, then you should replicate at the level of pond-habitat type. For instance, to examine whether amphibians can survive larval development in golf course ponds, Boone *et al.* (2008) used multiple golf course and reference ponds with replicate cages both within and between ponds. Using a single golf course and reference pond with replicate enclosures within each pond would not have clarified whether golf-course management affected larval amphibians; instead, it would only confirm that the two ponds are different for some reason, perhaps due to golf course management, historically different abiotic and biotic factors present in the pond, or chance. Replicate cages in the latter case would constitute "pseudoreplicates" (Hurlbert 1984) and would increase the probability of making a type I error or determining there is a significant difference related to treatment when there is not. However, if you are interested in how overwintered bullfrog tadpoles may influence the amphibian community, then manipulating bullfrog abundance in replicate cages in a single pond would be acceptable. Replication in field cage studies requires replicating at the level of your question and factor of influence (e.g. golf course), which will likely be at separate locations. Replication is also necessary within ponds to determine variation among specific cage locations (e.g. aspect, slope, or shading along a shoreline). Split-plot or nested designs are especially useful in studies using cages in ponds and may allow the testing of hypotheses not testable with less sophisticated designs (Gunzburger 2007).

The number of replicates you should use is also an important consideration. Typically there is a trade-off between the number of replicates and the total size of the study related to logistical and financial constraints. However, the more divergent your replicates, the greater the variation within your treatments, and the less your ability to detect significant differences between treatments. Therefore, consideration of the level of variation inherent in your replicates will influence the number of replicates that you need. While laboratory studies frequently have five or more replicates because of the ease of replication, mesocosm tank studies will often be limited to three to five replicates (Boone and James 2005), which is typically sufficient to detect significant differences in behavior, morphology, development, and survival. You can estimate *a priori* how many replicates you should use by determining your desired statistical power $(1-\beta;$ a power of 0.8 or greater is desirable) based on your α criteria and the effect size (size of the difference). Using a statistical program like SAS (Proc Power) or using online calculators (e.g. www.math.yorku.ca/SCS/Online/power/) you can determine the ideal number of replicates to use based on the system you are studying.

The random assignment of treatments is one of the most critical steps of all true experiments and should be applied to mesocosm studies. Randomizing your treatments is extremely important to avoid confounding variation among replicate tanks or cages with variation among treatments. You can use a random number chart or a random number generator in SAS (Proc Plan). Treatments may be randomly assigned across all tanks or cages (completely random design), or they may be blocked so that experimental units are grouped along a known gradient that may cause variation (e.g. shading) not relevant to the experimental question and which can be removed statistically (randomized complete block).

6.7 Case studies

6.7.1 Community ecology

The study of community ecology has had a long history of experimental studies to help disentangle the complex interactions among species. Morin (1981) used three species of anuran tadpoles (*Scaphiopus holbrookii*, *Pseudacris crucifer*, and *Bufo terrestris*) as competitors and adult newts (*Notophthalmus viridescens*) as a predator to address the role of interspecific competition. Tadpoles were reared together from hatching to metamorphosis in 16 cattle tanks (epoxy-coated galvanized steel, 1.52 m diameter, 1000 L) at constant initial densities in all tanks. Newts were added in four densities (none, two, four, and eight per tank) to vary the intensity of predation and replicated four times each. The results showed that in the absence of newt predators *Scaphiopus* (62%) and *Bufo* (33%) metamorphs were relatively abundant, followed by *Pseudacris* (5%). However, in the presence of a high density of newts, *Pseudacris* (68%) metamorphs were most abundant followed by *Scaphiopus* (20%) and *Bufo* (12%). The rarity of *Pseudacris* in tanks without newts was best explained by intense interspecific competition with *Scaphiopus* and *Bufo*. The increased survival and larger body mass of *Pseudacris* in tanks with newts was best explained by release from interspecific competition due to newt predation on the preferred prey of *Scaphiopus* and *Bufo* tadpoles. Morin suggested that newts acted as a 'keystone' predator to prevent exclusion of competitively inferior species by reducing the survival and density of competitively superior species. These results demonstrate that predation and competition could interact to produce deterministic patterns of community structure in amphibians.

In a subsequent experiment, Wilbur (1987) tested the interaction effects of competition, predation, and disturbance on the structure of a similar aquatic amphibian community. He used four species of anuran tadpoles (*Rana sphenocephala*, *S. holbrookii*, *P. crucifer*, and *B. terrestris*) as competitors and adult newts (*Notophthalmus viridescens*) as a predator. Tadpoles of the four species were reared together in 36 cattle tanks (epoxy-coated galvanized steel, 1.52 m diameter, 1000 L) at a constant high or low density. Four newts were added to half the tanks and ponds were drained at three rates; either 50 days, 100 days, or control tanks kept at 50 cm deep. All 12 treatment combinations were replicated in three spatial blocks to account for physical differences across a field were tanks were positioned. The results showed that *Scaphiopus* was least sensitive to intraspecific density, metamorphosed quickly in fast-drying tanks, and was the competitive dominant in tanks without newt predators, but suffered highest predation from newts. *Rana* was most successful at low density but could not

grow fast enough to metamorphose in fast-drying tanks and suffered high predation by newts. *Bufo* were very sensitive to high density and produced few metamorphs without newt predators. This effect was reversed by newts selectively
preying on *Scaphiopus* and *Rana* tadpoles, especially in rapidly drying tanks.
Pseudacris performed poorly in all slow-drying tanks but showed moderate success in high-density tanks where newts had removed most competitors. The
overall conclusion of this study was that all three factors interacted with each
other and with species' life history to determine community structure. Further,
it suggested that competition dominates simple habitats of short duration and
predation becomes increasingly important in ponds with longer hydroperiods
that permit the establishment of diverse predators. Both of the studies described
above were influential in focusing attention on the complex interactions of predation, competition, and pond drying in aquatic communities rather than historical single-factor arguments.

6.7.2 Evolutionary ecology

The "common-garden" experimental approach to evolutionary studies has been
important for clarifying genetic and environmental interactions by measuring
genotype performance in a simplified controlled environment (Clausen *et al.*
1947). Parris *et al.* (1999) examined the larval performance of hybrid and parental genotypes of the southern and plains leopard frog (*Rana sphenocephala*
and *Rana blairi*) in a common-garden approach using cattle tanks. They established 56 experimental populations in cattle tanks (polyethylene, round, 1.83 m
diameter) by adding 960 L of tap water, 1.0 kg of deciduous leaf litter, and multiple additions of plankton from natural ponds. Six genotypes of hatchling tadpoles were added to tanks at a low (20 per tank) or high (60 per tank) density to
simulate a competitive gradient. The 12 treatment combinations were replicated
either three or five times in the 56 tanks (Parris *et al.* 1999). Survival was the
same among all genotypes, but decreased with increasing density. Body mass
at metamorphosis was the same among genotypes, but decreased with increasing density. Tadpoles of *R. sphenocephala* had the shortest larval period while
the other parental species *R. blairi* had the longest larval period. Ponds with
hybrid genotypes produced a greater proportion of metamorphs than those with
R. blairi tadpoles but a smaller proportion than ponds with *R. sphenocephala*
tadpoles. Their results demonstrate equal or increased larval performance of
hybrids relative to parental genotypes in cattle tank environments. Studies such
as the one described above are critical for determining whether or not natural
hybridization could potentially lead to the production of recombinant genotypes possessing novel adaptations.

6.7.3 Ecotoxicology

The field of toxicology historically has taken a reductionist approach by using single species and chemical testing in the laboratory. However, there is growing effort to develop multispecies, community, and ecosystem-level experimental approaches that better mimic real-world problems (Kimball and Levin 1985; Rowe and Dunson 1994; Boone and James 2005; Semlitsch and Bridges 2005). A model study by Boone *et al.* (2007) examined the role that chemical contamination, competition, and predation play singly and in combination in an aquatic amphibian community. They established replicate aquatic communities in 64 polyethylene cattle tanks (round, 1.85 m diameter, 0.6 cm height, 1480 L volume) by adding 1000 L of tap water, 1.0 kg of leaf litter, plankton from natural ponds, 60 *Bufo americanus* tadpoles, 30 *R. sphenocephala* tadpoles, and 10 *Ambystoma maculatum* salamander larvae to each. Four factors were manipulated and replicated four times in a fully factorial design: no or two bluegill sunfish, no or six overwintered bullfrog tadpoles, 0 or 2.5 mg/L carbaryl (an insecticide), and 0 or 10 mg/L ammonium nitrate fertilizer. The results showed that bluegills had the largest impact on the community by eliminating *B. americanus* and *A. maculatum* and reducing the abundance of *R. sphenocephala*. Chemical contaminants had the second strongest effect on the community with the insecticide reducing *A. maculatum* abundance by 50% and increasing the mass of anurans (*R. sphenocephala* and *B. americanus*) at metamorphosis; the fertilizer positively influenced time and mass at metamorphosis for both anuran and salamander larvae. Presence of overwintered bullfrog tadpoles reduced mass and increased time to metamorphosis of the anurans. Although both bluegill and overwintered bullfrog tadpoles had negative effects on the amphibian community, they performed better in the presence of one another and in contaminated ponds. These results indicate that predicting complex interactions from single-factor effects may not be straightforward. Further, the research supports the hypothesis that combinations of factors can contribute to population declines.

6.7.4 Land use and management

Only recently have ecologists begun to use mesocosms to address questions concerning human land use and habitat fragmentation. Hocking and Semlitsch (2007) tested the effects of timber harvest on oviposition-site selection by gray treefrogs (*Hyla versicolor*) in four experimental forest treatments. Five plastic wading pools (1.5 m diameter, 30 cm deep) were placed above ground in each of the four forest treatments: (1) a clear-cut with high coarse woody debris (CWD), (2) a clear-cut with low CWD, (3) a partial cut with 50% canopy removal, and (4) a control. The five pools were placed at three distances from the natural breeding pond in each treatment (see Figure 1 in Hocking and Semlitsch 2007): one pool at 10 m, two pools at 64 m, and two pools at 115 m from a natural breeding pond. The four forest treatments were positioned around three replicate natural breeding ponds for a total of 60 wading pools. Pools filled naturally with rainwater by April. Following the first oviposition event, all pools were checked every 24–48 h for treefrog eggs. Eggs were removed and counted. The results indicated that treefrogs laid significantly more eggs in the clear-cuts (low-CWD 77185 eggs; high-CWD 51990 eggs) than in either the partial (13553 eggs) or the control (14068 eggs) treatments. Further, the interaction of forest × isolation treatment showed that oviposition in both clear-cut treatments decreased with increasing isolation whereas oviposition was the same at all levels of isolation in the partial and control treatments. Hocking and Semlitsch (2007) suggested that the strong preference for open-canopy pools likely benefits the larval stage because of higher-quality food and warmer water temperatures but that isolation of breeding pools beyond 50–100 m can inhibit oviposition by female treefrogs. They also add that the possible benefits to one stage (tadpoles) might be diminished by the risk of desiccation to the terrestrial juvenile stage after metamorphosis. Thus, mesocosms can be used to address land-use questions beyond the boundary of aquatic ecosystems.

6.8 Conclusion

Although we are both advocates for the use of mesocosms in amphibian ecology and conservation, we also agree that a pluralistic approach to any question is likely necessary for complete understanding. The use of laboratory experiments to understand mechanisms as well as field studies to anchor mesocosm results back to the real world are critical. The use of cattle watering tanks, wading pools, and cages is a powerful technique in any researcher's toolbox. The choice of mesocosm should reflect knowledge of the natural history of the species and system being studied. If used properly, at the correct scale for the questions one is asking, many questions can be clearly answered and results can have strong inferences back to natural systems. However, we caution readers that neither type of mesocosm described here can address all ecological and conservation questions alone. One has only to read Schindler (1998) to realize that most large-scale ecosystem processes cannot be replicated in cattle tanks or cages. But, for many population- and community-level studies of amphibians, mesocosms offer an excellent approach.

6.9 References

Boone, M. D. and Bridges-Britton, C. M. (2006). Examining multiple sublethal contaminants on the gray treefrog, *Hyla versicolor*: effects of an insecticide, herbicide, and fertilizer. *Environmental Contamination and Chemistry*, **25**, 3261–5.

Boone, M. D. and James, S. M. (2003). Interactions of an insecticide, herbicide, and natural stressors in amphibian community mesocosms. *Ecological Applications*, **13**, 829–41.

Boone, M. D. and James, S. M. (2005). Use of aquatic and terrestrial mesocosms in ecotoxicology. *Applied Herpetology*, **2**, 231–57.

Boone, M. D., Semlitsch, R. D., Fairchild, J. F., and Rothermel, B. B. (2004). Effects of an insecticide on amphibians in large-scale experimental ponds. *Ecological Applications*, **14**, 685–91.

Boone, M. D., Semlitsch, R. D., Little, E. E., and Doyle, M. C. (2007). Multiple stressors in amphibian communities: interactive effects of chemical contamination, bullfrog tadpoles, and bluegill sunfish. *Ecological Applications*, **17**, 291–301.

Boone, M. D., Semlitsch, R. D., and Mosby, C. (2008). Suitability of golf course ponds for amphibian metamorphosis when bullfrogs are removed. *Conservation Biology*, **22**, 172–9.

Brockelman, W. Y. (1969). An analysis of density effects and predation in *Bufo americanus* tadpoles. *Ecology*, **50**, 632–43.

Carpenter, S. R. (1996). Microcosm experiments have limited relevance for community and ecosystem ecology. *Ecology*, **77**, 677–80.

Clausen, J., Keck, D. D., and Hiesey, W. M. (1947). Heredity of geographically and ecologically isolated races. *American Naturalist*, **81**, 114–33.

Diamond, J. (1986). Overview: laboratory experiments, field experiments, and natural experiments. In J. Diamond and T. Case (eds), *Community Ecology*, pp. 3–22. Harper and Row Publishers, New York.

Drake, J. A., Huxel, G. R., and Hewitt, C. L. (1996). Microcosms as models for generating and testing community theory. *Ecology*, **77**, 670–7.

Garcia, T. S., Stacy, J., and Sih, A. (2004). Larval salamander response to UV radiation and predation risk: color change and microhabitat use. *Ecological Applications*, **14**, 1055–64.

Gascon, C. (1992). Aquatic predators and tadpole prey in central Amazonia: field data and experimental manipulations. *Ecology*, **73**, 971–80.

Gascon, C. and Travis, J. (1992). Does the spatial scale of experimentation matter? A test with tadpoles and dragonflies. *Ecology*, **73**, 2237–43.

Griffis-Kyle, K. L. and Ritchie, M. E. (2007). Amphibian survival, growth, and development in response to mineral nitrogen exposure and predator cues in the field: An experimental approach. *Oecologia*, **152**, 633–42.

Gunzburger, M. S. (2007). Habitat segregation in two sister taxa of hylid treefrogs. *Herpetologica*, **63**, 301–10.

Hairston, Sr, N. G. (1989a). Hard choices in ecological experimentation. *Herpetologica*, **45**, 119–22.

Hairston, Sr, N. G. (1989b). *Ecological Experiments*. Cambridge University Press, Cambridge.

Hocking, D. J. and Semlitsch, R. D. (2007). Effects of timber harvest on breeding-site selection by gray treefrogs (*Hyla versicolor*). *Biological Conservation*, **138**, 506–13.

Hurlbert, S. H. (1984). Pseudoreplication and the design of ecological field experiment. *Ecological Monographs*, **54**, 187–211.

Ireland, D. H., Wirsing, A. J., and Murray, D. L. (2007). Phenotypically plastic responses of green frog embryos to conflicting predation risk. *Oecologia*, **152**, 162–8.

Jaeger, R. G. and Walls, S. C. (1989). On salamander guilds and ecological methodology. *Herpetologica*, **45**, 111–19.

Kimball, K. D. and Levin, S. A. (1985). Limitations of laboratory bioassays: the need for ecosystem-level testing. *Bioscience*, **35**, 165–71.

Loman, J. (2004). Density regulation in tadpoles of *Rana temporaria*: a full pond field experiment. *Ecology*, **85**, 1611–18.

Mills, N. E. and Semlitsch, R. D. (2004). Competition and predation mediate indirect effects of an insecticide on southern leopard frogs. *Ecological Applications*, **14**, 1041–54.

Morin, P. J. (1981). Predatory salamanders reverse the outcome of competition among three species of anuran tadpoles. *Science*, **212**, 1284–6.

Morin, P. J. (1983). Predation, competition, and the composition of larval anuran guilds. *Ecological Monographs*, **53**, 119–38.

Parris, M. J., Semlitsch, R. D., and Sage, R. D. (1999). Experimental analysis of the evolutionary potential of hybridization in leopard frogs (Anura: Ranidae). *Journal of Evolutionary Biology*, **12**, 662–71.

Relyea, R. A. (2004). Fine-tuned phenotypes: tadpole plasticity under 16 combinations of predators and competitors. *Ecology*, **85**, 172–9.

Relyea, R. A. and Auld, J. R. (2005). Predator- and competitor-induced plasticity: how changes in foraging morphology affect phenotypic trade-offs. *Ecology*, **86**, 1723–9.

Rowe, C. L. and Dunson, W. A. (1994). The value of simulated pond communities in mesocosms for studies of amphibian ecology and ecotoxicology. *Journal of Herpetology*, **28**, 346–56.

Rubbo, M. J., Belden, L. K., and Kiesecker, J. M. (2008). Differential responses of aquatic consumers to variations in leaf-litter inputs. *Hydrobiologia*, **605**, 37–44.

Rubbo, M. J. and Kiesecker, J. M. (2004). Leaf litter composition and community structure: translating regional species changes into local dynamics. *Ecology*, **85**, 2519–25.

Schindler, D. W. (1998). Replication versus realism: the need for ecosystem-scale experiments. *Ecosystems*, **1**, 323–34.

Scott, D. E. (1990). Effects of larval density in *Ambystoma opacum*: an experiment in large-scale field enclosures. *Ecology*, **71**, 296–306.

Semlitsch, R. D. (1987). Relationship of pond drying to the reproductive success of the salamander *Ambystoma talpoideum*. *Copeia*, **1987**, 61–9.

Semlitsch, R. D. and Bridges, C. M. (2005). Amphibian ecotoxicology. In M. Lannoo (ed.), *Amphibian Declines: the Conservation Status of United States Species*, pp. 241–3. University of California Press, Berkeley, CA.

Semlitsch, R. D., Scott, D. E., Pechmann, J. H. K., and Gibbons, J. W. (1996). Structure and dynamics of an amphibian community: evidence from a 16-year study of a natural pond. In M. L. Cody and J. A. Smallwood (eds), *Long-term Studies of Vertebrate Communities*, pp. 217–48. Academic Press, San Diego, CA.

Skelly, D. K. (2002). Experimental venue and estimation of interaction strength. *Ecology*, **83**, 2097–2101.

Skelly, D. K. and Kiesecker, J. M. (2001). Venue and outcome in ecological experiments: manipulations of larval anurans. *Oikos*, **94**, 198–208.

Vonesh, J. R. and Buck, J. C. (2007). Pesticide alters oviposition site selection in gray treefrogs. *Oecologia*, **154**, 219–26.

Wilbur, H. M. (1972). Competition, predation, and the structure of the *Ambystoma-Rana sylvatica* community. *Ecology*, **53**, 3–20.

Wilbur, H. M. (1987). Regulation in complex systems: experimental temporary pond communities. *Ecology*, **68**, 1437–52.

Wilbur, H. M. and Travis, J. (1984). An experimental approach to understanding pattern in natural communities. In D. R. Strong, Jr, D. Simberloff, L. G. Abele, and A. B. Thistle (eds), *Ecological Communities: Conceptual Issues and the Evidence*, pp. 113–22. Princeton University Press, Princeton, NJ.

Williams, B. K., Rittenhouse, T. A. G., and Semlitsch, R. D. (2008). Leaf litter input mediates tadpole performance across forest canopy treatments. *Oecologia*, **155**, 377–86.

7

Water-quality criteria for amphibians

Donald W. Sparling

7.1 Introduction

Most amphibian species spend portions or even their entire life cycle in water. Whether their habitats are streams, wetlands, vernal ponds, farm ponds, or larger bodies of water, amphibians are directly affected by several natural and anthropogenic chemical and physical characteristics. Dissolved oxygen, temperature, pH, salinity and water conductivity, organic carbons, and pollutants are important factors of their habitats that can affect survival, growth, maturation, and physical development. In addition to direct effects, these characteristics interact with other factors such as predators, prey, parasites, and competitors to affect populations. Just as one example, the incidence of infestation of tadpoles by *Ribeioria ondatrae*, a trematode parasite that causes limb malformations, has been linked to the nutrient status of ponds (Johnson and Chase 2004). Field studies of amphibians should at least include a description of naturally occurring environmental conditions. In some cases, this description should also include analysis of environmental contaminants. This chapter discusses chemical and physical factors that are most important to aquatic life stages of amphibians and directs the reader to reviews for more complicated issues dealing with water quality.

Because of space limitations, the descriptions of these factors and the methods used in analyzing them are abbreviated. More complete descriptions of physiological requirements can be found in Feder and Bruggren (1992) and McDiarmid and Altig (1999). I recommend US Geological Survey (USGS) (2008a) and Eaton *et al.* (2005) for standardized protocols of water-quality analysis.

7.2 Dissolved oxygen

Whereas free oxygen in the atmosphere occurs as O_2 and is a relatively abundant gas, it must be dissolved in water to be useful to aquatic aerobic organisms.

These organisms take in oxygen through moist membranes such as gill epithelium, dermis, or other structures. Aside from for rare exceptions, only aquatic life stages of amphibians potentially face oxygen problems; atmospheric oxygen is sufficient for terrestrial forms. Some species of amphibian, such as tailed frogs (*Ascaphus* spp.), live in swift-moving lotic environments where water is constantly mixed with air and oxygen concentrations are usually near saturation. Others, however, inhabit quiescent, warm water, lentic environments where oxygen concentrations can fluctuate or persist at low concentrations. Under laboratory conditions, oxygen concentrations below 4 mg/L are deemed stressful to amphibian larvae and other aquatic organisms (ASTM 1988) and prolonged exposure to such hypoxic conditions in the field can be considered adverse. However, there are differences among species and animals can acclimate over time to low oxygen concentrations.

Many amphibians have anatomical features, behaviors, and physiological processes that allow survival under temporary hypoxic conditions. For example, members of the families Ranidae and Ambystomatidae assimilate oxygen through dermal uptake, gills, and lungs. Under normoxic conditions (non-stressful oxygen concentrations), approximately 70% of oxygen uptake by these larvae is through the dermis, 20% via the gills, and 10% through the lungs (Gatten *et al.* 1984). In contrast, *Bufo* spp. and *Ascaphus* spp. do not inflate their lungs until just before metamorphosis (Feder 1984) and Plethondontids lack lungs entirely.

In hypoxic conditions, larvae with lungs often swim to the surface and gulp air into the lungs and pulmonary uptake of oxygen becomes more important. Surfacing, however, incurs metabolic costs and exposes tadpoles to predators such as turtles (Feder 1983). However, in an experiment using aquariums, McIntyre and McCollum (2000) found that the alternative behavior of not coming to the surface could also incur risks. Young, pre-lunged bullfrog (*Rana catesbeiana*) tadpoles experienced a higher degree of predation in high-oxygen tanks than in low-oxygen ones because predacious tiger salamanders (*Ambystoma tigrinum*) spent a greater proportion of time searching for tadpoles at the bottom of the tanks under high-oxygen conditions than when the dissolved oxygen was low.

Hypoxia can induce similar physiological responses in amphibians as in other vertebrates including changes in blood pH, build-up of lactate in muscles, lethargy, and, under severe conditions, mortality (Ultsch *et al.* 1999). Chronic and intermittent hypoxia can reduce growth and developmental rates and delay hatching of salamander embryos (Valls and Mills 2007).

Several factors affect the concentration and measurement of dissolved oxygen. Oxygen concentrations can vary widely through the course of a day, especially in warm, eutrophic bodies of water. In ponds, a major source of oxygen comes from photosynthesis by algae and other green plants. Because photosynthesis is driven by water temperature and sunlight, dissolved oxygen concentrations are frequently lowest at dawn, increase several-fold during the course of the day, reach maximum concentrations in mid afternoon, and decrease sharply during the night due to biological oxygen demand (Figure 7.1). Cloudy or cool weather may reduce the extremes in concentrations. These daily variations demonstrate the need to record measurements under similar times of day and sunlight when comparing oxygen concentrations among ponds or through time. Measurements taken in early morning may be most meaningful with regard to environmental stress.

In lotic conditions, especially in shallow headwater streams inhabited by many species of salamander, dissolved oxygen concentrations are relatively uniform throughout the water column. In lentic bodies, however, there is often a declining gradient from the surface, where mixing with air is maximal, to the bottom, where anaerobic decay is greatest. Because larvae of different species may occupy different water depths, oxygen concentrations should be taken at least several centimeters below the surface or at the depth commonly used by the species of concern.

The measurement of dissolved oxygen is relatively straightforward with one of the many commercially available electronic dissolved-oxygen meters. For

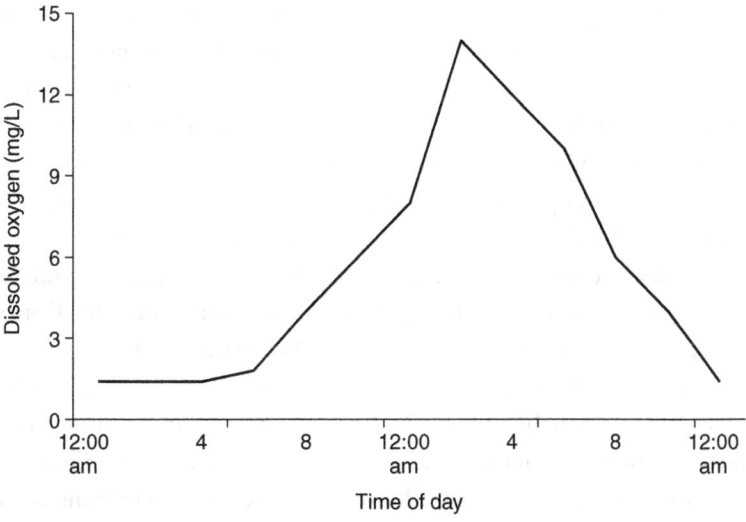

Fig. 7.1 Hypothetical variation in dissolved oxygen concentration in a warm, semi-eutrophic, lentic body of water.

example, YSI Instruments, Orion, Accumet, and Oakton are a few of the companies that manufacture portable field meters. These almost always come with temperature sensors and may be bundled with conductivity and pH sensors. Colorimeter kits (e.g. LaMotte, Orion, Chemetrics) are available and tend to be less expensive than electronic meters but require reagents for every test and may be less sensitive. When collecting water for oxygen analysis it is important to fill plastic or glass bottles to the very top without adding air and to keep the bottle cool until analysis. Ideally, oxygen should be analyzed on site to obtain the most accurate values.

7.3 Temperature

Because amphibians are poikilotherms they have limited ability to regulate their body temperature and are greatly affected by the temperature of their surrounding environment. Water temperature, therefore, is extremely important in affecting metabolic rates, other physiological processes, and behavior.

As a class, amphibians demonstrate a wide tolerance to temperature regimens: some species are able to withstand "temperature changes of 30°C on a daily basis and up to about 35°C on a seasonal basis" (Rome *et al.* 1992, p. 183). Wood frogs (*Rana sylvatica*) have the most northern distribution of any North American species (Conant and Collins 1998) and live above the Arctic Circle. Couch's spadefoot (*Scaphiophus couchii*) inhabits very arid, hot desert areas in the west where temperatures may exceed 45°C. Some species are eurythermic; for example, American bullfrogs range from central Mexico to northern Maine and Nova Scotia and tiger salamanders can be found from southern Texas into Alberta and Saskatchewan. Given time to acclimate, adult and larval amphibians in temperate and northern climes can survive near and even subzero temperatures by becoming dormant and greatly reducing their metabolic rates (Voituron *et al.* 2003). Those that inhabit dry, hot regions often burrow into the ground, enmesh themselves in cocoons, and estivate (Pinder *et al.* 1992). Some species, such as northern leopard frogs (*Rana pipiens*), can shunt water from vital organs to prevent tissue damage from ice crystal formation and concentrate glucose in organs to serve as a type of anti-freeze (Tattersall and Ultsch 2008).

In general, between 10 and 40°C, each 10° increase in ambient temperature increases metabolism by 1.4–2.4 times (Rome *et al.* 1992). Higher metabolic rates require greater oxygen; however, oxygen concentrations decrease as water temperatures increase (see below). At temperatures near 0°C most amphibians are very sluggish. At high temperatures, say above 30–35°C for some species but as low as 25–30°C for less tolerant animals, thermal stress can result in reduced mobility, abnormally high heart rates, and eventually death. Subtle interactions between temperature and other stressors can also occur. For example, Anderson *et al.* (2001) demonstrated complex relationships between predators and temperature in Pacific treefrogs (*Pseudacris regilla*). At 25.7°C tadpoles grew faster and exceeded the size limitation of their insect predators more quickly than tadpoles raised at 9.9°C. However, tadpoles of similar size encountered greater predation rates at the warmer temperature. Temperature can have a strong influence on life history progress (Camp and Marshall 2000) and microhabitat use (Crawford and Semlitsch 2008) in both aquatic and terrestrial amphibians.

There is a direct relationship between water temperature and dissolved oxygen concentrations. At sea level, where the partial pressure of oxygen, P_O, is highest, the concentration of dissolved oxygen is around 14 ppm at 0°C and declines as water temperatures increase (Figure 7.2). The same general relationship exists at higher elevations except that P_O saturation concentrations decline. Because oxygen demand increases as water increases in temperature, physiological stress can be very high in warm, low oxygenated waters.

Measuring water temperature is straightforward with any digital or analog thermometer. Many electronic meters used for other water chemistry parameters such as pH, conductivity, or dissolved oxygen come equipped with thermometers. Some thermometers automatically record minimum and maximum temperatures between readings so that a range can be ascertained. More sophisticated systems include data loggers that keep a continuous record of temperatures.

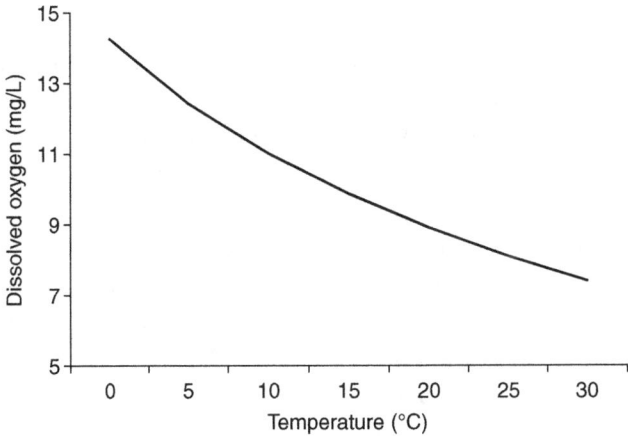

Fig. 7.2 Relationship between water temperature and dissolved oxygen concentration at sea level.

7.4 pH

pH, the negative log of the hydrogen ion concentration in water, is another key characteristic in describing amphibian habitats. The pH scale ranges from 0 to 14 which corresponds to a solution of 1 M H^+ (10^0) to 1 M OH^- with all H^+ bound with oxygen as a hydroxide. A pH of 7 defines neutral solutions, those with lower pH values are acidic, and those with higher values are alkaline. For ecological purposes, pH ranges of 6.0–7.5 are generally considered to be circumneutral or within a range that should present no harm to most aquatic organisms. A corresponding term, alkalinity, is the ability of water to buffer changes in pH.

A key factor in determining pH is the type of bedrock found within the watershed. Sedimentary rock such as sandstone, limestone, and dolomite are high in carbonates and bicarbonates that bind with hydrogen ions and form natural buffers. Alkalinity is typically high in these waters. Watersheds with a bedrock of igneous rock such as granite or basalt typically have low alkalinity due to the absence of carbonates.

Another natural factor affecting pH is the concentration of organic acids including tannic, fulvic, and humic acids. These organic compounds are weak acids in that they have low dissociation constants compared to mineral acids. Organic acids can be important in maintaining a low pH in isolated water bodies with high organic matter such as fens and bogs, but they also serve to some degree as buffers against anthropogenic sources of acidity.

Two major sources of anthropogenic acidity are acid deposition and acid mine drainage. Acid deposition comes from the production of nitrous oxides (NO_x) and sulfates (SO_4^{2-}) during the combustion of fossil fuels. These molecules combine with hydrogen in the air and are deposited on to land and water either with rain or attached to dust and organic particles that precipitate in dry form. Once in water the hydrogen and anions dissociate and free hydrogen ions increase acidity. Acid deposition can be of major concern in wetlands and streams that lie downwind of major industrial centers or cities and in watersheds with low alkalinity. The US Environmental Protection Agency (USEPA) (2008) is a good reference for current aspects of acid rain and deposition. Acid mine drainage is most prevalent in unreclaimed mined areas. Inversion of soil during mining places layers high in sulfides towards the surface and oxidation of these soils results in acidic seeps and leaching. The USGS (2008b) provides a good review of acid mine drainage.

Many review articles have been written on the adverse effects of acid deposition and anthropogenic acidity on amphibians (e.g. Sparling 1995; Rowe and Freda 2000). Variation in sensitivity to low pH occurs at the life stages, population, and species levels. At pH values equal to or below 4.5, embryonic development may cease entirely. At somewhat higher pH values ($\approx$4.5–5.0) development continues but hatching is curtailed (Dunson and Connell 1982). Interspecific differences to low pH have been found in several studies (e.g. Gosner and Black 1957; Sadinski and Dunson 1992). The critical pH or that which can cause significant increases in mortality for amphibian embryos ranges from 5.0 to 3.5 (Freda 1990). Tolerant species include carpenter's frogs (*Rana virgatipes*), wood frogs, and pine barren's treefrog (*Hyla andersoni*). Tolerance to low pH may be acquired in that animals living in acidic environments for many generations may have higher tolerance to reduced pH than those inhabiting circumneutral waters (Andren *et al.* 1989). The primary mechanism of toxicity due to low pH is interference with ion transport. Other effects include compromised immune systems (Brodkin *et al.* 2003), an inability of embryos to hatch, reduced growth, and delayed metamorphosis.

Acidic waters can be treated to increase pH. In watersheds with low alkalinity the addition of limestone to streams can substantially increase pH. Hard water with high calcium and magnesium concentrations can affect the toxic effects of acidity on embryos and larvae. Calcium can increase amphibian tolerance to low pH by reducing the loss of plasma ions. For example, adding 10 mg/L dissolved calcium to water increased hatching of Jefferson salamander (*Ambystoma jeffersonianum*) embryos at pH 4.5 compared with 1 mg/L (Freda and Dunson 1985). Conversely, calcium reduced hatching in wood frogs at pH 4.5. Freda (1990) explained that the difference in response to calcium was related to a curling effect seen in embryos that is elicited by H^+ and by calcium. Embryos that can hatch even though curled are benefited by calcium whereas those that cannot hatch when curled suffer.

Metals are more soluble with reduced pH and may interact with acidity to influence toxicity. Aluminum is particularly of concern because it is the most common metal in the Earth's crust, is soluble with low pH, and changes its speciation form as pH varies (reviewed by Sparling and Lowe 1996). Aluminum interacts with acidity in complex ways (e.g. Skei and Dolmen 2006) and a complete review of this interaction is beyond the scope of this chapter.

A common way of measuring pH is through one of the many commercially available pH meters. A typical meter will accurately measure pH to $\pm$0.01 pH units. If budgets are limited, however, pH papers with different ranges can be purchased and they are accurate to about $\pm$0.2 pH units. As with other water-quality measurements, it is usually preferable to assess water several centimeters below the surface. Sealed probes can be submerged to the desired depth or water can be collected in narrow-mouthed bottles for open or pH papers. The measurement of pH is temperature sensitive and most meters are set to 25°C for optimal accuracy. For field measurements, therefore, I recommend meters with automatic temperature compensation (ATC); these simultaneously record temperature and adjust the pH values accordingly.

7.5 Conductivity, hardness, and salinity

Conductivity is the ability of water to carry an electrical current and is due to the total concentration of anions and cations in water, collectively called total dissolved solids (TDS) in water. It is measured in units of micro-Siemens per centimeter or µS/cm. For comparison, ultrapure water has a conductivity of 5.5×10^{-3} µS/cm, drinking water around 5–50 µS/cm and sea water around 5000 µS/cm. Waters that have conductivity ranging between 150 and 500 µS/cm are adequate for aquatic organisms (USEPA; www.epa.gov/volunteer/stream/vms59.html).

Hardness can be measured as either the concentration (mg/L) of calcium ions or the sum of calcium and magnesium ions in water. These cations help buffer the effects of acidity and alter the toxic effects of dissolved metals (Horne and Dunson 1995b). Hard water is traditionally defined as a calcium-equivalent concentration of 80–120 mg/L and soft water as 0–20 mg/L. Moderate to hard water tends to be better for amphibians than soft water because calcium and magnesium can ameliorate the toxic effects of metals and pH, improve osmoregulation, and are required for proper formation of bones and other physiological processes (Ultsch et al. 1999).

Salinity is the concentration of chloride salts in water and is usually measured in parts per thousand (g/L or ‰), molarity (mM), or osmoles (mOsm). Most amphibians have low tolerance to salinity because they are not good osmoregulators (Gomez-Mestre et al. 2004). Balinksy (1981) listed 64 species of anurans and 13 species of urodeles that are tolerant to brackish water, although none are truly marine. Among the most tolerant species are cane toads (Bufo marinus), crab-eating frog (Rana cancrivora), African clawed frog (Xenopus laevis), European green toad (Bufo viridis), and some Batrachoseps spp. salamanders (Shoemaker et al. 1992). Euryhaline species can live in water that is 600–800 mOsm (70–93 ppm NaCl) but adults of most others do best at 200–300 mOsm (23–34 ppm NaCl; Shoemaker et al. 1992). Relatively little data are available on tadpoles but, in general, the lower the salinity the better. Road salts may pose problems for amphibians inhabiting ponds located near roads (Karraker et al. 2008).

As with other water-quality factors mentioned above, there are electronic meters and colorimeter kits available to measure these parameters.

7.6 Total and dissolved organic carbon

Organic carbon comes from organic molecules as contrasted to inorganic carbon that is derived from carbon dioxide and carbonates. Total organic carbon includes all sizes of organic molecules, both dissolved and non-dissolved. Dissolved organic carbon consists of organic molecules smaller than 0.45 μm in diameter. This carbon comes from many sources including surface run-off, aerial deposition, and decomposition of aquatic organisms. Elevated concentrations of organic carbon, along with suspended soil particles, are principal constituents of total suspended solids (TSS) and indicate potential problems with run-off into streams and lakes. Organic carbons are used as nutrients by bacteria and can increase biological oxygen demand. However, moderate levels of organic carbon are important because these molecules are a source of food for microorganisms. They also bind with certain pollutants such as metals and reduce their availability to aquatic organisms.

Dissolved organic carbon (DOC) is ecologically important to amphibians. Banks *et al.* (2007) showed a positive relationship between DOC concentration in wetlands of Acadia National Park (range 2–18 mg/L) and the concentration of methylated mercury (MeHg) in sediments and northern green frog (*Rana clamitans*) and bullfrog tadpoles. Because MeHg is more bioavailable and toxic than inorganic mercury, DOC in this situation can be detrimental to amphibians. In contrast, some forms of DOC, such as tannins, cause a brownish coloration to the water. These DOCs reduce light penetration into water which may curtail primary productivity (Carpenter *et al.* 2001). They also significantly suppress penetration of ultraviolet-B radiation which has been linked to malformations. Diamond *et al.* (2005) determined that DOC reduced ultraviolet-B penetration in the top 1 cm of water up to 87% and Calfee *et al.* (2006) determined that DOC protected salamander embryos from harmful ultraviolet-B exposure. Freda (1991) demonstrated that DOC can mitigate the toxicity of aluminum and other metals in acidified waters. However, humic and fulvic acids in excess of 10–20 mg/L can be toxic to amphibians (Freda *et al.* 1989; Horne and Dunson 1995a).

The analysis of TOC and DOC is a bit more complex than the other analytes so far described. Briefly, the first step is to separate inorganic and organic carbons in the water. This can be done by acidifying the water, thus converting carbonates to carbon dioxide which is then sparged from the water. The remaining carbon is considered organic. An aliquot of this is filtered through a 0.45 μm membrane and the carbon in the filtrate is DOC. Both TOC and DOC can be measured on a carbon analyzer or other instruments.

Colored DOCs (tannic, humic, and fulvic acids) can be measured indirectly by comparing the color of water to a set of standards available in kit form. Turbidity, which is affected by the total amount of inorganic and organic suspended particles, can be measured with a turibidimeter that measures light penetration through a standardized vial in units called nephelometric turbidity units (NTUs).

7.7 Pollutants

The number of types of pollutants or contaminants that can affect amphibian populations is enormous and beyond the scope of this brief chapter. Sparling (2003) surveyed the potential effects of contaminants on amphibians and Sparling et al. (2000) provided an extensive review of contaminant effects on amphibians and reptiles. Here I list a brief summary of contaminant classes and their significance.

7.7.1 Fertilizers and nitrogenous compounds

There are several ways of measuring ammonia and nitrates in water, including: (1) ion chromatography (for nitrates); (2) ion-specific probes coupled with a pH meter that registers millivolt output; (3) colorimetry; (4) test papers; and (5) other methods for sewage treatment operations and large volumes of material. Of the first four methods, ion-specific probes are less expensive than purchasing an ion chromatograph and are more sensitive and accurate than colorimetry or test papers. Test papers, although least expensive, are more limited in their resolution and detection limits than the other methods.

7.7.2 Pesticides

Pesticides are chemicals that are used to control plant and animal species considered to be noxious or unwanted by humans. The types of pesticides can be distinguished by intended target and by chemical class, which usually corresponds to primary mode of action. By target the primary classes are insecticides, fungicides, herbicides, and rodenticides.

Very roughly and with exceptions, insecticides are most acutely toxic to amphibians and other aquatic organisms of these classes. However, insecticides, fungicides, and herbicides exert a wide variety of sublethal or chronic effects that affect individual and population health. Rodenticides are not of much concern to amphibians, simply because of how they are used. There is a large and growing number of chemical families used as pesticides (see Cowman and Mazanti 2000 for a more complete review). Of greatest concern due to their widespread usage are organophosphates, carbamates, pyrethrins, and a few organochlorine insecticides. These are neurotoxins and either function at the synapse between neurons (organophospates and carbamates) or by affecting the transmission of action potentials down axons. In addition to direct mortality, pesticides can reduce growth, inhibit development, affect escape and predatory behavior, cause genotoxicity, affect gonad development, and alter other physiological processes (see Cowman and Mazanti 2000; Sparling 2003) and interact with other stressors (e.g. McIntyre and McCollum 2000; Relyea and Mills 2001). Aerial spraying of pesticides results in airborne particles that can travel many kilometers from the point of application to affect remote populations of amphibians (Davidson *et al.* 2001; Sparling *et al.* 2001).

There is no field method to determine the presence or concentration of pesticides in that their analyses are complex and often expensive. Matrix samples (sediment, water, or organisms) need to be collected and brought to a laboratory where the pesticides are extracted and analyzed using high-performance liquid chromatography (HPLC), gas chromatography, or other methods, all following recognized protocols.

7.7.3 Metals

Metals and metalloids such as zinc, copper, lead, chromium, arsenic, cadmium, mercury, selenium, and others are naturally occurring elements but are also released through many different industrial processes at concentrations that can be toxic to amphibians. Linder and Grillitsch (2000) reviewed the ecotoxicology of metals on amphibians and reptiles and showed that effects can range from direct mortality to a host of sublethal and potentially debilitating effects. Toxicity is affected by pH in that metal solubility increases with acidity and soluble forms of metals are more bioavailable than non-soluble forms.

Ion-specific probes are available for lead, silver, and copper, but the method for measuring is more complicated than inserting the probe directly into a water sample and should be done under laboratory conditions. Fortunately, metals are very stable, and once samples have been acidified to pH < 2 they can be stored for several months before analysis. Colorimetric methods exist for iron, manganese, aluminum, molybdenum, and copper, but most metals are analyzed through atomic absorption spectrophotometry or ion-coupled spectrophotometry (ICP) in a laboratory.

7.7.4 Organic pollutants and halogenated hydrocarbons

For this chapter, I include polycyclic aromatic hydrocarbons (PAHs), polychlorinated biphenyls (PCBs), dioxins, furans, and chlorinated pesticides (e.g. DDT, toxaphene, dieldrin) in this category. All of these chemicals are organically based and some have chlorine or other halogens incorporated into the molecular structure. PAHs include benzene, toluene, benzo[a]pyrene, and hundreds of other molecules that form through natural processes of decomposition, but also come from combustion of fossil fuels, processing of petroleum, and other sources. PCBs were used as lubricating fluids and in electrical transformers until the 1970s when their use in North America was banned. Dioxins and furans are formed during forest fires but also through the combustion of fossil fuels and are strong carcinogens. Similarly, many of the chlorinated hydrocarbon pesticides have been banned because they are extremely persistent in the environment and cause cancer, genotoxicity, endocrine disruption, and other harmful effects to humans and wildlife. Nevertheless, because they are extremely persistent, they can still be found throughout the world. At most environmentally realistic concentrations these chemicals produce a variety of sublethal effects including genotoxicity, inhibition of growth and development, endocrine disruption, skeletal defects and malformations, and reduced hatching success (reviewed by Sparling 2000). Sophisticated laboratory methods including HPLC or gas chromatography are required for the analysis of these chemicals.

7.7.5 Pharmaceuticals

Ecotoxicologists are just becoming aware of a myriad of chemicals that pass through our waste-treatment plants and enter natural bodies of waters essentially intact. These include prescription and non-prescription medicines, antibiotics, caffeine, and other products that pass through human bodies or are disposed of into toilets and down drains. These chemicals can exert numerous effects such as endocrine disruption and many physiological and behavioral changes that have yet to be defined. Because our awareness of these substances is so new, methods for determining their concentrations in field-collected matrices are often not available. Others can be determined through gas chromatography/mass spectrophotometry.

7.8 Summary and conclusions

In summary, students of amphibian ecology and behavior must be aware of various chemical and physical factors. These, in addition to vegetation, substrate composition, structural elements in the environment, water depth, stream flow, and the presence of other vertebrate and invertebrate species constitute the basic elements of habitat assessment for amphibians. There are many different methods of measuring these factors and the choice of method should be based on desired accuracy and precision as well as economics.

7.9 References

Anderson, M. T., Kiesecker, J. M., Chivers, D. P., and Blaustein, A. R. (2001). The direct and indirect effects of temperature on a predator-prey relationship. *Canadian Journal of Zoology*, **79**, 1834–41.

Andren, C., Henrikson, L. Olsson, M., and Nilson, G. (1989). Effects of pH and aluminium on embryonic and early larval stages of Swedish brown frogs *Rana arvalis, R. temporaria* and *R. dalmatina. Holoarctic Ecology*, **11**, 127–35.

ASTM (1988). *Standard Practice for Conducting Acute Toxicity Tests with Fishes, Macroinvertebrates, and Amphibians*. ASTM, West Conshohocken, PA.

Balinsky, J. B. (1981). Adaptation of nitrogen metabolism to hyperosmotic environment in Amphibia. *Journal of Experimental Zoology*, **215**, 335–50.

Banks, M. S., Crocker, J., Connery, B., and Amirbahman, A. (2007). Mercury bioaccumulation in green frog (*Rana clamitans*) from Acadia National Park, Maine, USA. *Environmental Toxicology and Chemistry*, **26**, 118–25.

Brodkin, M., Vatnick, I., Simon, M., Hopey, H., Butler-Holston, K., and Leonard, M. (2003). Effects of acid stress in adult *Rana pipiens. Journal of Experimental Zoology*, **298A**, 16–22.

Calfee, R. D., Bridges, C. M., and Little, E. E. (2006). Sensitivity of two salamander (*Ambystoma*) species to ultraviolet radiation. *Journal of Herpetology*, **40**, 35–42.

Camp, C. D. and Marshall, J. L. (2000). The role of thermal environment in determining the life history of a terrestrial salamander. *Canadian Journal of Zoology*, **78**, 1702–11.

Carpenter, S. R., Cole, J. J., Hodgon, J. R., Kitchell, J. F., Pace, M. L., Bade, D., Cottingham, K. L., Essington, T. E., Houser, J. N., and Schindler, D. E. (2001). Trophic cascades, nutrients, and lake productivity: whole-lake experiments. *Ecological Monographs*, **71**, 163–86.

Conant R. and Collins, J. T. (1998). *Reptiles and Amphibians. East/Central North America. Peterson Field Guides*. Houghton-Mifflin, Boston, MA.

Cowman, D. F. and Mazanti, L. E. (2000). Ecotoxicology of "New Generation" pesticides to amphibians. In D. W. Sparling, G. Linder, and C. A. Bishop (eds), *Ecotoxicology of Amphibians and Reptiles*, pp. 233–68. SETAC Press, Pensacola, FL.

Crawford, J. A. and Semlitsch, R. D. (2008). Abiotic factors influencing abundance and microhabitat use of stream salamanders in southern Appalachian forests. *Forest Ecology and Management*, **255**, 1841–7.

Davidson, C., Shaffer, H. B., and Jennings, M. R. (2001). Declines of the California red-legged frog: climate, UV-B, habitat, and pesticides hypotheses. *Ecological Applications*, **11**, 464–79.

Diamond, S. A., Trenham, P. C., Adams, M. J., Hossack, B. R., Knapp, R. A., Stark, S. L., Bradford, D., Corn, C. S., Czarnowski, K., Brooks, P. D. *et al.* (2005). Estimated ultraviolet radiation doses in wetlands in six national parks. *Ecosystems*, **8**, 462–77.

Dunson, W. A. and Connell, J. (1982). Specific inhibition of hatching in amphibian embryos by low pH. *Journal of Herpetology*, **16**, 314–16.

Eaton, A. D., Clesceri, L. S., Rice, E. W., and Greenberg, A. E. (2005). *Standard Methods for the Examination of Water and Waste Water*, 21st edn. American Public Health Association Press, Washington DC.

Feder, M. E. (1983). The relation of air breathing and locomotion to predation on tadpoles, *Rana berlandieri*, by turtles. *Physiological Zoology*, **56**, 522–31.

Feder, M. E. (1984). Consequences of aerial respiration for amphibian larvae. In R. S. Seymour (ed.), *Respiration and Metabolism of Embryonic Vertebrates*, pp. 71–86. Junk, Dordrecht.

Feder, M. E. and Burggren, W. W. (1992). *Environmental Physiology of the Amphibians*. University of Chicago Press, Chicago, IL.

Freda, J. (1990). Effects of acidification on amphibians. In J. P. Baker (ed.), *State of Science and State of Technology*, SOS/T report no. 13, *Biological Effects of Changes in Surface Water Acid-Base Chemistry*, pp. 2-185–2-202. National Acid Precipitation Assessment Program, Washington DC.

Freda, J. (1991). The effects of aluminum and other metals on amphibians. *Environmental Pollution*, **71**, 305–28.

Freda, J. and Dunson, W. A. (1985). The influence of external cation concentration on the hatching of amphibian embryos in water of low pH. *Canadian Journal of Zoology*, **63**, 2649–56.

Freda, J., Cavdek, V., and McDonald, D. G. (1989). Role of organic complexation in the toxicity of aluminum to *Rana pipiens* embryos and *Bufo americanus* tadpoles. *Canadian Journal of Fisheries and Aquatic Sciences*, **47**, 217–24.

Gatten, Jr, R. E., Caldwell, J. P., and Stockard, M. E. (1984). Anaerobic metabolism during intense swimming by anural larvae. *Herpetologica*, **40**, 164–9.

Gomez-Mestre, I., Tejedo, M. R., and Estepa, J. (2004). Developmental alterations and osmoregulatory physiology of a larval anuran under osmotic stress. *Physiological and Biochemical Zoology*, **77**, 267–74.

Gosner, K. L. and Black, J. H. (1957). The effects of acidity on the development and hatching of New Jersey frogs. *Ecology*, **38**, 256–62.

Horne, M. T. and Dunson, W. A. (1995a). The interactive effects of low pH, toxic metals, and DOC on a simulated temporary pond community. *Environmental Pollution*, **89**, 155–61.

Horne, M. T. and Dunson, W. A. (1995b). Effects of low pH, metals, and water hardness on larval amphibians. *Archives of Environmental Contamination and Toxicology*, **29**, 500–5.

Johnson, P. T. J. and Chase, J. M. (2004). Parasites in the food web: linking amphibian malformations and aquatic eutrophication. *Ecology Letters*, **7**, 521–6.

Karraker, N. E., Gibbs, J. P., and Vonesh, J. R. (2008). Impacts of road deicing salt on the demography of vernal pool-breeding amphibians. *Ecological Applications*, **18**, 724–34.

Linder, G. and Grillitsch, B. (2000). Ecotoxicology of metals. In D. W. Sparling, G. Linder, and C. A. Bishop (eds), *Ecotoxicology of Amphibians and Reptiles*, pp. 325–408. SETAC Press, Pensacola, FL.

McDiarmid, R. W. and Altig, R. (1999). *Tadpoles: The Biology of Anuran Larvae*. University of Chicago Press, Chicago, IL.

McIntyre, P. B. and McCollum, S. A. (2000). Responses of bullfrog tadpoles to hypoxia and predators. *Oecologia*, **125**, 301–8.

Ortiz-Santaliestra, M. E., Marco, A., Fernandez, M. J., and Lizana, M. (2006). Influence of developmental stage on sensitivity to ammonium nitrate of aquatic stages of amphibians. *Environmental Toxicology and Chemistry*, **25**, 105–11.

Pinder, A. W., Storey, K. B., and Ultsch, G. R. (1992). Estivation and hibernation. In M. E. Feder and W. W. Burggren (eds), *Environmental Physiology of the Amphibians*, pp. 250–74. University of Chicago Press, Chicago, IL.

Relyea, R. A. and Mills, N. (2001). Predator-induced stress makes the pesticide carbaryl more deadly to gray treefrog tadpoles (*Hyla versicolor*). *Proceedings of the National Academy of Sciences USA*, **98**, 2491–6.

Rome, L. C., Stevens, E. D., and John-Alder, H. B. (1992). The influence of temperature and thermal acclimation on physiological function. In M. E. Feder and W. W. Burggren (eds), *Environmental Physiology of the Amphibians*, pp. 183–205. University of Chicago Press, Chicago, IL.

Rowe, C. L. and Freda, J. (2000). Effects of acidification on amphibians at multiple levels of organization. In D. W. Sparling, G. Linder, and C. A. Bishop (eds), *Ecotoxicology of Amphibians and Reptiles*, pp. 545–72. SETAC Press, Pensacola, FL.

Sadinski, W. J. and Dunson, W. A. (1992). A multilevel study of effects of low pH on amphibians of temporary ponds. *Journal of Herpetology*, **26**, 413–22.

Shoemaker, V. H., Hillman, S. S., Hillyard, S. D., Jackson, D. C., McClanahan, L. L., Withers, P. C., and Wygoda, M. L. (1992). Exchange of water, ions, and respiratory gases in terrestrial amphibians. In M. E. Feder and W. W. Burggren (eds), *Environmental Physiology of the Amphibians*, pp. 125–50. University of Chicago Press, Chicago, IL.

Skei, J. K. and Dolmen, D. (2006). Effects of pH, aluminum, and soft water on larvae of the amphibians *Bufo bufo* and *Triturus vulgaris*. *Canadian Journal of Zoology*, **84**, 1668–77.

Sparling, D. W. (1995). Acidic deposition: a review of biological effects. In D. Hoffman, B. A. Rattner, G. A. Burton, and J. Cairns Jr (eds), *Handbook of Ecotoxicology*, pp. 301–29. Lewis Publishers, Boca Raton, FL.

Sparling, D. W. (2000). Ecotoxicology of organic contaminants to amphibians. In D. W. Sparling, G. Linder, and C. A. Bishop (eds), *Ecotoxicology of Amphibians and Reptiles*, pp. 461–94. SETAC Press, Pensacola, FL.

Sparling, D. W. (2003). A review of the role of contaminants in amphibian declines. In D. Hoffman, B. A. Rattner Jr, and J. Cairns (eds), *Handbook of Ecotoxicology*, pp. 1099–1128. Lewis Publishers, Boca Raton, FL.

Sparling, D. W. and Lowe, T. P. (1996). Environmental hazards of aluminum to plants, invertebrates, fish, and wildlife. *Reviews Environmental Contamination Toxicology*, **145**, 1–127.

Sparling, D. W., Linder, G., and Bishop, C. A. (2000). *Ecotoxicology of Amphibians and Reptiles*. SETAC Press, Pensacola, FL.

Sparling, D. W., Fellers, G. M., and McConnell, L. L. (2001). Pesticides and amphibian population declines in California, USA. *Environmental Toxicology and Chemistry*, **20**, 1591–5.

Tattersall, G. J. and Ultsch, G. R. (2008). Physiological ecology of aquatic overwintering in ranid frogs. *Biological Review*, **83**, 119–40.

Ultsch, G. R., Bradford, D. F., and Freda, J. (1999). Physiology: coping with the environment. In R. W. McDiarmid and R. Altig (eds), *Tadpoles. The Biology of Anuran Larvae*, pp. 189–214. University of Chicago Press, Chicago, IL.

USEPA (US Environmental Protection Agency) (2008). *Acid Rain.* www.epa.gov/
acidrain/.

USGS (United States Geological Survey) (2008a). *National field manual for the collection of
water-quality data: U.S. Geological Survey Techniques of Water-Resources Investigations*,
book 9, chapters A1–A9. http://pubs.water.usgs.gov/twri9A.

USGS (United States Geological Survey) (2008b). *Acid mine drainage: introduction.*
http://energy.er.usgs.gov/health_environment/acid_mine_drainage/.

Valls, J. H. and Mills, N. E. (2007). Intermittent hypoxia in eggs of *Ambystoma maculatum*: embryonic development and egg capsule conductance. *Journal of Experimental
Biology*, **210**, 2430–5.

Voituron, Y., Eugene, M., and Barré, H. (2003). Survival and metabolic responses to
freezing by the water frog (*Rana ridibunda*). *Journal of Experimental Zoology*, **299A**,
118–26.

Part 3
Juveniles and adults

8

Measuring and marking post-metamorphic amphibians

John W. Ferner

8.1 Introduction

In this chapter, I review marking and measuring techniques for post-metamorphic amphibians; larvae are covered in Chapter 4 in this volume. Measuring the size of individuals is important for determining age, categorizing life history stages (juvenile or adult), and calculating growth rates (larval measurements are covered in Chapter 3). More field and behavioral studies requiring marking are now being conducted than in the past, and often investigators need very specific marking techniques and non-invasive ways of identifying individuals. Additionally, ethical issues relative to certain techniques, such as radioactive tags and mutilation procedures, are now more commonly considered as, for example, when considering toe-clipping. Some older techniques are used much less often than they used to be. Major advancements in biotelemetry (see Chapter 11) and passive integrated transponder (PIT) tags have added needed flexibility to the choice of marking methods.

Recent reviews of marking and identification techniques include Baker and Gent (1998) and Ferner (2007) for amphibians and reptiles, and Donnelly *et al.* (1994) for amphibians. Additional references are found in a number of papers published in a special edition of *Mertensiella* (Henle and Veith 1997).

Some criteria for ideal marks or tags are as follows (after Ferner 2007):

- they should not affect the survivorship or behavior of the organism;
- they should allow the animal to be as free from stress or pain as possible;
- they should identify the animal as a particular individual, if desirable;
- they should last indefinitely, or at least the duration of the study;
- they should be easily read and/or observable;

- they should be adaptable to organisms of different sizes;
- they should be easy to use in both laboratory and field, and use easily obtained material at minimal cost.

There are no techniques that satisfy all of these criteria. Selecting a technique requires deciding which are the most feasible and scientifically accurate in meeting the objectives of the particular study. Two or more marks often are used to meet multiple study objectives or to serve as a back-up to the primary mark (e.g. simultaneous toe-clipping and photographing of dorsal patterns). Techniques

Table 8.1 *Sources of materials for marking amphibian species (as reviewed in Ferner 2007 unless otherwise indicated).*

Technique item	Source of materials
Adhesive (small wounds)	Vetbond®
Bandage (New Skin®)	Medtech (www.newskinproducts.com)
Beads, balls, and elastic for waistbands	Ball Chain Manufacturing Co. (www.ballchain.com) Bead Industries (www.beadindustries.com) B. Toucan (www.btoucan.com)
Fluorescent pigments for branding	Scientific Marking Materials, PO Box 23122, Seattle, WA 98124, USA Radiant Color Company, 2800 Radiant Ave., Richmond, CA 94804, USA (+1 800 Radiant; fax +1 510 233 9138)
Fluorescent elastomer dyes; visual implant elastomers (VIEs)	Northwest Marine Technology, PO Box 427, 976 Ben Nevis Road, Shaw Island, WA 98286, USA (www.nmt.us)
PIT tags	Destron Fearing Corporation/Digital Angel, St. Paul, MN, USA (www.destronfearing.com, www.digitalangel.com) Also see www.biomark.com
PIT pack	Battery powered Destron Fearing transceiver (model FS 2001 A-ISO) and custom-built antenna (Blomquist *et al.* 2008)
Scanner (for PIT tags)	AVID Marketing, Norco, CA USA
Fluorescent tags (VIAlpha Numeric Tags)	Northwest Marine Technology (see address above)
Ultraviolet lamp for detecting fluorescents	UVP, San Gabriel, CA, USA (www.uvp.com)

PIT, passive integrated transponder.

described for use with one group of amphibians might be adapted for a different group. Therefore, investigators might benefit by considering all of the techniques available, regardless of how originally used. Further, there may be a need for standardizing marking techniques in comparative studies, or those that might be carried on by other researchers in the future.

In the following text, I emphasize the use of marking techniques for adult amphibians which have proved most successful. When available, information on the advantages and disadvantages of each technique has been given. This chapter provides information adequate for considering techniques without consulting the original source. However, once a technique has been tentatively selected, it is advisable to consult the primary literature if time and facilities permit. The use of complex techniques, such as telemetry or PIT tags, requires careful review of the original sources. Sources for materials used with various techniques are given in Table 8.1, although availability and contact information are likely to change constantly.

8.2 Toe-clipping

8.2.1 Anurans

Although several toe-clipping methods are proposed in the literature, that of Martof (1953) is most widely used. He cut toes of *Rana clamitans* with scissors, trying to avoid damage to webbing between them. Little blood loss or loss of swimming power was observed and no regeneration of the digits was found. This system assigns serial numbers to the digits of the feet. No more than two toes were removed from any one foot. The left hind foot indicated units, and for those larger than five, combinations of two digits (the fifth and one other) were excised. For example, clipping the fifth and third toes designated tens and the front feet denoted hundreds. With this arrangement, 6399 individuals can be marked in series; the greatest number can be indicated by removing the two outer digits on each foot. The only concern Martof expressed about this procedure was that confusion might arise where only a single toe on a single hand foot was removed with recapture of a frog with an injured foot. He suggested alleviating this problem with a special designation, such as a zero marking by removal of the second and fourth toes on the other hind foot. Waichman (1992) also proposed a numbering system for toe-clipping that letters the four limbs (A through D) and numbers the toes. This scheme makes all 959 combinations available using two and three toes, but not removing more than two toes for each limb (see Table 8.2).

Table 8.2 *Alphanumeric code for toe-clipping amphibians and reptiles from Waichman (1992)*

One toe	Two toes					
A1	A1A2	A2A5	A3D2	A5B5	B1D4	B4C1
A2	A1A3	A2B1	A3D3	A5C1	B1D5	B4C2
A3	A1A4	A2B2	A3D4	A5C2	B2B3	B4C3
A4	A1A5	A2B3	A3D5	A5C3	B2B4	B4C4
A5	A1B1	A2B4	A4A5	A5C4	B2B5	B4C5
B1	A1B2	A2B5	A4B1	A5C5	B2D1	B4D1
B2	A1B3	A2C1	A4B2	A5D1	B2D2	B4D2
B3	A1B4	A2C2	A4B3	A5D2	B2D3	B4D3
B4	A1B5	A2C3	A4B4	A5D3	B2D4	B4D4
B5	A1C1	A2C4	A4B5	A5D4	B2D5	B4D5
C1	A1C2	A2C5	A4C1	A5D5	B3B4	B5C1
C2	A1C3	A2D1	A4C2	B1B2	B3B5	B5C2
C3	A1C4	A2D2	A4C3	B1B3	B3C1	B5C3
C4	A1C5	A2D3	A4C4	B1B4	B3C2	B5C4
C5	A1D1	A2D4	A4C5	B1B5	B3C3	B5C5
D1	A1D2	A2D5	A4D1	B1C1	B3C4	B5D1
D2	A1D3	A3A4	A4D2	B1C2	B3C5	B5D2
D3	A1D4	A3A5	A4D3	B1C3	B3D1	B5D3
D4	A1D5	A3C1	A4D4	B1C4	B3D2	B5D4
D5		A3C2	A4D5	B1C5	B3D3	B5D5
		A3C3	A5B1	B1D1	B3D4	C1C2
		A3C4	A5B2	B1D2	B3D5	C1C3
		A3C5	A5B3	B1D3	B4B5	C1C4
		A3D1	A5B4			

C1C5	C5D3	A1A4B2	A1A5D1	A2A4B5	A2A5D4
C1D1	C5D4	A1A4B3	A1A5D2	A2A4C1	A2A5D5
C1D2	C5D5	A1A4B4	A1A5D3	A2A4C2	A3A4B1
C1D3	D1D2	A1A4B5	A1A5D4	A2A4C3	A3A4B2
C1D4	D1D3	A1A4C1	A1A5D5	A2A4C4	A3A4B3
C1D5	D1D4	A1A4C2	A2A3B1	A2A4C5	A3A4B4
C2C3	D1D5	A1A4C3	A2A3B2	A2A4D1	A3A4B5
C2C4	D2D3	A1A4C4	A2A3B3	A2A4D2	A3A4C1
C2C5	D2D4	A1A4C5	A2A3B4	A2A4D3	A3A4C2
C3C4	D2D5	A1A4D1	A2A3B5	A2A4D4	A3A4C3
C3C5	D3D4	A1A4D2	A2A3C1	A2A4D5	A3A4C4
C3D1	D3D5	A1A4D3	A2A3C2	A2A5B1	A3A4C5
C3D2	D4D5	A1A4D4	A2A3C3	A2A5B2	A3A4D1
C3D3	A1A2C3	A1A4D5	A2A3C4	A2A5B3	A3A4D2
C3D4	A1A2C4	A1A5B1	A2A3C5	A2A5B4	A3A4D3
C3D5	A1A2C5	A1A5B2	A2A3D1	A2A5B5	A3A4D4
C4C5	A1A2D1	A1A5B3	A2A3D2	A2A5C1	A3A4D5
C4D1	A1A2D2	A1A5B4	A2A3D3	A2A5C2	A3A5B1
C4D2	A1A2D3	A1A5B5	A2A3D4	A2A5C3	A3A5B2
C4D3	A1A2D4	A1A5C1	A2A3D5	A2A5C4	A3A5B3
C4D4	A1A2D5	A1A5C2	A2A4B1	A2A5C5	A3A5B4
C4D5	A1A3B1	A1A5C3	A2A4B2	A2A5D1	A3A5B5
C5D1	A1A3B2	A1A5C4	A2A4B3	A2A5D2	A3A5C1
C5D2	A1A3B3	A1A5C5	A2A4B4	A2A5D3	A3A5C2
Three toes	A1A3B4				and so on
A1A2B1	A1A3B5				
A1A2B2	A1A3C1				
A1A2B3	A1A3C2				
A1A2B4	A1A3C3				
A1A2B5	A1A3C4				
A1A2C1	A1A3C5				
A1A2C2	A1A3D1				
	A1A3D2				
	A1A3D3				
	A1A3D4				
	A1A3D5				
	A1A4B1				

Codes A5 and B5 are available only for animals with five foretoes.

Regardless of the numbering system used, the potential for toe regeneration must be considered (see Table 8.3). In addition, the thumbs of the forefeet are required by males during amplexus, and they should not be clipped (for example, Briggs and Storm 1970). Another potential problem is understanding how toe-clipping may stress an animal and affect resource acquisition. Daugherty (1976) indicated a problem with weight loss in toe-clipped *Rana pipiens*. The most serious criticism of toe-clipping anurans was raised by Clarke (1972), who noted that the probability of recapturing *Bufo fowleri* decreased as the number of digits excised increased; researchers today do not clip as many toes as Clarke, however. Halliday (1995) reported that studies of *Atelopus elegans*, *Atelopus carbonerensis*, *Bufo marinus*, *Bufo granulosus*, and *Bufo bufo* provided little or no evidence for any impact on survival after toe-clipping. An assessment and concern about use of toe-clipping of *Hyla labialis* was provided by Lüddecke and Amezquita (1999). Fewer than 1% of toe-clipped *Rana pretiosa* showed any sign of infection, and no mortality was found (Reaser and Dexter 1996). They stressed that hygienic, sterile techniques must be used, and that impacts should be carefully monitored in

Table 8.3 *Digit-regeneration occurrence and times for selected toe-clipped amphibian species*

Species	Time interval for regeneration
Anurans	
Bufo hemiophrys	Several months but recognizable
Hyla regilla	Only slight after 1 year
Hyperolius viridiflavus ferniquei	Complete in newly metamorphosed
Rana catesbeiana	None in 3-year study
Rana erythraea	Slow or nonexistent over 4 years
Caudates	
Ambystoma opacum	Some in short period
Taricha spp.	Several years
Taricha granulosa	Indefinitely, if kept below or at 10°C
Triturus cristatus	Slow compared to tail clips
Triturus vulgaris	Up to 10 months; little winter regeneration (Griffiths 1984)
Plethodon cinereus	7 months in laboratory
Plethodon glutinosus	At least 2 years
Plethodon wehrlei	50% after 100 days; regenerated digits identifiable by lack of pigmentation
Cryptobranchus alleganiensis	1 year
Batrochoseps spp.	Slow; regenerated toes recognizable

After review by Ferner (2007) unless otherwise indicated.

studies using toe-clipping. Despite these potential problems, toe-clipping is still the most common marking technique used for anurans.

8.2.2 Salamanders

Toe-clipping has been the common method of marking salamanders in recent years, and in most cases something similar to Martof's (1953) system of marking anurans is used. Another scheme (as above) was provided by Waichman (1992) (Table 8.2). The potential of toe regeneration needs to be considered in salamanders, but whether or not it is a deterrent to using toe-clipping depends upon the species and study objectives. Table 8.3 summarizes reports in the literature concerning the regeneration of digits in salamanders.

Some salamanders, such as juvenile (< 2.5 cm snout–vent length, or SVL) dusky salamanders (*Desmognathus fuscus*), have toes too small for precise clipping. In such cases, animals have been marked successfully by clipping small pieces of the tail at either right angles to the longitudinal axis of the body or in the transverse plane at different angles (Orser and Shine 1972). This technique allowed groups of marked juveniles to be distinguished for at least 1 month before regeneration became a problem.

Mutilation techniques, such as toe-clipping, may use some form of local or general anesthesia. Peterman and Semlitsch (2006) stressed the importance of properly buffering solutions of the anesthetic MS-222 in plethodontid salamanders, and the effect of various concentrations on the time to become anesthetized and recover. They found induction time usually decreased, and the time for recovery increased, with increasing concentrations of MS-222. This study should be consulted before using this anesthetic.

Davis and Ovaska (2001) found that toe-clipping in *Plethodon vehiculum* resulted in less weight gain than animals marked with fluorescent tags or used as controls. These authors used unique combinations of three clipped toes per salamander, removing no more than one toe per foot. The number of ambiguous marks in toe-clipped individuals increased dramatically in frequency compared with the fluorescent tagged animals after 50 days, but the majority of both retained useful marks throughout the 87-week field study.

8.2.3 Ethical issues related to toe-clipping of amphibians

In recent years there has been increasing concern over the possible impact of various marking techniques on animals. Mutilation techniques such as toe-clipping have been reviewed in several studies to determine their impact on survival. In addition, ethical issues relative to the possibility of inflicting pain on organisms are a concern. University Institutional Animal Care Committees (IACUC) in the USA may request documentation on marking techniques and consideration of several alternatives. For guidelines see www.ssarherps.org/pages/ethics.php. In Europe legislation is often even more strict and a license is required before any invasive surgical procedure including scientific studies using PIT tagging.

Parris and McCarthy (2001) reported evidence that toe-clipping anurans decreases the rate of return of animals in mark–recapture studies. They encouraged clipping as few toes as possible, and stated that researchers must control for the impact of clipping on the rate of recapture. They further pointed out that these results have "important implications for the ethical treatment of animals, the continued use of toe-clipping to mark species of conservation concern, and the removal of multiple toes from an individual frog or toad" (McCarthy and Parris 2004). The debate on toe-clipping anurans continues (summarized by Ferner 2007 and Phillott *et al.* 2008).

8.3 Branding

8.3.1 Anurans

Kaplan (1958) described a branding technique for frogs involving the incorporation of India ink into scarified skin of the venter. Numbers were etched in the skin with a hypodermic needle and then filled with India ink. Erythema disappeared quickly, no infection was observed, and the numerals remained distinguishable for more than 3 months. Kaplan also described an electric tattooing technique as an improvement upon the scarification method. Higgins India Ink™ was found most effective if mixed with a drop of glycerin to aid its spreading into the skin. Surplus ink was wiped away, leaving a clear, permanent mark. Some initial inflammation was observed in the frogs, but was only rarely prolonged. Additional branding techniques are provided in Table 8.4.

Table 8.4 *Branding techniques used for marking amphibian species (as reviewed in Ferner 2007)*

Species	Description of branding technique
Anurans	Scarring with needle or electric tattoo marker then filled with India ink
	Heated chromium nickel wire
	Silver nitrate (75%), potassium nitrate (25%)
Ascaphus truei	Freeze branding with copper wire cooled in dry ice
Bufo bufo	Alcian Blue dye solution using a "Panjet®" innoculator
Eleutherodactylus podiciferus	Pressurized fluorescent granular powder

8.3.2 Salamanders

Salamander branding techniques involve heat or freeze brands, or the application of pigments or dyes into scarified skin. Taber *et al.* (1975) branded *Cryptobranchus alleganiensis* with heated 1.5-cm-high numerical brands of thin nickel-chrome wire. These remained clear through a 2-year study, but some faded and required rebranding. Bull *et al.* (1983) tested freeze-branding application times while marking *Ambystoma macrodactylum*. Letter brands of 5 mm high made with silver-tipped copper rods were immersed for 5 min in liquid nitrogen. A 0.75-s application produced the best results, with maximum clarity occurring at 3 months. Woolley (1962) used a black Carter's felt ink pencil to mark salamanders. Dilute acetic acid or ammonium hydroxide was used to remove mucus on the salamander's tail before application. These marks lasted at least a month.

Taylor and Deegan (1982) injected a fluorescent pigment into red eft (*Notophthalmus viridescens*) dermis using a compressed air spray gun with minimal mortality. After 39 days, individuals began succumbing to water mold (*Saprolegnia*) infections, but these could not be linked to the use of fluorescent branding.

Nishikawa and Service (1988) described a technique using dry fluorescent dust applied with pressurized air on *Plethodon jordoni* and *Plethodon glutinosus*. The technique required four to 10 colours of inert fluorescent pigments, canisters, a spray gun with 6.35 mm nozzle, a hose, a single-stage regulator, and pressurized air. Animals were placed on a dry enamel pan, and the spray gun was held 1 cm away from the skin surface. Marks were applied using lower pressures (25 psi) for smaller animals and greater pressures (40 psi) for larger animals, depending on the size of the fluorescent particles used. Three to four marks could be applied to each animal on either side of the body (cranial and caudal to the forelimb, mid-body, and cranial and caudal to the hind limb) for a maximum of 10 possible locations. Nishikawa and Service (1988) reported higher recapture rates than for any salamander toe-clipping studies performed to that time.

8.4 Tagging and banding

The use of any external tag or band must be considered carefully, as the device may impede the movement of the animal or snag on the substrate and vegetation.

8.4.1 Anurans

Kaplan (1958) used aluminum toe bands to tag frogs ("butt-end" bird band, #1242, size 2½). These numbered, cylindrical bands were placed around a toe, and the two ends were pressed together with pliers. The bands were tightened so as not to restrict circulation, but they pierced the webbing of the foot. These remained fixed indefinitely and caused no apparent hindrance in the movement of the frog.

A variety of tags and bands used for anurans is summarized in Table 8.5. McAllister *et al.* (2004) recommended against the knee tags used by Watson *et al.* (2003) in *R. pretiosa* due to skin and muscle lacerations in 33% of the recaptured animals. Emlen (1968) reported no differences in behavior, mortality, emigration rates, or weight loss between tagged and untagged American bullfrogs (*Rana catesbeiana*). Due to soiling and staining, seasonal replacement of waistbands was necessary. However, Robertson (1984) reported a negative impact using Emlen's (1968) waistband design on 10 species of Australian hylids and leptodactylids.

Windmiller (1996) suspended yarn tags in a plastic bag filled with fluorescent pigment which was pressed into the yarn, and attached to the dorsal integument of juvenile ranid frogs. The pigment trails were followed at night by illuminating them with a longwave (366 nm) ultraviolet lamp (Blak-Ray® model ML-49) while wearing safety/ultraviolet-enhancing glasses. Buchan *et al.* (2005) used soft VIAlpha Numeric Tags (US $1.00 per tag) and an injector (US$120) to insert the tag. Buchan *et al.* (2005) found high survivorship, retention, and readability of these tags in both the laboratory and field.

Table 8.5 *Tagging and banding techniques used for marking amphibian species (as reviewed in Ferner 2007 unless otherwise specified)*

Species	Description of tagging/banding technique
Anurans	Eppendorf microcentrifuge tubes filled with Cyalume™ luminescent fluid glued to dorsum
	Soft VIAlpha Numeric Tags® (fluorescent) with imprinted letters and numbers inserted into sartorius muscle
Bufo bufo	Knee-tagging (Elmberg 1989; Kuhn 1994)
Rana catesbeiana	Painted nylon waistband around pelvic girdle
Rana catesbeiana and Rana clamitans	Fluorescent acrylic yarn held between squares of self-adhering verterinary bandage and glued to dorsum
Rana clamitans	Waistband using surgical thread sewn through piece of surveyor's flagging
Rana pretiosa	Numeric-coded fingerling tags attached over knee with elastic thread
Uperoleia rugosa	1-mm squares Scotchlite® reflective sheeting attached with cyanoacrylate tissue cement
Xenopus laevis	Glass beads attached to forelimb with surgical wire

Visual implant elastomers (VIEs), as described for salamanders below, are a very useful mark for anuran larvae. Their potential for use in marking adult anurans may not be fully appreciated.

8.4.2 Salamanders

Woolley (1973) tagged *Eurycea lucifuga* and *Eurycea longicauda* with a sub-cutaneous injection of two parts Liquitex acrylic polymer to one part distilled water. This mixture was injected into the lateral proximal caudal region with a 22-gauge hypodermic needle, leaving a mark 7–10 mm in diameter. Woolley observed no adverse effects with this procedure and found slight fading in very few individuals. The author suggested that a series of acrylic colours be used to differentiate individuals.

Subcutaneous injections of fluorescent, elastomer dyes were used in *Plethodon vehiculum* by Davis and Ovaska (2001). Red, orange, or yellow dyes (VIEs) were injected into six locations in various combinations to create 816 unique marks. The elastomer was mixed with a hardener following the manufacturer's instructions, and 0.1 mL was placed in a 0.3 mL syringe. The authors recommend fluorescent marking or pattern mapping over toe-clipping in studies of plethodontid salamanders.

Bailey (2004) evaluated VIE marking in *Eurycea bislineata* by documenting mark retention, salamander survivorship, and growth rates. No adverse effects and no loss of marks were found over the 11-month study. Misreading marks in the field by observers was found to be a factor in some cases, so a training period for investigators is recommended to minimize observer bias. Rittenhouse *et al.* (2006) used the same technique with powdered fluorescent pigments with newly metamorphosed *Ambystoma maculatum,* as described for *Rana sylvatica* (see Table 8.4). No temporal effects of powder-dusting were detected.

8.4.3 Caecilians

Gower *et al.* (2006) reviewed the use of alphanumeric fluorescent tags (VIAlpha tags) on the Indian caecilian *Gegeneophis ramaswamii.* They found that making an incision with a scalpel blade before using the injector increased the efficiency of tagging, and that anesthesia was needed to quiet specimens. Equipment sterilization between uses minimizes the risk of spreading pathogens.

8.5 Trailing devices

As with bands and tags, the use of trailing devises must be carefully considered because of the potential to impair the movement of animals. They may be very

Table 8.6 *Trailing devices used for marking amphibian species (as reviewed in Ferner 2007 unless otherwise indicated)*

Species	Description of trailing technique
Ambystoma tigrinum	Aquatic forms tagged with trailing plastic float on monofilament line sutured in tail
Bufo bufo	Sewing thread on bobbin fastened around waist with elastic band (Sinsch 1988)
Bufo valliceps	Cotton thread bobbin on rigid plastic tubing holder on elastic band
Leptodactylus labyrinthicus	Cotton thread spool-and-line quilting cocoon on 1 cm inguinal elastic band
Rana pipiens	Nylon sewing thread on rigid plastic tubing holder on elastic band
Rana sylvatica	Metamorphs dipped in fluorescent powder pigments leaving trail

difficult to attach, and they may snag on the substrate and vegetation. Still, they are useful in certain applications. Table 8.6 reviews these techniques.

8.6 Pattern mapping

The patterns and markings of many species are distinctive and yet have a high degree of individual variation. Photographing or sketching these patterns may be useful for individual recognition on subsequent recaptures (Meyer and Grosse 1997; Streich *et al.* 1997; Winkler and Heunisch 1997).

8.6.1 Anurans

While more commonly used in salamanders, the patterns of anurans may be unique enough to make individual recognition possible. Wengert and Gabrial (2006) used photographs of chin-spot patterns in *Rana mucosa* and found a high success rate in reidentification over a 3-month study; changes in spot pattern over time were observed so regular recaptures might minimize misidentifications using this technique. The effectiveness of using photographic identification on long-term studies of adult anurans has yet to be determined.

Pointing out concerns in using artificial or invasive marks with the protected *Leiopelma archeyi* (Archey's frog) in New Zealand, Bradfield (2004) developed a highly accurate photographic identification technique using natural markings. This study is a model on how to develop and test a photographic pattern mapping technique. A series of six digital photographs were taken from dorsal, ventral, cranial (facial), caudal, right lateral, and left lateral views. Photographs using a flash were most successful in discerning usable characteristics of the various black markings. A filing system for the photos used subgroupings which substantially reduced the amount of time needed to identify recaptures. Intra- and interobserver consistency of identification was tested and found to be high. Overall, 99.2% of identifications were successful once recaptures were assigned to their subgroups.

8.6.2 Salamanders

Naturally occurring variation in morphology was used for *Triturus cristatus* and *Triturus vulgaris* by photographing belly patterns (Hagstrom 1973). With this technique, one must be sure that the patterns are both recognizably different and constant through time. Ontogenetic belly pattern variation of *T. cristatus* was documented by Arntzen and Teunis (1993). Healy (1974) found the variation in dorsal spot patterns of *Notophthalmus viridescens* useful in identifying individuals. This technique requires recapture and handling which may not be consistent with the animals' behavior.

Digital photographs of dorsal spot patterns were tested for their effectiveness in the identification of *Eurycea bislineata* by Bailey (2004). Observers were given color printouts of the dorsal view of each animal with the SVL written below the image for size reference. The observers compared these photographs to test animals and were timed as to how long it took them to identify each animal. Photo-identification rates were very high (about 94% correct), but were potentially lower if observers were also given un-marked (photographed) individuals in the attempted comparisons. Use of Adobe Photoshop® was effective in comparing qualities of integument patterns of vertebrates in digital photographs and may be useful in matching images (L. Rifai, personal communication). Sweeney *et al.* (1994) describe the use of computer analysis for the belly patterns of *T. cristatus*. Computer applications for animal pattern analysis can now be found on websites such as www.conservationresearch.co.uk/.

Loafman (1991) photographed the natural variation of spot patterns in adult *Ambystoma maculatum* to successfully identify about 97% of the recaptures. Adult *A. opacum* were photographed in a box and successfully reidentified using their distinctive bar pattern over a 1-year period; using head patterns alone, 80% of the adults could be distinguished from one another (Doody 1995).

The use of spot patterns for identification of *A. maculatum* is reviewed by Grant and Nanjappa (2006) with special attention to possible errors in this technique. They quantified error rates in specific mapping techniques, modified techniques to address sources of error, compared methods relative to observer bias in both the field and laboratory, and determined the effort needed to search databases for specific individuals. They provided very specific suggestions for researchers dealing with large samples of specimens identified by patterns.

8.7 Passive integrated transponder (PIT) tags

An extensive review and evaluation of a microchip marking system was given by Camper and Dixon (1988). The 10 mm × 2.1 mm PIT tags are encased in glass and encoded with an alpha-numeric code which is read by a portable reader with a hand wand. The authors tested tags on 95 individuals from a wide range of species, and implanted tags with a metal syringe having a 12-gauge needle. This baseline study experienced only one failed PIT due to a cracked glass cover, a 6% error rate in the readings, a reading success on the first pass of the wand of 92% and a migration rate of a tag within the animal of 36% (with possible reduction of first-pass reading-rate success) (Camper and Dixon 1988). Examples of species of amphibians studied with PIT tags are given in Table 8.7.

Table 8.7 *Use of PIT tags for marking amphibian species (as reviewed in Ferner 2007 unless otherwise indicated)*

Species	Description of technique
Anurans	
Bufo bufo	Needle injection into pinched dorsum, massaged caudally to base of spine
Limnodynastes peronii and *Litoria aurea*	Inserted with needle behind front limb along side of body
Rana catesbeiana, Rana clamitans, and *Rana pipiens*	Needle injected into pinched ventral side
Rana pretiosa	Inserted through 2 mm incision on dorsum 1–2 cm caudal to eyes, then massaged back along spine
Rana sylvatica	Inserted through 2 m incision above the scapula; "PIT pack" used for detection (Blomquist *et al.* 2008)
Rana temporaria	Needle injection into pinched dorsum, massaged caudally to base of spine
Caudates	
Ambystoma opacum	Inserted into body cavity at 3 mm incision 5 mm cranial to hind limb
Taricha torosa	Needle insertion into abdomen

8.7.1 Anurans

All anurans in the Camper and Dixon (1988) project had the PIT tags implanted intra-abdominally using the implanter syringe with the needle dipped in 70% ethanol before implantation. Wounds were cleaned after injection with 70% ethanol and sealed with Krazy Glue®. Synthetic liquid bandages can also be used.

Ireland *et al.* (2003) found that the tag migrated on its own to the caudal lymph space between the hind legs in all but 5% of the frogs they studied. A field scanner was used for identification and the authors believed the technique to be the best available for marking medium-to-large-sized anurans. The primary drawback is the cost of the AVID microchips ($\approx$US\$8.00) and scanner ($\approx$\$1000). McAllister *et al.* (2004) inserted 12-mm tags (125 kHz model) on frogs $\geq$42 mm and massaged them back along the body to the base of the spine to minimize the chance of their being lost through the incision prior to healing (about 2 weeks).

Pyke (2005) provides a good review on the use of PIT tags and reports on the marking of 3000 individuals of nine species. The tags were supplied by Trovan and are "individually-packaged needles inside hermetically sealed packages." The resulting small wound was sealed with Vetbond® (*n*-butyl cyanoacrylate adhesive) and fewer than 1% of the animals exhibited any sort of distress call during the procedure. As with previous studies, Pyke (2005) found little immediate impact or effect on reproduction and long-term survival.

8.7.2 Salamanders

Ott and Scott (1999) anesthetized salamanders (see Table 8.6) for about 10 min with 2-phenoxy ethanol (30 drops/500 ml distilled H_2O) and sealed the incision with New Skin® liquid bandage. Recovery from anesthesia was in containers of distilled H_2O with the head above water for a 3–4-h period until the animals appeared active when prodded gently. Where no adverse impacts of PIT tags were reported, Ott and Scott (1999) pointed out the disadvantage of cost and marking time.

8.7.3 Caecilians

PIT tags probably hold the most promise for marking caecilians, but it is only appropriate for larger specimens and requires perfecting the injection technique. Inasmuch as the long salamanders *Siren* and *Amphiuma* have been marked successfully over a period of more than 5 years (C. K. Dodd Jr, personal communication), PIT tagging holds promise for studies of caecilians. Since caecilians may be less likely to be recaptured than some other amphibians, the cost of lost tags should be considered.

8.8 Taking measurements

It is important to age specimens, place them in a class size (i.e. juvenile or adult), and document growth rates that precise SVL and tail length measurements be taken. Salamanders and caecilians may be placed on flat transparent surfaces or restrained in clear plastic tubes with a millimeter scale or metric stick used for measuring. Snout–vent distances are taken from the tip of the snout to the caudal end of the cloacal aperture. Tail lengths are taken from the caudal end of the cloacal aperture to the tip of the tail. Any regenerated portion of the tail should be noted and measured separately. Anurans are usually measured from the tip of the snout to the caudal projection of the urostyle as the vent is often not easily found. Calipers are useful in measuring the lengths of anurans.

Body mass should also be measured in grams after the specimen is patted dry of any surface moisture. A specimen bag or customized bin on a scale may be used to contain the animal. Use of an electronic or spring scale (such as Pesola®) should be selected to obtain a weight to the nearest 0.10 g if possible. As many amphibians evacuate the bladder or cloaca upon capture, investigators should allow this to happen before weighing the specimen.

8.9 Recommendations

After extensive review of the literature and considering the criteria for selecting a marking technique listed in the Introduction to this chapter, the following are recommended.

- Toe-clipping remains the method of choice for most studies of anurans with necessary consideration of the impact on the animals and long-term retention of the clip.
- The emerging technique of choice with caudates is the use of VIE which may also be used with anurans in the future. Non-invasive techniques, particularly pattern recognition, are to be considered as a possibility for all amphibians.
- When financial considerations and practicality dictate, PIT tagging is an increasingly popular method for long-term studies.

In summary, it is essential that all techniques be considered and understood thoroughly before use (the advantages, disadvantages, sources of bias), be practiced with specimens in captivity using good hygiene, and be monitored for the impact of the mark on the survival and behavior of the amphibian.

8.10 References

Arntzen, J. W. and Teunis, S. F. M. (1993). A six year study on the population dynamics of the crested newt (*Triturus cristatus*) following the colonization of a newly created pond. *Herpetological Journal*, **3**, 99–110.

Bailey, L. L. (2004). Evaluating elastomer marking and photo identification methods for terrestrial salamanders: marking effects and observer bias. *Herpetological Review*, **35**, 38–41.

Baker, J. and Gent, T. (1998). Marking and recognition of animals. In T. Gent and S. Gibson (eds), *Herpetofauna Worker's Manual*, pp. 45–54. Joint Nature Conservation Committee, Peterborough.

Blomquist, S. M., Zydlewski, J. D., and Hunter, M. L. (2008). Efficacy of PIT tags for tracking the terrestrial anurans *Rana pipiens* and *Rana sylvatica*. *Herpetological Review*, **39**, 174–9.

Bradfield, K. S. (2004). *Photographic Identification of Individual Archey's Frogs, Leiopelma archeyi, from Natural Markings*. DOC Science Internal Series 191. New Zealand Department of Conservation, Wellington.

Briggs, J. L. and Storm, R. L. (1970). Growth and population structure of the cascade frog, *Rana cascadae* Slater. *Herpetologica*, **26**, 283–300.

Buchan, A., Sun, L., and Wagner, R. S. (2005). Using alpha numeric fluorescent tags for individual identification of amphibians. *Herpetological Review*, **36**, 43–44.

Bull, E. L., Wallace, R., and Bennett, D. H. (1983). Freeze-branding: a long term marking technique on long-toed salamanders. *Herpetological Review*, **14**, 81–2.

Camper, J. D. and Dixon, J. R. (1988). Evaluation of a microchip marking system for amphibians and reptiles. *Texas Parks and Wildlife Department, Research Publication*, **7100–59**, 1–22.

Clarke, R. D. (1972). The effect of toe clipping on survival in Fowler's toad (*Bufo woodhousei fowleri*). *Copeia*, **1972**, 182–5.

Daugherty, C. H. (1976). Freeze branding as a technique for marking anurans. *Copeia*, **1976**, 836–8.

Davis, T. M. and Ovaska, K. (2001). Individual recognition of amphibians: effects of toe clipping and fluorescent tagging on the salamander *Plethodon vehiculum*. *Journal of Herpetology*, **35**, 217–25.

Donnelly, M. A., Guyer, C., Juterbock, J. E., and Alford, R. A. (1994). Techniques for marking amphibians. In W. R. Heyer, M. A. Donnelly, R. W. McDiarmid, L. C. Hayek, and M. S. Foster (eds), *Measuring and Monitoring Biological Diversity, Standard Methods for Amphibians*, appendix 2, pp. 277–284. Smithsonian Institution Press, Washington DC.

Doody, J. S. (1995). A photographic mark-recapture method for patterned amphibians. *Herpetological Review*, **26**, 19–21.

Elmberg, J. (1989). Knee-tagging: a new marking technique for anurans. *Amphibia-Reptilia*, **10**, 101–4.

Emlen, S. T. (1968). A technique for marking anuran amphibians for behavioral studies. *Herpetologica*, **24**, 172–3.

Ferner, J. W. (2007). *A Review of Marking and Individual Recognition Techniques for Amphibians and Reptiles*. Herpetological Circular No. 35. Society for the Study of Amphibians and Reptiles, Salt Lake City, UT.

Gower, D. J., Oommen, O. V., and Wilkinson, M. (2006). Marking amphibians with alpha numeric fluorescent tags: caecilians lead the way. *Herpetological Review*, **37**, 302.

Grant, E. H. C. and Nanjappa, P. (2006). Addressing error in identification of *Ambystoma maculatum* (Spotted Salamanders) using spot patterns. *Herpetological Review*, **37**, 57–60.

Griffiths, R. A. (1984). Seasonal behaviour and intrahabitat movements in an urban population of smooth newts *Triturus vulgaris* (Amphibian: Salamandridae). *Journal of Zoology*, **203**, 241–51.

Hagstrom, T. (1973). Identification of newt specimens (Urodela, *Triturus*) by recording the belly pattern and a description of photographic equipment for such registrations. *British Journal of Herpetology*, **4**, 321–6.

Halliday, T. (1995). More on toe-clipping. *Froglog*, **12**, 2–3.

Healy, W. R. (1974). Population consequences of alternative life histories in *Notophthalmus v. viridescens*. *Copeia*, **1974**, 221–9.

Henle, K. and Veith, M. (eds) (1997). Naturschutzrelevante Methoden der Feldherpetologie. *Mertensiella*, **7**.

Ireland, D., Osbourne, N., and Berrill, M. (2003). Marking medium to large sized anurans with Passive Integrated Transponder (PIT) tags. *Herpetological Review*, **34**, 218–20.

Kaplan, H. M. (1958). Marking and banding frogs and turtles. *Herpetologica*, **14**, 131–2.

Kuhn, J. (1994). Methoden der Anuren-Marierung für Frielandstudien: Übersicht-Knie- Ringetiketten-Erfahrunger mit der Phalangenamputation. *Zeitscrift für Feldherpetologie*, **1**, 177–92.

Loafman, P. (1991). Identifying individual spotted salamanders by spot pattern. *Herpetological Review*, **22**, 91–2.

Lüddecke, H. and Amezquita, A. (1999). Assessment of disc clipping on the survival and behavior of the Andean frog *Hyla labialis*. *Copeia*, **1999**, 824–30.

Martof, B. S. (1953). Territoriality in the green frog, *Rana clamitans*. *Ecology*, **34**, 165–74.

McAllister, K. R., Watson, J. W., Risenhoover, K., and McBride, T. (2004). Marking and radiotelemetry of Oregon spotted frogs (*Rana pretiosa*). *Northwestern Naturalist*, **85**, 20–5.

McCarthy, M. A. and Parris, K. M. (2004). Clarifying the effect of toe clipping on frogs with Bayesian statistics. *Journal of Applied Ecology*, **41**, 780–6.

Meyer, F. and Grosse, W.-R. (1997). Populationsökologische Studien an Amphibien mit Hilfe der fotografischen Individualerkennung: Übersicht zur Methodik und Anwendung bei der Kreuzkröte (*Bufo calamita*). *Mertensiella*, **7**, 79–92.

Nishikawa, K. C. and Service, P. M. (1988). A fluorescent marking technique for individual recognition of terrestrial salamanders. *Journal of Herpetology*, **22**, 351–3.

Orser, P. N. and Shine, D. J. (1972). Effects of urbanization on the salamander *Desmognathus fuscus fuscus*. *Ecology*, **53**, 1148–54.

Ott, J. A. and Scott, D. E. (1999). Effects of toe-clipping and PIT-tagging on growth and survival in metamorphic *Ambystoma opacum*. *Journal of Herpetology*, **33**, 344–8.

Parris, K. M. and McCarthy, M. A. (2001). Identifying effects of toe clipping on anuran return rates: the importance of statistical power. *Amphibia-Reptilia*, **22**, 275–89.

Peterman, W. E. and Semlitsch, R. D. (2006). Effects of tricaine methanesulfonate (MS-222) concentration on anesthetization and recovery in four plethodontid salamanders. *Herpetological Review*, **37**, 303–4.

Phillott, A., Skerratt, L., McDonald, K., Lemckert, F., Hines, H., Clarke, J., Alford, R., and Speare, R. (2008). Toe clipping of anurans for mark-recapture studies: acceptable if justified. That's what we said! *Herpetological Review*, **39**, 149–50.

Pyke, G. H. (2005). The use of PIT tags in capture-recapture studies of frogs: a field evaluation. *Herpetological Review*, **36**, 281–5.

Reaser, J. K. and Dexter, R. E. (1996). *Rana pretiosa* (Spotted Frog). Toe clipping effects. *Herpetological Review*, **27**, 195–6.

Rittenhouse, T. A. G., Altnether, T. T., and Semlitsch, R. D. (2006). Fluorescent powder pigments as a harmless tracking method for ambystomids and ranids. *Herpetological Review*, **37**, 188–91.

Robertson, J. G. M. (1984). A technique for individually marking frogs in behavioral studies. *Herpetological Review*, **15**, 56–7.

Sinsch, U. (1988). Seasonal changes in the migratory behaviour of the toad *Bufo bufo*: direction and magnitude of movements. *Oecologia*, **76**, 390–8.

Streich, W. J., Beckmann, H., Schneeweiss, N., and Jewgenow, K. (1997). Computergestützte Bildanalyse von Fleckenmustern der Rotbauchunke (*Bombina bombina*). *Mertensiella*, **7**, 93–102.

Sweeney, M., Oldham, R. S., Brown, M., and Jones, J. (1994). Analysis of belly patterns for individual newt recognition. In T. Gent and R. Bray (eds), *Conservation and Management of Great Crested Newts Triturus cristatus. Symposium Proceeding at Kew Gardens*. English Nature Science Series 20, pp. 75–7. English Nature, Peterborough.

Taber, C. A., Wilkinson, R. F., and Topping, M. S. (1975). Age and growth of hellbenders in the Niangua River, Missouri. *Copeia*, **1975**, 633–9.

Taylor, J. and Deegan, L. (1982). A rapid method for mass marking amphibians. *Journal of Herpetology*, **16**, 172–3.

Waichman, A. V. (1992). An alphanumeric code for toe clipping amphibians and reptiles. *Herpetological Review*, **23**, 19–21.

Watson, J. W., McAllister, K. R., and Pierce, D. J. (2003). Home ranges, movements, and habitat selection of Oregon spotted frogs (*Rana pretiosa*). *Journal of Herpetology*, **37**, 292–300.

Wengert, G. M. and Gabrial, M. W. (2006). Using chin spot patterns to identify individual mountain yellow-legged frogs. *Northwestern Naturalist*, **87**, 192.

Windmiller, B. (1996). Tracking techniques useful for field studies of anuran orientation and movement. *Herpetological Review*, **27**, 13–15.

Winkler, C. and Heunisch, G. (1997). Fotografische Methoden der Individualerkennung bei Bergmolch (*Triturus alpestris*) und Fadenmolch (*T. helveticus*) (Urodela, Salamandridae). *Mertensiella*, **7**, 71–7.

Woolley, H. P. (1973). Subcutaneous acrylic polymer injections as a marking technique for amphibians. *Copeia*, **1973**, 340–1.

Woolley, P. (1962). A method of marking salamanders. *Missouri Speleology*, **4**, 69–70.

9

Egg mass and nest counts

Peter W. C. Paton and Reid N. Harris

9.1 Background: using egg mass and nest counts to monitor populations

There is a critical need to develop statistically sensitive monitoring programs to assess changes in the distribution and abundance of amphibians and determine underlying mechanisms affecting populations (Alford and Richards 1999; Marsh 2001; Storfer 2003; Petranka *et al.* 2004, 2006). However, assessing trends is challenging because amphibian populations may experience dramatic fluctuations (Pechmann *et al.* 1989) and all species have imperfect detection probabilities (Bailey *et al.* 2004). Biologists interested in monitoring long-term amphibian population trends have a number of methods available (Heyer *et al.* 1994), but selecting the most appropriate survey and analysis techniques can be difficult (Williams *et al.* 2002).

Techniques used to assess amphibian populations during the breeding season include surveys of calling anurans (Weir *et al.* 2005), dip-net sweeps to capture larvae in aquatic systems (Werner *et al.* 2007), drift fences with pit falls (Pechmann *et al.* 1989), egg-mass counts (Adams *et al.*, 1998; Lips *et al.* 2001; Dodd 2003; Rödel and Ernst 2004), and nest counts (Corser and Dodd 2004; Harris 2005). Advantages of calling surveys include the ability to survey many breeding wetlands rapidly and most anurans can be detected, but urodeles are not sampled because they do not vocalize. Dip-net surveys assess the occurrence and density of larval amphibians, but are labor-intensive, require knowledge of local breeding phenology, and require expertise in larval identification. Drift-fence arrays can accurately assess changes in adult population size and productivity for individual ponds; however, they are expensive because fences are difficult to install and traps need to be checked often.

Many amphibians oviposit spawn in clumped aggregations, rather than individual eggs. These spawn clumps are usually called "egg masses" in the

North America literature (e.g. Paton and Crouch 2002) and in the European literature as "spawn clumps" (Griffith and Raper 1994), "batches" (Barton and Rafinski 2006), "clutches" (Ficetola *et al.* 2006), or "egg masses" (Sofianidou and Kyriakopoulou-Sklavounou 1983; Waringer-Löschenkohl 1991; Hartel 2008).

Nest counts can monitor populations in a similar manner to egg-mass counts. Both types of count estimate annual breeding effort of populations and provide an index of population size. In addition, both counts can test for effects of climate on population-level breeding effort (Corser and Dodd 2004), habitat relationships (Chalmers and Loftin 2006), population presence, and life-history information (e.g. relationships between egg size and clutch size, and between egg size and embryonic survival). Comparisons of egg size and clutch size among populations can investigate maternal investment patterns as adaptations to local environmental conditions. For nest counts, behavioural information can often be gained, such as frequency of solitary and communal nesting and patterns of nest attendance or brooding.

In the following sections we describe characteristics that make a species suitable for egg mass or nest counts. We also discuss factors that need to be considered when designing studies to collect and analyze these types of survey data. Both egg mass and nest counts can be used to study habitat use over a variety of spatial scales, thus are good candidates for landscape and metapopulation studies designed to manage and conserve at-risk species.

9.2 Oviposition strategies

Amphibians exhibit up to 39 different reproductive modes based on factors such as oviposition site (e.g. aquatic, terrestrial, live-bearing), egg type, and mode of development (Wells 2007). Worldwide, at least 10 families of anurans and some urodeles have terrestrial oviposition (Wells 2007; see Appendix 9.1). Although terrestrial oviposition strategies for many species have been described, using egg counts to monitor their populations typically are not considered due to low detection probabilities and potential negative impacts on populations by disturbing eggs in nests. However, we review several case studies below where nest-count surveys were used to monitor populations because detection probabilities were high and negative impacts were low, which suggests that a similar approach might be used for other species.

About 80% of anuran families deposit eggs in standing water (Wells 2007). Species that oviposit in smaller wetlands, such as vernal pools (e.g. Ranidae, Microhylidae), are probably best suited for egg-mass counts because biologists can search sites completely. For example, Egan and Paton (2004) found that most pond-breeding amphibian sites were less than 0.05 ha in southern New England, USA. Oviposition sites selected by some tropical species are extremely small bodies of water, such as treeholes, aerial plants, or water-filled seedpods (Wells 2007). However, although their oviposition habitat has been described, to our knowledge no one has used egg-mass counts to monitor tropical amphibian populations. Searching for egg masses of species that breed in large lakes is probably not feasible due to low detection probabilities, and thus other monitoring strategies should be employed. For stream-breeding species, egg masses can be detected (e.g. Ascaphidae), although stream monitoring programs typically search for juveniles and adults (Adams and Bury 2002).

Some anurans lay eggs in foamy nests on the water's surface or in water-filled basins (e.g. members of the subfamily Leptodactylinae; Wells 2007), which can be highly visible on the surface of shallow, temporary pools. Also, rhacophorid and hyperoliid frogs build foam nests in trees and other substrates. Thus, one possible technique to monitor their populations might be to attempt to count foam nests, but given the difficulties of finding the nests of most foam-nesting species, this is probably not a cost-effective technique for most species.

There are many arboreal tropical species that deposit eggs on leaves, branches, or tree trunks above standing water (see summary in Wells 2007). Neotropical hylids oviposit on upper surfaces of leaves and develop rapidly, where tadpoles fall into the water below to complete metamorphosis. Phyllomedusine hylid frogs from Central and South America deposit eggs on leaves and other substrates. Although monitoring schemes have not incorporated egg mass or nest counts of these arboreal breeding species, it is something that tropical biologists could consider.

9.3 Egg-mass counts

Many species of pool-breeding amphibians oviposit egg masses attached to submerged vegetation or on the pond bottom, with adults not attending eggs during incubation. For these species, biologists have used egg-mass counts to assess population sizes and trends of anurans (Table 9.1) and urodeles (Table 9.2). The best long-term studies come from Europe, where several biologists have monitored *Rana temporaria* egg counts for decades (Cooke 1985; Meyer *et al.* 1998; Loman and Andersson 2007). In North America, egg-mass counts were initiated for *Bufo canorus* (Sherman and Morton 1993), *Rana sevosa* (Richter *et al.* 2003), *Rana sylvatica*, and *Ambystoma maculatum* (Petranka *et al.* 2004).

Table 9.1 *Examples of studies where anuran populations were assessed by counting egg masses for more than 2 years.*

Species	Country	Habitat	Years	Source
Hyla regilla	USA (Washington)	Ponds	4	Adams *et al.* (1998)
Bufo calamita	England	Ponds	20	Buckley and Beebee (2004)
Bufo canorus	USA (California)	Lakes	20	Sherman and Morton (1993)
Rana sevosa	USA (Mississippi)	Seasonal pools	13	Richter *et al.* (2003)
Bufo boreas	USA (Colorado)	Lakes	12	Muths *et al.* (2003)
Rana capito	USA (Alabama)	Pond	13	Jensen *et al.* (2003)
Rana sylvatica	USA (Wyoming)	Seasonal pools	10	Corn and Livo (1989)
R. sylvatica	USA (southeastern Kentucky)	Ponds	10	Petranka *et al.* (2004)
Rana aurora	USA (Washington)	Ponds	4	Adams *et al.* (1998)
Rana temporaria	England	Ponds	14	Cooke (1985)
R. temporaria	Sweden	Ponds	17	Loman and Andersson (2007)
R. temporaria	Switzerland	Fens	28	Meyer *et al.* (1998)
R. temporaria	Austria	Ponds	7	Gollmann *et al.* (2002)
R. temporaria	England	Ponds	23	Williams (2005)
Rana arvalis	Sweden	Ponds	17	Loman and Andersson (2007)
Rana dalmatina	Germany	Pond	7	Gollmann *et al.* (2002)
R. dalmatina	Romania	Pond	11	Hartel (2008)
R. dalmatina	Greece	Pond	3	Sofianidou and Kyriakopoulou-Sklavounou (1983)
R. dalmatina	Austria	Pond	7	Waringer-Loschenkohl (1991)
R. capito, Rana sphenocephala	USA (Mississippi)	Seasonal pools	3	Richter (2000)
Rana pipiens	USA (Wyoming)	Ponds	10	Corn and Livo (1989)
Rana japonica, Rana tsushinensis, Rana nigromaculata	Japan	Lenthic waters	1	Kuramoto (1978)
Hyperolius spp.	Uganda	Arboreal	1	Vonesh (2000)

Table 9.2 *Examples of studies where urodele populations were assessed by counting egg masses.*

Species	Country	Habitat	Years	Source
Ambystoma maculatum	USA (Rhode Island)	Seasonal pools	2	Egan and Paton (2004)
A. maculatum	USA (Maine)	Seasonal pools	2	Calhoun *et al.* (2003)
A. maculatum	USA (Ohio)	Semipermanent pond	4	Brodman (1995)
A. maculatum	USA (southeastern Kentucky)	Ponds	10	Petranka *et al.* (2004)
Ambystoma. jeffersonianum	USA (Ohio)	Semipermanent pond	4	Brodman (1995)
Ambystoma gracile	USA (Washington)	Lenthic	4	Adams *et al.* (1998)
Ambystoma macrodactylum	USA (Washington)	Lenthic	4	Adams *et al.* (1998)
Ambystoma barbouri	USA (Kentucky)	Stream	2	Kats and Sih (1992)
Taricha torosa	USA (California)	Stream	2	Gamradt and Kats (1997)

In contrast to drift-fence arrays, egg-mass counts only require biologists to survey a pond several times annually to accurately estimate annual breeding effort, and thus are much less labor-intensive (hence less expensive) than drift-fence arrays. In addition, many sites can be monitored annually using egg-mass counts, for example, Loman and Andersson (2007) monitored *Rana arvalis* and *R. temporaria* populations using egg-count surveys in 120 ponds in Sweden over 17 years. In Australia, Martin and Cooper (1972) conducted egg-mass counts of *Crinia victoriana*, and egg-mass counts are proposed for several vulnerable species in including *Geocrinia alba*, *Litoria littlejohni*, and *Litoria verreauxii alpina* (University of Canberra 2003).

9.4 Amphibian nests and nest counts

We use the term *nest* to mean an oviposition site associated with at least one of the following characteristics: habitat alteration, including reduction of pathogenic fungi, nest construction, and nest attendance. An amphibian nest is a specific location selected by a female for oviposition and is often associated with nest attendance by a parent. In some species, males or females may conduct some habitat alteration or nest construction. For example, male frogs construct basins in *Hyla rosenbergi* (Hobel 1999) and females do so in *Litoria lesueuri* (Richards and Alford 1992). Female four-toed salamanders *Hemidactylium scutatum* appear to compress soil or mat vegetation at their nesting site. In other species, and perhaps most species, males or females select specific oviposition sites, such as cavities in rotten logs or slight depressions in a rock, but probably do not obviously modify the location or engage in nest construction. However, nest attendants can modify the nest site to reduce the presence of pathogenic fungi, which requires specific techniques to detect (Banning *et al.* 2008).

Before initiating nest counts, we recommend studying published accounts and consulting with herpetologists who work or have worked on the species of interest. Detailed studies of four-toed salamander nests by three different research groups illustrate how nest counts can be incorporated into a study of population dynamics (Corser and Dodd 2004; Harris 2005) and habitat requirements (Chalmers and Loftin 2006; Wahl *et al.* 2008). The nests of this species are fairly easy to detect; however, nest-site detection probabilities vary among species. Salamanders that nest deep within talus, such as *Plethodon punctatus*, will probably prove impossible to monitor by nest-count surveys.

Female four-toed salamanders lay eggs in moss and grass on the banks of ponds and bogs. After hatching, gilled larvae disperse into aquatic habitats, where they metamorphose after a relatively short larval period. Females attend nests, communal nesting can be very common, and males are not associated with nests. Because nests are restricted to banks near the water, they are relatively easy to locate by gently separating mosses and grasses (Figure 9.1).

Corser and Dodd (2004) used nest counts to track population sizes of several populations in the Great Smoky Mountain National Park, Tennessee, USA, and determine that the population was not declining. Harris (2005) used nest counts to reach similar conclusions for populations in montane Virginia. Since Harris (2005) observed no decline in two populations over a 14-year period and he used the same nest-location techniques each year, nests can be located and embryos counted without harming populations. Chalmers and Loftin (2006) and Wahl *et al.* (2008) used nest counts to quantify habitat characteristics of ponds and bogs associated with successful nesting, which can be incorporated into habitat-restoration efforts. In addition, nest counts are the primary technique used to assess communal nesting and solitary brooding behaviour (Harris 2008).

Corser (2001) used nest counts to monitor population trends of green salamanders *Aneides aeneus* in the Appalachian Mountains, USA, by conducting

Fig. 9.1 Nests of the four-toed salamander (*Hemidactylium scutatum*) can be discovered or checked by carefully spreading grasses and reeds along pond edges. Nests were best located at this site by standing in the pond and carefully checking the margin of the pond for nests.

extensive surveys of nests and attending females in the same populations beginning in the early 1970s and then again in the 1990s. These data were critical in showing that at least three large populations had severely declined, assuming that each pond surveyed represented a separate population or deme.

We know of no studies that have used anuran nest counts to track changes in population size. However, nest sites of *Eleutherodactylus coqui* have been studied extensively in Puerto Rico (Stewart and Pough 1983; Townsend 1989; Townsend and Stewart 1994). Counting nests may track *E. coqui* population trends because researchers were able to confidently check all suitable potential nest sites. For *E. coqui*, preferred nest sites are elevated, enclosed, and often occur in curled dead leaves of *Cecropia peltata* and fronds of the palm *Prestoea montana*. In an experimental study, adding artificial nest sites increased population density, including the number of clutches, indicating that preferred nest sites are limited (Stewart and Pough 1983). One advantage of diurnal nest-count surveys is that they can be conducted under most weather conditions, whereas frog-call surveys in many areas are best done at night under rainy, moist, or humid conditions.

9.5 Clutch characteristics

Females of some explosive-breeding species deposit a single jelly spawn annually that can contain from 750 to 1750 eggs (e.g. *R. sylvatica*; Crouch and Paton 2000; Figure 9.2). Both *R. sylvatica* and *R. temporaria* are cold-water breeders that deposit egg masses in large communal aggregations near the surface (Seale 1982). For relatively short-lived species, such as *R. sylvatica*, females may only breed once or twice in their lifetime and breed every year. Crouch and Paton (2000) documented a 1:1 ratio between the number of egg masses counted in ponds and the number of female *R. sylvatica* entering ponds based on drift-fence captures. Thus, egg-mass counts can represent a precise estimate of annual breeding effort or the total number of females depositing spawn at the breeding site that year.

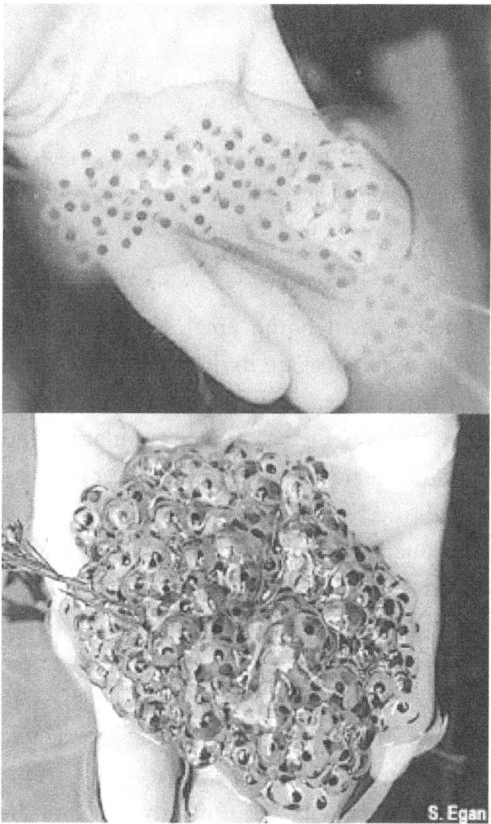

Fig. 9.2 Egg masses of spotted salamanders (*Ambystoma maculatum*; upper panel) and wood frogs (*Rana sylvatica*; lower panel) are commonly found in seasonal pools in eastern North America. Photographs by Scott Egan.

In other explosive-breeding species, such as *A. maculatum*, females may oviposit two to four clutches annually at a single wetland, with total reproductive production of 100–300 eggs annually (Petranka 1998; Figure 9.2). Because there is not a 1:1 ratio between number of egg masses and number of females ovipositing for these multiple clutch breeders, quantifying long-term population trends is more difficult. In fact, many temperate (12 species) and tropical anurans (nine species) can deposit from two to 12 clutches annually (Wells 2007, p. 501).

Many species are not suited for egg-mass counts. Some species deposit clutches of more than 1 m wide on the surface, but only one egg thick (e.g. *Rana catesbeiana* can deposit 12 000 eggs in 12 min). Because eggs in warm water tend to hatch rapidly (< 5 days), detection probabilities during occasional surveys are low. Therefore species that breed in warm water are probably best surveyed by techniques other than egg-mass counts. Toad (*Bufo*) and some pelobatid egg masses are often not distinctive because they are deposited in long intertwined strings, thus it is difficult to count individual egg masses (but see Muths *et al.* 2003). Finally, many explosive-breeders oviposit single eggs or small packets at the bottom of ponds rather than in aggregated clutches (e.g. *Pseudacris*, *Bombina*, *Discolglossus*, and myobatrachid frogs; Parmelee *et al.* 2002; Wells 2007), thus using egg counts to monitor those species is not practical. We suggest that alternative techniques such as calling surveys need to be used to monitor species with these oviposition strategies.

Another issue is reproductive frequency, as females of species that live longer, such as many *Ambystoma* salamanders (e.g. *A. maculatum* may live to be 20–30 years old), may only breed every 2–3 years (Petranka 1998). Females vary reproductive frequency based on their ability to accumulate adequate resources necessary for oviposition, which may vary with any number of factors, such as climatic conditions (Harris and Ludwig 2004). In four-toed salamanders, Corser and Dodd (2004) noted a positive correlation between the number of nests and precipitation during the nesting season. A similar relationship between rainfall and nests was found in frog *E. coqui* (Townsend and Stewart 1994). Females may skip reproduction when conditions do not favor migration to nesting sites, even if they have accumulated enough resources to oviposit. Environmental influences are probably common among amphibian species, suggesting that egg mass and nest counts could be used to reveal important life history and behavioural correlates but must be used cautiously to estimate population size. Thus egg and nest counts may represent an estimate of annual breeding effort, but not the total number of females in the vicinity of a breeding pond.

9.6 Spatial distribution of eggs

Knowledge of the spatial distribution of oviposition sites within a wetland is helpful when searching during monitoring programs. Anurans exhibit a variety of spatial distribution patterns when ovipositing in lentic habitats (still-water habitats such as ponds and lakes). Wells (2007) characterized nine different social organization strategies among anurans, with a combination of clumped oviposition sites (e.g. *R. sylvatica*, *Rana lessonae. R. catesbeiana*, *Bufo calamita*, *Scaphiopus couchii*), dispersed oviposition sites across the entire wetland (*Bombina bombina*, *Hyla versicolor*), and uniformly dispersed oviposition sites in shallow sections of wetlands (*Rana porosa*, *R. clamitans*).

Spatial distribution patterns of oviposition sites can vary within a species depending on a variety of characteristics (light, temperature, vegetation, predation). These may lead to shifts within habitats. For example, Crouch and Paton (2000) documented a clumped aggregation of *R. sylvatica* in ponds with more than 100 egg masses, whereas smaller aggregations were dispersed throughout a pond where fewer than 100 egg masses were detected. In addition, although communal aggregations of *R. sylvatica* egg masses often are located in the northeast corner of ponds where water temperatures are highest (Seale 1982), this can vary among ponds (P. Paton, unpublished results). Thus observers often need to search an entire pond to accurately count all egg mass at a site.

9.7 Breeding phenology

Amphibians exhibit an array of reproductive modes during oviposition that can be classified as either explosive (most individuals breeding in a single night) or prolonged (some tropical species breed year-round). The length and timing of the oviposition period is primarily related to abiotic factors such as temperature, precipitation, and hydroperiod at aquatic breeding sites (reviewed by Wells 2007). One of the most important considerations when conducting egg-mass counts is to conduct surveys when the number of clutches reaches peak abundance, although obviously for prolonged breeders this means multiple visits to breeding sites.

There is considerable interspecific and intraspecific variation in egg-deposition dates; thus, determining breeding phenology in your local area is critical. You should refer to local field guides or experts before initiating egg counts, or if no quantitative data are available for your region, then you have to gather initial breeding phenology data. Biologists in North America can use the North American Amphibian Monitoring Program (NAAMP; Weir *et al.* 2005) to help determine local breeding phenology.

Species in temperate areas typically breed in warmer months, with many ranids, bufonids, and ambystomatids breeding explosively in early spring. Some explosive breeders, such as *R. sylvatica* in North America (Crouch and Paton 2002) or *R. temporaria* in Europe, breed synchronously once annually within 2–3-week period (Elmberg 1990), whereas other species, such as southern leopard frog (*R. sphenocephala*) exhibit a more prolonged breeding season, with short breeding bouts separated by 1–2 months (Doody and Young 1995). Thus, local knowledge of the length of the oviposition period and hatching rates is critical when designing an egg-count monitoring program.

Tropical and subtropical species often breed throughout the year during long rainy seasons, with the number of breeding individuals varying as a function of rainfall (reviewed by Wells 2007). Some species that breed in permanent water may be continuously active throughout the rainy season (e.g. *Scinax boulengeri*), whereas species that breed in temporary aquatic habitats may oviposit explosively only after heavy rainfall and then undetectable for the rest of the year (e.g. *Hyla pseudopuma*). Finally, species that breed in the desert or savanna are explosive breeders that take advantage of ephemeral seasonal wetlands, thus oviposition and egg development in these species tends to be rapid, hence monitoring schemes for these species also have to be flexible.

9.8 Number of surveys needed

The goal of egg mass and nest counts is to determine the total number of females that deposited clutches that year at a particular site. Species that are explosive breeders are better suited for egg-mass counts. Species that breed relatively synchronously (e.g. *R. sylvatica*, Crouch and Paton 2000; *R. temporaria*, Meyer *et al.* 1998) are ideal because a pond only need to be surveyed a few times to determine the maximum number of egg masses. In Sweden, Loman and Andersson (2007) studied *R. temporaria* populations by conducting three to six surveys annually at each pond, with surveys separated by 3–7-day intervals. They stopped surveying ponds only after no new (< 5-day-old) spawn was detected. In contrast, species with prolonged breeding seasons may require multiple surveys to a site to estimate total annual reproduction (e.g. *R. sevosa*; Richter *et al.* 2003).

In addition, species whose eggs do not hatch rapidly (> 10 days from deposition to hatching) are also best for egg-mass counts because they have higher detection probabilities. In North America and Europe, those species that breed early in the breeding season tend to be the best species to monitor with egg-mass counts because their eggs can take over 3 weeks to hatch (e.g. *R. sylvatica*, *R. temporaria*, *R. arvalis*, *Rana dalmatina*, *A. maculatum*); thus, biologists have a longer survey window. For example, in Rhode Island, *R. sylvatica* eggs may take over 20 days to hatch in cold ponds (Crouch and Paton 2000). In contrast, *R. catesbeiana* eggs that are deposited in June may hatch in under 3 days.

9.9 Estimating egg-mass detection probabilities

Herpetologists have been criticized for assuming perfect detection probabilities when conducting surveys for individuals, egg masses, and nests (Mazerolle *et al.* 2007). Even trained observers often do not count all egg masses present at a site (Williams *et al.* 2002; Grant *et al.* 2005). Not adjusting counts for the proportion of egg masses that are not detected can lead to a bias in trend estimates. Therefore it is critical for observers to estimate detection probabilities of egg masses, which is the proportion of the true number of egg masses present at a site that are detected.

Grant *et al.* (2005) suggested that estimating detection probability using a double-observer sampling procedure or another capture–recapture method was essential if the primary objective was to estimate changes in population size. In this technique, two observers survey sites together, with the primary observer initially detecting egg masses and reporting them to the recorder. The recorder documents any egg masses that the primary observer missed. After half of a pond has been surveyed, the two biologists switch observer and recorder roles (Nichols *et al.* 2000). Grant *et al.* (2005) did not detect any variables (e.g. observer variation, pool area or depth, or vegetation structure) or covariates (e.g. observer experience or differences in visual abilities, variation between amphibian species, or egg mass density) that explained variation in detection probability of egg masses.

There are alternative methods that biologists could use to estimate detection probabilities and estimate population sizes. For example, one observer could count egg masses in a pond on multiple occasions within a short time interval, making sure to mark detected egg masses (e.g. Regester and Woosley 2005). These types of data could be modeled using a capture–recapture framework (Williams *et al.* 2002; Grant *et al.* 2005). Another potential technique is an independent double-observer method, where a second observer conducts a count separately from the primary observer without sharing information until after the survey has been completed.

9.10 Variation in counts among observers

Grant *et al.* (2005) found strong evidence of variation among observers in their ability to detect and count egg masses. Windmiller (1996) found less than 10% difference between surveys during independent double-observer counts of

A. maculatum egg masses in four seasonal ponds. Crouch and Paton (2000) estimated about 12% variation by two independent observers surveying *R. sylvatica*
egg masses. Egan (2001) found that independent double-observer counts of *A.
maculatum* egg masses varied by 25%, while *R. sylvatica* egg-mass counts varied
by 11%. All these species had relatively high detection probabilities (84–100%,
Grant *et al.* 2005). Based on work by Grant *et al.* (2005) and Williams *et al.*
(2002), studies using multiple observers need to incorporate an observer parameter when modeling trend estimates.

9.11 Marking eggs

Some methods require that egg masses are marked so that observers can tell
which egg masses have or have not been detected. Egg masses can be marked
with colored flagging tied to adjacent vegetation, but this can be confusing
and difficult if no vegetation is close to the eggs (P. Paton, personal observation). Alternatively, Regester and Woosley (2005) tested the field practicality
and effects of visible fluorescent elastomer (visible implant elastomer; VIE) on
embryo development and retention in the jelly matrix. There was no effect on
embryo viability, and marks remained intact for more than 35 days and could be
detected up to 0.45 m deep. Because VIE comes in multiple colors, egg masses
could be injected with a different color on each survey.

9.12 Situations in which nest counts are not practical

9.12.1 Nest destruction

Nests of many salamander and frog species are cryptic or difficult to detect
(e.g. underground, inside rotting logs, deep in talus, under logs). Nest discovery may result in damage to the nest, the nest environment, the embryos, and
to the nest attendant. For example, Peterson (2000) found that 40 red-backed
salamander *Plethodon cinereus* nests were damaged, while another 15 nests that
were not damaged. Merely turning a log to check for a nest can sometimes damage the nest. One can minimize nest destruction by altering search techniques.
However, if nests or nest attendants are typically damaged upon discovery, then
nest counts *cannot* be used to monitor changes in population size. Only use nest
counts when there is no harm to nests or nest attendants.

9.12.2 Desertion of attendant

Nest discovery may cause the nest attendant to desert the nest. Breitenbach (1982)
suggested that nest disturbance caused the desertion of female *H. scutatum*,

although Harris *et al.* (1995) found the females could be disturbed daily in meso-cosms without causing desertion. Some species, such as the anuran *Cophixalus ornatus*, appear to treat humans as predators and will lunge toward human fingers and not desert (Felton *et al.* 2006). Preliminary studies should be conducted to assess the effects of nest discovery and degree of nest disturbance. Quickly checking for nest presence will be less likely to cause desertion than temporarily removing and measuring the attendant and counting embryos. In many species, desertion by the nest attendant is associated with reduced embryonic success (Harris *et al.* 1995). Therefore, if the objective is to monitor or conserve population size, then it is important to not cause desertion.

9.13 How to count eggs in a nest

Valuable information can be obtained by counting embryos in a nest. Clutch size is a fundamental fitness component for studies ranging from natural selection and evolution to conservation biology. Obtaining an accurate count of embryos will depend on the number of embryos in the nest, how the embryos are configured within the nest, and the line of sight from observer to nest. When there are few eggs arranged in a monolayer in easy sight, counts should be accurate. Nest photographs can be used to count eggs when all embryos are visible. When embryos are laid in a three-dimensional cluster, accurate counts are more difficult, especially for large clutches. Amphibian embryos are often attached to each other in nests, so moving embryos must be done carefully to avoid damaging adjacent embryos. Once an investigator has experience with counting embryos, accurate estimates of the number of embryos without actually seeing each one can be obtained by understanding how clusters of embryos are arranged. Such a procedure should be verified by first estimating and then counting all embryos. Measurement error should be estimated by replicating counts within a nest. Nests in crevices or cavities can sometimes be surveyed with a strong light source and tools, such as a dental mirror (Corser 2001).

9.14 Estimating hatching success

Hatching success is calculated as number of embryos that hatch divided by the initial number of embryos, and requires that both measurements are made accurately. The number of embryos needs to be estimated before predation and oophagy compromise the count. The number of hatchings needs to be estimated just before hatching occurs and larvae or young have dispersed. A table of embryonic stages can be very helpful in determining when hatching is just about to occur. Care must be taken at this stage because movement of late-stage embryos can induce hatching. Estimates of hatching or embryonic success typically take several visits to nest sites.

9.15 Analysis of egg-mass count data

For species where egg-mass counts represent accurate estimates of annual breeding effort by females at a site, then population trends could be modeled using a variety of regression approaches (Meyer *et al.* 1998; Alford and Richards 1999; Loman and Andersson 2007). Information theoretic model selection based on the Akaike Information Criterion may help identifying the best models. Population growth rates and density dependence can be tested using relatively simple formulae: the population growth rate can be measured using ΔN method (Houlahan *et al.* 2000):

$$\Delta N = \log(N+1)_{t-1} - \log(N+1)_t$$

where N represents the population size (number of the egg masses) at time t. Density dependence can be tested by regressing ΔN (growth rate) against the number of egg masses (see also Hartel 2008 for *R. dalmatina* and Meyer *et al.* 1998 for *R. temporaria*).

If one is interested in annual changes in site occupancy (i.e. the proportion of ponds with breeding populations), then recent developments by MacKenzie *et al.* (2006) could be of interest. This is relevant for species with lower detection probabilities, particularly for species or areas where it is difficult to determine total population size based on total egg mass or nest counts.

9.16 Summary

Egg-mass counts and nest counts are both viable techniques that have been used successfully by herpetologists to monitor population trends of amphibian species. Egg mass and nest counts are particularly powerful because one can model annual breeding effort or true population sizes, rather than crude indices of population sizes that are provided by other techniques such as calling surveys. In addition, for conservation biologists interested in assessing factors that might affect populations, egg counts and nest counts are useful tools. In particular, biologists have been using egg counts recently to assess the effects of habitat structure at a variety of spatial scales on estimates of annual reproductive effort (e.g. Egan 2001; Egan and Paton 2004; Skidds *et al.* 2007). Biologists working in the tropics, where monitoring information is critical given recent disease events there, might be particularly interested in incorporating egg counts or nest counts in monitoring schemes as a supplement to existing calling surveys and visual encounter surveys.

9.17 References

Adams, M. H. and Bury, R. B. (2002). The endemic headwater stream amphibians of the American Northwest: associations with environmental gradients in a large forested preserve. *Global Ecology and Biogeography*, **11**, 169–78.

Adams, M. H., Bury, R. B., and Swarts, S. A. (1998). Amphibians of the Fort Lewis Military Reservation, Washington: Sampling techniques and community patterns. *Northwestern Naturalist*, **79**, 12–18.

Alford, R. A. and Richards, S. J. (1999). Global amphibian declines: a problem in applied ecology. *Annual Review of Ecology and Systematics*, **30**, 133–56.

Andreone, F., Cadle, J. E., Cox, N., Glaw, F., Nussbaum, R. A., Raxworthy, C. J., Stuart, S. N., Vallan, D., and Vences, M. (2005). Species review of amphibian extinction risks in Madagascar: conclusions from the global amphibian assessment. *Conservation Biology*, **19**, 1790–1802.

Bailey, L. L., Simons, T. R., and Pollock, K. H. (2004). Estimating site occupancy and species detection probability parameters for terrestrial salamanders. *Ecological Applications*, **14**, 692–702.

Banning, J. L., Weddle, A. L., Wahl III, G. W., Simon, M. A., Lauer, A., Walters, R. L., and Harris, R. N. (2008). Antifungal skin bacteria, embryonic survival, and communal nesting in four-toed salamanders, *Hemidactylium scutatum*. *Oecologia*, **156**, 423–9.

Barton, K. and Rafinski, J. (2006). Co-occurrence of Agile Frog (*Rana dalmatina* Fitz. in Bonaparte) with Common Frog (*Rana temporaria* L.) in breeding sites in southeastern Poland. *Polish Journal of Ecology*, **54**, 151–7.

Bickford, D. (2004). Differential parental care behaviors of arboreal and terrestrial microhylid frogs from Papua New Guinea. *Behavioral Ecology and Sociobiology*, **55**, 402–9.

Breitenbach, G. L. (1982). The frequency of joint nesting and solitary brooding in the salamander, *Hemidactylium scutatum*. *Journal of Herpetology*, **16**, 341–6.

Brodman, R. (1995). Annual variation in breeding success of two syntopic species of *Ambystoma* salamanders. *Journal of Herpetology*, **29**, 111–13.

Buckley, J. and Beebee, T. J. C. (2004). Monitoring the conservation status of an endangered amphibian: the natterjack toad *Bufo calamita* in Britain. *Animal Conservation*, **7**, 221–8.

Bury, R. B. and Adams, M. J. (2000). *Inventory and monitoring of amphibians in North Cascades and Olympic National Parks, 1995–1998*. Final Report. Forest and Rangeland Ecosystem Science Center, Corvallis, OR.

Calhoun, A. J. K., Walls, T. E., Stockwell, S. S., and McCollough, M. C. (2003). Evaluating vernal pools as a bases for conservation strategies: a Maine case study. *Wetlands*, **23**, 70–81.

Chalmers, R. J. and Loftin, C. S. (2006). Wetland and microhabitat use by nesting four-toed salamanders in Maine. *Journal of Herpetology*, **40**, 478–85.

Cooke, A. S. (1985). The deposition and fate of spawn clumps of the common frog *Rana temporaria* at a site in Cambridgeshire, 1971–1983. *Biological Conservation*, **32**, 165–87.

Corn, P. S. and Livo, L. J. (1989). Leopard frog and wood frog reproduction in Colorado and Wyoming. *Northwestern Naturalist*, **70**, 1–9.

Corser, J. D. (2001). Decline of disjunct green salamander (*Aneides aeneus*) populations in the southern Appalachians. *Biological Conservation*, **97**, 119–26.

Corser, J. D. and Dodd Jr, C. K. (2004). Fluctuations in a metapopulation of nesting four-toed salamanders, *Hemidactylium scutatum*, in the Great Smoky Mountains National Park, USA, 1999–2003. *Natural Areas Journal*, **24**, 135–40.

Crochet, P.-A., Chaline, O., Cheylan, M., and Guillaume, C. P. (2004). No evidence of general decline in an amphibian community of Southern France. *Biological Conservation*, **119**, 297–304.

Crouch, W. B. and Paton, P. W. C. (2000). Using egg mass counts to monitor wood frog populations. *Wildlife Society Bulletin*, **28**, 895–901.

Crouch, W. B. and Paton, P. W. C. (2002). Assessing the use of call surveys to monitor breeding anurans in Rhode Island. *Journal of Herpetology*, **36**, 185–92.

DeAlmeida Prado, C. P., Uetanabaro, M., and Lopes, F. S. (2000). Reproductive strategies of *Leptodactylus chaquensis* and *L. podicipinus* in the Pantanal, Brazil. *Journal of Herpetology*, **34**, 135–9.

Dodd Jr, C. K. (2003). *Monitoring amphibians in Great Smoky Mountains National Park*. Circular Number 1258. US Geological Survey, Tallahassee, FL.

Doody, J. S. and Young, J. E. (1995). Temporal variation in reproduction and clutch mortality of leopard frogs (*Rana utricularia*) in South Mississippi. *Journal of Herpetology*, **29**, 614–16.

Driscoll, D. A. (1998). Counts of calling males as estimates of population size in the endangered frogs *Geocrina alba* and *G. vitellina*. *Journal of Herpetology*, **32**, 475–81.

Egan, R. S. (2001). *Within-pond and Landscape-level Factors Influencing the Breeding Effort of Rana sylvatica and Ambystoma maculatum*. MSc thesis, University of Rhode Island, Kingston, RI.

Egan, R. S. and Paton, P. W. C. (2004). Within-pond parameters affecting oviposition by wood frogs and spotted salamanders. *Wetlands*, **24**, 1–13.

Elmberg, J. (1990). Long-term survival, length of breeding season, and operational sex ratio in a boreal population of common frogs, *Rana temporaria* L. *Canadian Journal of Zoology*, **68**, 121–7.

Felton, A., Alford, R. A., Felton, A. M., and Schwarzkopf, L. (2006). Multiple mate choice criteria and the importance of age for male mating success in the microhylid frog, *Cophixalus ornatus*. *Behavioral Ecology and Sociobiology*, **59**, 786–95.

Ficetola G. F., Valota, M., and de Bernardi, F. (2006). Temporal variability of spawning site selection in the frog *Rana dalmatina*: consequence for habitat management. *Animal Biodiversity and Conservation*, **29**, 157–63.

Freda, J., Sadinski, W. J., and Dunson, W. A. (1991). Long term monitoring of amphibian populations with respect to the effects of acidic deposition. *Water, Air, and Soil Pollution*, **55**, 445–62.

Gamradt, S. C. and Kats, L. B. (1997). Impact of chaparral wildfire-induced sedimentation on oviposition of stream-breeding California newts (*Taricha torosa*). *Oecologia*, **110**, 546–9.

Gollmann, G., Gollmann, B., Baumgartner, C., and Waringer–Löschenkohl, A. (2002). Spawning site shifts by *Rana dalmatina* and *Rana temporaria* in response to habitat change. *Biota*, **3**, 35–42.

Gottsberger, B. and Gruber, E. (2004). Temporal partitioning of reproductive activity in a neotropical anuran community. *Journal of Tropical Ecology*, **20**, 271–80.

Grant, E. H. C., Jung, R., Nichols, J., and Hines, J. (2005). Double-observer approach to estimating egg mass abundance of pool-breeding amphibians *Wetlands Ecology and Management*, **13**, 305–20.

Griffith, R. A. and Raper, S. J. (1994). How many clumps are there in a mass of frog spawn? *British Herpetological Bulletin*, **50**, 14–17.

Harris, R. N. (2005). *Hemidactylium scutatum*. In M. Lannoo (ed.), *Amphibian Declines: the Conservation Status of United States Species*, pp. 780–1. University of California Press, Berkeley, CA.

Harris, R. N. (2008). Body condition and order of arrival affect cooperative nesting behaviour in four-toed salamanders *Hemidactylium scutatum*. *Animal Behaviour*, **75**, 229–33.

Harris, R. N. and Ludwig, P. M. (2004). Resource level and reproductive frequency in female four-toed salamanders, *Hemidactylium scutatum* (Caudata: Plethodontidae). *Ecology*, **85**, 1585–90.

Harris, R. N., Hames, W. W., Knight, I. T., Carreno, C. A., and Vess, T. J. (1995). An experimental analysis of joint nesting in the salamander *Hemidactylium scutatum* (Caudata: Plethodontidae): the effects of population density. *Animal Behaviour*, **50**, 1309–16.

Hartel, T. (2008). Weather conditions, breeding date and population fluctuation in *Rana dalmatina* from Central Romania. *Herpetological Journal*, **18**, 40–4.

Heyer, W. R., Donnelly, M. A., McDiarmid, R. W., Hayek, L.-A. C., and Foster, M. S. (1994). *Measuring and Monitoring Biological Diversity: Standard Methods for Amphibians*. Smithsonian Institution Press, Washington DC.

Hobel, G. (1999). Facultative nest construction in the gladiator frog *Hyla rosenbergi* (Anura: Hylidae). *Copeia*, **1999**, 797–801.

Houlahan, J. E., Findlay, C. S., Schmidt, B. R., Meyer, A. H., and Kuzmin, S. L. (2000). Quantitative evidence for global amphibian population declines. *Nature*, **412**, 499–500.

Icochea, J., Quispitupac, E., Portilla, A., and Ponce, E. (2004). Framework for assessment and monitoring of amphibians and reptiles in the Lower Urubamba Region, Peru. *Environmental Monitoring Assessment*, **76**, 55–67.

Jensen, J. B., Bayley, M. A., and Blankenship, E. L. (2003). The relationship between breeding by the gopher frog *Rana capito* (Amphibia: Ranidae) and Rainfall. *American Midland Naturalist*, **150**, 185–90.

Kats, L. B. and Sih, A. (1992). Oviposition site selection and avoidance of fish by streamside salamanders (*Ambystoma barbouri*). *Copeia*, **1992**, 468–73.

Kuramoto, M. (1978). Correlations of quantitative parameters of fecundity in amphibians. *Evolution*, **32**, 287–96.

Kusano, T., Sakai, A., and Hatanaka, S. (2005). Natural egg mortality and clutch size of the Japanese treefrog *Rhacophorus arboreus* (Amphibia: Rachophoridae). *Current Herpetology*, **25**, 79–84.

Lips, K. R., Reaser, J. K., Young, B. E., and Ibáñez, R. (2001). Amphibian monitoring in Latin America: a protocol manual. Society for the Study of Amphibians and Reptiles, *Herpetological Circular* No. **30**, 1–115.

Loman, J. and Andersson, G. (2007). Monitoring brown frogs *Rana arvalis* and *Rana temporaria* in 120 south Swedish ponds 1989–2005. Mixed trends in different habitats. *Biological Conservation*, **35**, 46–56.

MacKenzie, D. I., Nichols, J. D., Royle, J. A., Pollock, K. H., Bailey, L. L., and Hines, J. E. (2006). *Occupancy Estimation and Modeling: Inferring Patterns and Dynamics of Species Occurrence.* Elsevier Press, Amsterdam.

Marsh, D. M. (2001). Fluctuations in amphibian populations: a meta-analysis. *Biological Conservation*, **101**, 327–35.

Martin, A. A. and Cooper, A. K. 1972. The ecology of terrestrial anuran eggs, genus *Crinia* (Leptodactylidae). *Copeia*, **1972**, 163–9.

Mazerolle, M. J., Bailey, L. L., Kendall, W. L., Royle, J. A., Converse, S. J., and Nichols, J. D. (2007). Making great leaps forward: accounting for detectability in herpetological field studies. *Herpetologica*, **41**, 672–89.

Meyer, A. H., Schmidt, B. R., and Grossenbacher, K. (1998). Analysis of three amphibian populations with quarter-century long time-series. *Proceedings of the Royal Society of London Series B Biological Sciences* **265**, 523–8.

Muths, E., Corn, P. S., Pessier, A. P., and Green, D. E. (2003). Evidence for disease-related amphibian decline in Colorado. *Biological Conservation*, **110**, 357–65.

Neckel-Oliveira, S. and Wachlevski, M. (2004). Predation on the arboreal eggs of three species of *Phyllomedusa* in Central Amazônia. *Journal of Herpetology*, **38**, 244–8.

Nichols J. D., Hines, J. E., Sauer, J. R., Fallon, F., Fallon, J., and Heglund, P. J. (2000). A double-observer approach for estimating detection probability and abundance from avian point counts. *Auk*, **117**, 393–408.

Parmelee, J. R., Knutson, M. G., and Lyon, J. E. (2002). *A Field Guide to Amphibian Larvae and Eggs of Minnesota, Wisconsin, and Iowa.* Information and Technology Report USGS/BRD/ITR-2002-0004. US Geological Survey, Washington DC.

Paton, P. W. C. and Crouch, W. B. (2002). Using the phenology of pond-breeding amphibians to develop conservation strategies. *Conservation Biology*, **16**, 194–204.

Pechmann, J. H. K., Scott, D. E., Gibbons, J. W., and Semlitsch, R. D. (1989). Influence of wetland hydroperiod on diversity and abundance of metamorphosing juvenile amphibians. *Wetland Ecology and Management*, **1**, 3–11.

Peterson, M. G. (2000). Nest, but not egg, fidelity in a terrestrial salamander. *Ethology*, **106**, 781–94.

Petranka, J. W. (1998). *Salamanders of the United States and Canada.* Smithsonian Institution Press, Washington DC.

Petranka, J. W. and Holbrook, C. T. (2006). Wetland restoration for amphibians: should local sites be designed to support metapopulations or patchy populations? *Restoration Ecology*, **14**, 404–11.

Petranka, J. W., Smith, C. K., and Scott, A. F. (2004). Identifying the minimal demographic unit for monitoring pond-breeding amphibians. *Ecological Applications*, **14**, 1065–78.

Regester, K. J. and Woosley, L. B. (2005). Marking salamander egg masses with Visible Fluorescent Elastomer: retention time and effect on embryonic development. *American Midland Naturalist*, **152**, 52–60.

Richards, S. J. and Alford, R. A. (1992). Nest construction by an Australian rainforest frog of the *Litoria lesueuri* complex (Anura: Hylidae). *Copeia*, **1992**, 1120–3.

Richter, S. C. (2000). Larval caddisfly predation on the eggs and embryos of *Rana capito* and *Rana sphenocephala*. *Journal of Herpetology*, **34**, 590–3.

Richter, S. C., Young, J. E., Johnson, G. N., and Seigel, R. A. (2003). Stochastic variation in reproductive success of a rare frog, *Rana sevosa*: implications for conservation and for monitoring amphibian populations. *Biological Conservation*, **111**, 171–7.

Roberts, W. E. (1994). Explosive breeding aggregations and parachuting in a neotropical frog, *Agalychis saltator* (Hylidae). *Journal of Herpetology*, **28**, 193–9.

Rödel, M. O. and Ernst, R. (2004). Measuring and monitoring amphibian diversity in tropical forests. I. An evaluation of methods with recommendations for standardization. *Ecotropica*, **10**, 1–14.

Seale, D. (1982). Physical factors influencing oviposition by the wood frog, *Rana sylvatica*, in Pennsylvania. *Copeia*, **1982**, 627–35.

Sherman, C. K. and Morton, M. L. (1993). Population declines of Yosemite Toads in the eastern Sierra Nevada of California. *Journal of Herpetology*, **27**, 186–98.

Skidds, D. E., Golet, F. C., Paton, P. W. C., and Mitchell, J. C. (2007). Habitat correlates of reproductive effort in wood frogs and spotted salamanders in an urbanizing watershed. *Journal of Herpetology*, **41**, 439–50.

Sofianidou, T. S. and Kyriakopoulou-Sklavounou, P. (1983). Studies on the biology of the frog, *Rana dalmatina*, Bonap. during the breeding season in Greece. *Amphibia-Reptilia*, **4**, 125–36.

Stewart, M. M. and Pough, F. H. (1983). Population density of tropical forest frogs: relation to retreat sites. *Science*, **221**, 570–2.

Storfer, A. (2003). Amphibian declines: future directions. *Diversity and Distributions*, **9**, 151–63.

Summers, K., McKeon, C. S., and Heying, H. (2006). The evolution of parental care and egg size: a comparative analysis in frogs. *Proceedings of the Royal Society of London Series B Biological Sciences* **273**, 687–92.

Thurley, T. and Bell, B. D. (1994). Habitat distribution and predation on a western population of terrestrial *Leiopelma* (Anura: Leiopelmatidae) in the northern King Country, New Zealand. *New Zealand Journal of Zoology*, **21**, 431–6.

Townsend, D. S. (1989). The consequences of microhabitat choice for male reproductive success in a tropical frog (*Eleutherodacylus coqui*). *Herpetologica*, **45**, 451–8.

Townsend, D. S. and Stewart, M. M. (1994). Reproductive ecology of the Puerto Rican frog *Eleutherodactylus coqui*. *Journal of Herpetology*, **28**, 34–40.

University of Canberra (2003). *Survey Standards for Australian Frogs. Final Report*. Applied Ecology Research Group, Canberra.

Veith, M., Lötters, S., Andreone, F., and Rödel, M.-O. (2004). Measuring and monitoring amphibian diversity in tropical forests. II. Estimating species richness from standardized transect censing. *Ecotropica*, **10**, 85–99.

Vonesh, J. (2000). Dipteran predation on the arboreal eggs of four *Hyperolius* frog species in western Uganda. *Copeia*, **2000**, 560–6.

Wahl III, G. W., Harris, R. N., and Nelms, T. (2008). Nest site selection and embryonic survival in four-toed salamanders, *Hemidactylium scutatum* (Caudata: Plethodontidae). *Herpetologica*, **64**, 12–19.

Waringer-Löschenkohl, A. (1991). Breeding ecology of *Rana dalmatina* in lower Austria: a 7-year study. *Alytes*, **9**, 121–34.

Weir, L. A., Royle, J. A., Nanjappa, P., and Jung, R. E. (2005). Modeling anuran detection and site occupancy on North American Amphibian Monitoring Program (NAAMP) routes in Maryland. *Journal of Herpetology*, **39**, 627–39.

Wells, K. D. (2007). *The Ecology and Behaviour of Amphibians*. University of Chicago Press, Chicago, IL.

Werner, E. E., Yurewicz, K. L., Skelly, D. K., and Relyea, R. A. (2007). Turnover in an amphibian metacommunity: the role of local and regional factors. *Oikos*, **116**, 1713–25.

Williams, L. R. (2005). Restoration of ponds in a landscape and changes in Common Frog (*Rana temporaria*) populations, 1983–2005. *Herpetological Bulletin*, **94**, 22–9.

Williams, B. K., Nichols, J. D., and Conroy, M. J. (2002). *Analysis and Management of Animal Populations: Modeling, Estimation and Decision Making*. Academic Press, San Diego, CA.

Windmiller, B. S. (1996). *The Pond, the Forest, and the City: Spotted Salamander Ecology and Conservation in a Human-dominated Landscape*. PhD dissertation, Tufts University, Medford, MA.

Zheng, Y. and Fu, J. (2006). Making a doughnut-shaped egg mass: oviposition behaviour of *Vibrissaphora boringiae* (Anura: Megophryidae). *Amphibia-Reptilia*, **28**, 309–11.

Appendix 9.1 *Oviposition strategies of anuran families of the world, with comments on whether egg counts could be used monitor their populations (modified from Wells 2007).*

Family	Distribution	Deposition site	Eggs could be counted?
Allophrynidae	South America	Ponds	No?[1]
Arthroleptidae	Africa	Terrestrial	No?
Ascaphidae	Northwestern USA	Strings attached to rocks in streams	Maybe?[2]
Astylosternidae	Africa	Rivers and streams	No?
Bombinatoridae	Europe and Asia	Ponds/streams	Yes?
Brachycephalidae	Northern South America	Terrestrial	No
Bufonidae	Worldwide, except Australia	Long strings in still water	Small populations of some species
Centrolenidae	Central America	Arboreal over water	Yes?
Dendrobatidae	South and Central America	Terrestrial eggs carried to water by adults	No?
Discoglossidae	Europe, Middle East, North Africa	Terrestrial (males carry eggs), some breed in ponds	No
Helephrynidae	Africa	Under rocks in streams	No?
Hemiphractinae	South America	Females carry eggs	No
Hemisotidae	Africa	Terrestrial near ponds	No
Hylidae	Worldwide	In ponds and on vegetation	Some species[3, 4, 5]
Hyperoliidae	Africa	Terrestrial and aquatic eggs	No?
Leiopelmatidae	New Zealand	Terrestrial and water-filled depressions	Some species[6]
Leptodactyidae	Southern USA, Central and South America	Ponds, foam nests in water or near water	Some species?[7]
Mantellidae	Madagascar	In ponds or streams	No?[8]
Megophryidae	Asia	Pools, under rocks in streams	No?[9]

Appendix 9.1 *Continued*

Family	Distribution	Deposition site	Eggs could be counted?
Microhylidae	Worldwide	Variety of terrestrial, arboreal, and aquatic habitat	Yes?[10,11]
Myobatrachidae	Australia and New Guinea	Aquatic foam nests or burrows or terrestrial or ponds	No?[12]
Pelodytidae	Europe and Asia	Strings in ponds	Yes[13]
Petropedetidae	Africa	Ponds and terrestrial	No?[14]
Pipidae	South American and Africa	Strictly aquatic; in some species females incubate eggs on back	No?[15]
Ranidae	Worldwide	Aquatic	Yes (see Table 9.1)[16]
Rhacophoridae	Asia, Africa	Foam nests in trees, on rocks, and surface of surface	Yes?[17]
Rhinodermatidae	South America	Terrestrial,	No
Rhinophrynidae	Central America	Shallow ponds	No?
Sooglossidae	Seychelles	Terrestrial	No?

[1]Gottsberger and Gruber (2004); [2]Bury and Adams (2000); [3]Adams *et al.* (1998); [4]Roberts (1994); [5]Neckel-Oliveira and Wachlevski (2004); [6]Thurley and Bell (1994); [7]DeAlmeida Prado *et al.* (2000), [8]Andreone *et al.* (2005); [9]Zheng and Fu (2006); [10]Bickford (2006); [11]Summers *et al.* (2006); [12]Driscoll (1998); [13]Crochet *et al.* (2004); [14]Veith *et al.* (2004); [15]Icochea *et al.* (2004); [16]Freda *et al.* (1991); [17]Kusano *et al.* (2005).

10

Dietary assessments of adult amphibians

Mirco Solé and Dennis Rödder

10.1 Introduction

Dietary information is essential to understand amphibian life history, population fluctuations, and the impact of habitat modification, and to develop conservation strategies (Anderson 1991). In ecology, trophic relationships represent a functional connection between different taxa and are a key subject in many aut- and synecological studies. Studies on diets are essential for assessments of energy flow and food webs in ecological communities. Variables involved in an animal's diet are seldom obvious and include behavioral, physiological, and morphological features of both predator and prey (Simon and Toft 1991). Therefore, dietary information from locally imperilled species is a necessary component in designing management and conservation programs.

Traditionally, amphibians are described as generalist predators feeding mainly on arthropods, but mollusks, annelids, and even small vertebrates are also frequently consumed. Amphibians are able to distinguish between different prey types allowing for different degrees of feeding specialization (Freed 1982). Dietary specialization is often associated with morphological, physiological, and behavioral characteristics that facilitate location, identification, capture, ingestion, and digestion of prey items. Feeding dynamics can vary considerably between species, but differences between the three major amphibian lineages and between species feeding on land and in water become obvious. Therefore, the feeding mechanisms and dietary specialties of the different groups are discussed separately (for amphibian larvae see Chapter 5).

10.1.1 Adult anurans

Anurans are carnivores and their main diet consists of arthropods. Parts of plants or fruits are accidentally consumed with other food items by adult amphibians, although exceptions have been documented, such as the hylid frog *Xenohyla truncata* (Silva and Britto-Pereira 2006). In most anurans, prey capture is enhanced by the use of projectile tongues fixed at the anterior part of the mouth. The posterior part is commonly free, and the whole tongue can be thrust out in order to capture prey items. Among anurans, the family of the predominately aquatic Pipidae lack tongues and uses suction feeding. The species are characterized by broad elements of the hyobranchial skeleton, especially the ceratohyals and ceratobranchials. Sokol (1969) observed that the small pipid *Hymenochirus boettgeri* depressed its hyobranchium prior to mandibular depression, so that maximum suction is achieved as soon as the mouth is opened. The larger pipid genera *Pipa* in South America and *Xenopus* in Africa augment buccal suction by stuffing their prey into their mouth with their forelimbs.

Differences in feeding behavior among frogs with different diets have been observed, encompassing a continuum between sit-and-wait and opportunistic search strategies. The composition of the diet can reflect the feeding strategy to some degree. Whereas sit-and-wait foragers commonly consume larger, mobile prey and only a few food items per time unit, opportunistic searchers consume greater numbers of smaller food items. These behavioral patterns are also reflected in morphological properties of the species. Emerson (1985) found a relationship between the skull shapes of frogs and their diet, whereby frogs that eat relatively small, slow prey have relatively long jaws and an asymmetrical feeding cycle; time to capture prey is less than the time to bring prey into the mouth. On the other hand, snail-eating frogs have lower jaws with a relatively long distance from the insertion point of the adductor muscle to the jaw articulations. Two main diet patterns in anurans were identified by Toft (1980a, 1981, 1985). They comprise "ant specialists" eating slow-moving, strongly chitinized arthropods such as ants, termites, and mites, and "non-ant specialists" eating generally more variable and less chitinous arthropods (e.g. spiders and grasshoppers).

10.1.2 Adult caudata

Terrestrial salamanders typically feed by capturing their prey with a large, moist, sticky tongue. Although predominately carnivorous, several species of the genus *Siren* include larger plant parts in their diet (Duellman and Trueb 1994). Some neotropical salamanders exhibit lower prey diversity than most temperate-zone terrestrial and aquatic species, a difference that might be explained by the disproportionately high contribution of ants in the diets of the neotropical species (Anderson and Mathis 1999).

10.1.3 Adult gymnophiona

Despite being widely distributed in the tropics, only few dietary studies have been performed on caecilians (O' Reilly 2000). Most dietary studies of terrestrial and aquatic caecilians suggest that they are generalized carnivores. Kupfer *et al.* (2005) studied the diet of both larval and adult *Ichthyophis* cf. *kohtaoensis* from Thailand. The larvae feed exclusively on aquatic invertebrates, preferring benthic prey. The adults consume mostly soil invertebrates such as earthworms, ants, and termites. Ontogenetic dietary shifts and the trophic position of caecilians are more similar to salamanders than frogs with biphasic life cycles.

10.2 Methods to obtain prey items: historical overview

Most accounts of amphibian feeding are anecdotal and involve only a few taxa. As a result, much is unknown about prey selection and foraging strategies. The diet of amphibians has commonly been studied by sacrificing the animals and dissecting their stomachs. Technically, this is the easiest method and, when applied to series of preserved specimens housed in museum collections, ethically justified. Prey items can be obtained in good condition, facilitating identification and measurements. Today, the euthanasia of large samples of study animals for dietary analyses is ethically questionable because alternative methods are available. One of these alternative methods is stomach flushing. It avoids killing large number of animals, and has been used in herpetological studies since the 1970s. For example, Fraser (1976) used this method to study salamander diets, and it was used in later studies of both frogs and reptiles (Legler 1977; Legler and Sullivan 1979). Most researchers anesthetize the animals prior to stomach flushing, but this occasionally has resulted in mortality (Joly 1987). Despite early attempts at alternative study methods, most dietary studies continued to rely upon freshly killed specimens, probably because the description of stomach-flushing techniques proved rather inadequate and difficult to replicate. Solé *et al.* (2005) published a protocol clearly explaining which materials were needed, and which steps had to be followed to achieve successful stomach flushing. Their protocol has since been used in dietary research on different anurans (e.g. Miranda *et al.* 2006; Mahan and Johnson 2007; Solé and Pelz 2007; Wang *et al.* 2008). Wu *et al.* (2007) compared data on stomach dissection with those obtained by stomach flushing to assess dietary diversity, and found no significant differences, tending to validate the flushing method. They stated that "our study indicates that the result of stomach flush is accurate in the diet analysis. Stomach flush should be used more widely and stomach dissection should be thrown away ultimately." Since prey size, composition, and diversity are often correlated with the sex and size of the predator (Lima and Magnusson 2000), it is important to note the snout–vent length and the gender of each specimen.

10.2.1 Stomach flushing: materials

A complete flushing set consists of two containers, a spatula, forceps, two syringes (20 ml for small and 60 ml for large frogs), an infusion tube (for small frogs) or a rubber or PVC hose (commonly used for aquarium pumps), small airtight vials, and 70% ethanol solution (Figure 10.1). We recommend acquiring syringes with screw-on threading and infusion tubes with internal threading to prevent the detachment of the tube from the syringe during the flushing procedure. For use directly in the field, all materials can be stored in the larger container. Spring water or previously filtered pond water can be used as a flushing solution.

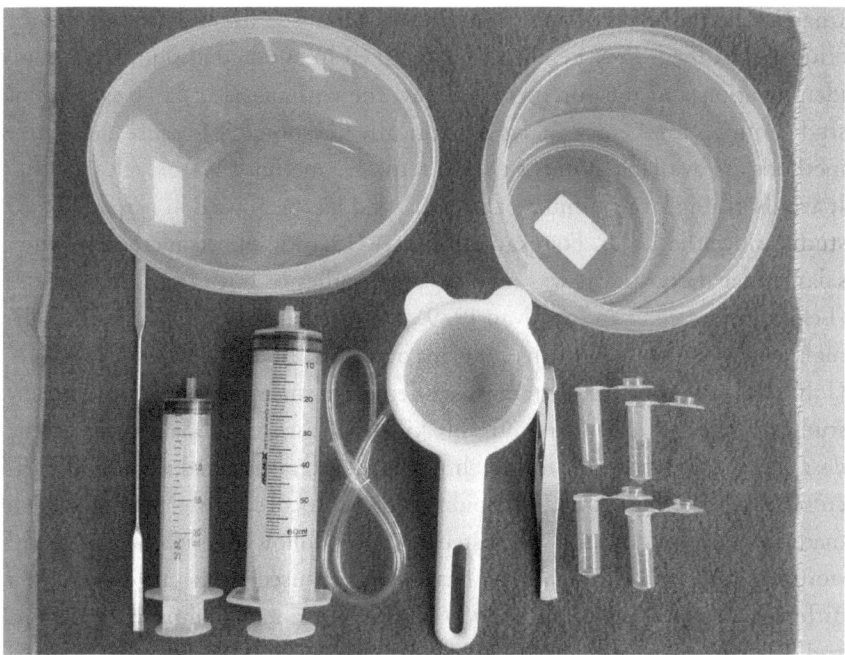

Fig. 10.1 Complete flushing set consisting of two containers, a spatula, forceps, two syringes, an infusion tube (for small frogs) or a rubber or PVC hose, small airtight vials, and 70% ethanol solution.

10.2.2 When should sampling be conducted?

Feeding-activity patterns of the target species should be considered to avoid a bias resulting from digestion, the rate of which is directly related to environmental temperature. Amphibians should be sampled during their foraging activity period or shortly thereafter, as digestion compromises accurate identification of prey items. For example, most frogs are crepuscular or nocturnal (Duellman and Trueb 1994), and searching for them is easier at night due to the reflective aspect of the eyes of most frogs when illuminated with a head lamp (Rocha *et al.* 2000). We recommend sampling frogs about 2–4 h after nightfall. During this time most frogs have already managed to find some food, and the chance to recover unfragmented prey items is much higher than if sampling and flushing are performed the next morning.

10.2.3 How to perform the flush

Before handling the frog, all material should be ready to be used. One container should be filled with flushing solution and the other positioned to collect the stomach contents. The syringe should be filled with flushing solution and attached to the tube. The specimen should be held safely by fixing its forelimb with one hand, while the syringe is held in the other. As both hands are in use, the spatula used to open the frog's mouth should be held between one's teeth. The spatula should be inserted gently at the hind part of the mouth and then turned at a 45° angle. The tube is then inserted into the mouth and down the esophagus to the stomach (Figure 10.2). The pyloric end of the stomach can easily be detected, but gentle care is highly recommended throughout the process.

Once the tube is in the correct position, the entire syringe contents are flushed into the stomach. The right amount of force is dependent on the size of the flushed frog. We recommend starting using little pressure and increase the pressure steadily until stomach contents start pouring out of the mouth. The empty syringe is then separated from the tube, which is kept inside the stomach, refilled with flushing solution, and attached to the tube again in order to allow a second flush. This procedure should be repeated as long as stomach contents are forced out. When no more stomach contents appear, the frog should be flushed one last time. The flushing solution containing the stomach contents is then decanted into a sieve. Items are then picked up by using a forceps and stored in a vial in 70% ethanol or formalin. The latter has the advantage that the prey items are not leached as much compared to when ethanol is used. On the other hand storage in ethanol allows subsequent DNA analyses and formalin does not. For specimens with very small prey, such as mites and Collembolans, gauze with a mesh size of at least 6 μm should be used instead of the sieve. Stomach contents

Fig. 10.2 Stomach flushing applied to *Trachycephalus mesophaeus*.

can be directly washed with 70% ethanol from the gauze into a small vial. Frogs should be released where they were captured as soon as possible after stomach flushing to facilitate a return to normal foraging activity and behavior.

10.3 Data analysis

Once the prey items are obtained, the most comfortable way to identify them is to spread each sample separately on a petri dish containing water and then examine them under a stereo microscope. Prey identification is commonly carried out to ordinal level, but prey categories must not all correspond to taxonomic orders. Most authors, for example, separate the Order Hymenoptera into the family Formicidae (ants) and non-Formicidae because in many amphibian species ants make up a large part of the diet. Depending on the degree of dietary specialization and the question at hand it might be useful to identify prey items to the lowest taxonomic level as possible. If identification keys for the local fauna exist, these can be used for taxonomic identification of prey. In areas with a lack of such data, general ordinal keys can be used (e.g. Triplehorn and Johnson 2005) or prey items can be sent to colleagues specialized in specific groups. Some authors categorize the prey items into ecological guilds, such as 'rather active' or 'sedentary' taxa, or flying, ground-dwelling, or burrowing species. These methods facilitate assessments of foraging strategies in the absence of behavioral information, since the quantity and volume of prey items belonging to different guilds allow researchers to draw conclusions about the feeding strategy of the amphibian predator.

Even if researchers manage to capture amphibians shortly after they have started foraging, and even if the flushing procedure is applied shortly thereafter, researchers frequently find that parts of the stomach contents have already been digested. Soft tissues generally are digested more rapidly than strongly chitinized parts, thus making prey items break into fragments. Hirai and Matsui (2001) noticed that certain specific parts of prey remain intact even after digestion has started. Examples are the heads of ants or wasps, abdomen and elytra of beetles, thoraxes of flies, moths, beetles, or Orthopterans (grasshoppers, crickets, katydids), forewings of flies and bees, and the saltatorial legs of Orthopterans. Hirai and Matsui (2001) calculated regression formulae between body parts and total body length, which can be used as a rough estimate of prey size. To be complete, however, information on the size of undamaged items is required for a thorough assessment of prey use.

In most dietary studies, information about the number, frequency, and volume and/or mass of the prey items of each taxonomic or ecological group is presented. These data allow direct comparison between different prey groups. The mass of prey items can be assessed with an analytic balance. Here, storage in formalin is superior to storage in ethanol since the later dehydrates the prey items. Furthermore, since stomach contents are commonly wet, it is important to remove excess moisture from the prey item before weighing to avoid bias. The best way would be to store the prey items in a drying oven at approximately 50°C until a constant dry mass is achieved. Information about prey number, frequency, and mass are relatively easy to obtain, as there are a number of different techniques used to measure the volume of prey items. Volume and mass of prey are directly correlated with the amount of energy they can provide; thus, this information is most appropriate in combination with the identification of important prey groups. Most researchers attempt to reconstitute volumes from linear measures using formulae for geometric shapes; body appendices of prey items should be ignored when measurements are taken. The most commonly used formula is the one for ellipsoid bodies (Colli and Zamboni 1999):

$$V = \frac{4\pi}{3} \frac{L}{2} \left(\frac{W}{2} \right)^2$$

where V is volume, L is length, and W is width of prey item. This formula is also referred to as prolate spheroid (Matori and Aun 1994; Vitt *et al.* 1999). Other formulas used to estimate volume are the formula for a parallelepiped ($V = LWH$) (Schoener 1967; Bergallo and Magnusson 1999) and for a cylinder (Werner *et al.* 1995):

$$V = \pi L \left(\frac{W}{2} \right)^2$$

The volume of prey items also can be measured directly by introducing them into a measuring cylinder graduated in intervals of 0.1 ml. Water is added from a pipette graduated in intervals of 0.01 ml until the item is completely covered and the water has reached the next highest graduation in the cylinder. The volume can then be taken as the water level in the cylinder minus the amount added with the pipette (Magnusson *et al.* 2003; Cuello *et al.* 2006).

10.3.1 Measuring prey availability and electivity

Although Schoener (1974) recommended that the study of a species' trophic niche should include prey availability, only a few studies have done so (Toft 1980a, 1981; MacNally 1983; Juncá and Eterovick 2007). There are many different methods used to measure prey availability, and techniques should be selected in keeping with the different foraging microhabitats, such as leaf litter and arboreal species. Traps are usually either mechanical or based on attractants such as food or pheromones; the latter should be avoided because of sampling biases.

To assess prey availability in a leaf-litter habitat, Maneyro and da Rosa (2004) used pitfall traps consisting of 300 ml cans with a solution of liquid soap and 10% formalin. This formulation may prove useful in sampling ground-dwelling, mobile prey items. Soap is added to prevent the escape of arthropods, and it reduces the surface tension of the solution, facilitating the drowning of small arthropods such as Collembolans. Furthermore, it makes the bucket walls of the trap slippery. Instead of formalin (which may be legally prohibited for use under natural settings and the use of which is often considered unethical in field research), ethanol may be substituted. Pitfalls also may be fitted with modified lids or funnels to prevent escape.

Pitfall traps may overestimate more mobile fauna (Ausden 1996). On the other hand, it is reasonable to assume that such bias would not be misleading in the assessment of prey availability since prey mobility may increase the detection probability by amphibians which use vision to locate prey (Duellman and Trueb 1994). Juncá and Eterovick (2007) recommended using Berlese–Tullgren funnel traps (Brower *et al.* 1998) in addition to pitfall traps and combining the results. These traps sample sedentary taxa in addition to mobile fauna. Both methods underestimate flying prey items.

In grassy or herbaceous habitats, the composition and abundance of prey items can be estimated along transects using linear sweeps with insect nets. Each sweep sample should be transferred into a Berlese–Tullgren trap for at least 30 min to separate prey items from litter (leaves, seeds, sticks). Properly conducted, this method can provide a comprehensive assessment of invertebrate availability. It should be noted that highly mobile flying insects such as Odonates (dragonflies) might be underrepresented. Furthermore, the time and temperature at which the sweeps are conducted can have a strong influence on the result depending on the activity periods of the insects.

Arboreal habitat can be sampled most easily using fogging techniques. Applying insecticides through fogging is widely used in pest control, and a wide range of commercial fogging machines is available. Fumigated insects falling from tree canopies are collected on plastic sheets laid out on the ground or by plastic trays or aluminum funnels (for a short review, see Adis *et al.* 1998). Insecticide fogging has become established as a standard technique since the 1980s for assessing arthropod diversity, especially in tropical forest canopies (Adis *et al.* 1998). A variety of "knockdown" insecticides have been tested, but natural pyrethrum is now established as a standard as it allows for collection of live specimens and its biodegradability facilitates application in ecologically sensitive habitats. Adis *et al.* (1998) suggested recommendations to standardize fogging procedures. When applying fogging procedures, researchers must consider that the entire vertical habitat that is sampled may overestimate the prey available to amphibians, which inhabit only parts of the vertical habitat structure.

Aquatic habitats can be sampled using a fine mashed net, such as the D-frame kick net or surber stream-bottom sampler. The D-frame kick net is either cone- or bag-shaped for the capture of organisms and commonly used by lying the spine of the net firmly on the ground and hauling it a specific number of times along a specific distance. This type of net is easy to transport and can be used in a variety of habitat types. However, the D-frame kick net must have a defined or delimited area that is sampled/kicked in order to compare results obtained from different sites or samples. Cao *et al.* (2005) investigated the performance and comparability between surber and D-frame kick-net samples and found that subsamples with the same number of individuals were highly and consistently comparable between sampling devices. Rocky stream habitats are best sampled by two persons. One collector should place a fine mashed net so that its bottom lies on the ground. Another collector should turn over as many rocks as possible upstream from the net for 1 min. Wiping off clinging insects from rocks and branches may enhance collection quantities and species diversity.

Once relative abundances of the prey items in the habitat and in the stomach contents are identified, an electivity index can be calculated, as proposed by Jacobs (1974):

$$D = \frac{R_k - P_k}{(R_k + P_k) - (2R_k P_k)}$$

whereby R_k is the proportion of prey category k in stomach contents, and P_k is the proportion of prey category k in the environment. D varies from +1 (complete selection of preference for prey category k) through 0 (prey category k is taken in the same proportion as found in the environment) to -1 (prey category k is present in the environment but absent in the diet). When interpreting the results, biases caused by sample methods need to be considered.

10.3.2 Comparing trophic niche structure

A number of indices have been proposed for the identification of (trophic) diversity, niche breadth, and niche overlaps between multiple species. Calculation of indices can be facilitated using the software EcoSim (N. J. Gotelli and G. L. Entsminger, version 7.0, available at http://together.net/~gentsmin/ecosim.htm, 2001). Trophic diversity can be assessed using Hurtubia (1973) or the Gladfelter–Johnson index (Gladfelter and Johnson 1983). A modification of this latter index was recently proposed by Cardona (1991) to measure trophic niche breadth using occurrence frequencies:

$$B' = \frac{\sum p_k - \sigma}{100R}$$

where p_k is the occurrence frequency of kth prey, σ is the standard deviation of the occurrence frequencies, and R is the number of prey species exploited by the guild.

In order to compare the niche breath between multiple species or changes in the niche breadth within different seasons, the Shannon–Weaver diversity index H' (Weaver and Shannon 1949; see Chapter 18) can be used:

$$H' = -\sum_i p_i \ln P_i$$

where p_i is the relative abundance of each prey category, calculated as the proportion of prey items of a given category to the total number of prey items.

The index of relative importance (*IRI*) is a measure that reduces bias in descriptions of animal dietary data. Introduced by Pianka *et al.* (1971) and Pianka (1973), it is mainly used currently to describe the diet of fish:

$$IRI_t = (PO_t)(PI_t + PV_t)$$

where PO_t is the percentage of occurrence (100 × number of stomachs containing *t* item/total number of stomachs), PI_t is the percentage of individuals (100 × total number of individuals of *t* in all stomachs/total number of individuals of all taxa in all stomachs), and PV_t is the percentage of volume (100 × total volume of individuals of *t* in all stomachs/total volume of all taxa in all stomachs). Hart *et al.* (2002) suggested that IRI should be applicable to a wide range of taxa. To calculate the trophic niche breadth, the index of Simpson (1949) can be used:

$$B = \frac{1}{\sum p_i^2}$$

where B is Simpson's index, p_i is the fraction of items in the food category i, and the range is 1 to n.

To compare various diets with a different number of food categories, the standardized Simpson's index (*BS*) form as proposed by Hulbert (1978) can be used:

$$BS = \frac{B-1}{n-1}$$

Here, n is the number of food categories and the index ranges from 0 to 1.

Overlap in trophic niches between two species can be assessed with an index proposed by May and MacArthur (1972) as modified by Pianka (1974):

$$O_{jk} = \frac{\sum_i^n P_{ij} P_{ik}}{\sqrt{\sum_i^n P_{ij}^2 \sum_i^n P_{ik}^2}}$$

Here, O_{jk} is niche overlap and p_{ij} and p_{ik} represent the proportions of the ith resource used by the jth and kth species.

10.3.3 Fatty acid and stable-isotope analyses

Analyses of isotopic composition and fatty acid profiles of tissues can provide reliable information on diets and trophic status because they reflect assimilation over long periods of time. Isotope signatures—that is, ratios of heavier (e.g. ^{15}N and ^{14}C) and lighter (^{14}N and ^{13}C) isotopes—of consumers are generally similar to those of their food and can thus provide insights into food-web structure. Fatty acids are essential energy storage lipids in consumers that cannot be readily synthesized and must be obtained from their food. Differences in fatty acid composition among available food types are reflected in fatty acid compositions in consumers; therefore, profiles can be good indicators of diets. For both techniques, tissue samples (or in the case of small individuals or food items, the whole organism) are simply collected and preserved in a manner that does not alter isotopic or fatty acid composition, depending on the analyses to be performed, and then sent to an analytical laboratory for analyses (for a more detailed description of these techniques see Chapter 5).

10.3.4 Further biochemical analyses

Some frog species in the neotropics and Madagascar secrete alkaloids for protection against predators. These so-called poison frogs do not produce the alkaloids, but instead obtain them from their insect-rich diet (Daly *et al.* 1994). While the diet of neotropical poison frogs is well studied (Daly *et al.* 1994, 2002), data on the Malagasy frogs was only provided recently by Clark *et al.* (2005). These authors extracted alkaloid samples from both Malagasy frogs and their food. Alkaloid fractions can be extracted and isolated from skin samples using methanol (one part skin/two parts methanol) (Daly *et al.* 1992). Possible methods for biochemical analyses comprise gas chromatography in conjunction with mass spectrometry and infrared spectroscopy to identify alkaloids (for detailed descriptions see Daly *et al.* 1992, 1993, 1994). They found that Malagasy frogs, like the neotropical species, acquire their alkaloids from the ants they prey upon. Thirteen of the 16 alkaloids detected in the Malagasy frogs also exist in insects and frogs in the neotropics. As neither the ants nor the frogs in these two regions are closely related, these authors suggested that the evolution of acquisition mechanisms for alkaloids in these ant species was likely responsible for the subsequent convergent evolution of the frogs that preyed on them. The researchers also found nicotine, a plant alkaloid, in one Malagasy frog species, suggesting a possible plant–ant–frog toxin food chain.

10.4 General considerations

Samples should be taken randomly throughout the habitat of the target species in order to avoid bias, such as the predominance of ants which would be expected if all samples were obtained next to an ant hill. Furthermore, sample sizes are an important issue for dietary analyses. Considering that diet may vary between the sexes, adequate samples from both genders should be obtained. When working with stomach contents, researchers should be aware that they will probably not get a picture of what prey items the focal species feeds upon, but rather a picture of the items that are hard to digest. Frogs are capable of digesting prey rapidly: Blanchard's cricket frogs (*Acris crepitans blanchardi*) fed dyed *Drosophila* (fruit flies) and killed 7 h later had no trace of prey remaining in their stomachs (Johnson and Christiansen 1976). In another study with *A. crepitans* immediately preserved frogs had a mean of seven prey items per stomach, whereas frogs preserved 6 h after capture revealed 3.6 prey items per stomach (Caldwell 1996). In a preliminary study with 20 *Physalaemus lisei*, frogs were fed three different prey categories (Oligochaeta, Curculionid beetles, and Aranea) at the same time, and four frogs were stomach-flushed every 4 h. Oligochaeta (worms) could not be detected after 4 h, whereas the pronotum of spiders was still undigested after 8 h and the beetle elytra were still present after 12 h (M. Solé, unpublished results). In conclusion, researchers will underestimate the prey categories that basically consist of soft tissue (worms, mollusks, and fly larvae) and overestimate those with strongly keratinized bodies (ants and beetles). One way in which biologists can mediate this problem is to carefully choose the sampling period according to the feeding period of each species.

10.4.1 Ontogenetic changes

Diet composition of amphibians may change during ontogeny, since larger individuals may be able to eat larger prey items, which significantly increases the variety of available prey types. However, Lima (1998) found changes in types and sizes of prey that were caused other than by a simple passive effect of selection for larger prey with growth of the frogs. Foraging modes and microhabitat use may also change during ontogeny, which can affect the composition of available prey items (Lima and Magnusson 2000). To assess ontogenetic changes in the composition of the diet, different snout–vent size classes of the target species can be compared. The best visualization provides a cluster analysis based on the relative frequency of the different prey categories. Data must be normally distributed to meet the assumptions of the statistical test (Zar 1999). If the data do not meet the required statistical assumptions, they can be transformed (see Chapter 18).

10.4.2 Seasonal changes

Availability and diversity of prey items can change between different seasons (Houston 1973). Although there are many records concerning feeding habits in amphibians, few include seasonal patterns. In Amazonian Peru, Toft (1980b) observed a decrease in arthropod biomass and in frog abundance in the dry season. However, she argued that frog species and their arthropod prey may respond differently to seasonal changes, since the principal feeding patterns of the litter anurans studied did not appear to change seasonally. Yilmaz and Kutrup (2006) analyzed seasonal changes in the diet of the European common frog, *Rana ridibunda*, and found changes in prey diversity between July (62 prey categories) and August (38). The highest mean prey number was observed in July. The mean prey size also changed during the year: frogs were more likely to consume large prey items in June than in the other months. These results suggested that the monthly variation in diet composition of this species was correlated with prey availability. Finally, large sample sizes are important when studying seasonal dietary changes.

10.5 Conclusions

Dietary information is pivotal for successful development of conservation strategies on species level and the understanding of ecosystem function. Unfortunately, this kind of information is not available for the vast majority of taxa and is often incomplete. Seasonal variations, ontogenetic shifts, and relationships between site-specific prey availability, presence of potential competitors, and diet composition are commonly not addressed. Future studies should focus on these issues. In the light of global amphibian declines an Amphibian Conservation Action Plan was formulated by the International Union for Conservation of Nature to prevent further biodiversity loss and captive breeding programmes are being developed for the most threatened species (Gascon *et al.* 2007). Such approaches depend crucially on autecological knowledge such as information on the diet of the target species. Here, every kind of information can be valuable.

10.6 Acknowledgments

We are grateful to Ken Dodd, Marian Griffey, and Matt Whiles for helpful comments on the manuscript and to Klaus Riede for fruitful discussions concerning quantitative insect sampling methods and help with the literature. The work of DR was funded by the Graduiertenförderung des Landes Nordrhein-Westfalen.

10.7 References

Adis, J., Basset, Y., Floren, A., Hammond, P. M., and Linsenmair, K. E. (1998). Canopy fogging on an overstorey tree- recommendations for standardization. *Ecotropica*, 4, 93–7.

Anderson, M. T. and Mathis, A. (1999). Diets of two sympatric Neotropical salamanders, *Bolitoglossa mexicana* and *B. rufescens*, with notes on reproduction for *B. rufescens*. *Journal of Herpetology*, **33**, 601–7.

Anderson, S. H. (1991). *Managing our Wildlife Resources*. Merrill Publishing Co., Columbus, OH.

Ausden, M. (1996). Invertebrates. In W. J. Sutherland (ed.), *Ecological Census Techniques*, pp. 139–77. Cambridge University Press, Cambridge.

Bergallo, H. G. and Magnusson, W. E. (1999). Effects of climate and food availability on four rodent species in southeastern Brazil. *Journal of Mammalogy*, **80**, 472–86.

Brower, J. E., Zar, J. H., and von Ende, C. N. (1998). *Field and Laboratory Methods for General Ecology*. WCB McGraw-Hill, Boston, MA.

Caldwell, J. P. (1996). The evolution of myrmecophagy and its correlates in poison frogs (family Dendrobatidae). *Journal of Zoology, London*, **240**, 75–101.

Cao, Y., Hawkins, C. P., and Storey, A. W. (2005). A method for measuring comparability of different sampling methods used in biological surveys: implications for data integration and synthesis. *Freshwater Biology*, **50**, 1105–15.

Cardona, L. (1991). Measurement of trophic niche breadth using occurrence frequencies. *Journal of Fish Biology*, **39**, 901–3.

Clark, V. C., Raxworthy, C. J., Rakotomalala, V., Sierwald, P., and Fisher, B. L. (2005). Convergent evolution of chemical defense in poison frogs and arthropod prey between Madagascar and the Neotropics. *Proceedings of the National Academy of Sciences USA*, **102**, 11617–22.

Colli, G. R. and Zamboni, D. S. (1999). Ecology of the worm-lizard *Amphisbaena alba* in the Cerrado of central Brazil. *Copeia*, **1999**, 733–42.

Cuello, M. E., Bello, T. M., Kun, M., and Ubeda, C. A. (2006). Feeding habits and their implications for the conservation of endangered semiaquatic frog *Atelognathus patagonicus* (Anura, Neobatrachia) in a northwestern Patagonian pond. *Phyllomedusa*, **5**, 67–76.

Daly, J. W., Secunda, S. I., Garraffo, H. M., Spande, T. F., Wisnieski, A., Nishihira, C., and Cover Jr, J. F. (1992). Variability in alkaloid profiles in Neotropical poison frogs (Dendrobatidae). *Toxicon*, **30**, 887–98.

Daly, J. W., Garraffo, H. M., and Spande, T. F. (1993). Amphibian alkaloids. In G. A. Cordell (ed.), *The Alkaloids*, vol. 43, pp. 185–288. Academic Press, San Diego, CA.

Daly, J. W., Garraffo, H. M., Spande, T. F., Jaramillo, C., and Rand, A. S. (1994). Dietary source for skin alkaloids of poison frogs (Dendrobatidae)? *Journal of Chemical Ecology*, **20**, 943–55.

Daly, J. W., Kaneko, T., Wilham, J., Garraffo, H. M., Spande, T. F., Espinosa, A., and Donnelly, M. (2002). Bioactive alkaloids of frog skin: combinatorial bioprospecting reveals that pumiliotoxins have an arthropod source. *Proceedings of the National Academy of Sciences USA*, **99**, 13996–14001.

Duellman, W. E. and Trueb, L. (1994). *Biology of Amphibians*. John Hopkins University Press, Baltimore, MD.

Emerson, S. B. (1985). Skull shape in frogs—correlations with diet. *Herpetologica*, **41**, 177–88.

Fraser, D. F. (1976). Coexistence of salamanders in the genus *Plethodon*, a variation of the Santa Rosalia theme. *Ecology*, **57**, 238–51.

Freed, A. N. (1982). A treefrog's menu: selection for an evening's meal. *Oecologia*, **53**, 20–6.

Gascon, C., Collins, J. P., Moore, R. D., Church, D. R., McKay, J. E., and Mendelson III, J. R. (2007). *Amphibian Conservation Action Plan - Proceedings: IUCN/SSC Amphibian Conservation Summit 2005*. IUCN, Gland.

Gladfelter, W. B. and Johnson, W. S. (1983). Feeding niche separation in a guild of tropical reef fish (Holocentridae). *Ecology*, **64**, 552–63.

Hart, R. K., Calver, M. C., and Dickman, C. R. (2002). The index of relative importance: an alternative approach to reducing bias in descriptive studies of animal diets. *Wildlife Research*, **29**, 415–21.

Hirai, T. and Matsui, M. (2001). Attempts to estimate the original size of partly digested prey recovered from stomachs of Japanese anurans. *Herpetological Review*, **32**, 14–16.

Houston, W. W. K. (1973). The food of the common frog, *Rana temporaria*, on high moorland in northern England. *Journal of Zoology*, **171**, 153–65.

Hulbert, S. H. (1978). The measurement of niche overlap and some relatives. *Ecology*, **59**, 67–77.

Hurtubia, J. (1973). Trophic diversity measurement in sympatric predatory species. *Journal of Ecology*, **549**, 885–90.

Jacobs, J. (1974). Quantitative measurement of food selection: a modification of the forage ration and Ivlev's electivity index. *Oecologia*, **14**, 413–17.

Johnson, B. K. and Christiansen, J. L. (1976). The food and food habits of Blanchard's cricket frog, *Acris crepitans blanchardi* (Amphibia, Anura, Hylidae), in Iowa. *Journal of Herpetology*, **10**, 63–74.

Joly, P. (1987). Le regime alimentaire des amphibiens: methods d'etude. *Alytes*, **6**, 11–17.

Juncá, F. A. and Eterovick, P. C. (2007). Feeding ecology of two sympatric species of Aromobatidae, *Allobates marchesianus* and *Anomaloglossus stepheni*, in Central Amazon. *Journal of Herpetology*, **41**, 301–8.

Kupfer, A., Nabhitabhata, J., and Himstedt, W. (2005). From water into soil: trophic ecology of a caecilian amphibian (Genus *Ichthyophis*). *Acta Oecologica*, **28**, 95–105.

Legler, J. M. (1977). Stomach flushing: a technique for chelonian dietary studies. *Herpetologica*, **33**, 281–4.

Legler, J. M. and Sullivan, L. J. (1979). The application of stomach-flushing to lizards and anurans. *Herpetologica*, **35**, 107–10.

Lima, A. P. (1998). The effects of size on the diets of six sympatric species of postmetamorphic litter anurans in central Amazonia. *Journal of Herpetology*, **32**, 392–9.

Lima, A. P. and Magnusson, W. E. (2000). Does foraging activity change with ontogeny? An assessment for six sympatric species of postmetamorphic litter anurans in Central Amazonia. *Journal of Herpetology*, **34**, 192–200.

MacNally, R. (1983). Trophic relationships of two sympatric species of *Ranidella* (Anura). *Herpetologica*, **39**, 130–40.

Magnusson, W. E., Lima, A. P., Silva, W. A., and Araújo, M. C. (2003). Use of geometric forms to estimate volume of invertebrates in ecological studies of dietary overlap. *Copeia*, **2003**, 13–19.

Mahan, R. D. and Johnson, J. R. (2007). Diet of the gray treefrog (*Hyla versicolor*) in relation to foraging site location. *Journal of Herpetology*, **41**, 16–23.

Maneyro, R. and da Rosa, I. (2004). Temporal and spatial changes in the diet of *Hyla pulchella* (Anura, Hylidae) in southern Uruguay. *Phyllomedusa*, **3**, 101–13.

Matori, R. and Aun, L. (1994). Aspects of the ecology of a population of *Tropidurus spinulosus*. *Amphibia-Reptilia*, **15**, 317–26.

May, R. M. and MacArthur, R. H. (1972). Niche overlap as a function of environmental variability. *Proceedings of the National Academy of Sciences USA*, **69**, 1109–13.

Miranda, T., Ebner, M., Solé, M., and Kwet, A. (2006). Spatial, seasonal and intrapopulational variation in the diet of *Pseudis cardosoi* (Anura: Hylidae) from the Araucária Plateau of Rio Grande do Sul, Brazil. *South American Journal of Herpetology*, **1**, 121–30.

O'Reilly, J. C. (2000). Feeding in caecilians. In K. Schwenk (ed.), *Feeding, Form, Function and Evolution in Tetrapod Vertebrates*. Academic Press, New York.

Pianka, E. R. (1973). The structure of lizard communities. *Annual Review of Ecology and Systematics*, **4**, 53–74.

Pianka, E. R. (1974). Niche overlap and diffuse competition. *Proceedings of the National Academy of Sciences USA*, **71**, 2141–5.

Pianka, E. R., Oliphant, M. S., and Iverson, Z. L. (1971). Food habits of albacore bluefin, tuna and bonito in Califorina waters. *California Department of Fish and Game Bulletin*, **152**, 1–350.

Rocha, C. F. D., Van Sluys, M., Alves, M. A. S., Bergallo, H. G., and Vrcibradic, D. (2000). Activity of leaf-litter frogs: when should frogs be sampled? *Journal of Herpetology*, **34**, 285–7.

Schoener, T. W. (1967). The ecological significance of sexual dimorphism in size in the lizard *Anolis conspersus*. *Science*, **155**, 474–7.

Schoener, T. W. (1974). The compression hypothesis and temporal resource partitioning. *Proceedings of the National Academy of Sciences USA*, **71**, 4169–72.

Silva, H. R. and Britto-Pereira, M. C. (2006). How much fruit do fruit-eating frogs eat? An investigation on the diet of *Xenohyla truncata* (Lissamphibia: Anura: Hylidae). *Journal of Zoology*, **270**, 692–8.

Simon, M. P. and Toft, C. A. (1991). Diet specialization in small vertebrates: mite-eating in frogs. *Oikos*, **61**, 263–78.

Simpson, E. H. (1949). Measurement of diversity. *Nature*, **163**, 688.

Sokol, O. M. (1969). Feeding in the pipid frog *Hymenochirus boettgeri* (Tornier). *Herpetologica*, **25**, 9–24.

Solé, M. and Pelz, B. (2007). Do male tree frogs feed during the breeding season? Stomach flushing of five syntopic hylid species in Rio Grande do Sul, Brazil. *Journal of Natural History*, **41**, 2757–63.

Solé, M., Beckmann, O., Pelz, B., Kwet, A., and Engels, W. (2005). Stomach-flushing for diet analysis in anurans: an improved protocol evaluated in a case study in Araucaria forests, southern Brazil. *Studies on Neotropical Fauna and Environment*, **40**, 23–8.

Toft, C. A. (1980a). Feeding ecology of thirteen syntopic species of anurans in a seasonal tropical environment. *Oecologia*, **45**, 131–41.

Toft, C. A. (1980b). Seasonal variation in populations of Panamanian litter frogs and their prey: a comparison of wetter and drier sites. *Oecologia*, **47**, 34–8.

Toft, C. A. (1981). Feeding ecology of Panamanian litter anurans: patterns in diet and foraging mode. *Journal of Herpetology*, **15**, 139–44.

Toft, C. A. (1985). Resource partitioning in amphibians and reptiles. *Copeia*, **1985**, 1–21.

Triplehorn, C. A. and Johnson, N. F. (2005). *Borror and Delong's Introduction to the Study of Insects*, 7th edn. Thomson Brooks/Cole, Belmont, CA.

Vitt, L. J., Zani, P. A., and Espósito, M. C. (1999). Ecological differences in tropical sympatric skinks (*Mabouya macrorhyncha* and *Mabouya agilis*) in southeastern Brazil. *Journal of Herpetology*, **30**, 60–7.

Wang, Y. P., Zhang, P., and Li, Y. (2008). Diet composition of post-metamorphic bullfrogs (*Rana catesbeiana*) in the Zhoushan archipelago, Zhejiang Province, China. *Frontiers of Biology in China*, **3**, 219–26.

Weaver, W. and Shannon, C. E. (1949). *The Mathematical Theory of Communication*. University of Illinois, Urbana, IL.

Werner, E. E., Wellborn, G. A., and McPeek, M. A. (1995). Diet composition in postmetamorphic bullfrogs and green frogs: implications for interspecific predation and competition. *Journal of Herpetology*, **29**, 600–7.

Wu, Z. J., Li, Y. M., and Wang, Y. P. (2007). A comparison of stomach flush and stomach dissection in diet analysis of four frog species. *Acta Zoologica Sinica*, **53**, 364–72.

Yilmaz, Z. C. and Kutrup, B. (2006). Seasonal changes in the diet of *Rana ridibunda* Pallas, 1771 (Anura: Ranidae) from the Gorele River, Giresun, Turkey. In M. Vences, J. Köhler, T. Ziegler, and W. Böhme (eds), *Herpetologia Bonnensis II. Proceedings of the 13th Congress of the Societas Europaea Herpetologica*, pp. 201–4. Bonn, Germany.

Zar, J. H. (1999). *Biostatistical Analysis*. Prentice-Hall, Englewood Cliffs, NJ.

11

Movement patterns and radiotelemetry

Dale M. Madison, Valorie R. Titus, and Victor S. Lamoureux

11.1 Introduction

There is no greater need for details on the year-round movements and refuges of amphibians in their natural habitats than today, mainly due to the pressure to develop lands for human habitation, agriculture, and commerce, and to define better wetland/buffer zones to protect amphibian communities. Unfortunately, there is a dearth of information on habitat use for most amphibians due to their relatively small size and secretive nature. Efforts to track individuals using radio-isotopes began decades ago (Madison and Shoop 1970), and other techniques have followed, including fluorescent powder, spool tracking, radiotelemetry, and harmonic radar, but for most amphibians details on habitat use through the annual cycle are still lacking. The recent surge in radiotelemetry use is providing new insights into habitat use and movements, but only a few studies have addressed whether the data might reflect transmitter transport, attachment, implantation, or the periodic disturbance of the investigator.

Favorable outcomes in the use of radiotelemetry require having a clear purpose for the study, obtaining reliable equipment and knowing how to use it, choosing the transmitter and external/internal tagging technique that best fit the need and the animal chosen for study, having the knowledge and skills necessary to use the attachment or implant techniques, anticipating the potential indirect and direct impacts of the radiotransmitter on the individual during and after tracking, and planning in advance how to analyze the anticipated data set. In this chapter, we discuss our experiences with amphibians to reduce duplication of effort and trauma to the animals studied, and to accelerate technical advances. We do not report all radiotelemetry studies, nor do we repeat the content of earlier reviews (see Heyer *et al.* 1994), but focus instead on more recent studies and models for future application of radiotelemetry procedure.

11.2 Equipment

The basic radiotelemetry equipment for hand-tracking amphibians includes radiotransmitters, a portable radio receiver, an antenna, and a cable. There are many international vendors of telemetry equipment but no single authority that has direct experience with all products and applications. Our advice is to rely on the recommendation of experienced users having the broadest biological and technical experience, and then choose the most reliable equipment available.

11.2.1 Receivers and antennas

The complexity and price of a receiver will vary with the scale of the study and the types of information being electronically recorded. There are basic receivers, such as the R-1000 (Communications Specialists), which can be programmed to 999 different animal frequencies between 148 and 174 MHz. Wider frequency ranges occur in more complex receivers, such as the ATS R5400S (Advanced Telemetry Systems), which can also upload information from data loggers in the transmitters. The cost of good-quality receivers is between US$700 and $2500. The small size of most amphibians usually restricts the transmitter size and application to the use of less expensive receivers like the R-1000.

The most commonly used antenna is a three- or four-element, hand-held Yagi antenna. Larger, five- (or more) element Yagi antennas are typically mounted on a vehicle for tracking wide-ranging animals, and give slightly better direction and distance information (Heyer *et al.* 1994). The main disadvantage of the five-element antenna in hand-tracking amphibians is that its larger size, necessitated by the fixed spacing between the elements for maximum frequency sensitivity, makes it more difficult to use in thick undergrowth. There are also H and loop antennas, but these somewhat reduce signal range/resolution. When ordering an antenna, tell the manufacturer what frequency range will be used so the antenna can be set up for optimal reception. Antennas typically range from US$100 to $300. We recommend a fold-up, three-element antenna for hand-held applications, and always carrying an extra antenna cable in case one fails in the field.

11.2.2 Transmitters

The types and sizes of transmitters, transmitter antennas and batteries vary, but the options grow fewer with smaller animals. In general, the weight of the transmitter/battery/attachment device should not exceed 10% of body weight on land (see Heyer *et al.* 1994), it should be neutrally buoyant in water, and the detection range should be at least 30–50 m line-of-sight. The smaller the transmitter/battery unit required to remain under the 10% weight rule, the more one has to compromise signal distance, lifetime, or pulse rate. There are three trade-offs in selecting a transmitter/battery combination at a given weight: (1) the greater the pulse rate (from about 50 beats/min to continuous), the more quickly and reliably you can determine signal direction, but the shorter the life of the transmitter/battery unit, (2) the larger (more powerful) the transmitter, the greater the detection distance, but the shorter the life of the transmitter/battery unit, and (3) the longer the transmitter antenna, the greater the detection distance, but the greater the impact on the animal. One should begin by determining the weight of the completed unit (including the waistband if used), and then choose the transmitter type/battery/attachment option that best serves the purpose of the study. In addition, in choosing transmitter frequencies one should avoid CB frequencies to ensure that there is no interference at the field site, especially near major highways or commercial properties. Finally, one may want to choose a vendor (e.g. Holohil Systems, Carp, Ontario, Canada) that will refurbish transmitters with spent batteries for a significant cost reduction.

11.2.2.1 External transmitters

In anurans, the transmitter is attached to a waist belt made out of several materials, few of which are satisfactory (Goldberg *et al.* 2002; Weick *et al.* 2005). Waist-belt trauma is frequent when the belts are attached snug enough so as not to slip off (Weick *et al.* 2005). External attachment has been achieved in cryptobranchid salamanders by sutures to the skin (Okada *et al.* 2006), but we agree with a similar study in hellbenders that doesn't recommend this procedure (Blais 1996).

Besides causing lesions and open wounds, especially in frogs with more delicate skin than toads, waist belts may cause trauma and death when the attachment bands or antennas become snagged in vegetation (e.g. McAllister *et al.* 2004). The whip antenna alone would likely affect movement in the thick grass and grass runways used by green frogs during summer foraging (Lamoureux *et al.* 2002). Slippage off the animal is also common, either following seasonal weight loss or being pulled off after snagging on vegetation (Rathbun and Murphey 1996; McAllister *et al.* 2004; Baldwin *et al.* 2006; Rittenhouse and Semlitsch 2007). Although some of these problems may be minimal over short time periods, many animals whose signals have failed or whose movements carry them outside the range of the tracking equipment will end up wearing these belts for the rest of their lives with increasing fitness reduction (e.g. weight loss, lower growth) and risk of injury/death. There are no known records to our knowledge of a transmitter-belt complex being worn successfully by an anuran spanning two breeding seasons. Even in one comprehensive and exceptionally well-documented study, including up to four re-beltings per frog, no mention of skin trauma or potential negative effect of transmitter belts was reported, and yet only five of 43 frogs had been tracked for more than 8 weeks (mean 25.6 days/frog) between 16 April and 12 November (Baldwin *et al.* 2006). Just one frog had its transmitter removed at the end of tracking, 18 had lost signals, 12 had been depredated, nine had "slipped out" of their waist bands, and three were found dead with no sign of predation. There is room here, and in essentially all other transmitter-belt studies lasting for more than a few days, for extensive direct and indirect negative effects of external belt attachment. It is prudent to consider any amphibian telemetry study relying on waist belts to be "long-term" unless the belts are removed after a week or so, or a release mechanism is built into the units. Benefits of waist bands include freedom from surgery, and being able to release an animal within moments of capture and transmitter attachment.

11.2.2.2 Internal transmitters

Whereas transmitter belts likely have value in certain circumstances (e.g. Fukuyama *et al.* 1988), preferably following laboratory-validated safety tests, we recommend implanted units for most amphibians, as have others (Madison 1997; Spieler and Linsenmair 1998; Lamoureux and Madison 1999; Lamoureux *et al.* 2002; Lemckert and Brassil 2003; Johnson 2006; Rittenhouse and Semlitsch 2006; McDonough and Paton 2007; Peterman *et al.* 2008). The implanted transmitter/battery unit has various antenna strands or coils embedded in the potting material to make a smooth-surfaced implant package. Alternatively, Jehle and Arntzen (2000) guided the 3–5 cm antenna out through the body wall (to increase signal distance reception) and sutured it to the body surface, an application we do not recommend. The potting material is an impervious coating of epoxy, dental acrylic, Elvax paraffin, beeswax, heat-shrink plastic, or non-reactive butyl rubber (Heyer *et al.* 1994). Care should be taken to use potting substances that do not elicit an immune response or release toxins upon implantation, which in the past were thought to stimulate cyst formation around the transmitter. Elongated implants should not have a length more than the diameter of the animal's greatest body or head width, because such units have caused an animal to get permanently stuck turning around in a narrow tunnel or passage (Madison 1997).

One variation of the "internal placement" of a radiotransmitter is down the esophagus as a gastric pill (Oldham and Swan 1992; Blais 1996; Schabetsberger *et al.* 2004), which in hellbenders were passed naturally after between 16 and

25 days, allowed short-term movement records, were recovered without the necessity of salamander recapture, and provided the guarantee of no long-term burden (Blais 1996). With this application, sizes just large enough to force regurgitation or small enough to be passed completely through the digestive system need to be considered carefully (Schabetsberger *et al.* 2004).

11.3 Surgical techniques

Surgery on amphibians requires knowing how to enter the coelomic cavity and properly close the wound. Individuals performing surgery should have training in the company of someone with previous surgical experience on amphibians. Amphibian skin, especially for most salamanders and smaller frogs, is much more delicate than mammalian skin, and has much slower healing rates in cool/cold environments. Transmitter collar and implant techniques have been developed for field studies of small mammals (see references in McShea and Madison 1992), but these techniques are inappropriate for, and even detrimental to, amphibians. The latest surgical advances for amphibians are incorporated in our recommendations below.

11.3.1 Surgery

Transmitters are typically implanted into the coelomic cavity. Some studies have implanted transmitters under the skin (Blais 1996; Lemckert and Brassil 2003), but sloughing of the transmitter through the skin (without edema or infection) can occur when using this technique (Blais 1996). Lumps on the body also promote abrasion/injury and suture failure as the animal moves through restrictive environments (Weick *et al.* 2005). We do not recommend subcutaneous transmitter implants for amphibians.

The following surgical procedure with minor variations has been used successfully (Colberg *et al.* 1997; Madison and Farrand 1998; Faccio 2003; Crook and Whiteman 2006; Rittenhouse and Semlitsch 2006; Cecala *et al.* 2007). For coelomic implants, all instruments and transmitters should be sterilized, such as in 95% ethanol followed by distilled water rinse and drying. For anesthetization, we recommend near total submersion in MS-222 (3-aminobenzoic acid ester methanosulfate salt) diluted with distilled water to a 0.25% solution and buffered to pH 7.0 with sodium bicarbonate (this concentration is specifically for adult *Ambystoma tigrinum*). The concentration should be adjusted to animal size, species, and habitat condition (see Lowe 2004; Crook and Whiteman 2006; Peterman and Semlitsch 2006), so that it takes about 10 min or more for the loss of the "righting" response or reaction to the poke of a probe. More rapid anesthesia, because either of higher temperatures or concentrations, increases the risk of death (Weick *et al.* 2005). Animals anesthetized too rapidly, or those over-anesthetized, should be rinsed continuously under cold tap water for at least 60 s to promote recovery. Animals should also be rinsed briefly under cool tap water immediately after anesthesia but before surgery to minimize the mild anticoagulant effect of MS-222 in the surgical field (Blais 1996). Benzocaine anesthetic has also been used (Crook and Whiteman 2006; Cecala *et al.* 2007), but MS-222 is preferred due to its wider margin of safety. For surgery, damp toweling placed under and over the animal, except for the surgical field, will reduce drying. A longitudinal incision should be made in the ventral posteriolateral abdominal wall with cutting iris scissors just long enough to insert the transmitter without tearing the tissue (Figure 11.1). Non-absorbable, braided, polyviolene sutures (5–0), applied by a 3/8 circle, reverse-cutting needle and secured by a standard surgeon's knot are recommended for wound closure. Other non-absorbable sutures are not recommended due to the lack of flexibility in the nylon. Just one layer of sutures is needed for most salamanders because the muscles are bonded to the skin. Otherwise, and for anurans in general, two layers of sutures are required, one for the inner musculature layer and the second for the skin (see Goldberg *et al.* 2002). A maximum of 1–1.5 mm spacing between all sutures is recommended to completely close the inner/outer incisions for prolonged healing and to put less stress on any single suture. Investigators routinely use too few sutures and consequently observe more dehiscence. Sutures drawn too tightly can also lead to tissue necrosis, severance, and dehiscence. Finally, some studies have used dissolvable sutures (Peterman *et al.* 2008), which we discourage, especially when healing is slowed, even for up to 6 months, during cooler times of the year (Colberg *et al.* 1997).

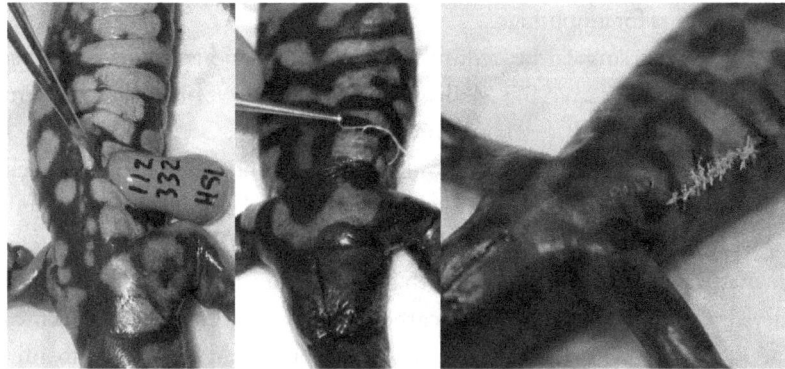

Fig. 11.1 Implanting of an internal transmitter into an eastern tiger salamander (*Ambystoma tigrinum tigrinum*). Photographs by Valorie R. Titus.

A final precaution regarding surgery involves gravid females. Amphibian eggs are hydrophilic, and exposing coelomic eggs to any water during implant surgery can cause coelomic egg-mass expansion and difficulty in suturing the wound closed. Too few sutures, or a permanent opening into the body cavity to accommodate an antenna, can also permit some water entry and possible injury or death. We also do not recommend implant surgery on females that will soon undergo amplexus. The physical stress on the sutures, even in anurans with generally tougher skins, may cause dehiscence.

11.3.2 Recovery and healing

The surgical field should not be treated before or after surgery with ethanol or other strong antiseptics, which will kill the animal's natural biota and promote bacterial/fungal infection soon after field release in a soiled environment. Probiotics have not been developed for amphibian skin. In addition, since the skin of amphibians is thin and permeable, ethanol or other antiseptics may be absorbed directly into the bloodstream and cause physiological stress, especially in aquatic organisms (see Crook and Whiteman 2006). Upon completion of surgery and rinsing in clean tap water, the animals should be placed under damp toweling or gauze for recovery. Animals should revive within 1 h and show full recovery in 3 h. Although complete healing may take several weeks when temperatures fall below 15°C, release soon after recovery from the anesthetic does not appear to have negative effects on survival or movements and is most likely to return the animal to its pre-capture activity (Madison 1997; Spieler and Linsenmair 1998; Johnson 2006; Rittenhouse and Semlitsch 2006; McDonough and Paton 2007; Peterman *et al.* 2008).

11.4 Tracking procedures

Radiotracking technique takes practice to glean the most information from a signal while minimizing any negative observer effect on the animal and its environment. Simply stated, a novice quickly becomes mesmerized by the signal and unaware of possible habitat disturbance and the risk of trampling the animal in trying to get an exact position, whether tracking in the water or on land. In this section, we discuss strategies and precautions relating to animal release, search, and data collection.

11.4.1 Animal release

Animals should normally be released under cover at the capture site at the beginning of the next active period, usually just after dusk the following evening. Just prior to release, check for transmitter functioning one last time, and tune the receiver to the strongest signal output of the transmitter. Tuning on an additional channel is also suggested (Baldwin *et al.* 2006). Re- or multi-programming the receiver to an exact frequency setting improves the chance of re-locating a lost signal. We also suggest that the investigator stay nearby after release and obtain positions every 30–60 min for several hours to determine the initial movement headings, which are generally straight and can often be used to find a lost signal the next day. Searching for a lost signal without any idea which direction to search can be time-consuming and frustrating.

11.4.2 Locating signal source

In order to obtain the best, most rapid directional information when searching for, or approaching, a distant signal, hold the antenna parallel to the ground and continually sweep it to the signal nulls right and left of the loudest signal intensity, rather than trying to aim the shaft in the direction of the loudest signal toward the animal. Signal information can be used to estimate the distance between the observer and the animal. When the initial signal is detected, adjust the volume on the receiver to be barely audible and continually re-adjust to this low volume as you approach the animal. At greater distances from the animal, white noise will dominate the signal background. The closer you are to the animal, the purer the signal will be. Once a relatively noise-free signal is obtained, we recommend a local triangulation procedure (about 3 m from the animal) to finally fix the animal's position (Mineau and Madison 1977). This procedure is also used to locate amphibians in a pond, stream, or up a tree (e.g. Blais 1996; Madison 1998); it involves the observer moving on a tangent to the signal and estimating two independent headings that define approximately a 90° angle of convergence to the signal source. Local triangulation minimizes the chance of disturbing the animal and/or its microenvironment, which might induce subsequent movement, and it reduces the chance of trampling the animal. To get an exact location, one should approach the last 3 m by holding the antenna shaft perpendicular to the ground (and the elements perpendicular to the signal source) and continually move the antenna tip away from your feet and up through an arc of about 135°, keeping the loudest signal in front of you. This procedure can alternate with a side-to-side scan to refine the directional heading. Soon the animal will become visible, or a location on the ground/leaf litter above the animal can be identified to within 10 cm (Madison and Farrand 1998; Faccio 2003). Avoid disturbing the animal unless it needs to be retrieved for weighing, replacing a transmitter, or checking on its condition. We generally do not search for an unseen animal until it occurs in exactly the same location for multiple position checks, especially following rain when animals are most likely to move. Removing an animal from a subterranean retreat should follow a circular excavation procedure to avoid injuring the animal and to enhance the description of the refuge (Madison and Farrand 1998).

Several factors can influence signal direction and detectability. Animals near or under metal objects, boulders, bluffs, small ridges and depressions, utility poles, or fences, as well as underground or underwater, will be more difficult to locate because of signal deflection and dampening. Madison and Farrand (1998) experienced about a 1 m distance reduction for each centimeter of transmitter depth, but this dampening can vary with different equipment. The reduction for water depth is much less than for soil. If an arboreal animal is being tracked, or if the observer and/or animal is on higher ground, the normal signal range will be increased and the animal will sound closer than it actually is. An observer looking for a lost signal should raise the antenna high overhead and, where possible, move to higher ground to improve detection distance.

11.4.3 Data collection

Most investigators use hand-held equipment to obtain positions, which minimally requires a GPS unit, flagging, a topographical map, and a waterproof field data book. Typically, 10–20 animals can be tracked concurrently, but this varies with topographic features, understory thickness, how dispersed the animals are, and how much data are taken during a position check. GPS data are less accurate on steep slopes and in ravines. Flagging should be used to mark each fix or key locations with the animal ID, date, and time of observation written in ballpoint pen on the flagging. A wide variety of environmental and refuge data can also be recorded for terrestrial or aquatic locations using data loggers, weather stations, and local field and laboratory techniques (e.g. Heyer *et al.* 1994; Madison 1997; Madison and Farrand 1998; Faccio 2003; Watson *et al.* 2003; Baldwin *et al.* 2006; Rittenhouse and Semlitsch 2007).

11.5 Analysis of movement data

Telemetry generates extensive good or poor data, depending on whether the study objectives, sampling methods and statistical procedures were planned in advance. There are several resources on analyzing directionality, space use and movement data, depending on the study objectives (cited in White and Garrott 1990; Kie *et al.* 1996; Madison 1997; Millspaugh and Marzluff 2001; Blomquist and Hunter 2007), including applications of advanced Geographic Information System (GIS) software, such as ArcGIS (Environmental Systems Research Institute, Redlands,

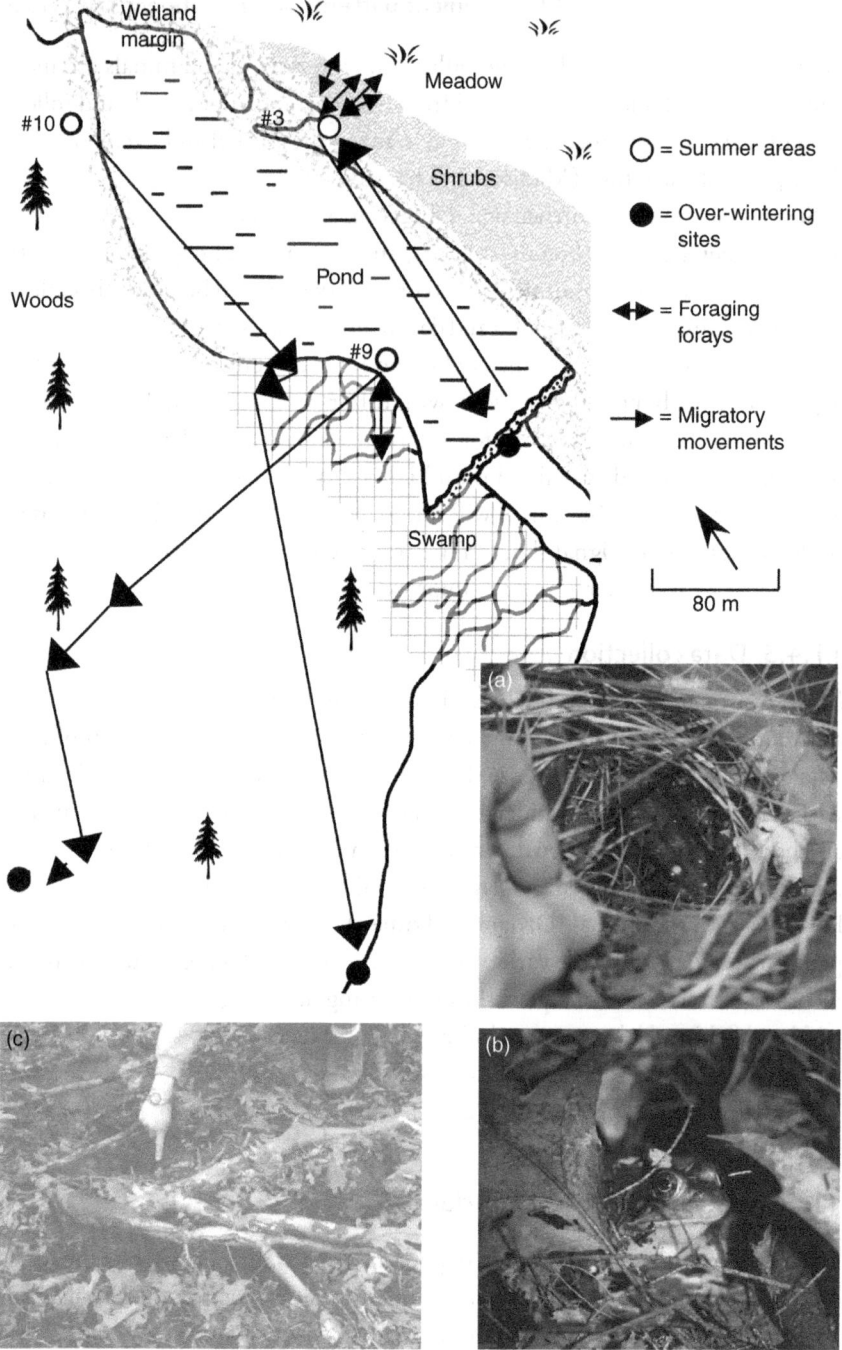

Fig. 11.2 Tracks for three radio-implanted northern green frogs (*Rana clamitans*), Broome County, NY, USA, showing vegetation types, habitat photographs, and major movement/refuge details. Wetland margins were structurally complex, consisting of *Leersia*, *Polygonum*, *Sparganium*, *Scirpus*, and *Carex*. The swamp had an

CA, USA). The latter permits detailed maps of space use to be overlaid with vegetation and soil maps for comprehensive studies of habitat use.

A common mistake in quantifying space use is rushing to label and quantify the "home range" for an amphibian, which should include all behaviors relating to reproduction and survival (Burt 1943), or calculating the average or maximum distance moved from a breeding pond without qualifying that the estimates are based on a small subset of individuals over a few weeks or months. Lacking for the most part is information on year-round movements in structurally different habitats. Results are seldom scaled or weighted according to the duration of a telemetry record, for example averaging a one-week distance record with a 4-month distance record can underestimate the potential for a species to move long distances (Madison and Farrand 1998). The potential for individuals to occupy widely spaced and diverse habitats for different purposes through the annual cycle must be kept in mind, and a good but preliminary descriptive model for these movements is that for green frogs (Figure 11.2), as described in Lamoureux and Madison (1999) and Lamoureux et al. (2002). More quantitative indices of multi-seasonal space use also occur (e.g. Spieler and Linsenmair 1998; Watson et al. 2003; Baldwin et al. 2006).

The most common home-range representation is the minimum convex polygon, and while this method may give fairly accurate short-term "daily"

Fig. 11.2 Continued

understory of ferns, various thick grasses, Scirpus, Juncus, and Carex, and a canopy consisting largely of Acer rubrum. The deciduous forest consisted of Fagus, Betula, Acer, Tsuga, and Quercus. Subject numbers at track origins are consistent with Lamoureux and Madison (1999) and Lamoureux et al. (2002). Frog #3 (a 31 g female released 12 September 1995) occupied a wetland margin, made four back-and-forth foraging forays to self-made forms or rodent runways in dense grasses during October (photograph a), migrated 240 m on 3 November to an over-wintering site in a large beaver dam, and returned to its release (breeding) location in April 2006. Frog #9 (a 45 g male released 10 August 1996) made one out-and-back foray into the swamp, migrated across the swamp and 240 m up a forested hillside on 8 September 1996, made additional movements with a succession of leaf litter refuges (e.g. photograph b) 200 m further up the hillside to its 19 October 1996 over-wintering location consisting of a small spring with approximately 10 cm-diameter tunnels under a large root (photograph c), and finally was recaptured there May 1997 prior to the spring migration. Frog #10 (a 63 g male released on 26 August 1996) moved 250 m on 30 September from its wetland/woodland-edge habitat to the swamp, moved 40 m further into the swamp on 3 October 1996, and then was lost until relocated on 11 October 1996 360 m into forested habitat buried in a small, seasonal stream under flat stones and small pebbles, where it over-wintered. Photographs by Victor S. Lamoureux.

or "activity" range estimates for a mobile individual (Madison 1985; McShea and Madison 1992), many migratory amphibians make only a few large movements with most activity being in small subterranean food or shelter patches strung out somewhat longitudinally through upland habitat, such as in *A. tigrinum* (Figure 11.3). Similar observations have been recorded by others

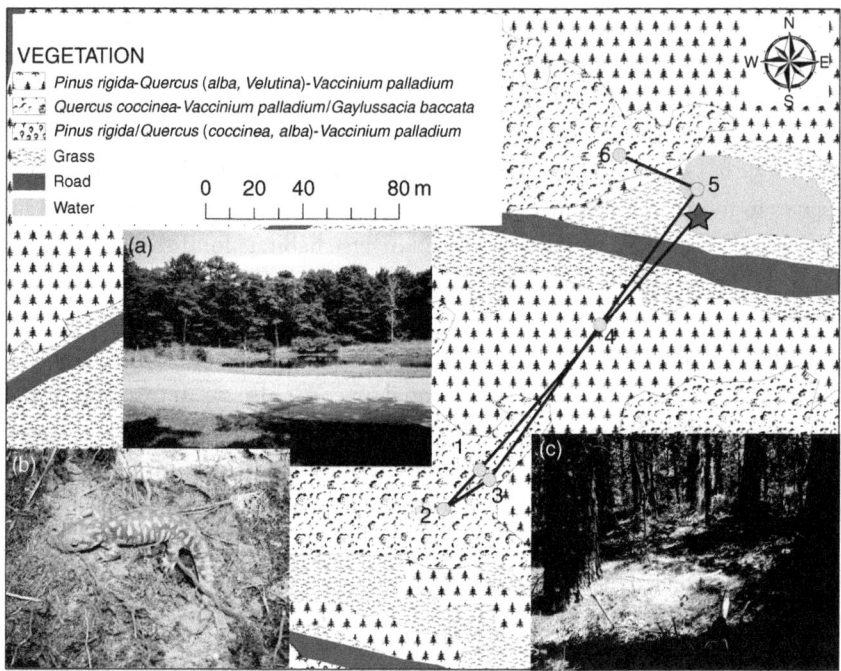

Fig. 11.3 Year-long telemetry track of an implanted 33 g male *Ambystoma tigrinum*, Suffolk County, NY, USA, showing vegetation types, habitat photographs, and major movement/refuge details as follows: Star (30 March 2005): initial position following breeding in the vegetation at the edge of the pond. Point 1 (4 April 2005): initial movement 137 m from pond edge into small self-made borrow 8 cm from surface. Point 2 (21 June 2005): movement 21 m further from pond into a small self-made borrow 10 cm from surface. Point 3 (2 July 2005): movement 23 m back toward pond into a small mammal borrow 8 cm from surface. Point 4 (1 October 2005): movement 78 m toward pond into a small mammal borrow 12 cm from surface. Point 5 (4 December 2005): movement 68 m back into the breeding pond the following season. Point 6 (15 March 2006): movement 33 m out of pond; later only the transmitter was found on the surface (possible predation). All movements were associated with rain events. Photographs include (a) grassland and road traversed at least twice during migrations, (b) salamander near burrow at point 3, and (c) typical woodland/understory at points 1, 2, 3, and 6. Photographs by Valorie R. Titus.

(Fukuyama *et al.*1988; Madison 1997; Madison and Farrand 1998; Lemckert and Brassil 2000; Faccio 2003; Lemckert and Brassil 2003; Rittenhouse and Semlitsch 2006, 2007). If an estimate of the home range or the buffer zone for a species is required, we recommend caution in whatever model is used, and a clear statement concerning the limitations of such values so they are not misinterpreted by others to represent the year-round resource needs or habitat buffer for a sustainable amphibian population (see McDonough and Paton 2007).

11.6 Validation of telemetry procedures

Radiotelemetry and drift fences with pitfall traps, although standard techniques for monitoring amphibian movements, have not been fully examined for shortcomings (see Heyer *et al.* 1994; Madison and Farrand 1998). The following data suggest that telemetry can be used to monitor amphibians with reduced impact on survival and reproduction, at least in short-term studies.

11.6.1 Internal condition and mass

In external attachment studies, no effect on body mass was shown for two free-ranging toad species, since carrying transmitters of different weight or over different time periods had no differential effect on body mass (Indermaur *et al.* 2008). In another study, 16 frogs (*Rana aurora*) gained an average of 5.7 g after 2 months of carrying an external transmitter on a beaded chain (Rathbun and Murphey 1996), but how much the frogs might have weighed without carrying transmitters wasn't known. In implant studies during the post-breeding season, frogs (*Rana clamitans*) with transmitter implants gained the same weight as frogs without implants (Lamoureux *et al.* 2002), and weight and growth gains also occurred before and during the breeding season in *Hoplobatrachus occipitalis* (Spieler and Linsenmair 1998). However, *Hyla versicolor* showed a 9% average weight loss (1.1 g) following 3 weeks of carrying an implant, and other individuals with long-term implants coated with dental acrylic showed fibrous connective tissue cysts or strands around the implants, but neither of these effects was considered detrimental (Johnson 2006). Similar connective tissue was associated with toad implants and was thought to anchor the implants away from body organs (Gray *et al.* 2005). Failure to form such cysts may result in cloacal expulsion, as observed in leopard frogs (Weick *et al.* 2005). In a laboratory study of male spotted salamanders (*Ambystoma maculatum*), implanted animals showed significantly depressed feeding compared to non-implanted salamanders for 2 weeks after surgery, but no significant decrease in feeding behavior or body weight relative to the controls for the remaining 8 months of the study (Madison 1997). Collectively, these data show no major effects on body weight or internal condition.

11.6.2 Movements

Little is known about the full extent of habitat use in free-ranging amphibians other than that obtained using radiotelemetry, and so it is difficult to test whether transmitters might be affecting movements and habitat use. In field enclosures, *Rana sylvatica* and *Rana pipiens* with and without belted transmitters with 15 cm antennas showed some species-specific differences in behavior in response to simulated predation threat, prompting some caution in the use of the attached units studied (Blomquist and Hunter 2007). Studies of a rainforest frog (*Litoria lesueuri*) in laboratory enclosures showed no difference in the distance or frequency of movement before and after transmitter attachment, except for a slight reduction the first day after attachment (Rowley and Alford 2007). In nature, casual observations on non-implanted *A. maculatum* salamanders during tracking of implanted individuals revealed comparable movements, occurrence, and mortality (Madison 1997). Cross-study comparisons show that wood frogs monitored by drift fences (Vasconcelos and Calhoun 2004) emigrated comparable distances to implanted individuals (Baldwin *et al.* 2006; Rittenhouse and Semlitsch 2007). In addition, implanted tiger salamanders showed courtship movements in a breeding pond similar to those observed for non-implanted salamanders in captivity (Madison 1998). The above studies, although quite incomplete, suggest no major effects of implanted or attached transmitters on short-term movements.

11.6.3 Reproduction

Reproduction is seldom verified for amphibians carrying transmitters, although there are exceptions. Six gravid female frogs (*Buergeria buergeri*) were given waist-band transmitters (with a 3 cm whip antenna) and monitored along a stream for up to 8 days during successful mating and oviposition (Fukuyama *et al.* 1988). In implant studies, Johnson (2006) recorded a female *Hyla versicolor* ovipositing a year after carrying an implant and undergoing two implant surgeries. Tiger salamanders showed normal breeding activity soon after implantation (Madison 1998), and an implanted male spotted salamander returned 190 m to the same location at a breeding pond after carrying an implant in upland habitat for a year (Madison 1997). Further studies on possible long-term reproductive effects are obviously needed.

11.6.4 Injury and survivorship

For external attachment, the recurrent problem has been skin injury or getting entangled in vegetation or caught in tight spaces, perhaps more so for frogs than toads (e.g. Rathbun and Murphey 1996; Weick *et al.* 2005; Indermaur *et al.* 2008), which increases with track duration (Rittenhouse and Semlitsch 2007). For implant procedures, the main concern has been the somewhat high rate of recovery of bare transmitters, suggesting a high population turnover rate due to predators (herons, shrews, snakes) and/or the cloacal expulsion of the implants. Predation is strongly supported (Madison 1997; Madison and Farrand 1998; McDonough and Paton 2007), but recent evidence also supports expulsion (Weick *et al.* 2005).

Little is known about the effects of transmitter implants or belts on the long-term survival of amphibians. And even during short-term tracking where predation is the cause of death, the attached/implanted transmitters may predispose animals to predation, as cautioned by Blomquist and Hunter (2007). Clearly, more data are needed to evaluate effects on long-term survivorship.

11.7 Conclusions

In order to fully understand amphibian movement patterns and habitat use, radiotelemetry is an indispensable tool, despite some shortcomings in methodology. Waist bands for external transmitters with short whip antennas (less than ≈3 cm), or antennas built into the waist band rather than trailing behind the animal, are tentatively recommended for short-term tracking of anurans under conditions where the animals can be inspected periodically and the waist bands removed at the end of study (e.g. Fukuyama *et al.* 1988). However, long-term tracking using waist-band transmitters is not recommended because of the increased likelihood of skin trauma and getting snagged or trapped by objects in post-breeding and over-wintering refuges. External transmitters and antennas sutured to amphibians are not recommended.

Transmitters with antennas coiled in the potting material are recommended for short-term studies as gastric pills (e.g. Schabetsberger *et al.* 2004), or for extended studies as coelomic implants in both anurans (e.g. Lamoureux and Madison 1999) and salamanders (Madison 1997; Madison and Farrand 1998; Faccio 2003; Rittenhouse and Semlitsch 2006; McDonough and Paton 2007). Especially needed, however, is information on potential weight effects on fitness from the long-term transport of transmitter implants.

The future challenge in the use of radiotelemetry in amphibians is threefold: to minimize duplication in method development and unnecessary trauma to amphibians by fostering apprenticeship prior to use by an investigator, to educate journal editors and academic units of the appropriate procedures, and to encourage publication of data on injury, mortality, or other problems resulting from responsible radiotelemetry procedure, a notable example being Weick *et al.* (2005).

11.8 References

Baldwin, R. F., Calhoun, A. J. K., and deMaynadier, P. G. (2006). Conservation planning for amphibian species with complex habitat requirements: A case study using movements and habitat selection of the Wood Frog *Rana sylvatica*. *Journal of Herpetology*, **40**, 442–53.

Blais, D. P. (1996). *Movement, Home Range, and Other Aspects of the Biology of the Eastern Hellbender (Cryptobranchus alleganiensis alleganiensus): a Radiotelemetric Study.* Master's thesis, State University of New York at Binghamton, Binghamton, NY.

Blomquist, S. M. and Hunter Jr, M. L. (2007). Externally attached radio-transmitters have limited effects on the antipredator behavior and vagility of *Rana pipiens* and *Rana sylvatica*. *Journal of Herpetology*, **41**, 430–8.

Burt, W. H. (1943). Territoriality and home range concepts as applied to mammals. *Journal of Mammalogy*, **24**, 346–52.

Cecala, K. K., Price, S. J., and Dorcas, M. E. (2007). A comparison of the effectiveness of recommended doses of MS-222 (tricaine methanesulfonate) and Orajel (R) (benzocaine) for amphibian anesthesia. *Herpetological Review*, **38**, 63–6.

Colberg, M. E., Denardo, D. F., Rojek, N. A., and Miller, J. W. (1997). Surgical procedure for radio transmitter implantation into aquatic, larval salamanders. *Herpetological Review*, **28**, 77–8.

Crook, A. C. and Whiteman, H. H. (2006). An evaluation of MS-222 and benzocaine as anesthetics for metamorphic and paedomorphic tiger salamanders (*Ambystoma tigrinum nebulosum*). *American Midland Naturalist*, **155**, 417–21.

Faccio, S. D. (2003). Postbreeding emigration and habitat use by Jefferson and spotted salamanders in Vermont. *Journal of Herpetology*, **37**, 479–89.

Fukuyama, K., Kusano, T., and Nakane, M. (1988). A radio-tracking study of the behaviour of females of the frog *Buergeria buergeri* (Rhacophoridae, Amphibia) in a breeding stream in Japan. *Japanese Journal of Herpetology*, **12**, 102–7.

Goldberg, C. S., Goode, M. J., Schwalbe, C. R., and Jarchow, J. L. (2002). External and implanted methods of radiotransmitter attachment to a terrestrial anuran (*Eleutherodactylus augusti*). *Herpetological Review*, **33**, 191–4.

Gray, M. J., Miller, D. L., and Smith, L. M. (2005). Coelomic response and signal range of implant transmitters in *Bufo cognatus*. *Herpetological Review*, **36**, 285–8.

Heyer, W. R., Donnelly, M. A., McDiarmid, R. W., Hayek, L.-A. C., and Foster, M. S. (eds) (1994). *Measuring and Monitoring Diversity: Standard Methods for Amphibians*, Smithsonian Institution Press, Washington DC.

Indermaur, L.B., Schmidt, B.R., and Tockner, K. (2008). Effect of transmitter mass and tracking duration on body mass of two anuran species. *Amphibia-Reptilia*, **29**, 263–9.

Jehle, R. and Arntzen, J. W. (2000). Post-breeding migrations of newts (*Triturus crista-tus* and *T. marmoratus*) with contrasting ecological requirements. *Journal of Zoology, London*, 251, 297–396.

Johnson, J. R. (2006). Success of intracoelomic radiotransmitter implantation in the tree-frog (*Hyla versicolor*). *Lab Animal*, 35, 29–33.

Kie, J. G., Baldwin, J. A., and Evans, C. J. (1996). CALHOME: a program for estimating animal home ranges. *Wildlife Society Bulletin*, 24, 342–4.

Lamoureux, V. S. and Madison, D. M. (1999). Overwintering habitats of radio-implanted green frogs, *Rana clamitans*. *Journal of Herpetology*, 33, 430–5.

Lamoureux, V. S., Maerz, J. C., and Madison, D. M. (2002). Premigratory autumn for-aging forays in the green frog, *Rana clamitans*. *Journal of Herpetology*, 36, 245–54.

Lemckert, F. and Brassil, T. (2000). Movements and habitat use of the endangered giant barred river frog (*Mixophyes iteratus*) and the implications for its conservation in timber production forests. *Biological Conservation*, 96, 177–84.

Lemckert, F. and Brassil, T. (2003). Movements and habitat use by the giant burrowing frog, *Heleioporus australiacus*. *Amphibia-Reptilia*, 24, 207–11.

Lowe, J. (2004). Rates of tricaine methanesulfonate (MS-222) anesthetization in rela-tion to pH and concentration in five terrestrial salamanders. *Herpetological Review*, 35, 352–4.

Madison, D. M. (1985). Activity Rhythms and Spacing. In R.H. Tamarin (ed.), *Biology of New World Microtus*, pp. 373–419. American Society of Mammalogists, Lawrence, KA.

Madison, D. M. (1997). The emigration of radio-implanted spotted salamanders, *Ambystoma maculatum*. *Journal of Herpetology*, 31, 542–51.

Madison, D. M. (1998). Habitat-contingent reproductive behaviour in radio-implanted salamanders: a model and test. *Animal Behaviour*, 55, 1203–10.

Madison, D. M. and Shoop, C. R. (1970). Homing behavior orientation and home range of salamanders tagged with tantalum-182. *Science*, 168, 1484–7.

Madison, D. M. and Farrand, L. (1998). Habitat use during breeding and emigration in radio-implanted tiger salamanders, *Ambystoma tigrinum*. *Copeia*, 1998, 402–10.

McAllister, K. R., Watson, J. W., Risenhoover, K., and McBride, T. (2004). Marking and radiotelemetry of Oregon spotted frogs (*Rana pretiosa*). *Northwestern Naturalist*, 85, 20–5.

McDonough, C. and Paton, P. W. C. (2007). Salamander dispersal across a forested land-scape fragmented by a golf course. *Journal of Wildlife Management*, 71, 1163–9.

McShea, W.J. and Madison, D.M. (1992). Alternative approaches to the study of small mammal dispersal: insights from radiotelemetry. In W. Z. Lidicker Jr and N. C. Stenseth (eds), *Dispersal: Small Mammal as a Model*, pp. 319–32. Chapman and Hall, New York.

Millspaugh, J. J. and Marzluff, J. M. (2001). *Radio Tracking and Animal Populations*, Academic Press, San Diego, CA.

Mineau, P. and Madison, D. M. (1977). Radio-tracking of *Peromyscus leucopus*. *Canadian Journal of Zoology*, 55, 465–8.

Okada, S., Utsunomiya, T., Okada, T., and Felix, Z. I. (2006). Radio transmitter attach-ment by suturing for the Japanese giant salamander (*Andrias japonicus*). *Herpetological Review*, 37, 431–4.

Oldham, R. S. and Swan, M. J. S. (1992). Effects of ingested radio transmitters on *Bufo bufo* and *Rana temporaria*. *Herpetological Journal*, **2**, 82–5.

Peterman, W. E. and Semlitsch, R. D. (2006). Effects of tricaine methanosulfonate (MS-222) concentration on anesthetization and recovery in four plethodontid salamanders. *Herpetological Review*, **37**, 303–4.

Peterman, W. E., Crawford, J. A., and Semlitsch, R. D. (2008). Productivity and significance of headwater streams: population structure and biomass of the black-bellied salamander (*Desmognathus quadramaculatus*). *Freshwater Biology*, **53**, 347–57.

Rathbun, G. B. and Murphey, T. G. (1996). Evaluation of a radio-belt for ranid frogs. *Herpetological Review*, **27**, 187–9.

Rittenhouse, T. A. G. and Semlitsch, R. D. (2006). Grasslands as movement barriers for a forest-associated salamander: migration behavior of adult and juvenile salamanders at a distinct habitat edge. *Biological Conservation*, **131**, 14–22.

Rittenhouse, T. A. G. and Semlitsch, R. D. (2007). Postbreeding habitat use of wood frogs in a Missouri oak-hickory forest. *Journal of Herpetology*, **41**, 645–53.

Rowley, J. J. L. and Alford, R. A. (2007). Techniques for tracking amphibians: the effects of tag attachment, and harmonic direction finding versus radio telemetry. *Amphibia-Reptilia*, **28**, 367–76.

Schabetsberger, R., Jehle, R., Maletzky, A., Pesta, J., and Sztatecsny, M. (2004). Delineation of terrestrial reserves for amphibians: Post-breeding migrations of the Italian crested newt (*Triturus c. carnifex*) at high altitudes. *Biological Conservation*, **117**, 95–104.

Spieler, M. and Linsenmair, K. E. (1998). Migration patterns and diurnal use of shelter in a ranid frog of a West African savannah: a telemetric study. *Amphibia-Reptilia*, **19**, 43–64.

Vasconcelos, D. and Calhoun, A. J. K. (2004). Movement patterns of adult and juvenile *Rana sylvatica* (LeConte) and *Ambystoma maculatum* (Shaw) in three restored seasonal pools in Maine. *Journal of Herpetology*, **38**, 551–61.

Watson, J. W., McAllister, K. R., and Pierce, D. J. (2003). Home ranges, movements, and habitat selection of Oregon Spotted Frogs (*Rana pretiosa*). *Journal of Herpetology*, **37**, 292–300.

Weick, S. E., Knutson, M. G., Knights, B. C., and Pember, B. C. (2005). A comparison of internal and external radio transmitters with northern leopard frogs (*Rana pipiens*). *Herpetological Review*, **36**, 415–421.

White, G. C. and Garrott, R. A. (1990). *Analysis of Wildlife Radio-Tracking Data*, Academic Press, San Diego, CA.

12

Field enclosures and terrestrial cages

Elizabeth B. Harper, Joseph H.K. Pechmann, and

James W. Petranka

12.1 Introduction: amphibians in the terrestrial environment

Amphibians exhibit many life-history patterns and occur in biomes ranging from deserts and grasslands to boreal and tropical forests (Chapter 1). Most species have complex life cycles with a free-living aquatic larval stage and a non-gilled post-metamorphic stage. Of the more than 6000 extant species that have been described, almost all use terrestrial habitats during some or all of their life cycle. Common terrestrial specializations include arboreal, fossorial, surface-dwelling, and riparian species (Chapter 1).

The aquatic larval stage, if present, is typically much shorter than the post-metamorphic stage. The juveniles and adults of most species live in semi-aquatic or terrestrial habitats. Despite these facts, researchers have conducted surprisingly few studies on the terrestrial ecology of amphibians, especially pond-breeding amphibians (Pechmann 1995). Even some of the most basic questions—such as the extent to which population regulation occurs during the aquatic and terrestrial stages of the life cycle—are poorly resolved because of a shortage of information on density dependence during the terrestrial stages.

Conservation biologists have identified factors that are contributing to the global declines of amphibians (e.g. environmental pollutants, forest fragmentation, urbanization, and road construction), but the relative extent to which these are affecting the aquatic and terrestrial stages is often uncertain (Biek *et al.* 2002). Widespread gaps in knowledge about the terrestrial ecology of amphibians continue to hamper our ability to develop effective conservation strategies for the hundreds of species that are in long-term decline. Experiments using field enclosures can help resolve critical issues concerning the terrestrial ecology of amphibians and the ability of animals to tolerate environmental stressors.

Given that the use of enclosures is often expensive and time-consuming, why should a researcher use them? Enclosures allow researchers to expose animals to experimental treatments that test explicit hypotheses and to recapture animals to measure response variables. More importantly, field experiments maximize realism relative to laboratory or mesocosm studies and provide the strongest inferences as to whether the factors under investigation operate in nature (Hairston 1989; Morin 1998; Boone and James 2005).

When designing enclosure experiments, researchers must balance trade-offs in experimental design because it is impossible to simultaneously maximize realism, precision, and generality (Hairston 1989). Precision is maximized by creating replicates of experimental treatments that are as similar as possible to one another. Generality, which is the ability to extrapolate to a broad area or to a number of conditions, is maximized by placing replicate enclosures in as many locations or conditions as possible. In general, experimental designs that maximize generality compromise precision (and visa versa), and designs that maximize realism also compromise precision.

Optimally balancing these trade-offs depends on the questions that are being addressed. For example, a researcher who wishes to address how density affects the growth of juvenile amphibians may elect to maximize precision and reduce experimental error by placing enclosures at one site and in close proximity to one another (Altwegg 2003). In contrast, a researcher who wishes to determine the rate at which air-borne agricultural pesticides are accumulating in amphibians in a nearby mountain range may want to establish test sites at numerous locations to maximize generality.

In the sections that follow we provide examples of how terrestrial enclosures have been used to address questions about the ecology and conservation biology of amphibians. We also provide information on cage designs, add cautionary notes about experimental artifacts, and discuss trade-offs that should be considered when designing experiments.

12.2 What are the purposes of terrestrial enclosures?

Terrestrial enclosures have been used by amphibian ecologists for four general purposes, as follows.

1) To confine amphibians so that they can be subjected to an experimental treatment. The treatment can be the characteristics of the area itself (e.g. forested or non-forested; Chazal and Niewiarowski 1998; Todd and Rothermel 2006), or one applied by the researcher (e.g. population density;

Pechmann 1995; Harper and Semlitsch 2007). Enclosures allow these experiments to be conducted under realistic conditions in the field on organisms that could otherwise migrate from their assigned treatment.

2) To facilitate recapture of amphibians subjected to experimental treatments before they were placed in the enclosures, for example investigating the post-metamorphic carryover effects of larval exposure to pesticides (Boone 2005). In this case, enclosures permit evaluation of treatment effects under natural climatic conditions and diets, and perhaps under natural competition and predation regimes, depending on the type and size of the enclosures.

3) To facilitate recapture of dispersing or migrating amphibians to measure habitat choice and movement distances (Rothermel and Semlitsch 2002; Croshaw and Scott 2006). Enclosures in this case consist of walled corridors. Although these topics can also be addressed by tracking unenclosed individuals (Chapter 11), enclosures allow large groups of animals to be followed and recaptured. Enclosures can also confine choices and movements to particular habitats. For example, Croshaw and Scott (2006) examined the elevations marbled salamanders chose for nesting in enclosures in which the amount of cover was equalized across elevations.

4) To estimate terrestrial densities of amphibians (Regosin *et al.* 2003a, 2005). Enclosing an area containing naturally occurring amphibians limits their movement and can improve the accuracy of density estimates obtained by removal sampling or mark–recapture.

12.3 Defining the research question

The first step in designing an effective experiment using terrestrial enclosures is to clearly define the research question. This is essential in deciding: (1) which species to use, (2) the location, size, and number of enclosures, (3) methods of pen construction, (4) the number of amphibians per enclosure, (5) the frequency and methods of recapturing individuals, (6) the duration of the experiment, (7) the response metrics, and (8) how the data will be analyzed. The majority of research questions fall into two general categories: questions with habitat types as the experimental treatment and questions with treatments that can be randomly assigned to enclosures or to individuals within them.

12.3.1 Questions related to habitat types

If the research question examines the effects of specific habitat types, then the treatment consists of the conditions inside each enclosure and the surrounding conditions. Examples of habitat-focused research questions include studies of amphibian responses to clear-cutting (Chazal and Niewiarowski 1998; Todd and Rothermel 2006), vegetation density (Denton and Beebee 1994), forest edges (Marsh and Beckman 2004; Rothermel and Semlitsch 2006), and exotic plants (Maerz *et al.* 2005). Studies of this sort often do not permit researchers to randomly assign treatments to experimental units, and particular attention should be given to issues concerning spatial autocorrelation and pseudoreplication (Hurlbert 1984). To maximize generality, enclosures should be widely spaced. The trade-off is increased variability within treatments, meaning that increased replication may be necessary to achieve adequate statistical power.

Fig. 12.1 Small cages can be used to confine amphibians to specific microhabitats during short-term experiments.

Measuring habitat variables for use as covariates in later analyses can be useful in understanding the mechanisms underlying this within-treatment variability (e.g. aspect and surface temperature; Harper 2007). Densities of predators, competitors, and prey may also be important explanatory variables.

Research questions focused on habitat type have important implications for enclosure design and construction. Because the treatment includes the surrounding habitat as well as the conditions within the enclosure, realism may be improved by using mesh rather than solid materials for the enclosure walls. Mesh allows movement of air, water, nutrients, and small organisms in and out of enclosures and creates conditions within the enclosure that are more typical of the habitat type as a whole. Large enclosures (> 200 m^2) are more likely to incorporate most microhabitats available within a larger habitat type and to provide the range of choices that would be available to free-roaming individuals, thereby increasing realism. Large enclosures are also likely to support a more representative community including competitors, predators, and small mammals that dig burrows used by some amphibians (Regosin *et al.* 2003b). Conversely, if the research question focuses on microhabitats, small enclosures (Figure 12.1) can purposefully limit choice to determine the effects of specific microhabitats and their associated organisms on amphibian performance (Maerz *et al.* 2005; Rothermel and Luhring 2005).

12.3.2 Questions with treatments that can be assigned within enclosures

A broad range of research questions can be addressed using treatments that are randomly assigned to enclosures. Examples include manipulations of density (Cohen and Alford 1993; Pechmann 1995; Altwegg 2003; Harper and Semlitsch 2007), conditions during the larval stage (Boone 2005), and interspecific interactions (Southerland 1986; Denton and Beebee 1994; Price and Shields 2002). Because these questions do not focus on the surrounding habitat type, enclosures can be constructed within close proximity of one another, perhaps even sharing walls, while still achieving interspersion of treatments. If walls are shared, pens should be constructed of solid material so that the individual enclosures remain independent and are not influenced by treatments applied to neighboring pens (Chalcraft *et al.* 2005). Building pens in close proximity minimizes variability within treatments and increases precision; however, it compromises generality. A good compromise is to construct pens in blocks that are spaced widely apart, which allows extension of the results beyond a single site (Hurlbert 1984). One major advantage of questions that allow treatments to be assigned randomly within enclosures is that enclosure arrays built for one experiment can later be used for a range of other experiments (Figure 12.2).

Fig. 12.2 Terrestrial enclosure (1 m × 2 m) at the University of Missouri's Research Park in Columbia, MO, USA. These enclosures were first built to test the terrestrial fitness of hybrids in the *Rana pipiens* complex and have since been used in experiments focusing on ecotoxicology and density dependence. A hole dug in the center of each pen and covered with a board provides a moist refuge.

12.4 Constructing enclosures

Decisions involved in enclosure construction include the location, number, dimensions, and construction materials. Options are typically constrained by funds, labor, and the logistical constraints unique to each study site.

12.4.1 Location of enclosures

Enclosures built in close proximity to one another maximize precision and increase statistical power, but compromise generality. Generality is increased by widely spacing blocks of enclosures hundreds of meters or several kilometers apart; however, it is important to consider the time and resources required to build and monitor pens that are widely spaced. Enclosures can be positioned randomly, haphazardly, or uniformly, across the entire study site or in a stratified fashion within it. Random or stratified random-site selection is the most statistically defensible, especially if habitat is the treatment. Completely random or uniform site selection may not be the best option if there are areas within the habitat that amphibians avoid or areas where it is impossible to build suitable pens. If the study animals prefer specific microhabitats, then stratified designs that place all or a disproportionately large number of pens in preferred habitats may be desirable. Areas that expose amphibians to high temperatures due to lack of canopy cover or aspect can result in high mortality (Harper 2007), as can lack of moisture, burrows, or cover objects (Rothermel and Luhring 2005). In some studies these factors may be part of the treatments, while in others they may be sources of experimental error that reduce statistical power. Haphazard placement of enclosures may be ideal for ensuring that pens are located in areas where installation is feasible and habitats seem suitable, but it often compromises generality and may introduce bias.

12.4.2 Size and number of enclosures

Limited resources explain the inverse relationship between the number and size of enclosures used in experiments and the prevalence of small (< 10 m^2) enclosures (Table 12.1). Striking the right balance between the number and size of enclosures depends largely on the research question and the duration of the experiment. In general, it is best to first determine the size of enclosure that best suits the research question and then ascertain the number of enclosures that can feasibly be built, stocked, and monitored given the circumstances. Then decide whether this number of enclosures is adequate to answer the research question without a high probability of type II error (i.e. failing to reject a null hypothesis when the alternate is true), preferably by doing a statistical power analysis. Keep in mind that some enclosures may be lost to falling trees or other misfortunes, especially during long-term experiments.

Small pens compromise realism because they rarely encompass the home range and associated microhabitats that are used by amphibians. They may also exclude or restrict the movements of competitors, predators, and prey more than large pens. This can result in large variation in the organisms included in replicate pens. Cage effects, including unnatural behaviors resulting from confinement, are expected to be more prevalent in small enclosures because the available behavioral options, including habitat selection and predator avoidance, are limited.

Some amphibian species migrate over 1 km among over-wintering habitats, breeding habitats, and summer foraging areas, and the area of habitat used can vary dramatically depending on life-history stage and time of year (Beebee 1996). Experiments using these species can still achieve a reasonable level of realism if the duration of the experiment is limited; for example, to the juvenile stage or to the summer foraging period. Realism can also be improved by

Table 12.1 *Summary of methods used in terrestrial enclosure experiments.*

Reference	Species	Enclosure size (m²)	Number of enclosures	Research focus	Stocking density (per/m²)	Duration of experiment	Response metrics
Pearson (1955)	*Scaphiopus holbrookii*	74	5	Behavior	0.14–0.54	3 years	Activity rates
Jaeger (1971)	*Plethodon richmondi shenandoah* and *Plethodon cinereus*	0.25 and 0.5	6	Competitive exclusion	20–40	2 months	Survival
Southerland (1986)	*Desmognathus quadramaculatus, Desmognathus monticola, Desmognathus ochrophaeus*	0.25 and 0.5	9	Competitive exclusion	4–16	2 summers	Change in mass and survival
Southerland (1986)	*Desmognathus quadramaculatus, Desmognathus monticola, Desmognathus ochrophaeus*	5	4	Competitive exclusion	4	28 days	Growth and habitat selection
Cohen and Alford (1993)	*Bufo marinus*	3	15	Density dependence	3.3–16.7	3 weeks	Growth and survival
Denton and Beebee (1994)	*Bufo bufo, Bufo calamita*	4.5	12	Niche separation	0.89	2 months	Growth, survival, and activity rates

Pechmann (1994)	*Ambystoma talpoideum*	100	16	Density dependence	0.28–0.84	6 years	Survival, size, and age at first reproduction
Pechmann (1994)	*Gastrophryne carolinensis*	100	12	Density dependence	0.6–2.4	6 years	Survival, size, and age at first reproduction
Pechmann (1995)	*Ambystoma opacum* and *Ambystoma talpoideum*	225	18	Density dependence, competition	0.31–0.62	5 years	Survival, size, and age at first reproduction
Chazal and Niewiarowski (1998)	*Ambystoma talpoideum*	100	8	Forestry practices	0.8	5–6 months	Growth, fecundity, age at maturity, lipid storage
Hopkins *et al.* (1998)	*Bufo terrestris*	Not reported	16	Contaminant exposure	Not reported	7–12 weeks	Body concentrations of trace elements
Beck and Congdon (1999)	*Bufo terrestris*	0.5	2	Size-dependent growth and survival	52	2 months	Growth and survival
Laposata and Dunson (1999)	*Ambystoma jeffersonianum*	0.1134	20	Contaminant exposure	8.8	1 month	Growth, body concentrations of water and elements
Parris (2001)	*Rana blairi, Rana sphenocephala* and hybrids	2	48	Hybrid fitness	3.5	1 year	Growth and survival
Beard *et al.* (2002)	*Eleutherodactylus coqui*	1	20	Nutrient cycling	7	4 months	Concentrations of carbon and nitrogen in the body urine and feces

Table 12.1 Continued

Reference	Species	Enclosure size (m²)	Number of enclosures	Research focus	Stocking density (per/m²)	Duration of experiment	Response metrics
Price and Shields (2002)	*Plethodon cinereus* and *Plethodon glutinosus*	0.26	52	Interspecific interactions	3.8–7.7[1]	8 weeks	Growth
Rothermel and Semlitsch (2002)	*Ambystoma maculatum*, *Ambystoma texanum*, *Bufo americanus*	125	8	Juvenile emigration	Variable (total n = 179)	2 months	Rate of movement and distance moved
Rothermel and Semlitsch (2002)	*Ambystoma maculatum*, *Ambystoma texanum*, *Bufo americanus*	0.025	8	Dehydration	40[1]	22–25 h	Change in mass
Altwegg (2003)	*Rana lessonae*	9	12	Density dependence	2.8–8.3	1 year	Growth and survival
Altwegg and Reyer (2003)	*Rana lessonae* and *Rana esculenta*	9	12	Carryover effects	2.8–8.3	1 year	Growth and survival
Regosin et al. (2003a)	*Rana sylvatica*	272	17	Over-wintering ecology	Not stocked	9 months	Sex ratio, density, distance from pond
Marsh and Beckman (2004)	*Plethodon cinereus*	0.81	24	Road effects	2.5	2 months	Detectibility and surface activity
Moseley et al. (2004)	*Ambystoma talpoideum*	20.88 and 9	9 and 18	Microhabitat use	Not reported	8 months	Movement and habitat selection
Regosin et al. (2004)	*Ambystoma maculatum*	3.8	10	Burrow-occupancy patterns	0.26–0.53[1]	8 days	Probability of burrow occupancy

Study		Species		Topic		Time	Proportion detected
Williams and Berkson (2004)	1.18	*Plethodon cinereus*	124	Detection probabilities	0.85–1.7[1]	2 months	
Yurewicz and Wilbur (2004)	0.81	*Plethodon cinereus*	28	Cost of reproduction	1.2[1]	2 years	Egg survival, female growth, production of ova
Boone (2005)	2	*Rana sphenocephala, Rana blairi, Rana clamitans, Bufo woodhousii*	48	Contaminant exposure	3–5	7–12 months	Survival and growth
Maerz et al. (2005)	0.06	*Rana clamitans*	38	Foraging success	17[1]	38 h	Change in mass
Rothermel and Luhring (2005)	0.025	*Ambystoma talpoideum*	48	Forestry practices	40[1]	72 h	Water loss and survival
Croshaw and Scott (2006)	112–223	*Ambystoma opacum*	4	Nest-site selection	10.76	1 month	Number of nests per area
Greenlees et al. (2006)	2.88	*Bufo marinus*	60	Effects on native invertebrates	Standard biomass	Not reported	Invertebrate abundance
Rothermel and Semlitsch (2006)	18	*Ambystoma maculatum* and *Ambystoma opacum*	12	Forest fragmentation	1.3	2 years	Survival to maturity
Todd and Rothermel (2006)	16	*Bufo terrestris*	8	Forestry practices	1.8	2 months	Growth and survival
Harper (2007)	9	*Rana sylvatica* and *Bufo americanus*	64	Forestry practices	2	3 months	Survival

Table 12.1 *Continued*

Reference	Species	Enclosure size (m²)	Number of enclosures	Research focus	Stocking density (per/m²)	Duration of experiment	Response metrics
Harper and Semlitsch (2007)	*Rana sylvatica, Bufo americanus*	2	48	Density dependence	1–10	1 year	Survival, growth, reproductive development
Todd *et al.* (2007)	*Ambystoma opacum and Ambystoma talpoideum*	0.025	96	Fire ant predation	40[1]	48 h	Survival
Blomquist (2008)	*Rana pipiens and Rana sylvatica*	14.4	28	Forestry practices	1.73 and 1.39	3 months	Survival and growth
Patrick *et al.* (2008)	*Rana sylvatica*	100 and 8	1 and 24	Density dependence and movement	2–7	2–3 weeks	Survival and habitat selection

[1] Densities represent one animal per enclosure.

choosing species that have relatively small home ranges, as is often the case for direct-developing species (Price and Shields 2002).

Large enclosures maximize realism, allowing increased movement and access to a greater range of microhabitats and interacting organisms. One important trade-off is that the probability of recapturing individuals decreases as enclosure size increases. On the other hand, large enclosures can hold more individuals and provide more precise estimates of mean responses.

Large enclosures can also allow densities to be at or near "natural levels". Unfortunately, estimates of natural densities for terrestrial life history stages are not available for most amphibian species, and density can have significant effects on growth and survival (Figure 12.3). It may be worthwhile to estimate the natural densities of study species if estimates are unavailable. Terrestrial densities can be extremely high following metamorphosis, but quickly decrease thereafter due to mortality and migration (Cohen and Alford 1993). Therefore, short-term experiments using recently metamorphosed individuals can use high

Fig. 12.3 The density of animals in enclosures can have profound effects on individual growth rates, probabilities of survival, and age at sexual maturity. These American toads were the same size at metamorphosis and were raised in similar enclosures, but at densities of $1/m^2$ (left) and $10/m^2$ (right). This photo was taken 3 months after the animals were introduced to enclosures.

densities without sacrificing realism (Beck and Congdon 1999), but experiments that are longer term or use adults may require larger enclosures to approximate natural densities. Because amphibians occur at higher densities in high-quality habitat (Patrick *et al.* 2008), ensuring high habitat quality within enclosures (e.g. sufficient moisture, cover, and prey) can improve realism when enclosures are small and densities are relatively high.

12.4.3 Building enclosures that minimize escapes and trespasses

Most amphibians, especially fossorial species, are notorious escape artists and enclosure designs should ensure that study animals do not escape. Very secure pens are often less permeable to biotic and abiotic components of the ecosystem, a feature which may or may not be desirable (see section 12.3). Walls made of mesh materials such as hardware cloth allow greater exchange with the surrounding environment but are more easily climbed by some species than smooth, solid materials such as aluminum flashing. Even taxa that do not usually climb, such as *Ambystoma* and *Bufo*, will climb walls to get out of enclosures (E.B. Harper and J.H.K. Pechmann, personal observations).

Enclosures for fossorial species may require a mesh or solid bottom (Marsh and Beckman 2004; Williams and Berkson 2004; Yurewicz and Wilbur 2004). In the case of caecilians, it is necessary to use solid materials because caecilians can easily penetrate through flexible plastic (J. Measey, personal communication). Pre-made plastic containers, ranging in size from shoe boxes to large rain barrels or cattle tanks, can serve as enclosures for fossorial species (Price and Shields 2002; Williams and Berkson 2004). Caecilians have successfully been enclosed in plastic rain barrels punctured with small drainage holes and filled with sifted soil. To prevent caecilians from escaping, the container should be filled to a level that leaves an unfilled top portion that is greater in length than the longest individual (J. Measey, personal communication).

For less effective burrowers, for example most ranids, enclosures do not require a bottom, but usually have sides that are buried in well-packed soil 5–50 cm below ground depending on the species and size of enclosure. Small enclosures, for example those constructed from buckets or coffee cans (Rothermel and Semlitsch 2002; Maerz *et al.* 2005; Rothermel and Luhring 2005), can be installed with a shovel or post digger. Large enclosures are typically built by first digging a trench, then driving wooden, metal, or PVC posts into the ground at each corner and at points in between. Walls are then attached to the posts, and can be built with sheet metal, aluminum flashing, galvanized hardware cloth, or silt fencing. These materials vary considerably in price and durability, but even the least expensive silt fencing will survive at least a full year under most conditions. After the walls are secured, soil is backfilled into the trench and compressed so that a portion of the wall is underground. If pitfall traps are used to recapture animals, they can be installed before the trench is filled. During construction, soil, leaf litter, and vegetation are often disturbed. To maximize realism, the time between the building of the enclosures and the introduction of animals should be sufficient to allow accumulation of leaf litter and invertebrate prey, and the regrowth of vegetation.

Animals are most likely to attempt to escape from the corners of the enclosures, by climbing up the posts, or through the overlapping ends of the material used for the walls. Rounded corners can be more difficult to climb, and placing the support posts 10cm or more below the top of the wall can also reduce escapes. Strips of additional material across the top edge of the walls can serve as baffles, both to keep enclosed animals in and to prevent animals from natural populations from entering (Figure 12.4). Quick detection and repair of damaged or worn enclosures is also necessary to prevent escapes. Enclosures with lids may be necessary for arboreal species (Beard *et al.* 2002), and can be useful for other species to exclude predators, provide shade, and reduce escapes (Harper and Semlitsch 2007). However,

Fig. 12.4 Array of terrestrial enclosures (10 m × 10 m each) used in studies of density dependence in *Ambystoma talpoideum* and *Gastrophryne carolinensis* at the Savannah River Ecology Laboratory in South Carolina, USA. Note baffles at the top of the walls to deter escapes and immigration.

lids typically influence light, temperature, rainfall, litterfall, and movement of prey, which can hinder realism and may confound the treatment.

12.4.4 "Standardizing" conditions among enclosures

One advantage of terrestrial enclosures is that they can expose animals to the range of conditions found within a study area, such as the densities of prey, competitors, and predators, as well as variation in leaf-litter depth, vegetation, and solar radiation. Unfortunately, this variability decreases statistical power by contributing to experimental error. One option to reduce this "noise" is to standardize conditions within enclosures. For example, quantities of leaf litter and woody debris can be equalized (Harper 2007). Devices that provide wet refuges, such as a pit with moist leaf litter, a bowl of water, or a covering object, can reduce the risk of desiccation (Greenlees *et al.* 2006), and reduce variation in survival within treatments, especially if amphibians must be released into enclosures during dry periods when they would not normally be active. Predators including snakes may have to be removed, especially from small pens, because of the difficulty of equalizing predation across enclosures without excessive mortality. Competition can also be standardized by removing competitors or by introducing an equal number to all enclosures. Standardization comes at the cost of reduced realism, and it is important to consider how this may compromise the interpretation of the results. This is especially important when biotic and abiotic factors comprise the treatments.

12.5 Study species

The choice of study species is an important consideration when designing enclosure experiments. It must be possible to confine the species under realistic conditions and densities and to have access to enough animals to design an experiment with high statistical power.

12.5.1 Choice of species

Terrestrial enclosures are most appropriate for species that are primarily terrestrial after metamorphosis. For species that are both aquatic and terrestrial (e.g. *Rana clamitans* and *Desmognathus monticola*), short-term terrestrial enclosure experiments that coincide with normal habitat use can be informative (Maerz *et al.* 2005). Enclosures that encompass both aquatic and terrestrial habitats can also be used (Southerland 1986). For arboreal species, enclosures must be completely covered to prevent escape, which decreases realism. Effective jumpers, such as *Acris*, are difficult to keep in enclosures without very high walls or covers.

12.5.2 Source and age of animals

Animals can be wild-caught or raised from eggs in captivity (e.g. in artificial ponds). Both methods have advantages and disadvantages in terms of logistics, experimental design, and conservation. Using wild-caught animals is often necessary for studies of adult amphibians when raising individuals to adulthood is unfeasible due to constraints of time, space, and/or expense. Research on the ecology of adult amphibians is important because amphibian populations are especially sensitive to changes in adult vital rates (Biek *et al.* 2002); for this same reason, however, the effects on wild populations should be carefully considered before adults are removed. Raising individuals through metamorphosis in outdoor tanks or experimental ponds is often the best method for studies of juvenile pond-breeders for several reasons. First, large numbers of individuals can be produced with minimal effects on wild populations. Survival to metamorphosis in predator-free aquatic enclosures and tanks often exceeds 50–80% compared with fewer than 5–10% in many wild populations (Petranka and Sih 1986). Secondly, the conditions experienced by larvae can be controlled, thereby maximizing precision. Experimental ponds can also be used in conjunction with terrestrial enclosures to study carryover effects by manipulating conditions during the larval period and observing their effects on terrestrial life-history stages (Pechmann 1994; Altwegg and Reyer 2003; Boone 2005). When collecting eggs for future experimental populations, one should consider the genetic make-up in the context of the experiment. In some instances it may be prudent to select eggs from multiple populations or from multiple clutches within a population so the results are more representative of the species or population of interest.

12.6 Census techniques

For most enclosure studies, gathering data requires recapturing amphibians during and/or at the end of the study. Gathering data at regular intervals rather than waiting until the end of the experiment is often a wise strategy because mortality can be higher than expected. Multiple census intervals are also useful in estimating capture probabilities, which are often necessary to calculate estimates of survival (Chapter 24).

12.6.1 Individual marks

Individually marking animals before they are released into enclosures allows monitoring of individual growth rates, estimation of capture probabilities, and analysis of factors such as initial mass or time to metamorphosis (Beck and Congdon 1999). There are many techniques that can be used to identify individuals, including toe-clipping, visible implant elastomer, passive integrative transponder tags (PIT tags), and photographing unique dorsal patterns (Chapter 8). Group marks may be necessary when it is impractical to mark all individuals. In general, the marking technique chosen should strive to minimize the increased probability of mortality and stress to the animals while maximizing the probability of successful identification.

12.6.2 Methods of capture

The most common methods of recapturing amphibians in enclosures are hand capture, pitfall traps, and coverboards (Chapter 13). A combination of methods can be used to maximize capture probability. Hand captures work well for animals that are active on the surface or that jump when the leaf litter is disturbed. However, this method can result in biased data if animals are easier to see and catch in some treatments than others. Pitfall traps (Chapter 13) may provide a more standardized method for recapturing amphibians and are especially useful for fossorial species like ambystomatid salamanders that are not easily captured by hand (Pechmann 1995). Traps can be covered or filled with leaf litter when not in use and opened on rainy nights when the animals are most likely to be active. A wet sponge in the bottom of the traps can reduce desiccation risks, although only sponges that are free of chemicals should be used. Cover objects such as rotting logs or boards can also be placed in enclosures as a method for recapturing animals (Chapter 13) as long as their addition does not confound the effects of the treatment. Similarly, burrows can be created within enclosures to provide retreats where animals can predictably be found (Regosin et al. 2004). For caecilians and other truly fossorial amphibians that are rarely active on the surface, it may be necessary to do a single census at the end of the experiment by removing and thoroughly sifting through the soil and leaf litter. Radiotransmitters or harmonic radar tags (Moseley et al. 2004) can make study animals easier to relocate, but the equipment may affect behavior and expense can be an issue. Animals marked with PIT tags can be detected up to 13 cm underground (Blomquist et al. 2008). Pond-breeding amphibians can be censused along the enclosure walls during breeding migrations by pitfall traps, funnel traps, or hand capture as they attempt to migrate out of the enclosure to find a breeding pond (Pechmann 1995; Regosin et al. 2003b).

12.6.3 Frequency of censuses

Because enclosure experiments usually require an initially large input of time, labor, and resources, it is worthwhile in longer-term experiments to collect data early and often to ensure that useful data are gathered even if natural disasters or unexpectedly high mortality occur before the end of the experiment. Frequent censuses also improve estimates of capture probability and survival (Chapter 25). However, it is important to consider that handling animals too frequently can affect survival and induce unnecessary stress. The optimal frequency of censuses for estimating capture probability and survival also depends on the statistical methods that will be used. Some methods require equal time between sampling intervals or a minimum number of sample periods. Other statistical methods rely on multiple consecutive samples spaced at regular intervals (e.g. 3 days of sampling every 2 weeks). These methods are discussed in greater detail in Chapter 25. It is important to become familiar with these statistical methods prior to gathering the data to ensure that the frequency and methods used in the censuses maximize statistical power.

12.7 Response metrics

Experiments conducted in enclosures can use a wide range of response variables to describe the effects of experimental treatments on amphibians. Vital rates, including growth and survival, are among the most commonly used metrics (Table 12.1), but more subtle effects can be quantified with physiological measurements such as lipid content (Chazal and Niewiarowski 1998) or stress-hormone concentrations (Cooperman *et al.* 2004). Behavioral observations, including habitat selection, can also yield useful response metrics (Patrick *et al.* 2008). Data that quantify conditions within enclosures can provide covariates for analyses or can be used to provide metrics describing the role of amphibians in the ecosystem, for example in nutrient cycling (Beard *et al.* 2002) or reducing prey densities (Greenlees *et al.* 2006). In most enclosure experiments the enclosure is the experimental unit, so for any response metric that is based on the individual (e.g. mass), means per enclosure (or the equivalent) should be used in the analysis (Pechmann 1995). Proportions can also be used (e.g. percentage survival or the percentage that selected a given habitat type; Patrick *et al.* 2008).

12.7.1 Vital rates: survival, growth, age at reproductive maturity, fecundity

Survival can be surprisingly difficult to measure in all but very small enclosures. The minimum number known alive (MNKA) can be used as a proxy for survival (Rothermel and Semlitsch 2006), but does not allow the estimation of error resulting from undetected individuals. Because capture probabilities are rarely 100%, capture–mark–recapture models can be useful for estimating survival (Altwegg 2003; Chapter 24). However, when capture probabilities are low, there is a large amount of error associated with these estimates. Methods for incorporating errors from mark–recapture estimates into other statistical analyses of multiple experimental populations, such as analysis of variance, have received limited attention. Survival estimates can be improved by using an end-point associated with breeding migrations, such as first reproduction, because migrations increase the likelihood of capture (Pechmann 1995; Regosin *et al.* 2003b).

Growth is usually measured as change in mass or length over time. Some researchers prefer measuring snout–vent length, often with additional measurements including head width or tibia length (Altwegg and Reyer 2003), because mass is prone to short-term fluctuations.

Reproductive maturity can be determined by dissection, candling, or external signs of maturity such as nuptial pads or a swollen cloaca (Pechmann 1995; Rothermel and Semlitsch 2006; Harper and Semlitsch 2007). Age at reproductive maturity or the proportion of individuals that have reached maturity at the conclusion of the experiment are demographically important response metrics. Although fecundity is a component of individual fitness and population growth, clutch size has not typically been used as a response metric for enclosure experiments (but see Yurewicz and Wilbur 2004).

12.7.2 Physiological responses

Response metrics that quantify physiological responses are especially useful when the effects of experimental treatments are likely to be sublethal. Many of the methods used to quantify physiological responses require that animals be killed, so it is necessary to gather these data at the end of the experiment, or by removing a subset of animals at points during the experiment. Lipids stored in the body or in eggs can be measured and used as indicators of body condition and parental investment (Chazal and Niewiarowski 1998). Whole-body concentrations of elements such as arsenic, selenium, and vanadium can indicate bioaccumulation of contaminants (Hopkins *et al.* 1998; Laposata and Dunson 1999). Heightened concentrations of corticosterone circulating in the blood indicate a stress response, and can be determined using non-lethal methods (Cooperman *et al.* 2004).

12.7.3 Behavioral responses

Measures of behavioral responses in enclosure experiments have typically focused on movement and habitat selection. Movement has been quantified as activity rate (Pearson 1955; Rothermel and Semlitsch 2002; Marsh and Beckman 2004;

Moseley *et al.* 2004) and distance moved (Rothermel and Semlitsch 2002). Studies of habitat selection typically quantify the proportion of individuals that select a particular habitat type (Southerland 1986; Moseley *et al.* 2004; Patrick *et al.* 2008). Regosin *et al.* (2004) used probability of burrow occupancy as a response metric.

12.8 Thinking outside the box

Amphibian studies using terrestrial enclosures and cages have so far focused on a relatively narrow range of species and experimental designs. However, there is great scope for creativity in the use of these techniques. Some of the most exciting and least studied aspects of amphibian ecology can be studied using terrestrial enclosures. For example, a study by Beard *et al.* (2002) used cages as "exclosures" to document the effects of both the presence and absence of a terrestrial frog on nutrient cycling. Creative techniques for answering complex amphibian research questions have also involved using enclosures in conjunction with other experimental techniques (Marsh and Beckman 2004; Todd and Rothermel 2006) or conducting multiple enclosure experiments at a range of spatial and temporal scales (Patrick *et al.* 2008).

Although most amphibian enclosure experiments have been conducted in temperate regions, there is enormous potential for the use of enclosures in tropical ecosystems. Many tropical amphibians may actually be far better suited for enclosure experiments than their temperate counterparts. Leaf-litter species such as the neotropical *Eleutherodactylus*, *Arthroleptis* of Africa, or many of the Asian microhylids, occur at remarkably high densities and are unlikely to have large home ranges. Well-designed enclosures can also be used to gain a better understanding of the ecology of caecilians, for which few data are currently available.

12.9 References

Altwegg, R. (2003). Multistage density dependence in an amphibian. *Oecologia*, **136**, 46–50.

Altwegg, R. and Reyer, H.-U. (2003). Patterns of natural selection on size at metamorphosis in water frogs. *Evolution*, **57**, 872–82.

Beard, K. H., Vogt, K. A., and Kulmatiski, A. (2002). Top-down effects of a terrestrial frog on forest nutrient dynamics. *Oecologia*, **133**, 583–93.

Beck, C. W. and Congdon, J. D. (1999). Effects of individual variation in age and size at metamorphosis on growth and survivorship of southern toad (*Bufo terrestris*) metamorphs. *Canadian Journal of Zoology*, 77, 944–51.

Beebee, T. J. C. (1996). *Ecology and Conservation of Amphibians*. Chapman & Hall, London.

Biek, R., Funk, W. C., Maxell, B. A., and Mills, L. S. (2002). What is missing in amphibian decline research: insights from ecological sensitivity analysis. *Conservation Biology*, **16**, 728–34.

Blomquist, S. M. (2008). *Relative Fitness and Behavioral Compensation of Amphibians in a Managed Forest*. PhD dissertation, University of Maine, Orono, ME.

Blomquist, S. M., Zydlewski, J. D., and Hunter, M. L. (2008). Efficacy of PIT tags for tracking the terrestrial anurans *Rana pipiens* and *Rana sylvatica*. *Herpetological Review*, **39**, 174–9.

Boone, M. D. (2005). Juvenile frogs compensate for small metamorph size with terrestrial growth: overcoming the effects of larval density and insecticide exposure. *Journal of Herpetology*, **39**, 416–23.

Boone, M. D. and James, S. M. (2005). Aquatic and terrestrial mesocosms in amphibian ecotoxicology. *Applied Herpetology*, **2**, 231–57.

Chalcraft, D. R., Binckley, C. A., and Resetarits Jr, W. J. (2005). Experimental venue and estimation of interaction strength: comment. *Ecology*, **86**, 1061–7.

Chazal, A. C. and Niewiarowski, P. H. (1998). Responses of mole salamanders to clear-cutting: using field experiments in forest management. *Ecological Applications*, **8**, 1133–43.

Cohen, M. P. and Alford, R. A. (1993). Growth, survival and activity patterns of recently metamorphosed *Bufo marinus*. *Wildlife Research*, **20**, 1–13.

Cooperman, M. D., Reed, J. M., and Romero, L. M. (2004). The effects of terrestrial and breeding densities on corticosterone and testosterone levels in spotted salamanders, *Ambystoma maculatum*. *Canadian Journal of Zoology*, **82**, 1795–1803.

Croshaw, D. A. and Scott, D. E. (2006). Marbled salamanders (*Ambystoma opacum*) choose low elevation nest sites when cover availability is controlled. *Amphibia-Reptilia*, **27**, 359–64.

Denton, J. S. and Beebee, T. J.C. (1994). The basis of niche separation during terrestrial life between two species of toad (*Bufo bufo* and *Bufo calamita*): competition or specialisation? *Oecologia*, **97**, 390–8.

Greenlees, M. J., Brown, G. P., Webb, J. K., Phillips, B. L., and Shine, R. (2006). Effects of an invasive anuran [the cane toad (*Bufo marinus*)] on the invertebrate fauna of a tropical Australian floodplain. *Animal Conservation*, **9**, 431–8.

Hairston Sr, N. G. (1989). *Ecological Experiments: Purpose, Design, and Execution*. Cambridge University Press, Cambridge.

Harper, E. B.H. (2007). *The Population Dynamics and Conservation of Pond-breeding Amphibians*. PhD dissertation, University of Missouri, Columbia, MO.

Harper, E. B. and Semlitsch, R. D., (2007). Density dependence in the terrestrial life history stage of two anurans. *Oecologia*, **153**, 879–89.

Hopkins, W. A., Mendonca, M. T., Rowe, C. L., and Congdon, J. D., (1998). Elevated trace element concentrations in southern toads, *Bufo terrestris*, exposed to coal combustion waste. *Archives of Environmental Contamination and Toxicology*, **35**, 325–9.

Hurlbert, S. H. (1984). Pseudoreplication and the design of ecological field experiments. *Ecological Monographs*, **54**, 187–211.

Jaeger, R.G. (1971). Competitive exclusion as a factor influencing the distributions of two species of terrestrial salamanders. *Ecology*, **52**, 632–7.

Laposata, M. M. and Dunson, W. A. (1999). Enclosure design for ecotoxicological studies of terrestrial salamanders. *Herpetological Review*, **30**, 28–30.

Maerz, J. C., Blossey, B., and Nuzzo, V. (2005). Green frogs show reduced foraging success in habitats invaded by Japanese knotweed. *Biodiversity and Conservation*, **14**, 2901–11.

Marsh, D. M. and Beckman, N. G. (2004). Effects of forest roads on the abundance and activity of terrestrial salamanders. *Ecological Applications*, **14**, 1882–91.

Morin, P. J. (1998). Realism, precision, and generality in experimental tests of ecological theory. In W. J. Resetarits and J. Bernardo (eds), *Issues and Perspectives in Experimental Ecology*, pp. 50–70. Oxford University Press, Oxford.

Moseley, K. R., Castleberry, S. B., and Ford, W. M. (2004). Coarse woody debris and pine litter manipulation effects on movement and microhabitat use of *Ambystoma talpoideum* in a *Pinus taeda* stand. *Forest Ecology and Management*, **191**, 387–96.

Parris, M. J. (2001). Hybridization in leopard frogs (*Rana pipiens* complex): terrestrial performance of newly metamorphosed hybrid and parental genotypes in field enclosures. *Canadian Journal of Zoology*, **79**, 1552–8.

Patrick, D. A., Harper, E. B., Hunter Jr, M. L., and Calhoun, A. J. K. (2008). Terrestrial habitat selection and strong density-dependent mortality in recently metamorphosed amphibians. *Ecology*, **89**, 2563–74.

Pearson, P. G. (1955). Population ecology of the spadefoot toad, *Scaphiopus h. holbrooki* (Harlan). *Ecological Monographs*, **25**, 233–67.

Pechmann, J. H. K. (1994). *Population Regulation in Complex Life Cycles: Aquatic and Terrestrial Density-dependence in Pond-breeding Amphibians*. PhD dissertation, Duke University, Durham, NC.

Pechmann, J. H. K. (1995). Use of large field enclosures to study the terrestrial ecology of pond-breeding amphibians. *Herpetologica*, **51**, 434–50.

Petranka, J. W. and Sih, A. (1986). Environmental instability, competition and density-dependent growth and survivorship of a stream-dwelling salamander. *Ecology*, **67**, 729–36.

Price, J. E. and Shields, J. A. S. (2002). Size-dependent interactions between two terrestrial amphibians, *Plethodon cinereus* and *Plethodon glutinosus*. *Herpetologica*, **58**, 141–55.

Regosin, J. V., Windmiller, B. S., and Reed, J. M. (2003a). Terrestrial habitat use and winter densities of the Wood Frog (*Rana sylvatica*). *Journal of Herpetology*, **37**, 390–4.

Regosin, J. V., Windmiller, B. S., and Reed, J. M. (2003b). Influence of abundance of small-mammal burrows and conspecifics on the density and distribution of spotted salamanders (*Ambystoma maculatum*) in terrestrial habitats. *Canadian Journal of Zoology*, **81**, 596–605.

Regosin, J. V., Windmiller, B. S., and Reed, J. M. (2004). Effects of conspecifics on the burrow occupancy behaviour of spotted salamanders. *Copeia*, **2004**, 152–8.

Regosin, J. V., Windmiller, B. S., Homan, R. N., and Reed, J. M. (2005). Variation in terrestrial habitat use by four poolbreeding amphibian species. *Journal of Wildlife Management*, **69**, 1481–93.

Rothermel, B. B. and Semlitsch, R. D. (2002). An experimental investigation of land-scape resistance of forest versus old-field habitats to emigrating juvenile amphibians. *Conservation Biology*, **16**, 1324–32.

Rothermel, B. B. and Luhring, T. M. (2005). Burrow availability and desiccation risk of mole salamanders (*Ambystoma talpoideum*) in harvested versus unharvested forest stands. *Journal of Herpetology*, **39**, 619–26.

Rothermel, B. B. and Semlitsch, R. D. (2006). Consequences of forest fragmentation for juvenile survival in spotted (*Ambystoma maculatum*) and marbled (*Ambystoma opacum*) salamanders. *Canadian Journal of Zoology*, **84**, 797–807.

Southerland, M. T. (1986). Coexistence of three congeneric salamanders: the importance of habitat and body size. *Ecology*, **67**, 721–8.

Todd, B. and Rothermel, B. B. (2006). Assessing quality of clearcut habitats for amphibians: effects on abundances versus vital rates in the southern toad (*Bufo terrestris*). *Biological Conservation*, **133**, 178–85.

Todd, B. D., Rothermel, B. B., Reed, R. N., Luhring, T. M., Schlatter, K., Trenkamp, L., and Gibbons, J. W. (2007). Habitat alteration increases invasive fire and abundance to the detriment of amphibians and reptiles. *Biological Invasions*, **10**, 539–46.

Williams, A. K. and Berkson, J. (2004). Reducing false absences in survey data: detection probabilities of red-backed salamanders. *Journal of Wildlife Management*, **68**, 418–28.

Yurewicz, K. L. and Wilbur, H. M. (2004). Resource availability and costs of reproduction in the salamander *Plethodon cinereus*. *Copeia*, **2004**, 28–36.

Part 4
Amphibian populations

13

Drift fences, coverboards, and other traps

John D. Willson and J. Whitfield Gibbons

13.1 Introduction

Many of the simplest yet most highly productive sampling methods in herpetological field research use some type of trap or attraction device to increase capture rates or target secretive species. These techniques fall into two general categories: those that actually trap animals, accumulating captures on their own over time (passive traps) and those that attract animals but require an observer to actively capture them at the moment of the census (active traps). The most popular examples of these two trap categories are drift fences (generally with pitfall and/or funnel traps) and coverboards, respectively. Both of these methods are inherently simple concepts, and their description and explanation need not be made complex or complicated. Both techniques are usually best modified by the investigator who can use common sense to focus on the needs of a particular project that involves capturing animals in a field situation. However, we provide a general discussion of some of the fundamental issues that investigators who use these techniques must face, with particular emphasis on how choice of capture method and sampling design influence interpretation of capture data.

13.2 Drift fences, funnel traps, and other passive capture methods

13.2.1 What are passive traps?

Passive capture methods are designed to restrain animals that enter the trap of their own accord, accumulating captures that are then assessed upon regular censuses by the observer. Passive traps are among the more intensive methods for sampling amphibians in terms of time and effort, but often yield higher capture rates and more standardized samples than opportunistic or visual searches. Perhaps most importantly, passive traps are the most effective ways to sample many rare or secretive amphibian species, many of which are of conservation concern.

A huge variety of passive amphibian traps have been developed for nearly any imaginable habitat or situation; however, nearly all are variants on two basic trap types, the funnel trap and the pitfall trap. Funnel traps consist of a tapering funnel-shaped entrance that guides animals into a larger holding chamber (Figure 13.1). Once within the chamber, animals are unable to find their way back out the small entrance hole, becoming trapped. Pitfall traps work on a similar principle, consisting of some type of container that is sunk into the ground, with the rim level with the surface (Figure 13.2). Animals that fall into pitfalls are unable to climb out, becoming trapped. First described for use with herpetofauna by Gibbons and Bennett (1974) and Gibbons and Semlitsch (1981), drift fences are vertical barriers that intercept the intended trajectory of amphibians moving from one location to another. The fence typically guides animals toward a pitfall bucket, funnel trap, or other capture device (Figure 13.2). In most terrestrial and some aquatic situations, drift fences dramatically increase amphibian capture rates (Friend *et al.* 1989) and, under some circumstances, have been responsible for more amphibian captures per day, month, year, or decade than any other method used in field studies of amphibians (Pechmann *et al.* 1991; Gibbons *et al.* 2006). Although drift

Fig. 13.1 Several varieties of funnel trap are commonly used to sample amphibians in terrestrial and aquatic habitats. Front: soft-drink bottle funnel trap; back (left to right): plywood and hardware cloth box trap, steel "Gee" minnow trap, plastic minnow trap, and collapsible nylon trap. Photograph by John Willson, Savannah River Ecology Laboratory.

fences have been used most extensively in the southeastern USA, they have proven effective for many amphibian species worldwide (e.g. Gittins 1983; Friend 1984; Bury and Corn 1987; Friend *et al.* 1989; Jehle *et al.* 1995; Weddeling *et al.* 2004). However, drift fences were relatively ineffective for sampling anurans in forests of Queensland, Australia (Parris *et al.* 1999) and recorded lower numbers of anuran species than automated recording devices or nocturnal line-transect surveys in Taiwan (Hsu *et al.* 1985).

13.2.2 How are passive traps constructed, aligned, and monitored?

Funnel traps can be nearly any size and have been constructed of a variety of materials including twine netting, hardware cloth, window screen, nylon mesh, plywood, PVC pipe, and plastic soft-drinks bottles (Figure 13.1; Griffiths 1985; Shaffer *et al.* 1994; Adams *et al.* 1997; Buech and Egeland 2002; Willson *et al.* 2005). The wide range of funnel trap variants makes them effective for most species in nearly any habitat, aquatic or terrestrial. Likewise, pitfalls can be any size from small coffee cans to multi-gallon drums, but must be sufficiently deep to prevent escape of the target species. Pitfall traps are typically metal or plastic. Several studies have been conducted weighing the strengths and weaknesses of different trap types for various amphibian species (e.g. Vogt and Hine 1982; Friend 1984; Friend *et al.* 1989; Mitchell *et al.* 1993; Greenberg *et al.* 1994; Enge 2001; Ryan *et al.* 2002; Stevens and Paszkowski 2005; Todd *et al.* 2007), and we refer readers to these sources rather than discussing the merits of each trap type in detail here.

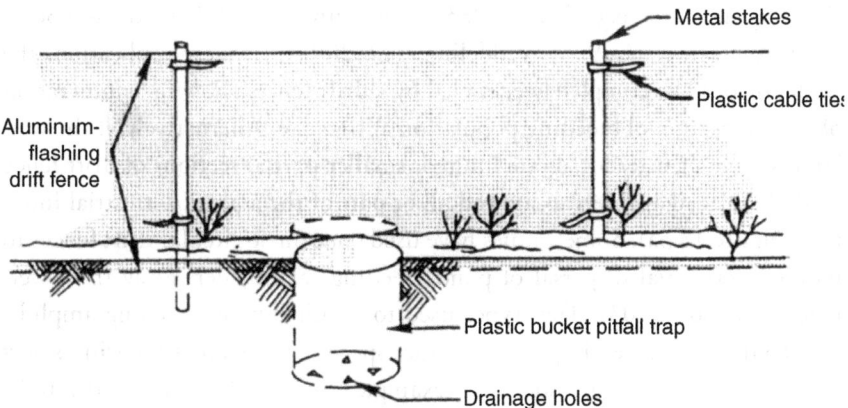

Fig. 13.2 Schematic of a terrestrial drift fence with large pitfall traps. Figure reprinted from Gibbons and Semlitsch (1981).

The evolution of the drift-fence technique in herpetology has resulted in the use of a variety of construction materials for the fence, including chicken wire, hardware cloth, aluminum flashing, stiff plastic, and erosion/silt fencing (Gibbons and Bennett 1974; Dodd and Scott 1994; Enge 1997a). Likewise, habitat constraints and the traits of particular target species have given rise to a variety of suggestions for the alignment of fencing within the habitat and an endless array of configurations of fences and traps is possible (see Corn 1994; Dodd and Scott 1994; Rice *et al.* 2006). No single construction material or trap type is universally "the best" because of several factors whose importance will vary depending upon the particular project. Among the issues that must be considered are goals of the study and how the data will be analyzed, the cost and availability of materials, anticipated longevity of the project and maintenance effort required, the size and behavior of target species, potential safeguards against predators, and the terrain and topography of the habitat itself. Mechanisms to minimize trespass by climbing species, drowning of captured individuals in buckets or dehydration in buckets or funnel traps, and predation by a variety of species continue to be developed to address specific situations. Rather than dictate what the most effective materials and trap configuration should be, we recommend that the investigator tailor drift-fence applications on the basis of budgets, time availability, and the general goals and specific objectives of the project. Following are examples of how differing goals can temper the style of fencing and type of traps.

First, perhaps the situation for which drift fences are most commonly employed is to intercept wetland-breeding amphibians as they undertake seasonal breeding migrations between terrestrial refugia and aquatic breeding sites (Figure 13.3a; e.g. Gittins 1983; Pechmann *et al.* 1991; Dodd and Scott 1994; Arntzen *et al.* 1995; Weddeling *et al.* 2004). In these applications the wetland could be completely encircled by a drift fence, allowing enumeration of the entire annual breeding population at that site. Alternatively, if the wetland is large or if resources are limited, smaller partial sections of drift fence could be placed at intervals around all or part of the aquatic/terrestrial interface (Figure 13.3a). Some studies have used concentric circular drift fences to monitor terrestrial dispersal of pond-breeding salamanders away from wetlands (Johnson 2003). Trap types used to monitor pond-breeding amphibians could vary depending on the target species. For studies focusing solely on ambystomatid salamanders, for example, small, coffee-can-sized pitfalls would be sufficient to capture individuals of all species, including the largest (*Ambystoma tigrinum*; Gibbons *et al.* 2006). Large plastic buckets, however,

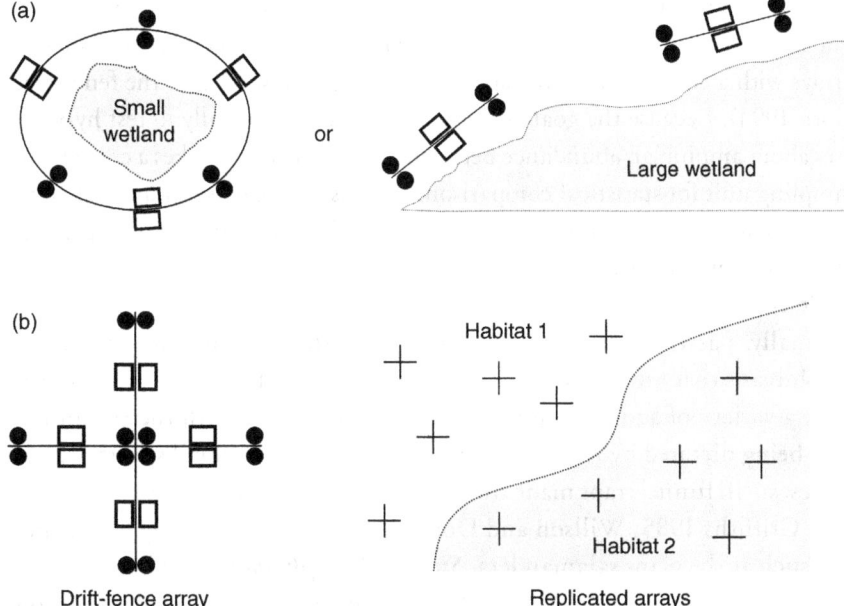

Fig. 13.3 Examples of drift-fence and passive-trap configurations used to address different research questions. Solid lines represent sections of drift fence, filled circles represent pitfalls, and open rectangles represent funnel traps. (a) Drift-fence configurations used to sample pond-breeding amphibians migrating to and from wetland breeding sites; (b) a drift-fence array pattern designed to compare amphibian abundance across two habitat types.

would be necessary to capture high numbers of most species, especially larger anurans (e.g. ranid frogs and toads) that could jump out of a smaller can (Mitchell *et al.* 1993). Larger buckets also increase the time that climbing hylids and microhylids are likely to remain within the trap. In some situations, funnel traps are the best single trap type for capturing many amphibian species (Enge 2001; Todd *et al.* 2007); however, funnel traps are generally more costly to construct, more time-consuming to maintain, and more prone to desiccating captured amphibians than are pitfalls. In nearly all situations, a combination of trap types (e.g. alternating large pitfalls and funnel traps) will produce the best possible assessment of the entire amphibian community (Vogt and Hine 1982; Greenberg *et al.* 1994; Todd *et al.* 2007).

Second, another common application of drift fences in terrestrial habitats is to compare abundances of amphibians in different areas (Figure 13.3b; e.g. habitat types, experimental treatments). In such cases, fences are generally located far from any obvious breeding wetland, hibernaculum, or other habitat focal point. For such applications, drift fences are often constructed in cross- or X-shaped arrays with a central trap and traps placed along each section of the fence (see Corn 1994). Because the goal of this type of study is generally to test hypotheses about amphibian abundance between areas, each array makes a convenient sampling unit for statistical comparisons. In this case, care should be taken to ensure that arrays are comparable (same length of fencing, number of traps, etc.) and that arrays are located randomly or systematically across the treatments or habitats.

Finally, passive traps can be used for quantitative sampling of aquatic amphibians that are not easily captured by other methods. For this application, a variety of aquatic funnel traps can be effective, with the size of the trap being dictated by the size of the target species. For small species or life stages, small funnel traps made from plastic soft-drinks bottles would suffice (e.g. Griffiths 1985; Willson and Dorcas 2003), while targeting larger species such at the giant salamanders, *Siren* and *Amphiuma*, would require larger traps such as commercially available minnow or crawfish traps (Johnson and Barichivich 2004; Willson *et al.* 2005). In most cases, aquatic traps should be set in water shallow enough to allow captured animals access to air and care should be taken to monitor fluctuations in water level that could submerge traps. Specific microhabitats where traps are set will vary by species, but heavily vegetated shallow areas are often preferable. Capture rates may also be increased by setting aquatic traps along natural barriers such as submerged logs or the shoreline or by the use of aquatic drift fences to direct amphibians into traps (Enge 1997b; Willson and Dorcas 2004; Palis *et al.* 2007). Finally, as in the previous example, traps may be set in any number of spatial configurations. Often a simple linear transect along a shoreline is sufficient. However, if the goal of the study is to compare captures statistically across different treatments (habitats, wetlands, etc.) standardized arrays of traps can be set that will serve as the sampling units in statistical comparisons.

Passive traps restrain captured animals, so frequent (generally at least daily) monitoring of traps is necessary to avoid mortality of captured animals. During hot or dry periods it is often advisable to provide access to moisture (e.g. a water bowl or damp sponge) within traps to avoid desiccation of captured animals. Finally, natural amphibian predators such as mid-sized mammals, birds, and large snakes often learn to target drift fences, and predator-control measures (raised covers for pitfalls, live-trapping and removal, or wide-width steel mesh trap covers, etc.) may be necessary to avoid undue predation.

13.2.3 What can passive traps tell you?
What can they not tell you?

Passive traps, especially when used in conjunction with drift fences, have proved highly effective for determining the distribution and abundance of amphibians both spatially and temporally. Many examples exist in which drift-fence captures revealed the presence of species that were rare or not even known to be present in an area, as well as providing a comparative assessment of annual and seasonal activity among species. Drift fences have also been used to capture large numbers of specific life stages of study species for laboratory experiments. The application of drift fences to a conservation effort was aptly demonstrated by Aresco (2005), who used silt fencing to create a barrier to prevent amphibians and reptiles from crossing a busy highway. Despite considerable hand-waving about the statistical approaches that could and should be applied to drift-fence data, as long as an investigator is aware of potential biases in the effectiveness of the fence and what is being revealed, the technique remains one that unquestionably can reveal natural history information about amphibians that may be unobtainable or unlikely to be discovered in any other way.

Generally, amphibians are only captured in passive traps when they are actively moving through the area where traps are deployed. Thus, the number of amphibians captured (often expressed as a rate, such as captures per trap per night) is ultimately a function of three major factors: (1) the density of animals within the area sampled, (2) the activity (movement) levels of those animals, and (3) the probability that an individual animal encountering a trap will be captured and not escape. Although trap capture rates can be extremely informative, consideration of all three factors is critical to interpreting capture data.

In situations where the population density of amphibians can be assumed to be relatively constant (e.g. within a single population over a fairly short time), most of the variation in capture rates can be assumed to be due to shifts in activity. Thus, drift fences have been instrumental in allowing investigators to identify seasonality and orientation of amphibian breeding migrations and environmental correlates of migratory activity (e.g. Semlitsch 1985; Semlitsch and Pechmann 1985; Phillips and Sexton 1989; Todd and Winne 2006). Likewise, when trap captures are pooled over relatively long intervals (e.g. seasons or years), thus minimizing short-term variation in activity due to environmental conditions, long-term shifts in abundance or activity can be assessed (e.g. Jehle *et al.* 1995). Ideally, studies wishing to use trap capture rates as abundance indices should test the assumption of equal catchability through mark–recapture, occupancy modeling, or similar methods (Chapter 24; Mazerolle *et al.* 2007). Additionally, marking captured animals allows the researcher to distinguish between novel individuals and recaptures, improving census counts by eliminating multiple counts of the same individual animal (Weddeling *et al.* 2004).

Although it is often tempting to attempt interspecific comparisons of abundance using capture-rate data, such comparisons are nearly always tenuous, given that species often differ in seasonal timing of activity levels and vary in their susceptibility to being captured by a particular trap type. For example, some species, such as many hylid and microhylid frogs, are adept at climbing out of buckets and over fences, whereas even small pitfalls are highly effective for capturing many salamanders and terrestrial anurans (Friend *et al.* 1989; Dodd 1991; Enge 2001; Todd *et al.* 2007). Thus, it would clearly be inappropriate to conclude that because more salamanders were captured in drift fences, they were actually more abundant than treefrogs within the habitat. Some conservative interspecific comparisons of abundance are possible, however, given careful consideration of potential biases. For example, if a wetland were to be completely surrounded by a drift fence with large pitfalls for an entire breeding season, it would probably be safe to compare total annual breeding population sizes of ambystomatid salamander species using that wetland. However, Arntzen *et al.* (1995) demonstrated that even for relatively small terrestrial species (*Bufo bufo* and *Triturus cristatus*), drift-fence efficiencies (proportion of breeding population captured) were often low and varied between species.

13.3 Coverboards and other traps that require active capture

13.3.1 What are coverboards and other active traps?

Unlike passive traps such as funnel traps or pitfalls, some so-called traps do not actually restrain or capture animals, but instead concentrate free-ranging amphibians to facilitate their capture by an active observer (usually by hand). For example, a herpetologist may lay down boards or other artificial cover objects that attract amphibians and allow them to be captured without disturbing natural cover such as logs, rocks, or vegetation. Such active traps generally operate on the principle of creating optimal microhabitats for the target species, attracting animals that can then be collected more easily than would otherwise be possible.

Because many amphibians are partially or exclusively fossorial for much of their lives and most prefer moist habitats, coverboards and other artificial cover objects are the most widely used active traps for amphibians. Coverboards simply consist of sections of cover material, most commonly wood or metal, which are placed on the ground in habitats preferred by target species. Coverboards act in the same way as natural-cover objects, trapping moisture and providing refugia for a variety of amphibian species. Amphibians are captured when an observer gently lifts the cover object and collects any animals observed hiding beneath. Because coverboards generally create moist subterranean microhabitats, this technique is most frequently used for amphibian species that prefer such habitats and include most terrestrial salamander species and some of the more fossorial anurans such as toads (e.g. *Bufo* and *Scaphiopus*) and microhylid frogs (e.g. *Gastrophryne*).

Other active capture methods have been developed, many for specific species or situations. One example is the use of PVC pipes for collecting hylid treefrogs (Moulton *et al.* 1996; Boughton *et al.* 2000). Pipes are placed vertically within the habitat, creating a moist arboreal microclimate favored by treefrogs for diurnal refugia. Pittman *et al.* (2008) used an extensive grid of PVC refugia within a wetland and surrounding upland habitats to document seasonal activity patterns, habitat use, and site fidelity in a North Carolina population of gray treefrogs (*Hyla chrysoscelis*). Additional examples of active traps include leaf-litter bags to aid in capture of aquatic salamanders and their larvae (Pauley and Little 1998; Waldron *et al.* 2003) and artificial pools to assess breeding activity of various anuran species (Resetarits and Wilbur 1991; Gascon 1994).

13.3.2 How are active traps constructed, aligned, and monitored?

Coverboards may be constructed from nearly any material and the most suitable material likely varies depending on the target species, research budget, habitat, and other characteristics of the study site (e.g. proximity to roads, terrain). For most amphibians, wooden boards are probably the best all-round option as they create moist conditions preferred by many species. Indeed, wood coverboards have been used in studies of a variety of woodland and stream-dwelling salamander species (Figure 13.4; Degraaf and Yamasaki 1992; Fellers and Drost 1994; Houze and Chandler 2002; Moore 2005; Luhring and Young 2006). Moore (2005) suggested using boards cut *in situ* from native trees for studying redback salamanders (*Plethodon cinereus*), providing a cost-effective coverboard option that can be used in habitats that are difficult to traverse. Although coverboards consisting of roofing tin or other metals are frequently used in reptile studies, these materials heat quickly and create conditions too hot and dry for amphibians in most situations. Indeed, Grant *et al.* (1992) found that amphibian captures in South Carolina were highest under plywood coverboards, while reptiles preferred tin. As most amphibian species are small,

Fig. 13.4 Example of a coverboard used to monitor woodland salamanders. Note that the low barrier around the board is part of an experiment and would normally not be used with coverboards used for general monitoring purposes. Photograph by Thomas Luhring, Savannah River Ecology Laboratory.

boards generally need not be large, but at least one study reported a positive correlation between the size of coverboards and the number of salamanders captured per board (Moore 2005). Moreover, larger or thicker boards may be preferable in warm or dry habitats as they generally hold moisture better than smaller boards. One study conducted in southern Georgia, USA, noted lower amphibian captures under coverboards than natural-cover objects, presumably resulting from warmer and more variable temperatures under boards (Houze and Chandler 2002). As some time is often necessary for suitable microhabitats (e.g. rotten leaf litter, burrows) to develop under refugia, the investigator should consider allowing boards to "weather" for several weeks or months before amphibian censuses are initiated.

Construction of other active traps varies, but the general goal is to create microhabitat conditions that attract target amphibian species. For example, sections of PVC pipe may be inserted vertically into the ground or affixed to tree trunks to provide arboreal refugia for hylid treefrogs (Moulton *et al.* 1996; Boughton *et al.* 2000). Johnson (2005) described several modifications to the so-called hylid tube technique, maximizing standardization of microhabitat within tubes and increasing ease of census and frog capture.

As uses of coverboards and other active traps vary, so will the designs for their placement. In general active traps should be placed in habitats that are favorable for amphibians, including well-shaded areas with abundant moisture. For species that breed in aquatic habitats, breeding sites may be the most appropriate places to maximize captures. For example, an obvious location to deploy PVC tubes for hylid frogs would be around the periphery of wetland breeding sites.

The spatial distribution of active traps also depends on the goals of the study. For simple amphibian inventories (documenting species presence) or collecting individuals for use in the laboratory, placing devices haphazardly in the most optimal habitats is the most cost-effective method. For studies where statistical comparisons are to be made, replicated sampling units must be designated. In general, because captures per individual trap are low, arrays of several boards or other traps are usually designated as the sampling unit. An array can consist of any arbitrary number of traps, generally arranged in a systematic pattern (e.g. a grid or transect). Replicate arrays are then placed systematically or randomly within treatments. Ideally, a power analysis can be used to determine the number of arrays (sample size) that is needed to obtain sufficient statistical power for the analysis.

For example, a researcher might wish to compare salamander abundance across three forest types within a relatively small geographical area. Having determined that salamanders are generally fairly common (say, an average of one salamander found per five coverboards checked), an array of 10 coverboards, spaced 5 m apart in a linear transect, might be determined as the sampling unit. Geographic Information Systems (GIS) technology could then be used to generate randomized locations for five arrays to be placed within each of the three habitat types. Thus, the total number of boards to be used would be 150 (10 boards/array × 5 arrays per habitat treatment × 3 habitats) and the sample size would be five per treatment.

Unlike passive traps, active traps can be monitored on nearly any temporal schedule, and the timing of censuses will reflect the question of interest. Generally, as animals may remain within the same refuge for extended periods, allowing some time (e.g. a few days or a week) between censuses may minimize repeated counts of the same individual animal. Alternatively, animals may avoid boards that are disturbed too frequently; one study noted that salamander captures under coverboards that were checked daily were reduced compared to boards that were censused on longer time intervals (1 or 3 weeks; Marsh and Goicochea 2003). Similarly, when designing a monitoring scheme, care should be taken to make samples as repeatable as possible. This typically means that environmental conditions should be as comparable as possible between samples, and many researchers set up environmental criteria for determining census times based on the biology of the study animal. For example, coverboard arrays might be checked once-weekly at 7–9 am on days without precipitation or only on nights with a temperature greater than 15°C and at least 1 cm of rain.

13.3.3 What can active traps tell you? What can they not tell you?

Coverboards and other active traps have several advantages as sampling methods for amphibians. First, in many cases, these are among the best methods for colleting large numbers of target species and can be the only ways to collect highly fossorial species or those that do not form breeding aggregations. Moreover, unlike passive traps, which require high-intensity monitoring on a daily basis to avoid mortality of captured animals, active traps can be monitored on a low-intensity or periodic basis because animals are not restrained and are not prone to accidental mortality or unnaturally high levels of predation. Also, because active traps generally concentrate animals into a highly searchable area (e.g. under a board, or in a PVC tube), these methods minimize the effects of observer bias, which can be substantial in other active capture methods such as visual searches. Finally, because artificial refugia are standardized for size and material, they create more repeatable microhabitats than natural-cover objects, yielding more standardized measures of abundance, and can be censused with minimal disturbance to the habitat (Heyer *et al.* 1994).

Because active traps yield high capture rates of target species, they can be useful for assessing patterns of abundance over time or space (e.g. population trends over time or variation in abundance across habitats). Generally, studies designed to investigate these types of questions compare capture rates among arrays of traps placed in different locations (e.g. two habitat types) or capture rates within an array or set of arrays over time. In these types of studies, the dependent variable is generally an index of relative abundance, typically the number of animals captured over some unit of time and effort (e.g. captures per array per census). However, as with any abundance index, data collected using these methods may be biased in a variety of ways, all of which must be considered when interpreting capture data.

The key assumption when comparing indices of relative abundance is that the capture probability of an individual animal is the same across arrays or time intervals. Thus, differences in capture rate among arrays reflect true differences in population density between sampling units. Although this assumption may be met in some situations, it is relatively easy to imagine situations where this premise is violated. Perhaps the most obvious case where the equal catchability assumption is violated is when comparing capture rates among species, as noted for the drift-fence method. It is often the case, regardless of the sampling method, that some species are highly catchable, while others are seldom encountered. Thus, it is generally unreasonable to assume that simply because one species is captured more frequently than another, that it is truly more abundant. Likewise, differences in catchability across time are critical to consider in temporal comparisons and the potential for activity patterns to influence catchability must be considered when interpreting capture rates. For example, in warm climates, salamanders often retreat deep underground during the summer and are seldom captured using any method. In this case, low capture rates of salamanders in the summer are best explained by seasonal differences in catchability rather than by changes in actual abundance within the landscape. A growing body of literature uses mark–recapture or occupancy modeling to incorporate detection probability (catchability) in interpretations of amphibian abundance data (Chapter 24; Mazerolle *et al.* 2007). Ideally, any researcher wishing to use abundance indices as indicators of population size or density should consider using these methods to test the equal catchability assumption.

A final consideration when using active traps is the potential for use of these methods to actually improve the overall quality of the habitat for target species. For example, if one of the factors limiting population size in an amphibian species is availability of suitable refugia, adding boards, hylid tubes, or other artificial refugia to the habitat may permit an increase in population density. Although such a situation would probably be favorable in studies designed to conserve at-risk amphibian species, it is worth remembering that population densities estimated in areas with abundant artificial refugia may not necessarily be representative of those in unaltered habitats. To our knowledge the potential for artificial refugia to improve habitat quality for amphibians has not been addressed experimentally and warrants future investigation.

13.4 References

Adams, M. J., Richter, K. O., and Leonard, W. P. (1997). Surveying and monitoring amphibians using aquatic funnel traps. In D. H. Olson, W. P. Leonard, and R. B. Bury (eds), *Sampling Amphibians in Lentic Habitats, Northwest Fauna*, no. 4, pp. 47–54. Society for Northwestern Vertebrate Biology, Olympia, WA.

Aresco, M. J. (2005). Mitigation measures to reduce highway mortality of turtles and other herpetofauna at a north Florida lake. *Journal of Wildlife Management*, **69**, 549–60.

Arntzen, J. W., Oldham, R. S., and Latham, D. M. (1995). Cost effective drift fences for toads and newts. *Amphibia-Reptilia*, **16**, 137–45.

Boughton, R. G., Staiger, J., and Franz, R. (2000). Use of PVC pipe refugia as a sampling technique for hylid treefrogs. *American Midland Naturalist*, **144**, 168–77.

Buech, R. R. and Egeland, L. M. (2002). Efficacy of three funnel traps for capturing amphibian larvae in seasonal forest ponds. *Herpetological Review*, **33**, 182–5.

Bury, R. B. and Corn, P. S. (1987). Evaluation of pitfall trapping in northwestern forests: trap arrays with drift fences. *Journal of Wildlife Management*, **51**, 112–19.

Corn, P. S. (1994). Straight-line drift fences and pitfall traps. In W. R. Heyer, M. A. Donnelly, R. W. McDiarmid, and L. C. Hayek (eds), *Measuring and Monitoring Biological Diversity. Standard Methods for Amphibians*, pp. 109–17. Smithsonian Institution Press, Washington DC.

Degraaf, R. M. and Yamasaki, M. (1992). A nondestructive technique to monitor the relative abundance of terrestrial salamanders. *Wildlife Society Bulletin*, **20**, 260–4.

Dodd, Jr, C. K. (1991). Drift fence-associated sampling bias of amphibians at a Florida sandhills temporary pond. *Journal of Herpetology*, **25**, 296–301.

Dodd, Jr, C. K. and Scott, D. E. (1994). Drift fences encircling breeding sites. In W. R. Heyer, M. A. Donnelly, R. W. McDiarmid, and L. C. Hayek (eds), *Measuring and Monitoring Biological Diversity. Standard Methods for Amphibians*, pp. 125–30. Smithsonian Institution Press, Washington DC.

Enge, K. M. (1997a). Use of silt fencing and funnel traps for drift fences. *Herpetological Review*, **28**, 30–1.

Enge, K. M. (1997b). *A Standardized Protocol for Drift-fence Surveys*. Technical report no. 14. Florida Game and Fresh Water Fish Commission, Tallahassee, FL.

Enge, K. M. (2001). The pitfalls of pitfall traps. *Journal of Herpetology*, **35**, 467–78.

Fellers, G. M. and Drost, C. A. (1994). Sampling with artificial cover. In W. R. Heyer, M. A. Donnelly, R. W. McDiarmid, and L. C. Hayek (eds), *Measuring and Monitoring Biological Diversity. Standard Methods for Amphibians*, pp. 146–50. Smithsonian Institution Press, Washington DC.

Friend, G. R. (1984). Relative efficiency of two pitfall-drift fence systems for sampling small vertebrates. *Australian Zoologist*, **21**, 423–34.

Friend, G. R., Smith, G. T., Mitchell, D. S., and Dickman, C. R. (1989). Influence of pitfall and drift fence design on capture rates of small vertebrates in semi-arid habitats of western Australia. *Australian Wildlife Research*, **16**, 1–10.

Gascon, C. (1994). Sampling with artificial pools. In W. R. Heyer, M. A. Donnelly, R. W. McDiarmid, and L. C. Hayek (eds), *Measuring and Monitoring Biological Diversity. Standard Methods for Amphibians*, pp. 144–6. Smithsonian Institution Press, Washington DC.

Gibbons, J. W. and Bennett, D. H. (1974). Determination of anuran terrestrial activity patterns by a drift fence method. *Copeia* **1974**, 236–43.

Gibbons, J. W. and Semlitsch, R. D. (1981). Terrestrial drift fences with pitfall traps: an effective technique for quantitative sampling of animal populations. *Brimleyana*, **7**, 1–16.

Gibbons, J. W., Winne, C. T., Scott, D. E., Willson, J. D., Glaudas, X., Andrews, K. M., Todd, B. D., Fedewa, L. A., Wilkinson, L., Tsaliagos, R. N. *et al.* (2006). Remarkable amphibian biomass and abundance in an isolated wetland: Implications for wetland conservation. *Conservation Biology*, **20**, 1457–65.

Gittins, S. P. (1983). The breeding migration of the common toad (*Bufo bufo*) to a pond in mid-Wales. *Journal of Zoology, London*, **199**, 555–62.

Grant, B. W., Tucker, A. D., Lovich, J. E., Mills, A. M., Dixon, P. M., and Gibbons, J. W. (1992). The use of coverboards in estimating patterns of reptile and amphibian biodiversity. In D. R. McCullough, and R. H. Barrett (eds), *Wildlife 2001*, pp. 379–403. Elsevier Science, London.

Greenberg, C. H., Neary, D. G., and Harris, L. D. (1994). A comparison of herpetofaunal sampling effectiveness of pitfall, single-ended, and double-ended funnel traps used with drift fences. *Journal of Herpetology*, **28**, 319–24.

Griffiths, R. A. (1985). A simple funnel trap for studying newt populations and an evaluation of trap behavior in smooth and palmate newts, *Triturus vulgaris* and *Triturus helveticus*. *Herpetological Journal*, **1**, 5–10.

Heyer, W. R., Donnelly, M. A., McDiarmid, R. W., and Hayek, L. C. (eds) (1994). *Measuring and Monitoring Biological Diversity. Standard Methods for Amphibians*, Smithsonian Institution Press, Washington DC.

Houze, C. M. and Chandler, C. R. (2002). Evaluation of coverboards for sampling terrestrial salamanders in south Georgia. *Journal of Herpetology*, **36**, 75–81.

Hsu, M. Y., Kam, Y. C., and Fellers, G. M. (2005). Effectiveness of amphibian monitoring techniques in a Taiwanese subtropical forest. *Herpetological Journal*, **15**, 73–9.

Jehle, R., Hodl, W., and Thonke, A. (1995). Structure and dynamics of central European amphibian populations: a comparison between *Triturus dobrogicus* (Amphibia, Urodela) and *Pleobates fuscus* (Amphibia, Anura). *Australian Journal of Ecology*, **20**, 362–6.

Johnson, J. R. (2005). A novel arboreal pipe-trap designed to capture the gray treefrog (*Hyla versicolor*). *Herpetological Review*, **36**, 274–7.

Johnson, S. A. (2003). Orientation and migration distances of a pond-breeding salamander (*Notophthalmus perstriatus*, Salamandridae). *Alytes*, **21**, 3–22.

Johnson, S. A. and Barichivich, W. J. (2004). A simple technique for trapping *Siren lacertina, Amphiuma means*, and other aquatic vertebrates. *Journal of Freshwater Ecology*, **19**, 263–9.

Luhring, T. M. and Young, C. A. (2006). Innovative techniques for sampling stream-inhabiting salamanders. *Herpetological Review*, **37**, 181–3.

Marsh, D. M. and Goicochea, M. A. (2003). Monitoring terrestrial salamanders: biases caused by intense sampling and choice of cover objects. *Journal of Herpetology*, **37**, 460–6.

Mazerolle, M. J., Bailey, L. L., Kendall, W. L., Royle, J. A., Converse, S. J., and Nichols, J. D. (2007). Making great leaps in herpetology: accounting for detectability in field studies. *Journal of Herpetology*, **41**, 672–89.

Mitchell, J. C., Erdle, S. Y., and Pagels, J. F. (1993). Evaluation of capture techniques for amphibian, reptile, and small mammal communities in saturated forested wetlands. *Wetlands*, **13**, 130–6.

Moore, J. D. (2005). Use of native wood as a new coverboard type for monitoring red-backed salamanders. *Herpetological Review*, **36**, 268–71.

Moulton, C. A., Fleming, W. J., and Nerney, B. R. (1996). The use of PVC pipes to capture hylid frogs. *Herpetological Review*, **27**, 186–7.

Palis, J. G., Adams, S. M., and Peterson, M. J. (2007). Evaluation of two types of commercially-made aquatic funnel traps for capturing ranid frogs. *Herpetological Review*, **38**, 166–7.

Parris, K. M., Norton, T. W., and Cunningham, R. B. (1999). A comparison of techniques for sampling amphibians in the forests of south-east Queensland, Australia. *Herpetologica*, **55**, 271–83.

Pauley, T. K. and Little, M. (1998). A new technique to monitor larval and juvenile salamanders in stream habitats. *Banisteria*, **12**, 32–6.

Pechmann, J. H. K., Scott, D. E., Semlitsch, R. D., Caldwell, J. P., Vitt, L. J., and Gibbons, J. W. (1991). Declining amphibian populations: the problem of separating human impacts from natural fluctuations. *Science*, **253**, 892–5.

Phillips, C. A. and Sexton, O. J. (1989). Orientation and sexual differences during breeding migrations of the spotted salamander, *Ambystoma maculatum. Copeia*, **1989**, 17–22.

Pittman, S. E., Jendrek, A. L., Price, S. J., and Dorcas, M. E. (2008). Habitat selection and site fidelity of Cope's gray treefrog (*Hyla chrysoscelis*) at the aquatic-terrestrial ecotone. *Journal of Herpetology*, **42**, 378–85.

Resetarits, W. J. and Wilbur, H. M. (1991). Calling site choice by *Hyla chrysoscelis*: effects of predators, competitors, and oviposition sites. *Ecology*, **72**, 778–86.

Rice, A. N., Rice, K. G., Waddle, J. H., and Mazzotti, F. J. (2006). A portable non-invasive trapping array for sampling amphibians and reptiles. *Herpetological Review*, **37**, 429–30.

Ryan, T. J., Philippi, T., Leiden, Y. A., Dorcas, M. E., Wigley, T. B., and Gibbons, J. W. (2002). Monitoring herpetofauna in a managed forest landscape: effects of habitat types and census techniques. *Forest Ecology and Management*, **167**, 83–90.

Semlitsch, R. D. (1985). Analysis of climatic factors influencing migrations of the salamander *Ambystoma talpoideum. Copeia*, 1985, 477–89.

Semlitsch, R. D. and Pechmann, J. H.K. (1985). Diel patterns of migratory activity for several species of pond-breeding salamanders. *Copeia*, **1985**, 86–91.

Shaffer, H. B., Alford, R. A., Woodward, B. D., Richards, S. J., Altig, R. G., and Gascon, C. (1994). Quantitative sampling of amphibian larvae. In W. R. Heyer, M. A. Donnelly, R. W. McDiarmid, and L. C. Hayek (eds), *Measuring and Monitoring Biological Diversity. Standard Methods for Amphibians*, pp. 130–41. Smithsonian Institution Press, Washington DC.

Stevens, C. E. and Paszkowski, C. A. (2005). A comparison of two pitfall trap designs in sampling boreal anurans. *Herpetological Review*, **36**, 147–9.

Todd, B. D. and Winne, C. T. (2006). Ontogenetic and interspecific variation in timing of movement and responses to climatic factors during migrations by pond-breeding amphibians. *Canadian Journal of Zoology*, **84**, 715–22.

Todd, B. D., Winne, C. T., Willson, J. D., and Gibbons, J. W. (2007). Getting the drift: examining the effects of timing, trap type, and taxon on herpetofaunal drift fence surveys. *American Midland Naturalist*, **158**, 292–305.

Vogt, R. C. and Hine, R. L. (1982). Evaluation of techniques for assessment of amphibian and reptile populations in Wisconsin. In N. J. Scott, Jr (ed.). *Herpetological Communities*, pp. 201–17. Wildlife Research Report 13. United States Fish and Wildlife Service, Washington DC.

Waldron, J. L., Dodd, Jr, C. K., and Corser, J. D. (2003). Leaf litterbags: factors affecting capture of stream-dwelling salamanders. *Applied Herpetology*, **1**, 23–6.

Weddeling, K., Hachtel, M., Sander, U., and Tarkhnishvili, D. (2004). Bias in estimation of newt population size: a field study at five ponds using drift fences, pitfalls, and funnel traps. *Herpetological Journal*, **14**, 1–7.

Willson, J. D. and Dorcas, M. E. (2003). Quantitative sampling of stream salamanders: comparison of dipnetting and funnel trapping techniques. *Herpetological Review*, **34**, 128–30.

Willson, J. D. and Dorcas, M. E. (2004). A comparison of aquatic drift fences with traditional funnel trapping as a quantitative method for sampling amphibians. *Herpetological Review*, **35**, 148–50.

Willson, J. D., Winne, C. T., and Fedewa, L. A. (2005). Unveiling escape and capture rates of aquatic snakes and salamanders (*Siren* spp. and *Amphiuma means*) in commercial funnel traps. *Journal of Freshwater Ecology*, **20**, 397–403.

14

Area-based surveys

David M. Marsh and Lillian M.B. Haywood

14.1 Introduction: what are area-based surveys?

Area-based surveys are frequently used to estimate the relative abundance, density, or diversity of amphibians. While the methodological details of area-based surveys can vary considerably, the basic principles behind these surveys are usually the same. Small, defined units, referred to as plots, quadrats, or transects, are selected randomly from larger areas of interest. These small units are then sampled for amphibians, and the data collected are used to make inferences about the relative abundance or diversity of amphibians in the larger areas.

Area-based surveys can be employed in a variety of situations. For example, one might want to compare amphibians in primary forest to amphibians in secondary forest or pasture. Or, one might want to analyze changes in amphibian communities over time (i.e. monitoring) or across an environmental gradient like elevation or distance from a disturbance. Area-based surveys are also used in experimental studies, for instance when habitats are modified and amphibians are compared between treatment and control areas. Finally, area-based studies are commonly used to study the basic habitat relationships and community ecology of amphibians (e.g. Toft 1980; Hairston 1987).

14.1.2 Why use area-based surveys?

Area-based surveys are useful primarily because they are rigorous and repeatable and because they can be replicated to put confidence intervals on estimates of abundance or diversity. Although haphazard surveys (i.e. walking around a large area and looking for amphibians) can yield count data and species lists, data from haphazard surveys are limited in their utility. One problem with haphazard surveys is that there is no way to repeat them consistently. If, years later, one wants to re-survey the same site or compare the site to some other area, there is simply no way to collect comparable data. Second, haphazard surveys result in single estimates with no information on variation. As a result, any measure of abundance or diversity will lack confidence intervals and hypothesis testing will often be impossible.

Time-constrained surveys, in which a set amount of time is spent searching for amphibians, are one step more rigorous than haphazard surveys. However, when carried out in very large areas, time-constrained surveys lead to many of the same problems concerning rigor and repeatability as do haphazard surveys. Controlling the time period involved in a survey is important, but this can be done within the context of area-based surveys.

In this chapter, we first outline the different kinds of area-based surveys and illustrate their use in the amphibian research literature. We then discuss when each of the different kinds of area-based surveys is preferable. Finally, we consider the major methodological issues involved in designing area-based surveys. In the interest of brevity, we focus on uses of area-based surveys in amphibian conservation and management rather than basic population or community ecology. Most of these uses involve the collection of count data (which are assumed to reflect relative abundance; see Chapters 24 and 25) or else the estimation of true density. Throughout this chapter, we use the term abundance to indicate relative abundance or counts and density to signify an estimate of true density. Our recommendations concerning area-based sampling are based in part on a survey of approximately 100 studies that used area-based sampling to collect data on amphibian abundance or diversity (see Table 14.1). They are also based

Table 14.1 *Summary of a literature review of 89 studies that used area-based sampling. Studies are divided by whether the primary sampling units were plots/quadrats, transects, transects nested within plots, or quadrats nested within plots or transects. Summary data show the percentage of studies of each type that were: (1) terrestrial, aquatic, or both; (2) tropical or temperate; (3) single-species or multi-species studies, (4) targeting frogs, salamanders, caecilians, or multiple groups; and (5) conducted during daytime, at night, or both.*

Approach	N	Terrestrial/ aquatic/both	Tropical/ temperate	Single/ multiple species	Frogs/salamanders/ caecilians/multiple	Day/night/ both
Plots/quadrats	33	53/37/10	34/66	34/66	39/26/8/26	55/28/17
Transects	29	34/45/21	34/66	52/48	55/14/0/31	35/56/9
Nested transects	15	67/7/26	27/73	33/67	40/47/0/13	38/50/12
Nested quadrats	12	92/0/8	42/58	50/50	42/8/50/0	29/0/71

on our own experience using plots and transects to sample terrestrial salamanders, stream salamanders, and neotropical frogs.

14.2 Kinds of area-based survey

There are two basic types of area-based survey: plot or quadrat surveys and transect surveys. The terms plot and quadrat are often used interchangeably and refer to rectangular or square sampling units. Plots and quadrats can span a wide range of sizes, from 0.5 m^2 stream-bed quadrats to forest plots of several hectares. Our literature review found that when plots were used as a primary sampling unit, their median dimensions were 25 m (range 4–400 m) by 20 m (range 2–240 m). The great variation in plot size in the literature highlights the need to adapt plot designs specifically to the species and habitat of interest.

Transects are long, narrow plots, designed to be searched by one investigator in a single pass. Our review found that the median length of amphibian transects was 100 m (range 7–2000 m) with a median width of 2 m (range 1–8 m). The most common transect lengths were 50 m and 100 m and the most common widths were 1 and 2 m. Transects are often sampled visually, but frogs can also be sampled by listening for calls (see Chapter 16).

In addition to these basic designs, researchers often use nested designs in which one type of sampling unit is contained within another (e.g. transects or quadrats nested within large plots). In these designs, researchers randomly select plot locations within the areas of interest, and then randomly choose quadrat or transect locations within the plots. These nested quadrats or transects are used to make inferences about the abundance or diversity of amphibians within the plots, and the plots are used to make inferences about amphibians in the larger areas.

14.3 Specific examples of area-based surveys

Beyond the general classification of plots, quadrats, and transects, there are many specific uses for area-based surveys. In this section, we give several examples chosen to reflect the diversity of area-based surveys and some of their more common uses.

14.3.1 Leaf-litter plots

Anurans are abundant in the leaf-litter of tropical forests and can contribute substantially to amphibian diversity in these habitats. Taxa frequently encountered in the leaf litter include terrestrial genera such as *Eleutherodactylus* in the

Neotropics, *Arthroleptis* in Africa, and *Philautus* in South Asia. The leaf-litter amphibian fauna may also include Bufonids, Ranids, Leptodactylids, and other amphibians that breed in aquatic habitat but spend most of their adult lives on the forest floor.

Small plots have been frequently used to sample leaf-litter amphibians in tropical forests (Inger 1980; Vonesh 2001; Rocha *et al.* 2001). To carry out leaf-litter surveys, researchers mark off the boundaries of a plot and carefully sift through the leaf litter for amphibians until they have covered the entire plot. Searching leaf litter in this manner is time-consuming, so plots are typically small: usually 1–64 m^2. Rocha *et al.* (2001) argued from a direct comparison of methods in Atlantic rainforest that small plots (i.e. 2 m^2) surveyed very carefully were likely to be superior to larger plots (i.e. 64 m^2) surveyed more superficially. Leaf-litter quadrats are typically searched during the day for reasons of convenience, though Rocha *et al.* (2000) found that both counts and species richness were actually higher at night.

Because leaf-litter surveys disturb the habitat, researchers usually search each plot only once. However, litter plots can be highly replicated within any area of interest. Vonesh (2001), for example, searched a total of 150 25 m^2 quadrats in Kibale National Park, Uganda, 50 each in undisturbed forest, selectively logged forest, and pine plantation. Conducting a few trial surveys will give a good idea of how large a plot can be effectively surveyed, how many plots can be searched, and how abundant amphibians are likely to be in any given sample. Habitat data can also be an important component of leaf-litter surveys: variables measured commonly include leaf-litter mass or depth (often correlated with amphibian abundance: Allmon 1991; Vonesh 2001), herb cover, canopy cover, and soil moisture, temperature, and pH.

14.3.2 Natural-cover surveys for terrestrial salamanders

Terrestrial salamanders are regularly found underneath natural-cover objects (e.g. rocks and logs) on the forest floor. Natural-cover surveys for terrestrial salamanders can be carried out using plots or transects, though these are typically much larger than leaf-litter plots, encompassing 100–2000 m^2. In natural-cover surveys, one turns over all accessible rocks and logs and searches for salamanders underneath them. Although natural-cover surveys sample only a fraction of salamander population, the resulting counts have been shown to be highly correlated with density estimates from more extensive mark–recapture (Smith and Petranka 2000). However, salamanders move back and forth between cover objects, the leaf litter, and underground retreats, so proper timing of these surveys is critical. Generally, the best time for natural-cover surveys is when the soil underneath cover objects is wetter than the surrounding leaf litter. Cover-object surveys, like leaf-litter surveys, can disturb the microhabitats of interest, and Smith and Petranka (2000) recommend re-sampling plots once annually. Habitat data collected during cover surveys usually include herb cover, canopy cover, soil moisture, and soil temperature, as well as data on the number, type, and decay classes of cover objects searched. Salamander counts can be divided by the number of cover objects on a plot to correct for differences in search effort (e.g. Marsh and Beckman 2004); however, whether or not this yields more accurate estimates of relative abundance has not been tested.

The ability to resample salamanders more frequently without degrading cover objects is one big advantage for artificial cover objects (ACOs; see Chapter 13) over natural-cover surveys. However, counts from grids of ACOs may be only weakly correlated with counts from natural-cover surveys (Hyde and Simons 2001), and ACOs may under-represent smaller size classes of salamanders (Marsh and Goicochea 2003). Thus, natural-cover surveys are recommended over ACOs for relative abundance and diversity estimation until ACOs have been more thoroughly validated.

14.3.3 Nocturnal transects

Nocturnal transects have regularly been used to survey forest frogs in tropical habitats. At night, frogs may crawl up onto vegetation where they can be seen at or below eye level. Terrestrial-breeding frogs are commonly found with nocturnal transects. Additionally, researchers have reported low counts but a remarkably high richness of aquatic-breeding species (e.g. Pearman 1997).

Researchers typically set up transects by flagging vegetation in a straight line of 30–100 m. These transects are walked slowly, searching with a flashlight or headlamp for frogs. The width of transects is usually no more than 1–2 m to each side, as beyond this distance too many frogs are likely to be missed (Funk *et al.* 2003). Some frogs along transects may be found by localizing their calls, so in a sense these transects may be considered a form of aural survey (Chapter 16). However, female frogs of most species do not call, and many additional males will be seen that are not heard calling. Thus, counts from nocturnal transects will usually be higher than counts from aural surveys.

Nocturnal transects often involve only moderate disturbance to the vegetation, and in these cases surveys can be repeated multiple times and used in conjunction with mark–recapture density estimation (Funk *et al.* 2003). However, in habitats with dense understory vegetation, habitat degradation could lead to reduced counts in subsequent surveys. The optimal conditions for carrying out nocturnal transects can also vary widely among species and habitats.

Temperature and moisture are often positively associated with frog calling activity (Townsend and Stewart 1994; Duellman 1995), although very heavy rains can actually reduce activity. Practicing transects in a variety of weather conditions can allow one to determine the best times for carrying out nocturnal transects. Habitat data measured in conjunction with nocturnal transects often include herb cover, canopy cover, tree stems, and tree diameter at breast height.

14.3.4 Quadrats for stream amphibians

Amphibians with adult or larval stages that live in streams may be sampled underneath rocks in the stream bed or at the edge of the stream. Several researchers have used quadrat surveys to sample amphibians, particularly larvae and small adults, in rocky stream beds (Barr and Babbitt 2002). Quadrats of 0.25–4 m² are randomly selected within a channel unit (pool, run, or riffle) and then all rocks are carefully removed to uncover the amphibians present. Quadrat surveys of a stream bed are in several respects analogous to leaf-litter surveys for tropical frogs: they are time-consuming so quadrats are typically small, they are moderately destructive so each quadrat is only searched once, and many quadrats can be surveyed in the course of a study. Habitat data collected during these kinds of surveys usually includes information on the size and number of rocks turned, as well as stream characteristics such as water temperature, pH, and flow rate.

14.3.5 Soil quadrats for caecilians

Several recent studies suggest that excavating soil quadrats may be a useful method for sampling caecilians. Measey *et al.* (2003) counted caecilians in the Western Ghats of India with 1 m² quadrats that were nested within 100 m² plots. They used a bladed hoe to dig out the soil to a depth of 0.3 m and sift through it for caecilians. This technique was surprisingly effective: they detected from zero to 12 caecilians per quadrat, with a mean density of 0.51–0.63/m². The one downside of the technique was that some of the caecilians (≈13%) showed signs of injury from the sampling procedure, although using a forked digging tool in place of a bladed hoe could possibly reduce injury rates (G.J. Measey, personal communication). Measey (2006) further suggested that because most caecilians were found near the surface, sampling could be made more efficient by first searching a larger plot (e.g. 25 m²) to only a shallow depth and then searching one or more quadrats within each plot to a greater depth. It bears mentioning that not all caecilians are ecologically equivalent and some species may be more likely than others to be detected in soil quadrats. Gower *et al.* (2004), for example, found that timed digs in Tanzania resulted in high counts of one caecilian species, whereas a second species was detected almost exclusively on the surface during rainy nights.

14.4 Modifications

The previous examples highlight some of the basic uses of plot and transect surveys. However, several modifications to these approaches have also been proposed. We cover two in detail: distance sampling (a modification of transect surveys) and adaptive cluster sampling (a modification of quadrat surveys).

14.4.1 Distance sampling

In bird surveys, distance sampling is widely used to estimate both density and detection rates (Gregory *et al.* 2004). The general idea of distance sampling is to make counts along a transect, but also keep track of the distance between the center of the transect and each animal found. Any decrease in counts with distance from the center of the transect is ascribed to changes in detectability. By modeling the way that detectability decreases with distance, one can estimate density from transect counts (for example with the program DISTANCE; Thomas *et al.* 2006). Unfortunately, distance sampling has been judged only moderately effective for amphibians (Fogarty and Vilella 2001; Funk *et al.* 2003), except when estimating burrow abundance of the fossorial salamander *Phaeognathus hubrichti* (Dodd 1990). This may be in part because many amphibians typically go undetected even at the center of the transect, leading to imprecise density estimates. Still, distance sampling does allow one to test whether detectability differs among habitats when mark–recapture is not feasible.

14.4.2 Adaptive cluster sampling

Amphibians are often highly aggregated in space with many individuals in some areas but few to none in other areas. This can be frustrating because it means that many randomly located sites may contain no amphibians. Adaptive cluster sampling can help focus sampling efforts in areas where animals are actually being found, but do so in a repeatable way (Thompson and Seber 1996). In adaptive cluster sampling, one starts by searching a randomly chosen quadrat within the area of interest. If this plot satisfies some pre-determined criterion (e.g. more than one amphibian found) then one begins to search adjacent plots. This process continues until a network of plots has been searched and is surrounded by so-called edge plots that fail to meet the criteria. Noon *et al.* (2006) used adaptive cluster sampling to search leaf-litter plots in the Western Ghats but found that the technique was actually less efficient than simple random sampling for detecting individual species. They postulated that the technique might have worked better if amphibians had been more abundant and more highly aggregated; however, more research is needed before adaptive cluster sampling can be recommended as a first-line approach to sampling.

14.5 Design issues: choice of sampling unit

There are two basic issues in the design of area-based studies: the size and shape of the sampling units and the number of these units to sample. In terms of size, small plots or quadrats are generally effective in situations where amphibians are small and fairly abundant and sampling a small area will tend to lead to nonzero counts. They are also useful when habitat heterogeneity is high because they can be highly replicated across the study area. In contrast, larger plots will generally be more effective when amphibians are less abundant, but are visible enough that larger areas can be efficiently searched. Plot surveys have been commonly used for searching low vegetation for frogs and searching coarse woody debris for salamanders (Woolbright 1991; Bailey *et al.* 2004). In both of these cases, small, randomly chosen quadrats might well contain no amphibians, and some would not even contain appropriate vegetation or woody debris for sampling.

With respect to the shape of the sampling units, linear transects sometimes offer advantages over rectangular plots. For one, transects are narrow and thus allow the researcher to do all sampling in a single pass. Because plots are fairly wide, it can be difficult for a single investigator to keep track of which areas have already been searched, particularly during nocturnal surveys. Second, transects cut across a good length of habitat, which may help to average out patchiness in the environment. For example, when we conducted transect searches to determine the effects of forest roads on salamanders, we found a highly significant relationship between distance from roads and salamander counts (Marsh and Beckman 2004). However, when we later surveyed 8 m × 8 m plots in similar habitats we found no significant relationship. In these plot surveys, the patterns in counts were almost identical, but the standard deviation in salamander counts was more than twice as high because of patchiness in salamander distributions.

A further use of transects is when one wants to test the effects of an environmental gradient such as elevation or distance from habitat disturbance. Transects can be oriented parallel to the gradient to allow for a continuous analysis of the effects of the gradient on amphibian abundance (e.g. Lehtinen *et al.* 2003). One can also set up transects perpendicular to the gradient of interest (e.g. Marsh and Pearman 1997); however, this approach has the disadvantage of restricting analysis to a few specific points along the gradient. Sampling perpendicular to the gradient may be reasonable when there are particular points of interest in the gradient (Jaeger and Inger 1994), but should probably be avoided when the gradient and its effects on amphibians are poorly understood.

Although transects may present some advantages over plots, there are also some situations where plots are likely to be superior. First, when the area to be sampled is small and round or rectangular, plots are far more practical than very short transects. Also, plots may be useful when many observers are available for sampling. Observers can cover a rectangular plot in parallel, making these more convenient to sample than several distinct transects. Finally, plots should generally be selected when one is sampling amphibians in conjunction with mark–recapture analysis. Even if amphibians move only a few meters between captures, they will tend to wander off of transects and small quadrats, leading to low detection rates and difficulty with estimating population parameters (Schaub *et al.* 2004).

14.5.1 Design issues: how many replicates?

After deciding between plots and transects the next question is usually "how many?" Strictly speaking, the number of replicates needed depends on the magnitude of differences in amphibian abundances one is hoping to detect and on how much background variation exists between survey dates, sites, and observers. For example, to compare amphibians between two habitats in which the standard deviation in counts is equal to 25% of the mean, one would need about nine replicates per habitat to detect a 50% difference in counts between habitats (at 90% power) and almost 50 replicates to detect a 20% difference.

In theory, larger plots or transects will tend to average out some habitat variation, so one would tend to need fewer plots than small quadrats. Typical levels of replication in the literature are five to 30 for large forest plots, 10–50 for transects, and 20–100 for small quadrats. To estimate how many replicates are needed, one can do a few practice plots to calculate variance. From these estimates, the R library *pwr* (R Development Core Team 2005) has a number of modules for doing power analysis and the downloadable program Monitor.exe (Gibbs *et al.* 1998) can perform power calculations for detecting changes over time.

In some cases, time may be limiting and the more realistic answer to the question of "how many plots?" is simply "as many as you can." Thus, an alternative approach is to do a few practice sites and estimate the time needed for one replicate. Then consider how much total time can be committed to the project. Dividing the latter by the former yields an estimate of how many replicates can be surveyed.

14.5.2 Reducing variation among replicates

More variation among replicates means surveying more sites, so anything that reduces variation will usually be helpful. Standardization of sampling methods is one critical aspect of reducing variation. For area-based surveys, standardization can include things like always searching plots for the same amount of time, always searching the same microhabitats within a plot, and always using the same surveyors for each plot.

These suggestions are valid for all animals, but for amphibians there are additional considerations. Because amphibian activity is highly dependent on weather conditions, amphibian counts can vary by orders of magnitude from day to day or even from hour to hour. There are two main approaches to dealing with this kind of variation. The first is to restrict surveys to periods of time when amphibians are most likely to be observed. This requires flexibility with scheduling surveys but can dramatically reduce the amount of survey effort needed. It also requires some initial surveys to get a sense of amphibian activity patterns. The second approach is to carry out surveys in a variety of conditions, but to collect data on weather-related variables that may be associated with activity levels (e.g. rainfall within the past week, rainfall at the time of the survey, temperature, humidity). These variables can then be included as covariates in any analysis of amphibian counts, thereby controlling some of the background variation.

In conjunction with either of these two approaches, surveys can be paired or grouped in a way that ensures that background variation will not confound a hypothesis being tested. For example, in a comparison of forest to pasture, each forest plot could be paired with a pasture plot so that one of each (or two or three of each) is surveyed on the same date. The pairing of these surveys ensures that any day-to-day variation affects both habitats equally. Additionally, by using survey date as a random effect in the statistical analysis, one can remove most of this variation from the variable of interest.

14.5.3 How many times to survey each replicate?

The number of times each replicate should be surveyed will depend on how destructive a method is to the microhabitats where amphibians are found. If a sampling method tends to degrade the habitats of interest, repeat surveys will result in lower counts and the appearance of population declines where none would otherwise exist. In these cases (e.g. leaf-litter plots) one can sample a different, randomly chosen set of quadrats in each survey period. This will tend to increase sampling error (since different plots will be surveyed each time) but will eliminate any bias associated with habitat degradation from surveys.

However, even when there is little disturbance associated with sampling, one is still faced with a trade-off between the number of times each site can be surveyed and the total number of sites that can be covered. If amphibian abundance is highly variable within an area, but there is not much variation from one survey to the next, more replicate plots that are surveyed less frequently is likely to be optimal (e.g. Smith and Petranka 2000). The exception to this is when using repeat surveys with mark–recapture to estimate detection rates (see Chapter 24).

14.6 An example of study design

Suppose one wanted to compare counts of a single amphibian species in forest habitat to counts in pasture. Assuming that only one forest area is available for sampling and that it is entirely surrounded by pasture, one would have at least four choices for area-based surveys: large plots, small quadrats, transects, or some nested design. If a total of 1000 m² could be surveyed in each of the two habitats, one could sample five 20 m × 10 m plots, 10 50 m × 2 m transects, or 40 5 m × 5 m quadrats (Figure 14.1a–c). Alternatively, one could use a nested design

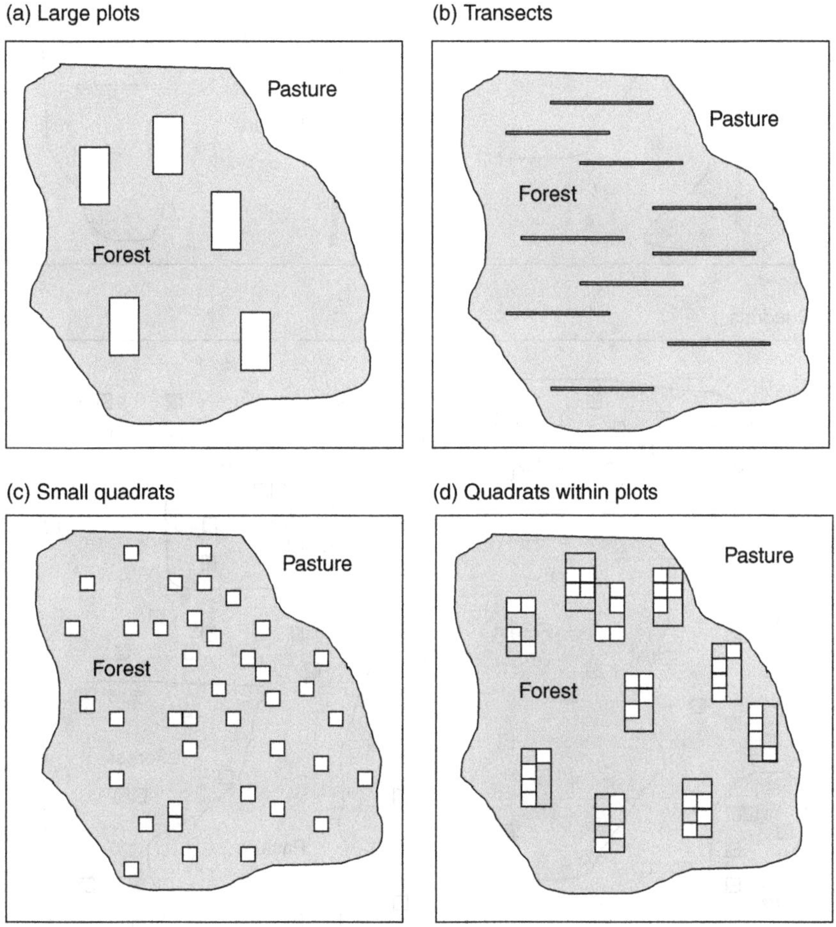

Fig. 14.1 Four possible designs for area-based sampling to compare a single forest area to surrounding pasture. Equal numbers of samples would be randomly located within pasture but these are not shown. (a) Five plots of 200 m². (b) Ten transects of 50 m × 2 m. (c) Forty 5 m × 5 m quadrats. (d) Four 5 m × 5 m quadrats nested within each of ten 200 m² plots.

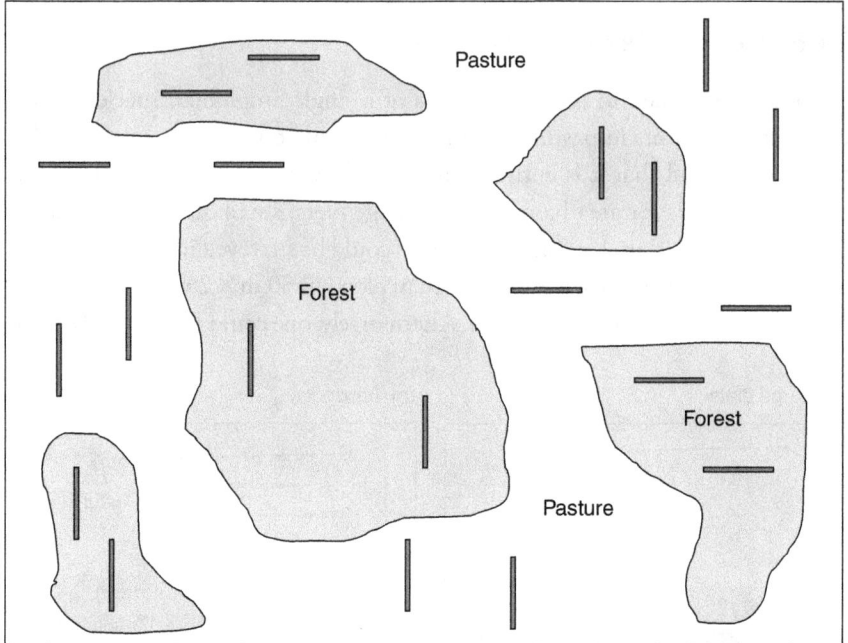

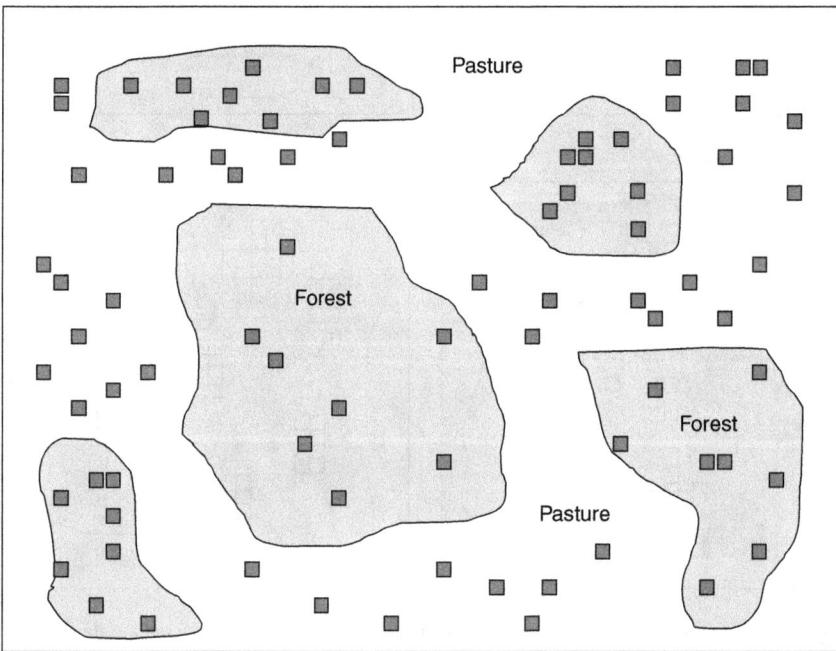

Fig. 14.2 Two possible designs for area-based sampling to compare five forest patches to surrounding pasture. (a) Two transects of 50 m × 2 m within each forest patch. (b) Eight 5 m × 5 m quadrats within each forest patch.

with 10 20 m × 10 m plots and four 5 m × 5 m quadrats chosen for sampling from within these (Figure 14.1d).

Any of these designs would permit a comparison of amphibian counts between the single forest patch and the surrounding pasture. However, because only one forest patch is sampled, inferences can only be made about that particular site. Thus, an even better design would use replicated forest patches rather than the single forest area, which would allow general inferences about amphibian counts in forest within this region. Assuming that five forest patches could be located, an area-based survey could employ a single 20 m × 10 m plot in each forest patch, two 50 m × 2 m transects, or eight 5 m × 5 m quadrats (Figure 14.2) Each of these could be paired with a parallel set-up in the surrounding pasture.

14.7 Assumptions of area-based surveys

Area-based surveys, as with any other research technique, require some assumptions to draw inferences from the data collected. First, plots or transects are assumed to be representative of the larger area of interest. In theory, a large sample of randomly chosen plots will closely approximate the larger area. However, smaller numbers of plots, even if randomly chosen, might be spatially aggregated or might under-represent particular microhabitats. When the larger area is known to vary, stratifying the area into different habitat types, and then randomly choosing plots within strata may be a better approach (Cochran 1977).

Second, area-based surveys generally yield count data with unknown detection rates. This is particularly true with more destructive techniques (e.g. leaf-litter plots, stream-bed quadrats) since one cannot use repeat surveys and mark–recapture to directly estimate detectability. Issues related to detectability and the interpretation of count data are covered in Chapters 24 and 25. Here, we simply point out that most uses of area-based surveys are aimed at comparing relative abundance or species richness among sites or time periods. Thus, the key assumption is not that all individuals are detected, but that detection rates are relatively constant among habitats or surveys. For instance, if vegetation differs between primary and secondary forest, counts from nocturnal transects in the two habitats may not be directly comparable since detection rates will likely depend on vegetation density. Furthermore, detectability is pretty much always an issue if one wants to compare abundances of different species. Distinct species tend to use different microhabitats, so area-based surveys will almost always detect some species at higher rates than others. Thus, characterizing amphibian communities usually requires multiple search techniques and some formal estimate of detectability.

Finally, with area-based surveys one generally assumes that amphibians do not move around much during the course of a survey. If amphibians are disturbed by the survey techniques and consistently escape the sampling area, these surveys will not be particularly effective.

14.8 Summary and recommendations

There is no single best approach for conducting area-based surveys. Which approach is optimal will depend on the site, the species, and the hypothesis being tested. That said, there are some general recommendations to keep in mind when conducting area-based surveys.

1) Before beginning a study, always practice the survey techniques. Testing out techniques is essential and usually leads to adjustments and improvements. Ultimately, trial surveys save time, because data will be of higher quality and there will be no need to repeat any surveys. Even two or three trial plots or transects can go a long way towards identifying potential problems and suggesting the necessary changes.

2) When in doubt, include more replicates. Almost nothing is as frustrating as coming up a few plots or transects short of an answer. When faced with a trade-off between the size of sampling units and the number of replicates, we believe that a greater number of smaller replicates will usually be optimal (the exception is when plots are so small that zero counts frequently result). Again, practicing plots of various sizes can help steer one to an appropriate-sized sampling unit.

3) Carefully consider the research question and how the survey design will answer that question. Going to the extreme of writing out a sample data set and attempting to analyze it can help identify errors or unstated assumptions in a sampling approach.

4) Know your amphibians. While this chapter has focused on the general concepts of area-based surveys, the details of a species' natural history will ultimately determine whether a particular sampling approach is successful. Understanding the basics of activity patterns and movement behavior are critical for effective sampling.

14.9 References

Allmon, W. D. (1991). A plot study of forest floor litter frogs, central Amazon, Brazil. *Journal of Tropical Ecology*, 7, 503–22.

Bailey, L. L., Simons, T. R., and Pollock, K. H. (2004). Spatial and temporal variation in detection probability of *Plethodon* salamanders using the robust capture-recapture design. *Journal of Wildlife Management*, **68**, 14–24.

Barr, G. E. and Babbitt, K. J. (2002). Effects of biotic and abiotic factors on the distribution and abundance of larval two-lined salamanders (*Eurycea bislineata*) across spatial scales. *Oecologia*, **133**, 176–85.

Cochran, W. G. (1977). *Sampling Techniques*. Wiley, New York.

Dodd, Jr, C. K. (1990). Line transect estimation of Red Hills salamander burrow density using a Fourier series. *Copeia*, **1990**, 555–7.

Duellman, W. E. (1995). Temporal fluctuations in abundances of anuran amphibians in a seasonal Amazonian rainforest. *Journal of Herpetology*, **29**, 13–21.

Fogarty, J. H. and Vilella, F. J. (2001). Evaluating methodologies to survey *Eleutherodactylus* frogs in montane forests of Puerto Rico. *Wildlife Society Bulletin*, **29**, 948–55.

Funk, W. C., Almeida-Reinoso, D., Nogales-Sornosa, F., and Bustamante, M. R. (2003). Monitoring population trends of *Eleutherodactylus* frogs. *Journal of Herpetology*, **37**, 245–56.

Gibbs, J. P., Droege, S., and Eagle, P. (1998). Monitoring populations of plants and animals. *Bioscience*, **48**, 935–40.

Gower, D. J., Loader, S. P., Moncrieff, C. B., and Wilkinson, M. (2004). Niche separation and comparative abundance of *Boulengerula boulengeri* and *Scolecomorphus vittatus* (Amphibia: Gymnophiona) in an East Usambara forest, Tanzania. *African Journal of Herpetology*, **53**, 183–90.

Gregory, R. D., Gibbons, D. W., and Donald, P. F. (2004). Bird census and survey techniques. In W. J. Sutherland, I. Newton, and R. E. Green (eds), *Bird Ecology and Conservation: a Handbook of Techniques*, pp. 17–55. Oxford University Press, Oxford.

Hairston, Sr, N. G. (1987). *Community Ecology and Salamander Guilds*. Cambridge University Press, New York.

Hyde E. J. and Simons, T. R. (2001). Sampling plethodontid salamanders: sources of variability. *Journal of Wildlife Management*, **65**, 624–32.

Inger, R. F. (1980). Densities of floor–dwelling frogs and lizards in lowland forests of SE Asia and Central America. *American Naturalist*, **115**, 761–70.

Jaeger, R. and Inger, R. F. (1994). Standard techniques for inventory and monitoring: quadrat sampling. In W. R. Heyer, M. A. Donnely, R. W. McDiarmid, L. C. Hayek, and M. S. Foster (ed.), *Measuring and Monitoring Biological Diversity. Standard Methods for Amphibians*, pp. 97–102. Smithsonian Institution Press, Washington DC.

Lehtinen, R. M., Ramanamanjato, J. B., and Raveloarison, J. G. (2003). Edge effects and extinction proneness in a herpetofauna from Madagascar. *Biodiversity and Conservation*, **12**, 1357–70.

Marsh, D. M. and Pearman, P. B. (1997). Effects of habitat fragmentation on the abundance of two species of *Leptodactylid* frogs in an Andean montane forest. *Conservation Biology*, **11**, 1323–8.

Marsh, D. M. and Goicochea, M. A. (2003). Monitoring terrestrial salamanders: biases due to frequent sampling and choice of cover objects. *Journal of Herpetology*, **37**, 460–6.

Marsh, D. M. and Beckman, N. G. (2004). Effects of forest roads on the abundance and activity of terrestrial salamanders in the Southern Appalachians. *Ecological Applications*, **14**, 1882–91.

Measey, G. J. (2006). Surveying biodiversity of soil herpetofauna: towards a standard quantitative methodology. *European Journal of Soil Biology*, **42**, S103–10.

Measey, G. J., Gower, D. J., Oommen, O. V., and Wilkinson, M. (2003). Quantitative surveying of endogeic limbless vertebrates – a case study of *Gegeneophis ramaswamii* (Amphibia: Gymnophiona: Caeciliidae) in southern India. *Applied Soil Ecology*, **23**, 43–53.

Noon, B. R., Ishwar, N. M., Vasudevan, K. (2006). Efficiency of adaptive cluster and random sampling in detecting terrestrial herpetofauna in a tropical rainforest. *Wildlife Society Bulletin*, **34**, 59–68.

Pearman, P. B. (1997). Correlates of amphibian diversity in an altered landscape of Amazonian Ecuador. *Conservation Biology*, **11**, 1211–25.

R Development Core Team (2005). *R: a Language and Environment for Statistical Computing, Reference Index Version 2.2.1.* www.R-project.org. R Foundation for Statistical Computing, Vienna.

Rocha, C. F. D., Van Sluys, M., Alves, M. A.S., Bergallo, H. G., and Vrcibradic, D. (2000). Activity of leaf-litter frogs: when should frogs be sampled? *Journal of Herpetology*, **34**, 285–7.

Rocha, C. F. D., Van Sluys, M., Alves, M. A. S., Bergallo, H. G., and Vrcibradic, D. (2001). Estimates of forest floor litter frog communities: a comparison of two methods. *Austral Ecology*, **26**, 14–21.

Schaub, M., Gimenez, O., Schmidt, B. R., and Pradel, R. (2004). Estimating survival and temporary emigration in the multistate capture–recapture framework. *Ecology*, **85**, 2107–13.

Smith, C. K. and Petranka, J. W. (2000). Monitoring terrestrial salamanders: repeatability and validity of area-constrained cover object searches. *Journal of Herpetology*, **34**, 547–57.

Thomas, L., Laake, J. L., Strindberg, S., Marques, F. F. C., Buckland, S. T., Borchers, D. L., Anderson, D. R., Burnham, K. P., Hedley, S. L., Pollard, J. H. *et al.* (2006). *Distance 5.0. Release 2.* www.ruwpa.st-and.ac.uk/distance/. Research Unit for Wildlife Population Assessment, University of St Andrews, St Andrews.

Thompson, S. K. and Seber, G. A. F. (1996). *Adaptive Sampling.* John Wiley and Sons, New York.

Toft, C. A. (1980). Feeding ecology of thirteen syntopic species of anurans in a seasonal tropical environment. *Oecologia*, **45**, 131–41.

Townsend, D. S. and Stewart, M. M. (1994). Reproductive ecology of the Puerto Rican frog *Eleutherodactylus coqui*. *Journal of Herpetology*, **28**, 34–40.

Vonesh, J. R. (2001). Patterns of richness and abundance in a tropical African leaf-litter herpetofauna. *Biotropica*, **33**, 502–10.

Woolbright, L. L. (1991). The impact of Hurricane Hugo on forest frogs in Puerto Rico. *Biotropica*, **23**, 462–7.

15

Rapid assessments of amphibian diversity

James R. Vonesh, Joseph C. Mitchell, Kim Howell, and

Andrew J. Crawford

15.1 Background: rapid assessment of amphibian diversity

More than 6400 amphibian species are known worldwide, with more than 50 new species being described in just the first half of 2008 (AmphibiaWeb 2008). Many of these species are threatened or declining and more than 150 may have recently become extinct (IUCN 2006). Such rates of species loss are far greater than the historic background extinction rate for amphibians (e.g. McCallum 2007; Roelants *et al.* 2007). Amphibians play diverse roles in natural ecosystems, and their decline may cause other species to become threatened or may undermine aspects of ecosystem function (Matthews *et al.* 2002; Whiles *et al.* 2006). Anthropogenic habitat loss and degradation, disease, introduced species, and pollution or combinations of these factors are at the root of most declines. As awareness of declines has increased, conservation groups, governments, and land managers have become more interested in protecting amphibian diversity. However, the lack of accurate data on amphibian distributions, particularly for tropical regions where diversity and declines are concentrated (IUCN 2006), is often a roadblock to effective conservation and management. Ideally, lack of information on the amphibian fauna for a particular area would result in a thorough inventory. Unfortunately, this is usually not realistic. Exhaustive inventories are costly and may take decades to compile (e.g. Timm 1994). Given the urgency of the current global biodiversity crisis, finite resources, and dynamic socio-economic environments with respect to conservation, an approach for rapidly gathering preliminary data on biodiversity is sometimes required.

Rapid biodiversity-assessment methods were developed in response to this need. A Rapid Assessment (RA) is an accelerated, targeted, and flexible biodiversity survey, often focusing on species associated with particular vegetation types or topographical features (Sayre *et al.* 2000). RAs consist of planning, field sampling, and analytical and writing stages, and are implemented by teams of biologists and resource managers with expertise in the region and taxonomic group(s) of interest. Most RAs aim to be completed, from planning to report, in less than a year. Although a variety of RA strategies have been proposed, two of the better documented are the Rapid Ecological Assessment methodology developed by the Nature Conservancy (Sayre *et al.* 2000) and the Rapid Assessment Program methodology developed by Conservation International (Roberts 1991; www.conservation.org). RA approaches share a number of features in common (modified from Sayre *et al.* 2000).

- *Speed.* Reducing project duration reduces costs and delivers results to decision-makers quickly.
- *Careful initial planning and training.* Effective planning saves time and money and increases the consistency of the information gathered.
- *Use of mapping technologies.* The use of Geographic Information Systems (GIS), remote sensing, and global positioning systems (GPS) can greatly facilitate planning, sampling design, and geo-referencing of data.
- *Careful scientific documentation.* Selecting classification, sampling, and surveying methods that are most effective for the habitat, season, and taxa being sampled can maximize assessment effectiveness.
- *Capacity-building and partnerships.* Developing collaborative relationships among conservation, government, and academic partners in host countries and internationally helps ensure that information generated will have a local impact.

An RA is not an exhaustive faunal inventory, monitoring program, environmental impact assessment, or management plan, nor does it seek to elucidate species interactions or ecological processes, although an RA could serve as an important prelude to any of these activities. Instead, an RA is a conservation planning tool that can provide an efficient initial characterization of diversity in areas for which relatively little is known (Sayre *et al.* 2000). Here we describe in brief the steps involved in RA and discuss sampling methodologies likely to be effective for assessing amphibian diversity. Due to the integrative nature of RA, we often refer the reader to other chapters in this volume that cover material important to the topic.

15.1.1 When is an RA needed?

Determining whether an RA approach is called for depends on how much is already known about the area of interest and the urgency to obtain additional information. The best candidate sites are those that are only marginally surveyed and are highly threatened (Sayre and Roca 2000). It is in these cases that data resulting from an RA, though typically limited in scope, can have important implications for decision-making.

For example, the Atewa Range Forest Reserve of Ghana contains some of the last remaining upland evergreen forest in the country and is home to endemic plant and insect species (McCullough *et al.* 2007). In 2006, Conservation International initiated a Rapid Assessment Program of amphibians (and other taxa) within the Atewa Range Forest Reserve in response to proposed mineral exploration in the reserve. Sampling sites were selected in areas suspected to support high biodiversity that also had large mineral deposits (McCullough *et al.* 2007). Field surveys were conducted over just 18 days at the beginning of the rainy season. Nearly one-third of the 32 amphibian species observed were listed as threatened on the IUCN Red List. One species was considered Critically Endangered (Kouamé *et al.* 2007). Based in part on the amphibian assessment results, the authors recommended that Atewa Range Forest Reserve be elevated to national park status and that mineral exploitation be prohibited (Kouamé *et al.* 2007; McCullough *et al.* 2007).

15.2 Planning an RA

When an RA is indicated, the next steps involve planning and preparation. Because RAs may involve government and non-government participants from multiple regions, are often conducted in remote, logistically challenging field localities, and investigate unknown or poorly described faunas, substantial advanced planning is necessary for success.

15.2.1 Developing objectives

Establishing clear and realistic objectives is the most important step of the planning process, as the objectives become the yardstick for subsequent activities and resource allocation (Hayek 1994; Sayre and Roca 2000). Often amphibian assessments will be conducted as part of a larger project involving specialized teams looking at a variety of taxonomic groups, and objectives will need to be coordinated (e.g. Kouamé *et al.* 2007). The objectives should define the temporal, spatial, and taxonomic scope of the project, and in doing so determine the sampling methods and intensity required. For example, the objective might be to provide a preliminary species list of stream-associated taxa from a particular watershed based on snapshot sampling of several sites during the rainy season. Or perhaps the goal is to provide an initial inventory of all amphibian species within an area of conservation interest based on sampling effort spread over multiple seasons. Ideally, one's objectives are focused, realistic, and quantifiable (Chapter 2). Initial effort in reviewing and refining objectives is always worth the investment.

15.2.2 Costs and funding

Consideration of the costs and time necessary to successfully complete an RA is important and the decision on whether or not to proceed with an RA is often affected by financial and time investments. Costs will depend upon the scope of the objectives and logistics of field implementation, and can include salary expenditures, field equipment and supplies, landscape imagery purchases, international and in-country travel expenses, collecting and research permits, training expenses, contracts, laboratory equipment, museum supplies, software, digital storage media, and publication costs (Sayre and Roca 2000). Advanced consideration of these costs allows careful evaluation of trade-offs in the use of finite resources. For example, the use of satellite images, aerial photographs, or aerial videography to identify and map different habitat types and topographical features can greatly facilitate selection of sampling localities and is a key feature of some RA approaches (e.g. Rapid Ecological Assessment; Sayre *et al.* 2000). However, acquisition and processing of remote imagery can be very expensive, putting it beyond the scope of projects with more limited resources. Similarly, additional field sampling across seasons may increase the number of species observed, but also increases costs associated with returning to the site or maintaining a team in the field. Once costs have been estimated and a short project proposal developed, potential financial supporters can be approached. Options for securing financial support include development banks, governments, development agencies, conservation organizations, foundations, corporations, military landowners, and private individuals (Sayre *et al.* 2000).

15.2.3 Team selection and training

Depending upon the scope of the project, RA teams range from a few to many people, and may include senior scientists, students, technicians, field assistants, guides, rangers, data managers, cartographers, and administrators from multiple collaborating institutions (Sayre and Roca 2000; McCullough *et al.* 2007). For larger teams, an initial planning workshop is advisable, with the goal of developing a work plan that clearly identifies the roles and responsibilities of participating organizations and individuals. An example outline for a planning workshop and an example working plan for Rapid Ecological Assessment is given in Sayre and Roca (2000). International collaborations are recommended, whenever possible. Foreigners planning an RA should collaborate with local scientists to increase the chances of a successful project, and to make lasting contributions to the host country. Nationals planning an RA should consider involving foreign experts in important methodologies or taxonomic groups, especially from neighboring countries with related faunas.

A training workshop specifically for field personnel can also be an important tool, by providing training related to animal identification, field sampling techniques and data collection, use of GPS for geolocation, mapping and navigation, and safety protocols. All team members in charge of data collection should have at least a basic appreciation of the sampling design and statistical analyses planned (Magnusson and Mourão 2004; Chapter 18). In addition, amphibian pathogens and parasites can be carried between habitats on hands, footwear, or field equipment, and could be spread to new localities. Thus, it is important that initial training for those involved in RA include how to minimize the spread of disease and parasites between study sites (Declining Amphibians Population Task Force (DAPTF) code of practice, Chapter 26; Aguirre and Lampo 2006). Typically, this training occurs at the study area at the commencement of fieldwork. An overview of basic qualifications for field biologists, training tips, project supervision, and quality assurance are given in Fellers and Freel (1995).

15.2.4 Permits

Conducting an RA may require permits from local, regional, national, and international governing agencies. Most national governments, and many state and local governments, require permits to study amphibians within their boundaries and public lands. Many protected areas (e.g. national parks) require additional permission. National agencies and international treaties (e.g. Convention on International Trade in Endangered Species of Wild Fauna and Flora (CITES), U.S. Fish and Wildlife Service, Convention on Biological Diversity) also regulate export and import of specimens and tissue samples. There is considerable variation among regions in permit fees and the times required to process and receive permits, ranging from weeks to well over a year. Thus, the permitting process should be initiated well in advance of the planned fieldwork.

15.2.5 Data management

Although short in duration, RA projects can produce large amounts of data of different types, include *quantitative* data that can be represented as numbers, *qualitative* data that are not easily represented in numerical form, *spatial* data that are linked to geographic coordinates, *metadata* documentation that accompanies other data sets, and images or video or audio clips. Data management is the process that helps ensure that diverse data sets can be efficiently collected, integrated and analyzed, and archived so that they can be easily retrieved in the future (Gotelli and Ellison 2004). Planning for good data management begins early in project development by making a table of the types of data to be produced in the field; for example habitat variables (Chapter 17), conditions, counts, specimens, and other media such as photos, audio (Chapter 16), or video. This information can be used to design protocols (e.g. field forms, labels, code numbers) for collection and management of disparate types of data. Once data are being generated it is important to review data in the field, and verify that team members are using identical protocols. Data should be transcribed regularly and transferred to a database as soon as possible. If a portable computer is brought to the field, raw data should still be maintained on paper in case of damage to the computer. Widely used software for data management include Microsoft Access and the non-proprietary OpenOffice BASE. Protocols need to be established for reviewing, cleaning, and backing up data on a regular basis. Museum specimen data should be stored in a manner compliant with the Distributed Generic Information Retrieval (DiGIR) protocols, such as the database management software, Specify (http://www. specifysoftware.org).

15.2.6 Developing the sampling plan

The sampling plan is a document that identifies areas to be sampled during field work, the time that each area is to be sampled, the sampling techniques to be used and designate the individuals or teams responsible for conducting the field work. The sampling plan lays out the strategy for obtaining data that are representative of the target area of interest (Sayre 2000).

15.2.6.1 Selecting sampling sites

RA, at its most basic level, involves the collection and characterization of taxonomic and spatial information about biodiversity (Sayre *et al.* 2000). Although identifying sampling localities is one of the first steps in developing a sampling plan for any RA, protocols differ in their emphasis of initial landscape characterization. In some cases, sampling sites may be identified based upon perceived high biodiversity potential (e.g. perhaps as indicated by previous sampling of other taxa), specific conservation threats, or access logistics (e.g. Kouamé *et al.* 2007). In other cases, sites may be selected to capture clines in important abiotic features (e.g. elevation, rainfall, vegetation) that may determine the boundaries of particular amphibian assemblages (e.g. Menegon and Salvidio 2005). Some projects use remote sensing imagery to visually identify specific habitats of interest prior to sampling. Finally, GIS (Chapter 19) coupled with ecological niche modeling provide a new tool for remotely identifying potential habitat for species of concern or biodiversity hotspots in general (e.g. Raxworthy *et al.* 2003; Pawar *et al.* 2007).

Development and accessibility of new mapping technologies over the past two decades have had important impacts on the way RAs are conducted. Spatial technologies, including GIS, remote sensing, and GPS are commonly used in defining project scope and establishing sampling localities. The Rapid Ecological Assessment approach developed by Conservation International places strong emphasis on using satellite and aerial imagery to classify landscapes of interest into vegetation or land-use cover categories (Sayre *et al.* 2000). Sampling sites are spread across these vegetation categories to establish the framework within which field sampling is conducted and to facilitate linkages between fine-scale sampling by field teams with landscape-level assessment of biodiversity (Nagendra and Gadgil 1999; Sayre *et al.* 2000). Sayre *et al.* (2000) provide a detailed example of how natural-colour and colour-infrared satellite and aerial fly-over imagery were used to develop an initial landscape characterization of Parque Nacional del Este, Dominican Republic, which was subsequently used to identify specific sampling sites and generate detailed site maps. While the utility of remote sensing imagery when planning an RA is readily apparent, costs can be considerable. Commercial satellite imagery may cost US$3000–5000 per scene and an aerial photo acquisition mission may cost between $20 000 and 120 000 (Sayre *et al.* 2000). For projects with limited resources, Google Earth™ maps integrate data from satellite imagery, aerial photography, and a GIS 3D globe to provide resolution of at least 15 m for most terrestrial areas.

We draw a distinction between the ways sampling sites are determined in most RA studies compared to traditional ecological inventory methods. Traditional ecological inventories emphasized objective field sampling based on randomized selection of replicated samples and substantial effort may be given to issues such as defining sampling coverage and determining detection probabilities (e.g. Buckland *et al.* 1993; 2001, 2004; Heyer *et al.* 1994; Williams *et al.* 2002). This emphasis on randomization, replication, and estimating detection functions greatly increases the kinds of inferences that can be made. However, for many RAs, sampling may be opportunistic, or determined by issues such as logistics, access, efficiency, and *a priori* perceptions of areas likely to be highest in diversity or under the greatest threat. Although replicate units may be sampled in some cases, the resulting data may not be rigorously applied to all questions of interest to the ecologist or biogeographer (Sayre 2000). As a result, the results from many RAs are best viewed as qualitative or semi-quantitative (e.g. Kouamé *et al.* 2007).

15.2.6.2 Selecting sampling time

Scheduling sampling times will depend on latitude, elevation, seasonal patterns of rainfall, and knowledge of species breeding phenology. For example, in Kibale National Park, Uganda, pond-breeding *Hyperolius* reed frogs are most active and breed during the two rainy seasons (Vonesh 2000). In contrast, *Leptopelis* treefrogs breed during the dry seasons (J.R. Vonesh, unpublished results). Successfully sampling both of these taxa might require fieldwork during wet and dry seasons. However, when resources limit sampling to a single season, more amphibians are likely to be sampled during wet months.

15.2.6.3 Selecting sampling techniques

A variety of techniques can be used to assess amphibian diversity, including drift fences, pitfall traps, and cover boards (Chapter 13; Menegon and Salvidio 2005), quadrat sampling (aquatic: Chapter 4; terrestrial: Chapter 14), call surveys (Chapter 16), road and trail censuses (Pearman *et al.* 1995; Rödel and Ernst 2004), aquatic sweep-net sampling (Chapter 4), bottle or minnow traps, and visual surveys (e.g. Crump and Scott 1994). References on amphibian sampling include Campbell and Christman (1982), Corn and Bury (1990), Heyer *et al.* (1994), Olson *et al.* (1997), Lips *et al.* (2001), Howell (2002), Rueda-Almonacid *et al.* (2006), and Williams *et al.* (2002).

Selecting the sampling methodologies for an RA requires careful thought. Above all an RA must be rapid, yet our objective of a complete species list would suggest that we employ multiple techniques because techniques differ in the set of species they sample successfully. Often the strengths of different techniques are readily apparent, for example litter quadrat sampling will be more effective at sampling direct-developing species than visual encounter surveys (VESs) around aquatic habitats (Vonesh 2000, 2001a, 2001b; J.R. Vonesh, unpublished results). In other cases the trade-offs among techniques are more subtle and may change seasonally (K. Howell, unpublished results). Techniques also vary in terms of cost, effort, and other logistical requirements. Drift-fence arrays can be effective at sampling amphibians in a variety of situations (Chapter 13; K. Howell, unpublished results) but they are relatively expensive and labor-intensive to set up and maintain. Therefore, each potential field method should be evaluated for its relative capture success, speed, bias, and the extent to which it can facilitate the objectives of the RA (e.g. see Chapter 6, Table 4 in Heyer *et al.* 1994).

15.2.6.3.1 Visual encounter surveys (VESs)

VESs involves field personnel searching the focal habitat systematically for a known period of time. The clock is stopped when not actively searching (e.g. animals are being processed). The number of animals observed can then be expressed in terms of animals observed per area (or distance) per person searching, or per unit time per person. VES is an effective technique for building species lists rapidly (Crump and Scott, 1994; Rodda *et al.* 2007), requires little equipment (e.g. minimally a headlamp and field notebook), and can be implemented in a variety of habitat types. For these reasons it is perhaps the most widely employed sampling method in RA of amphibian diversity, so we provide a brief overview of the method here. For additional information see Crump and Scott (1994), Corn and Bury (1990), Campbell and Christman (1982), Rödel and Ernst (2004), and Rueda-Almonacid *et al.* (2006).

There is considerable variation among past studies in how VES is conducted. In some instances (e.g. Mitchell 2006; Kouamé *et al.* 2007) experienced herpetologists selectively search areas and microhabitats determined most likely to yield amphibians. This approach has the benefit of potentially yielding more animals and species per effort than randomized sampling approaches. However, it is most subject to variation in the skill levels of the field personnel and is limited in its ability to generalize about relative abundances and habitat associations because (micro-)habitats are searched in a biased manner. Alternatively, an area may be searched via VES using a randomized-walk design (Crump and Scott 1994). In this case, prior to going to the field site the researcher generates a random sequence of compass headings as well as a random distance to be searched along each heading. A starting heading and distance is selected at random and field personnel simply work through the list of headings and distances, searching for animals, for a specified time. Since the path to be sampled is determined randomly, this approach has the advantage that replicated random walks from different areas could be compared statistically. However, it is important to appreciate that differences in the observed number of animals among sites may be as much due to differences in detection among sites as much as differences in true abundance. If the sampling design requires greater effort per area, VES can be conducted using a quadrat design (Crump and Scott 1994). In this case, quadrats of a known area are established and each is then sampled systematically by searching parallel transects across the plot (e.g. Hairston 1980; Aichinger 1987; Donnelly 1989). Crump and Scott (1994) recommend plot sizes of 10 m × 10 m or 25 m × 25 m, depending upon amphibian densities. To statistically compare areas of interest (e.g. different habitats) plots must be replicated and randomly located within areas to be compared (e.g. using GPS coordinates or a trail grid system). Finally, transect designs are often used in coordination with VES for sampling across habitat gradients that may affect amphibian diversity and abundance (e.g. moisture, elevation). Although these considerations may determine transect direction, distances to be sampled and starting points are best determined *a priori* and multiple randomly located replicate transects will be needed if areas are to be compared statistically.

It is also important to determine and carefully define the intensity of field sampling in advance. Crump and Scott (1994) suggest three levels of sampling intensity for sampling amphibians. The least intensive surveys are counts of animals that are active on the surface (Hairston 1980; Vonesh 2000). This approach is minimally invasive and thus is suitable for search areas with sensitive faunas and also requires the least amount of time, increasing the amount of area that can be covered. Intermediate-level searches count exposed animals but also turn over surface cover objects (note that cover objects must be returned to their original position to minimize disturbance). This level of VES can yield species often overlooked by low-intensity VES because of their secretive and semi-fossorial life histories (e.g. many salamanders and caecilians). When the most complete inventory possible is desired, all possible microhabitats are searched; surface objects are turned, decaying logs are torn apart, the leaf litter is systematically raked, epiphytes and tree holes are searched (Crump and Scott 1994), and aquatic habitats are searched for adults and larvae (Chapter 4). Such intensity requires considerable labor per unit area and causes the greatest disturbance to the habitat but may be more effective at sampling some rare species.

15.2.6.3.2 Assumptions and limitations of VESs

The analysis of VES data is based on the following assumptions: (1) all individuals are equally detectable; (2) individuals are recorded once during a survey; and (3) there are no observer-related biases in sampling (Crump and Scott 1994). In most instances, one or more of these assumptions will be violated. Amphibians of different species, sizes, or life stages are generally not equally conspicuous. Even the probability of detecting the same individual may vary due to temporal (e.g. diel, seasonal) variation in behavior (Bailey *et al.* 2004; Schmidt 2005). Similarly, differences in observer experience can often result in differences in detection. Thus, the assumption of equal detection probability across time, space, and individuals is unlikely to hold for most amphibian taxa or studies (Mazerolle *et al.* 2007). This has important implications for simple count-based methods like VES. Consider that the expected number of individuals or taxa counted (C) in a survey area is $E(C) = PN$, where P is the probably of detection and N is the true number in the population (Williams *et al.* 2002). Without an estimate of P, for any count of animals sampled, there is an infinite number of possible true population sizes. Mazerolle *et al.* (2007) and Schmidt (2004, 2005) provide lucid reviews of these issues and specifically focus on their application to her-petological studies. Given these limitations, VES is best viewed as a qualitative or semi-quantitative approach and should be used only when qualitative results are acceptable (e.g. threatened species were observed; Kouamé *et al.* 2007). VES may be a suitable sampling approach given the objectives of many RA projects. However, when resources and objectives allow, other more powerful study designs (e.g. distance sampling methods; Thomas *et al.* 2002; Funk *et al.* 2003; Fogarty and Viletta 2001; Viletta and Fogarty 2005) should be considered.

15.2.6.4 Determining what data to collect

Generating lists of observed species is usually the principle type of data-collection method used during an RA. In some cases, animals encountered in the field may be identified in the field. In other cases, it may not be possible to identify all species, requiring that specimens be held temporarily for identification (e.g. to consult a key) before being released at the site of collection, or it may be necessary to preserve voucher specimens that can be identified by taxonomic specialists (Jacobs and Heyer 1994; McDiarmid 1994). Voucher specimens are what ultim-ately define the identity and distribution of any amphibian species. Depending on collecting permits obtained, a series of one or more voucher specimens per sex, life stage, and species should be collected during any give RA, especially for remote geographic locations, and deposited in a recognized natural history museum. Voucher specimens allow independent researchers to confirm speci-men identification at a later date, a very important consideration given the highly unstable state of amphibian taxonomy. The systematic value of a voucher speci-men increases as more ancillary data are included.

Depending upon the project's goals, it may be important to collect add-itional data for animals observed or collected. GPS can be used to provide a geo-reference for each sample. Basic size measurements (snout–vent length, mass), sex, reproductive status based on secondary sexual characteristics, notes or photographs indicating coloration, and basic activity (e.g. calling, resting, mov-ing, etc.) may be recorded. Tissue samples for DNA analyses may be obtained from a toe clip, liver sample, or a buccal swab preserved in 95% ethanol or other buffer (Seutin *et al.* 1991; Gonser and Collura 1996; Goldberg *et al.* 2003). The benefits of integrating RA data with the DNA barcode of life campaign are potentially enormous but so far have been under-utilized (Vences *et al.* 2005; Fouquet *et al.* 2007; Ficetola *et al.* 2008; Smith *et al.* 2008). Additional types of sample, such as skin swabs, may be required for disease detection (Chapter 26). It is also important to record environmental characteristics, such as weather conditions, lunar cycle, and vegetation or habitat characteristics (Chapter 17).

However, collecting additional data on individuals is time-consuming and thus additional data on individuals may involve trading-off the amount of area that can be searched. Field forms are useful to remind the field team what data are to be collected and can help ensure consistency of data collection among team members. Crump and Scott (1994) provide an example data sheet for VES that includes some basic individual-level data. If designed with foresight, forms can also expedite data entry.

15.2.6.5 Field logistics

Poorly organized trips typically result in poor-quality data collection. Careful advance planning for the transportation of the team, sampling equipment, and basic supplies will be required, particularly for large teams and remote localities. Logistics also includes coordinating the safety of the team. Injuries and illness in remote locations can be dangerous and hinder project progress. In advance of fieldwork the field team should prepare a well-stocked medical kit, review vaccines and allergies of each participant, and plan for various possible medical emergencies that could arise.

15.3 In the field

After the planning and preparation has been completed, the team goes to the field and begins sampling in the manner indicated by the sampling plan. Situations will naturally arise that prevent the sampling protocols from being followed exactly. RAs are intended to be flexible enough to allow field teams to respond to unforeseen roadblocks as they arise. The field team should keep the primary objectives of the project in mind if they are required to alter plans in the field (Young *et al.* 2000).

15.4 Compiling data and interpreting results

Producing information that is useful to conservation biologists, land managers, and policy-makers requires skillful synthesis of the large volume of field data generated. If data-management strategies were developed and the data and metadata have been entered into the database, work can now focus on analysis and interpretation. RA data summaries typically focus on species lists (Kouamé *et al.* 2007). Species lists may be organized to highlight species sampled in different primary sampling localities, particular habitat types (e.g. stream, pond, litter), points along environmental gradients (e.g. elevation; Menegon and Salvidio 2005), indicator taxa (e.g. disturbance-tolerant versus -intolerant; Kerr *et al.* 2000), endemicity or conservation status (e.g. IUCN Red List status). The best summary approach will depend on the goals of the RA. Because sampling effort often varies among sites and may have relied to some degree on opportunistic sampling, it is often difficult to make direct comparisons between species lists among sites across or even within RA studies. In some cases, species accumulation or rarefaction curves can be used to show how rapidly the number of species sampled at a specific site or habitat increased with sampling effort (e.g. Chapter 18; Gotelli and Colwell 2001; Magurran 2003) to facilitate comparison at a given sampling effort. Similarly, in some cases it may be possible to use total diversity estimators to extrapolate from incidence sampling data to estimate the total number of amphibian species at a site (e.g. Chapter 18; Chao *et al.* 2005). When using VES data, however, investigators should clearly define their sampling unit and be aware how their choice could influence the diversity analyses.

15.5 Recommendations and reporting

Because RAs are typically motivated from a conservation or management perspective, recommendations will focus on how the information generated in the RA can be used to maintain long-term faunal diversity of the study sites. However, recommendations should be firmly based on the study results, not on preconceived ideas about the sites importance, and should consider the target audience and their ability to follow through on the recommendations. Little benefit comes from recommendations for which the target audience has no ability to implement (Young *et al.* 2000). Ways in which RA results are typically used to guide decision-making include providing direction for future land acquisition, identifying monitoring needs, establishing priorities for future research, suggestions for how to zone for multiple use, and for identifying potential threats to species of interest (Young *et al.* 2000).

15.6 Summary

Rapid assessment is an approach that can be useful for conservation planning because it can provide an efficient preliminary characterization of the amphibian fauna. However, because the methods are often qualitative or semi-quantitative, the RA approach is of most value when the fauna of the study areas is unknown and when even qualitative information is needed urgently to help decision-making. When resources, time, and project objectives allow, researchers should consider more quantitative approaches.

15.7 References

Aguirre, A. A. and Lampo, M. (2006). Protocolo de bioseguridad y cuarentena para prevenir la transmisión de enfermedades en anfibios. In A. Angulo, J. V. Rueda-Almonacid, J. V. Rodríguez-Mahecha, and E. La Marca (eds), *Técnicas de Inventario y Monitoreo para los Anfibios de la Región Tropical Andina*, pp. 73–92. Conservación Internacional, Bogotá.

Aichinger, M. (1987). Annual activity patterns of anurans of anurans in a seasonal Neotropical environment. *Oecologia*, **71**, 583–92.

AmphibiaWeb (2008). *Information on Amphibian Biology and Conservation*. AmphibiaWeb, Berkeley, CA. http://amphibiaweb.org.

Bailey, L. L., Simons, T. R., and Pollock, K. H. (2004). Estimating site occupancy and species detection probability parameters for terrestrial salamanders. *Ecological Applications*, **14**, 692–702.

Buckland, S. T., Anderson, D. R., Burnham, K. P., and Laake, J. L. (1993). *Distance Sampling: Estimating Abundance of Biological Populations*. Chapman and Hall, London.

Buckland, S. T., Anderson, D. R., Burnham, K. P., Laake, J. L., Borchers, D. L., and Thomas, L. (2001). *Introduction to Distance Sampling: Estimating Abundance of Biological Populations*. Oxford University Press, New York.

Buckland, S. T., Anderson, D. R., Burnham, K. P., Laake, J. L., Borchers, D., and Thomas, L. (2004). *Advanced Distance Sampling: Estimating Abundance of Biological Populations*. Oxford University Press, Oxford.

Campbell, H. W. and Christman, S. P. (1982). Field techniques for herpetofaunal community analysis. In N. J. Scott, Jr (ed.), *Herpetofaunal Communities*, pp. 193–200. Wildlife Research Report 13. U.S. Department of the Interior, Fish and Wildlife Service, Washington DC.

Chao, A., Chazdon, R. L., Colwell, R. K., and Shen, T.-J. (2005). A new statistical approach for assessing similarity of species composition with incidence and abundance data. *Ecology Letters*, **8**, 148–59.

Corn, P. S. and Bury, R. B. (1990). *Sampling Methods for Terrestrial Amphibians and Reptiles*. General Technical Report, PNW-GTR-256. U.S. Forest Service, Fort Collins, CO.

Crump, M. L. and Scott, Jr, N. J. (1994). Visual encounter surveys. In W. R. Heyer, M. A. Donnelly, R. W. McDiarmid, L. A. C. Hayek, and M. S. Foster (eds), *Measuring and Monitoring Biological Diversity, Standard Methods for Amphibians*, pp. 84–92. Smithsonian Institution Press, Washington DC.

Donnelly, M. A. (1989). Demographic effects of reproductive resource supplementation in a territorial frog, *Dendrobates pumilio*. *Ecological Monographs*, **59**, 207–21.

Fellers, G. M. and Freel, K. L. (1995). *A standardized protocol for surveying aquatic amphibians*. Technical Report NPS/WRUC/NRTR-95-01. U.S. National Park Service, Davis, CA.

Ficetola, G. F., Miaud, C., Pompanon, F., and Taberlet, P. (2008). Species detection using environmental DNA from water samples. *Biology Letters*, **4**, 423–5.

Fogarty, J. H. and Vilella, F. J. (2001). Evaluating methodologies to survey *Eleutherodactylus* frogs in montane forests of Puerto Rico. *Wildlife Society Bulletin*, **29**, 948–55.

Fouquet, A., Gilles, A., Vences, M., Marty, C., Blanc, M., and Gemmell, N. J. (2007). Underestimation of species richness in Neotropical frogs revealed by mtDNA analyses. *PLoS ONE*, **2**, e1109.

Funk, W., Almeida-Reinoso, D., Nogales-Sornosa, F., and Bustamante, M. (2003). Monitoring population trends of *Eleutherodactylus* frogs. *Journal of Herpetology*, **37**, 245–56.

Goldberg, C. S., Kaplan, M. E., and Schwalbe, C. R. (2003). From the frog's mouth: buccal swabs for collection of DNA from amphibians. *Herpetological Review*, **34**, 220–1.

Gonser, R. A. and Collura, R. V. (1996). Waste not, want not: toe-clips as a source of DNA. *Journal of Herpetology*, **30**, 445–7.

Gotelli, N. J. and Colwell, R. K. (2001). Quantifying biodiversity: procedures and pitfalls in the measurement and comparison of species richness. *Ecology Letters*, **4**, 379–91.

Gotelli. N. J. and Ellison, A. M. (2004). Chapter 8: Managing and curating data. *A Primer of Ecological Statistics*, pp. 207–36. Sinauer Associates, Sunderland, MA.

Hairston, N. G. (1980). The experimental test of an analysis of field distributions: Competition in terrestrial salamanders. *Ecology*, **61**, 817–26.

Hayek, L.-A. (1994). Research design for quantitative amphibian studies. In W. R. Heyer, M. A. Donnelly, R. W. McDiarmid, L. A. C. Hayek, and M. S. Foster (eds), *Measuring and Monitoring Biological Diversity, Standard Methods for Amphibians*, pp. 21–38. Smithsonian Institution Press, Washington DC.

Heyer, W. R., Donnelly, M. A., McDiarmid, R. W., Hayek, L. A. C., and Foster, M. S. (eds) (1994). *Measuring and Monitoring Biological Diversity, Standard Methods for Amphibians*. Smithsonian Institution Press, Washington DC.

Howell, K. M. (2002). Amphibians and reptiles: Herptiles. In G. Davies, (ed.), *African Forest Biodiversity*, pp. 17–44. Earthwatch, London.

IUCN, Conservation International, and NatureServe (2006). *Global Amphibian Assessment*. http://globalamphibians.org.

Jacobs, J. F. and Heyer, W. R. (1994). Collecting tissue for biochemical analysis. In W. R. Heyer, M. A. Donnelly, R. W. McDiarmid, L. A. C. Hayek, and M. S. Foster (eds), *Measuring and Monitoring Biological Diversity, Standard Methods for Amphibians*, pp. 103–7. Smithsonian Institution Press, Washington DC.

Kerr, J. T., Sugar, A., and Packer, L. (2000). Indicator taxa, rapid biodiversity assessment, and nestedness in an endangered ecosystem. *Conservation Biology*, **14**, 1726–34.

Kouamé, N. G., Bonteng, C. O., and Rödel, M.-O. (2007). A rapid survey of the amphibians from Atewa Range Forest Reserve, Eastern Region, Ghana. In J. McCullough, L. E. Alonso, P. Naskrescki, H. E. Wright, and Y. Osei-Owusu (eds), *A Rapid Biological Assessement of the Atewa Range Forest Reserve, Eastern Ghana*. RAP Bulletin of Biological Assessment 47, pp. 76–83. Conservation International, University of Chicago Press, Chicago, IL.

Lips, K. R., Reaser, J. K., Young, B. E., and Ibáñez, R. (2001). Amphibian monitoring in Latin America: a protocol manual/Monitoreo de Anfibios en América Latina: Manual de protocolos. *SSAR Herpetological Circular* no. 30.

Magnusson, W. E. and Mourão, G. (2004). *Statistics without Math*. Sinauer Associates, Sunderland, MA.

Magurran, A. E. (2003). *Measuring Biological Diversity*. Blackwell Publishing, Oxford.

Matthews, K. R., Knapp, R. A., and Pope, K. L. (2002). Garter snake distributions in high-elevation aquatic ecosystems: is there a link with declining amphibian populations and non-native trout introductions? *Journal of Herpetology*, **36**, 16–22.

Mazerolle, M. J., Bailey, L. L., Kendall, W. L., Royle, J. A., Converse, S. J., and Nichols, J. D. (2007). Making great leaps forward: Accounting for detectability in herpetological field studies. *Journal of Herpetology*, **41**, 672–89.

McCallum, M. L. (2007). Amphibian declines or extinction? Current declines dwarf background extinction rate. *Journal of Herpetology*, **41**, 483–91.

McCullough, J., Alonso, L. E., Naskrecki, P., Wright, H. E., and Osei-Owusu, Y. (eds) (2007). *A Rapid Biological Assessment of the Atewa Range Forest Reserve, Eastern Ghana*. RAP Bulletin of Biological Assessment 47, 1–191. Conservation International, Arlington, VA.

McDiarmid, R. W. (1994). Preparing amphibians as scientific specimens. In W. R. Heyer, M. A. Donnelly, R. W. McDiarmid, L. A. C. Hayek, and M. S. Foster (eds), *Measuring and Monitoring Biological Diversity, Standard Methods for Amphibians*, pp. 103–7. Smithsonian Institution Press, Washington DC.

Menegon, M. and Salvidio, S. (2005). Amphibian and reptile diversity in the southern Udzungwa Scarp Forest Reserve, south-eastern Tanzania. In B. A. Huber, B. J. Sinclair, and K.-H. Lampe (eds), *African Biodiversity. Molecules, Organisms, Ecosystems*, pp. 205–12. Springer, New York.

Mitchell, J. C. (2006). *Inventory of Amphibians and Reptiles of Fredericksburg and Spotsylvania National Historic Battlefield*. Technical Report/NERCHAL/NRTR-200600/000. U.S. National Park Service, Fredericksburg, VA.

Nagendra, H. and Gadgil, M. (1999). Biodiversity assessment at multiple scales: linking remotely sensed data with field information. *Proceedings National Academy of Sciences USA*, **96**, 9154–8.

Olson, D. H., Leonard, W. P. and Bury, R. B. (eds) (1997). *Sampling Amphibians in Lentic Habitats: Methods and Approaches for the Pacific Northwest: Northwest Fauna No. 4*. Society for Northwestern Vertebrate Biology, Olympia, WA.

Pawar, S., Koo, M. S., Kelley, C., Ahmed, M. F., Chaudhuri, S., and Sarkar, S. (2007). Conservation assessment and prioritization of areas in Northeast India: priorities for amphibians and reptiles. *Biological Conservation*, **136**, 346–61.

Pearman, P. B., Velasco, A. M., and Lopez, A. (1995). Tropical amphibian monitoring: a comparison of methods for detecting inter-site variation in species composition. *Herpetologica*, **51**, 325–37.

Raxworthy, C. J., Martinez-Meyer, E., Nussbaum, R. A., Schneider, G. E., Ortega-Huerta, M. A., and Peterson, A. T. (2003). Predicting distributions of known and unknown reptile species in Madagascar. *Nature*, **426**, 837–41.

Roberts, L. (1991). Ranking the rainforests. *Science*, **251**, 1559–60.

Rodda, G. H., Campbell, E.W, Fritts, T. H., and Clark, C. S. (2007). The predictive power of visual searching. *Herpetological Review*, **36**, 259–64.

Rödel, M.-O. and Ernst, R. (2004). Measuring and monitoring amphibian diversity in tropical forests. I. An evaluation of methods with recommendations for standardization. *Ecotropica*, **10**, 1–14.

Roelants, K., Gower, D. J., Wilkinson, M., Loader, S. P., Biju, S. D., Guillaume, K., Moriau, L., and Bossuyt, F. (2007). Global patterns of diversification in the history of modern amphibians. *Proceedings of the National Academy of Sciences USA*, **104**, 887–92.

Rueda-Almonacid, J. V., Castro, F., and Cortez, C. (2006). Técnicas para el inventario y muestreo de anfibios: Una compilación. In A. Angulo, J. V. Rueda-Almonacid, J. V. Rodríguez-Mahecha, and E. La Marca (eds), *Técnicas de Inventario y Monitoreo para los Anfibios de la Región Tropical Andina*, pp. 135–71. Conservación Internacional, Bogotá.

Sayre, R. (2000). Overview: Rapid Ecological Assessment after ten years. In R. Sayre, E. Roca, G. Sedaghatkish, B. Young, S. Keel, R. L. Roca, and S. Sheppard (eds), *Nature in Focus Rapid Ecological Assessment*, pp. 1–18. Island Press, Washington DC.

Sayre, R. and Roca E. (2000). Careful planning: a key to success. In R. Sayre, E. Roca, G. Sedaghatkish, B. Young, S. Keel, R. L. Roca, and S. Sheppard (eds), *Nature in Focus Rapid Ecological Assessment*, pp. 33–44. Island Press, Washington DC.

Sayre, R., Roca, E., Sedaghatkish, G., Young, B., Keel, S., Roca, R. L., and Sheppard S. (eds) (2000). *Nature in Focus: Rapid Ecological Assessment*. Island Press, Washington DC.

Schmidt, B. (2004). Declining amphibian populations: the pitfalls of count data in the study of diversity, distributions, dynamics, and demography. *Herpetological Journal*, **14**, 167–74.

Schmidt, B. (2005). Monitoring the distribution of pond breeding amphibians when species are detected imperfectly. *Aquatic Conservation: Marine and Freshwater Ecosystems*, **15**, 681–92.

Seutin, G., White, B. N., and Boag, P. T. (1991). Preservation of avian blood and tissue samples for DNA analyses. *Canadian Journal of Zoology*, **69**, 82–90.

Smith, M. A., Poyarkov, Jr, N. A., and Hebert, P. D. N. (2008). CO1 DNA barcoding amphibians: take the chance, meet the challenge. *Molecular Ecology Resources*, **8**, 235–46.

Thomas, L., Buckland, S. T., Burnham, K. P., Anderson, D. R., Laake, J. L., Borchers, D. L., and Strindberg, S. (2002). Distance sampling. In A. H. El-Shaarawi and W. W. Piegorsch (eds), *Encyclopedia of Environmetrics*, vol. 1, pp. 544–52. John Wiley & Sons, Chichester.

Timm, R. M. (1994). The mammal fauna. In L. A. McDade, K. S. Bawa, H. A. Hespenheide, and G. S. Hartshorn (eds), *La Selva: Ecology and Natural History of a Neotropical Rainforest*, pp. 229–37. University of Chicago Press, Chicago, IL.

Vences, M., Thomas, M., Bonett, R. M., and Vieites, D. R. (2005). Deciphering amphibian diversity through DNA barcoding: chances and challenges. *Philosophical Transactions of the Royal Society of London Series B Biological Sciences*, **360**, 1859–68.

Vilella, F. and Fogarty, J. (2005). Diversity and abundance of forest frogs (Anura: Leptodactylidae) before and after Hurricane Georges in the Cordillera Central of Puerto Rico. *Caribbean Journal of Science*, **41**,157–62.

Vonesh, J. R. (2000). Dipteran predation on the arboreal eggs of four *Hyperolius* frog species in western Uganda. *Copeia*, **2000**, 560–6.

Vonesh, J. R. (2001a). Patterns of richness and abundance in a tropical African leaf-litter herpetofauna. *Biotropica*, **33**, 502–10.

Vonesh, J. R. (2001b). Natural history and biogeography of the amphibians and reptiles of Kibale National Park, Uganda. *Contemporary Herpetology*. http://www.nhm.ac.uk/hosted_sites/ch/.

Whiles, M. R., Lips, K. R., Pringle, C. M., Kilham, S. S., Bixby, R. J., Brenes, R., Connelly, S., Colon-Gaud, J. C., Hunte-Brown, M., Huryn, A. D. *et al.* (2006). The effects of amphibian population declines on the structure and function of Neotropical stream ecosystems. *Frontiers in Ecology and the Environment*, **4**, 27–34.

Williams, B. K., Nichols, J. D., and Conroy, M. J. (2002). *Analysis and Management of Animal Populations*. Academic Press, New York.

Young, B., Sedaghatkish, G., and Roca, R. (2000). Fauna surveys. In R. Sayre, E. Roca, G. Sedaghatkish, B. Young, S. Keel, R. L. Roca, and S. Sheppard (eds), *Nature in Focus Rapid Ecological Assessment*. pp. 93–117. Island Press, Washington DC.

16

Auditory monitoring of anuran populations

Michael E. Dorcas, Steven J. Price, Susan C. Walls, and

William J. Barichivich

16.1 Introduction

Because anurans rely on vocalization for most communication, detection of species-specific calls provides relatively efficient mechanisms for studying and evaluating the status of anuran populations. Consequently, most amphibian monitoring programs focus on anurans and use call detection as their sole or primary monitoring technique. The overall goal of these programs is to determine and monitor the status of populations over time. Some monitoring programs may actually attempt some form of quantification of anuran population size or density, but this is often difficult when using only calling data. In this chapter we describe approaches for monitoring anuran populations based solely on auditory techniques. These approaches include manual calling surveys (MCS), automated recording systems (ARS), or some combination of the two. MCS can be used by researchers to monitor multiple amphibian populations, often over large spatial scales. ARS can be used by researchers interested in intensively monitoring populations of anurans at a single or a few locations. In this chapter, we describe both approaches and the potential costs and benefits of each. We also explain how data resulting from ARS can be used to optimize manual survey protocol and to interpret data resulting from MCS. Our goal is to provide an overview of these techniques and questions that can be addressed using them.

16.2 MCS

Generally, MCS simply involve observers listening to the vocalizations of male frogs and recording all species detected for the duration of the survey or for a given area. In many surveys, observers also score abundance of each species heard vocalizing. Researchers have long used MCS to conduct inventories of anurans in a given area. For example, Wright and Wright (1949) understood the importance of MCS when they stated that "If one knows the frog notes, he can in one night do more work on frog distribution than he might otherwise do in years." However, only within the last 30 years, when concerns of declining amphibian populations prompted the development of formal monitoring programs, have MCS been widely used for large-scale investigations of anuran populations. The widespread use of MCS for monitoring anurans has led numerous investigations into survey protocol, study design, and efficiency as a technique to detect anurans.

16.2.1 History and current status of MCS

Many historic and recent small-scale investigations have used MCS to detect anurans for ecological, behavioral, and conservation-related investigations (i.e. Martof 1953; Blair 1961; Woolbright 1985; Knutson *et al.* 1999). In this chapter, we emphasize MCS approaches that have specific objectives to monitor the proportion of sites where a given anuran species is observed, accounting for the potential effects of imperfect detection and other biases (site and sampling covariates such as habitat type and weather conditions). This metric is known as occupancy (MacKenzie *et al.* 2002, 2006).

Most MCS programs use surveys in which observers listen to anuran vocalizations at several points along a roadside transect or at designated sampling locations. Since the early 1980s, the use of roadside surveys to detect calling anurans has become used widely by many local and regional environmental monitoring programs. At least five Canadian provinces and 28 US states conduct, or have conducted, roadside surveys for calling anurans (Weir and Mossman 2005). If similar protocols are used, MCS data collected at the local or regional level (e.g. state or province) can be incorporated into a centralized database allowing for the evaluation of anuran populations at the national level. Two US programs initiated by the Department of Interior's US Geological Survey, the North American Amphibian Monitoring Program (NAAMP; www.pwrc.usgs.gov/naamp/; Weir and Mossman 2005), and the Amphibian Research and Monitoring Initiative (ARMI; http://armi.usgs.gov/), are good examples of this. Many US Geological Survey scientists use both MCS and ARS (see section 16.3) to meet their monitoring objectives, primarily on federal lands. Additional examples of North American MCS include the Marsh Monitoring Program in the Great Lakes region (Weeber and Vallianatos 2000), Ontario Backyard Frog Survey (de Solla *et al.* 2005), and FrogWatch USA (Inkley 2006). MCS are also used to monitor trends in anuran populations in Australia (Australia Frog Census; Walker 2002), as well as some European (e.g. Anthony 2002; Pellet and Schmidt 2005; Schmidt 2005; Scott *et al.* 2008), and Central American countries (e.g. Kaiser 2008).

16.2.2 Study objectives: what can MCS tell us?

The objective of any contemporary MCS is typically to conduct an inventory of an area of interest to determine the status of anurans in that region and/or to monitor populations and communities, over time, to address how species are responding to habitat change, climatic variation, or some other ecological or management issue. An objective of many monitoring programs, including those using MCS that are rapidly gaining recognition, is that of determining site occupancy (Marsh and Trenham 2008).

16.2.3 Survey design

Establishment of sampling sites is a critical component of all monitoring programs for which it is essential that sites are sampled according to a probabilistic scheme, so that statistical inference may be extrapolated to a defined area of interest. According to the NAAMP protocol (Weir and Mossman 2005), surveys are conducted along routes that generally are established along roads, dikes, waterways, or other points of access. Routes consist of 10 stops that are either placed at least 0.8 km apart or are stratified by habitat. Although NAAMP is often used as a template for designing MCS, roads are not typically established in a randomized fashion, thus illustrating one of the weaknesses of the NAAMP approach. One compromise for conducting road surveys and meeting this criterion is to, for example, randomly select stops along a given route that are at least 0.8 km apart. In this modification the area of inference would be limited to the route, rather than to a larger area of interest (e.g. a large protected area). Determining the number of routes and their associated sites/stops is another critical element if the resulting data are destined for occupancy analysis. The number of sites to be sampled varies depending upon the distribution and natural history of the anuran species being monitored, the question of interest, and other factors. But, generally speaking, a good rule of thumb in determining the number of sites to sample is usually as many as possible, typically at least 50, assuming that sites are sampled for a minimum of four seasons (see Figure 7.3 of MacKenzie *et al.* 2006).

MCS are usually conducted at night, starting one-half hour after sunset, and are completed by 01:00 h. For many anuran species, especially those in temperate regions, peak calling does occur within this sampling time. Bowers *et al.* (1998) found that the majority of species in their surveys ended peak calling at midnight. To minimize the potential effect of anthropogenic noise (see section 16.2.4), researchers typically allow 1 min between arriving at each site and beginning the chorus survey. All anuran species heard in a specified time frame (usually 5 min, but see section 16.2.4) are recorded; each calling anuran species is given a score, known as an amphibian calling index (ACI), ranging from 1 to 3, where 1 means distinct calls of individuals that can be counted and have no overlapping calls, 2 means calls of individuals that can be distinguished but have some overlapping calls, and 3 means a full chorus, with calls of individuals indistinguishable. Generally, MCS should not be conducted during heavy rain, high wind, or other inclement weather that could affect the detection of calling anurans. At each site, air temperature, relative humidity, barometric pressure, wind speed, and other variables are often measured immediately after recording call data to be used as sample covariates in occupancy analysis. These covariates can be used to help explain variation in site occupancy or detection probabilites.

In regions where anurans vocalize in a somewhat predictable manner, many MCS, such as NAAMP, recommend sampling each stop along a roadside route during a specified sampling period, such as during early spring, late spring, and summer. The NAAMP protocol recommends a single survey, based on convenience to volunteer observers, despite studies that have shown that significant variation in calling behavior does occur within the specified sampling periods (e.g. Todd *et al.* 2003; Gooch *et al.* 2006; Kirlin *et al.* 2006). For example, Gooch *et al.* (2006) conducted three MCS during the NAAMP-specified "summer" sampling period in the western Piedmont region of North Carolina, USA, and found that detection probabilities increased for some species (e.g. *Acris crepitans*, *Rana catesbeiana*) and decreased for other species (e.g. *Rana clamitans*) from survey 1 to survey 3 (Figure 16.1). Variation in calling behavior with prescribed sampling periods may cause the observer to not detect a species if single sampling occasions are employed. At least two surveys are required to calculate detection probabilities (preferably a minimum of three; J. Nichols, personal communication) during a sampling period (see MacKenzie *et al.* 2002 and Table 6.1 of MacKenzie *et al.* 2006 for more details).

16.2.4 Other survey-design issues to consider: the efficiency of MCS

The widespread use of MCS has led to numerous investigations that focus on the efficiency of MCS to detect anurans. In fact, few if any amphibian survey methodologies have been scrutinized to the extent of MCS. A central issue in all wildlife-monitoring programs, including MCS, is that of imperfect detection (MacKenzie *et al.* 2002, 2006). For example, even during the peak breeding season for a given anuran species, calling does not occur each night; variations

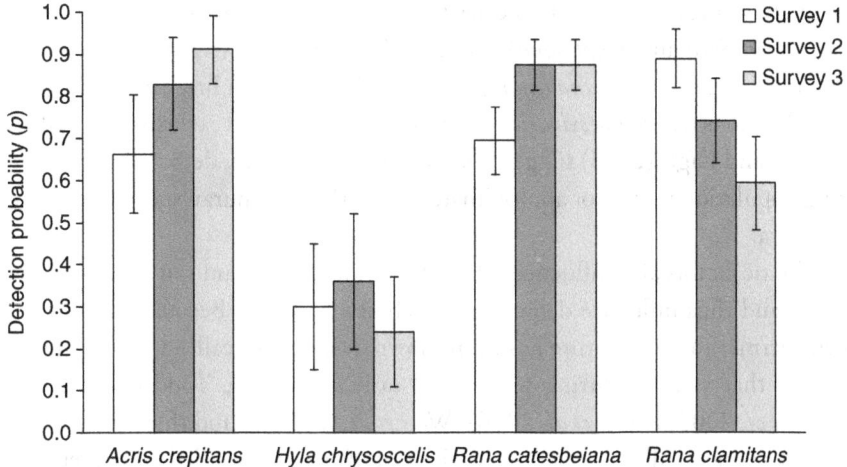

Fig. 16.1 Detection probability, p (± 1 standard error), for four summer-breeding anurans in the western Piedmont region of North Carolina, USA. MCS were conducted from 10 June through 13 July 2004. Detection probabilities were calculated using a model with survey-specific p and constant occupancy estimate, ψ (i.e. $\psi \cdot p(t)$), allowing for the calculation of p for each species during each of the three surveys within the sampling period. Note the influence of sampling occasion on p for *Acris crepitans* and *Rana clamitans*. Adapted from Gooch *et al.* (2006).

in species-specific calling behaviors and abiotic and biotic conditions can lead to the inference of absence despite the presence of a species. Aspects of survey protocol and observer bias can also influence detection probabilities. Fortunately, recent advances in statistical techniques have allowed for calculation of species-specific detection probabilities (e.g. program PRESENCE; MacKenzie *et al.* 2002, 2006; Chapter 24), which can greatly aid in inferring population status and potentially long-term population trends.

Inter- and intraspecific variation in anuran calling behavior may affect detection probability and should be considered when conducting a MCS or evaluating MCS data. Anurans exhibit a vast array of acoustic properties (Duellman and Trueb 1986) which influence probability of detection by observers. Some species have calls that can carry long distances (e.g. 1 km), whereas calls of other species cannot be detected until the observer is 100 m or less from the breeding site. Many species, such as *R. catesbeiana*, may call sporadically every few minutes. Other species (e.g. *Pseudacris crucifer*) call more continuously. In species-rich communities, louder, higher-pitch calls of one species may interfere with the detection of other, quieter species (Droege and Eagle 2005) or inhibit calling in another sympatric species (Littlejohn and Martin 1969). In general, MCS are best suited for regions where all species vocalize during a somewhat predictable breeding season. Species that breed in response to localized heavy rains (e.g. *Spea* and *Scaphiopus*), call quietly or not at all (e.g. *Ascaphus truei*), call infrequently (e.g. *Rana capito*), or have relatively short breeding (i.e. vocalizing) seasons (e.g. *Rana sylvatica*) should not be monitored exclusively by MCS. Droege and Eagle (2005) suggest that MCS are suitable for detecting and inferring population trends for approximately 55 of the 103 anuran species in North America.

Abiotic factors also influence calling behavior in many anuran species (Blair, 1961) and thus influence detection probability via MCS. Because anurans are ectotherms, air temperature has been shown to influence calling, especially for species that vocalize during winter and early spring (e.g. Todd *et al.*, 2003; Weir *et al.* 2005; Kirlin *et al.* 2006). Weir *et al.* (2005) found that air temperature explained calling variation in eight of 10 species studied; however, five species displayed preference for particular temperatures, indicating an optimal temperature for detection. Additionally, Pellet and Schmidt (2005) found that air temperature was a predictor of calling in *Hyla arborea* in Switzerland, with greater detections during warm temperatures. For winter-breeding species in temperate zones, calling behavior may be more sporadic and limited to daylight or early evening hours when temperatures are moderate (Todd *et al.* 2003; Kirlin *et al.* 2006; Saenz *et al.* 2006).

Abiotic factors other than air temperature have also been shown to influence anuran calling. For example, Oseen and Wassersug (2002) suggested that water temperature was the overall most important predictor of calling behavior of some Canadian anurans. Precipitation can also influence calling because many species are known to call more intensely after periods of heavy rain (Blair 1961; Oseen and Wassersug 2002). However, heavy rainfall during MCS is not recommended as it may obfuscate anuran vocalizations and decrease probability of detection. Other factors, such as wind speed (Weir *et al.* 2005; Oseen and Wassersug 2002; Johnson and Batie 2001), humidity of air (Oseen and Wassersug 2002), barometric pressure (Oseen and Wassersug 2002), and moonlight (Weir *et al.* 2005), may also influence anuran detection probability. However, anuran species respond differently to abiotic factors (Saenz *et al.* 2006). Knowledge of the relationship between anuran breeding strategies and abiotic factors prior to conducting MCS may allow for increased efficiency and/ or detection.

Anthropogenic noise also likely impacts anuran detection probability. Most calling anurans will respond to anthropogenic noise disturbance. For this reason, some MCS programs recommend that after arriving at a stop the observer must wait for a few minutes prior to conducting the survey. However, few studies have been conducted on the effects of noise on anuran calling behavior. Weir *et al.* (2005) found that only three of 10 species decreased calling behavior due to increased traffic. Regardless, anthropogenic noise can affect the observer's ability to detect frogs and should be recorded during MCS. Additionally, anthropogenic noise can also reduce a species proclivity to call, thus lowering its detectability (Sun and Narins 2005; C. Steelman, personal communication)

Several studies have investigated the effects of survey length on anuran detection. The majority of MCS range from 3 to 10 min per stop. Pierce and Gutzwiller (2004) found that 15 min was required to detect 90% of all species known to be present, whereas Shirose *et al.* (1997) found that 3 min surveys were adequate to detect most species. Gooch *et al.* (2006) found that 94% of summer-breeding anurans in the North Carolina Piedmont region were detected within the first 5 min of the MCS and detection probabilities were slightly higher as observers spent longer listening (3 min compared to 10 min). However, these and other investigations highlight that some species may go undetected even if surveys are extended up to an hour. The length of survey should be determined based on the specific objectives of the MCS; however, for detecting long-term population trends for most species, 3–5 min appears to be adequate.

Variation among observers can also substantially influence the quality of MCS data. Some large-scale MCS (e.g. NAAMP) rely heavily on volunteers, who may vary in experience and/or hearing ability. As a result, observers may not detect all species present, include species not present, or incorrectly identify vocalizations, resulting in flawed assessments of anuran populations (Lotz and Allen 2007). Weir *et al.* (2005) found that volunteer experience may influence species detection. However, studies by Genet and Sargent (2003), Shirose *et al.* (1997), and Lotz and Allen (2007) suggest that even relatively inexperienced volunteers were reliable in their abilities to determine species, yet estimating abundance categories often differed among observers. Pierce and Gutzwiller (2007) showed 79% agreement among nine observers conducting MCS in central Texas, USA, and stressed the importance of accounting for interobserver variation in data analysis. NAAMP currently requires all volunteers to pass an online anuran call test (available at www.pwrc.usgs.gov/frogquiz/) prior to conducting MCS. In general, training of volunteers will likely increase the probability of ensuring quality data.

16.2.5 Limitations of MCS data

MCS are excellent tools for monitoring changes in anuran occupancy or for inventorying which species occur in an area. At a given site, however, these surveys only yield qualitative data on abundance as obtained through the use of the ACI (section 16.2.3). These types of abundance data are of limited utility in assessing population densities. However, Nelson and Graves (2004) compared population estimates (via mark–recapture) of male *R. clamitans* with ACI collected at the same sites and found abundance to be correlated positively with ACI. In contrast, Corn *et al.* (2000) found that another measure of call frequency failed to reflect "a relatively large population" of *Bufo woodhousii* and only weakly distinguished among different-sized populations of *Pseudacris maculata*. The need to be able to use indices of relative abundance, such as the calling index, in occupancy models has been conceptualized and is currently an active area of research (Royle and Nichols 2003; Dorazio 2007).

Because call surveys are based upon the vocalizations of adult male anurans, they do not provide complete information on population structure; that is, non-calling females and subadults are not assessed by this method (Stevens and Paszkowski 2004). Similarly, this method is not useful for other non-calling amphibians, such as salamanders and caecilians. The males of many species of anurans will vocalize in contexts not necessarily related to breeding; moreover, the MCS does not consider the presence of egg masses, tadpoles, or metamorphosing juveniles in the population. Thus, the MCS provides no information about whether there was successful reproduction at a given site. Depending on the questions of interest in a given study, the MCS is therefore most useful when used in conjunction with other survey methods, such as visual encounter surveys and the use of traps or dipnets to assess the larval component of the population.

16.3 ARS

In addition to MCS, automated systems (ARS) can be used to detect anuran vocalizations and can be useful in monitoring many anuran populations (Peterson and Dorcas 1994). Typically, ARS (or so-called frogloggers) are used to collect data intensively at a single or a few locations, whereas MCS provide more superficial data but for a larger number of sites. ARS can be used to survey for anuran species in places difficult to access for MCS and can be left in the field for extended periods of time, thus increasing the probability of detecting a given species. ARS may be the only practical way to reliably detect species that have very short or unpredictable breeding seasons, such as *R. capito*. ARS minimize disturbance to calling anurans and provide a permanent sampling record that can be evaluated by multiple experts if required (Mohr and Dorcas 1999; Todd *et al.* 2003). When combined with information on environmental variation, data from ARS can be incorporated into models that can be used to optimize monitoring programs based on MCS (Bridges and Dorcas 2000; Oseen and Wassersug 2002).

16.3.1 Sources for ARS

At the time Peterson and Dorcas (1994) published on building and using ARS, there were few options for their procurement. Although not specifically intended for anuran monitoring, the Cornell Laboratory of Ornithology builds robust ARS units that have been used in a wide variety of tasks ranging from the detection of ivory-billed woodpeckers (*Campephilus principalis*) to monitoring whales around the world's oceans. Bedford Technical (www.frogloggers.com) and Wildlife Acoustics (www.wildlifeacoustics.com) have both produced ARS systems with a variety of options that make them attractive for monitoring frogs. Although their software has not been tested on anuran vocalizations, both Cornell Lab of Ornithology and Wildlife Acoustics produce products, Raven and Song Scope respectively, which may prove useful in automated call detection.

16.3.2 Construction, deployment, and retrieval of data

Although sources for ARS now exist, some biologists may still choose to build their own. This may be a logistical hurdle for some investigators; however, construction of an ARS can have the added benefit of familiarizing the user with the inner workings of their equipment. Construction of an ARS generally requires combining several simpler components: a recorder, timer/controller, microphone, power supply, and housing (Peterson and Dorcas 1994; Barichivich 2003). If the investigator is not familiar with electronics, we recommend working with someone skilled in building electrical devices.

The core component of an ARS is the recorder. Devices that have been successfully used include analog cassette and various digital (e.g. MPEG-1 Audio Layer 3, or MP3) recorders. Selection of a recorder is dependent on the required recording quality, capacity, and budget. Recordings produced on inexpensive analog cassette recorders are sufficient for manual listening, but higher sampling rates and frequency responses may be required in some circumstances. Additionally, storage capacity can usually be greatly increased when using digital recordings.

In most cases, researchers choose to make recordings at specific times of day (e.g. dusk) and on set intervals rather than continuously. Some recording devices (e.g. PDAs and digital voice recorders) have built-in timers or clocks. The addition of a timer/controller allows programming of the desired recording schedule. A wide variety of devices, from mechanical (K. Wharton, personal communication) to computer microprocessors (Acevedo and Villanueva-Rivera 2006), have been used for this purpose. Timers generally work in one of two ways: controlling the function of the recorder or interrupting the power supply to the recorder. In addition to time-based triggers, environmental triggers can be used to activate an ARS. For example, to detect explosive breeders like *Spea* or *Scaphiopus*, a tipping-bucket rain gauge with a reed valve could be used so a rainfall event would trip an ARS into service.

The choice of microphones is a crucial decision, as recordings will only be as good as the microphone, regardless of the recording device. At least nine types of microphone are available but condenser or dynamic varieties are those most commonly used. Unlike dynamic microphones, condenser microphones require a power supply. Condenser microphone power sources are often small (e.g. a single AA battery) and may not be sufficient if the ARS is deployed for extended periods of time without maintenance. This issue can be addressed by using the timer/controller to control the microphone as well as the recorder. Another consideration in selecting a microphone is the directional or acceptance cone. Cones can range from a 360° circle around the microphone (omnidirectional) to just a few degrees in front of the microphone (unidirectional or shotgun). To capture the vocalizations of species that call while partially or completely submerged (e.g. *Rana sevosa* or *Rana subaquavocalis*), a hydrophone would be more appropriate to use than a microphone (Platz 1993).

The final considerations of building an ARS are power supply and protection. Rechargeable batteries generally provide the best option for most researchers. Cost, power, size, and weight should be considered when selecting batteries. In environments with sufficient sunlight, rechargeable batteries can be supplemented by a solar panel and can generally operate without interruption. With the exception of the microphone, all components of an ARS should be firmly mounted inside a protective case. The case should provide adequate environmental protection and be large enough to hold all the components, yet small enough to be reasonably portable for ease of field deployment. For most field deployments, waterproof boxes (e.g. Otter or Pelican brands) work well, but less costly options exist. These include surplus ammunition cans, polyethylene coolers, plastic tool boxes, and plastic watertight marine boxes. Consideration should be given to the environment in which the ARS will be deployed. In areas frequented by people, the ARS can be locked, hidden, or even buried. Animals, such as raccoons and bears, may damage equipment and thus more rugged cases might be needed in some situations (Corn *et al.* 2000). Microphones should also be protected by a windscreen to reduce wind noise, along with a cover to shield the windscreen and microphone from the environment.

Before deploying an ARS, careful consideration should be given to placing and programming the ARS to capture the data germane to the question(s) of interest. Although the program should be restrictive enough to minimize the collection of superfluous data, sampling only during perceived peak times could lead to erroneous interpretation of data (Figure 16.2). Typically an ARS is placed near anuran breeding habitats with the microphone mounted facing the water. After the recordings have been retrieved, data need to be transcribed, either manually (i.e. by human ear) or by computer recognition. Manual assessment of recordings requires a trained observer to listen to and review the recordings while transcribing the data. Calling activity can be scored much like that in MCS and is typically semi-quantified as an ACI. Depending on observer experience and the complexity of the calling choruses, 1.5 h per hour of recording are required to review analog cassette tapes (Corn *et al.* 2000). Slightly less time

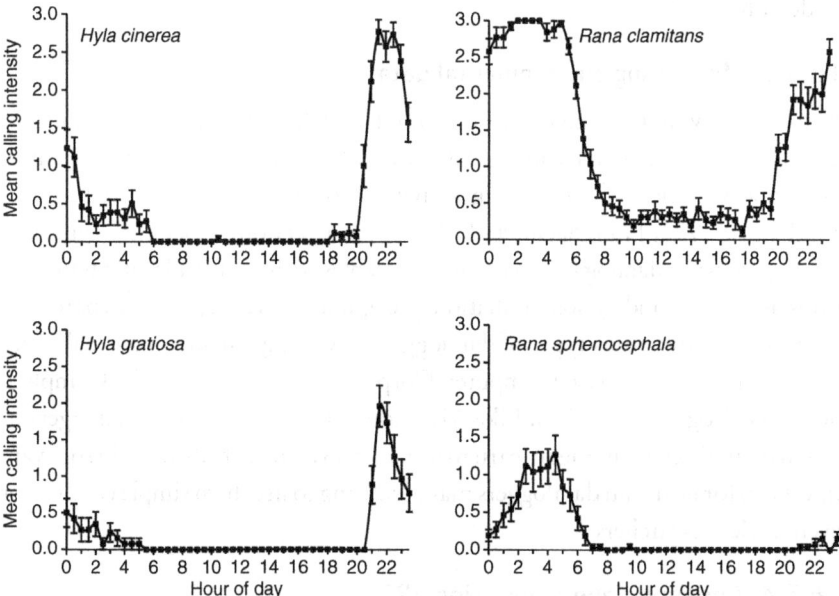

Fig. 16.2 Daily calling pattern of *Hyla cinerea*, *Hyla gratiosa*, *Rana clamitans*, and *Rana sphenocephala* recorded using an ARS at Carolina Bay near Aiken, SC, USA. Mean calling activity was calculated by averaging the recorded calling activity levels for each 30 min time recording period over all days of the study (from 16 June to 12 July 1997). Error bars denote ± 1 standard deviation. Note that calling in both species of treefrog peaked during the time period when manual calling surveys are recommended (dusk to midnight), but peak calling in *Rana* peaked well after midnight and *R. sphenocephala* called almost exclusively after midnight, and would thus likely be missed on most calling surveys. Adapted from Bridges and Dorcas (2000).

is usually required to review digital recordings. While in its infancy, computer recognition of high-quality recordings shows great promise for decreasing the time, and cost, of manually listening to ARS recordings and potentially increasing the accuracy of call recognition. Though not an automated process, one study showed that in a mixed-species chorus, including the acoustically dominant coqui (*Eleutherodactylus coqui*), several species were omitted or significantly underrepresented using only a manual data review (Villanueva-Rivera 2007). By adding a simple visual analysis of the spectrograms of digital call recordings using Adobe Audition, the calls of the less acoustically dominant species were no longer eclipsed (Villanueva-Rivera 2007). Research on birds demonstrates the potentially significant time savings when using computer recognition software. In 12 h, Agranat (2007) scanned more than 250 h of field recordings for bird vocalizations using Song Scope software. This process allowed a human observer to review 1552 potential calls of the cerulean warbler (*Dendrocia cerulea*) in under 1 h.

16.3.3 Monitoring environmental data

Automatically monitoring environmental data, while simultaneously monitoring anuran calling activity using ARS, can allow interpretation of how environmental variation affects calling activity (Dorcas and Foltz 1991; Peterson and Dorcas 1992, 1994; Saenz *et al.* 2006) and detection probability. Typically, investigators use dataloggers to monitor variables such as air and water temperatures, relative humidity, solar radiation, precipitation, wind speed, and barometric pressure. Numerous types of dataloggers of varying quality and capabilities are available (e.g. Onset Computer Corp., Pocasset, MA, USA; Campbell Scientific, Logan, UT, USA). Like ARS, some investigators find dataloggers to be particularly challenging to learn to use. However, recent advances in software interfaces for nearly all dataloggers make learning to use them simple enough for even novice researchers.

16.3.4 Answering questions using ARS

ARS can be used by researchers to address basic questions such as whether a species is present at a given location or to conduct more detailed investigations of factors affecting calling activity and population fluctuations (Corn and Muths 2002; Todd *et al.* 2003). Generally, investigators using ARS for monitoring purposes are simply interested in detecting whether a species is present. As such, many ARS systems can be set to only come on and record at certain times of the day when a particular species is most likely to vocalize (e.g. dusk until midnight), thus minimizing the number of recordings that must be evaluated. However, one must be careful not to assume that anurans call primarily during times traditionally recognized as peak calling periods (Bridges and Dorcas 2000).

In some cases, researchers use ARS to attempt to detect a species that is either rare or has very unpredictable calling patterns. Anurans such as *Spea* and *Scaphiopus* that generally only call under certain conditions (i.e. during or after heavy rains) may be particularly hard to detect using MCS and ARS may offer the best opportunity to detect their populations. For some extremely rare species (e.g. *R. sevosa*) detecting every known population is important for proper management and thus, increasing detectability of populations using ARS can play a vital role in conservation efforts.

Because ARS can collect data at regular and precise intervals, mathematical models can be developed that allow prediction of when and under what conditions species are likely to call, thus providing data that can be used to optimize MCS. Typically, data collected simultaneously with anuran calling data, such as temperature, precipitation, wind speed, and time of day, are used as independent variables in a logistic regression to predict the optimal conditions to conduct MCS (i.e. the times when detectability is maximized; Oseen and Wassersug 2002). Anurans may respond differently to environmental variables across their ranges, and thus development of models should be done for particular regions as needed. Such models were developed for three species of winter-breeding anurans (*Pseudacris crucifer*, *Pseudacris feriarum*, and *Rana sphenocephala*) in the western Piedmont region of North Carolina (Steelman and Dorcas, in press). Models showed that for *P. crucifer* day of year, time, precipitation, and water temperature positively influenced calling and air temperature negatively influenced calling; for *P. feriarum* time, precipitation, air temperature, and water temperature positively influenced calling, and day of year negatively influenced calling; for *R. sphenocephala* day of year, time, precipitation, and air temperature positively influenced calling, and higher water temperature negatively influenced calling. The models described for the winter-breeding species above were tested using previously collected data from MCS along with spot measurements of environmental data (Kirlin *et al.* 2006). Models accurately predicted whether a species was calling approximately 70% of the time.

In addition to using models to predict the best times and conditions to manually sample anurans, ARS can be used to interpret data collected previously, assuming sufficient environmental data are collected at the time of the surveys. Fortunately, nearly all existing MCS programs require collection of at least some environmental data. To do this, an investigator would insert the spot measurements of environmental variables collected by volunteers at the time the surveys were conducted into the model equation for each species of interest to generate a likelihood of calling (e.g. ranging from 0 to 1). Calling likelihoods can then be used to interpret previously collected calling survey results. If a species was not heard at a particular location but the model indicates a high likelihood of calling if it was present (e.g. 0.9), then the investigator can have a higher confidence in concluding that the species was not present, rather than it simply not vocalizing and being undetectable.

16.3.5 Limitation of ARS

Despite the advantages of using ARS, there are some disadvantages when compared to surveys based on MCS (Corn *et al.* 2000; Penman *et al.* 2005). For many investigators, ARS can be expensive compared to volunteer-based MCS. Additionally, ARS can typically only be deployed at one or a few sites, thus decreasing the number of anuran populations that can be monitored. This can have major implications for statistical analyses of data when each site is a replicate. The possibility of wildlife damage, vandalism, or theft of equipment is another issue that may limit investigators' ability to use ARS in some localities (e.g. Corn *et al.* 2000). Although ARS can be hidden relatively easily, one of us (M.E.D.) has had equipment stolen even in remote locations. Some investigators find using automated systems particularly challenging, especially if they consider themselves technologically challenged. However, we have found that nearly anyone can be taught to use automated systems effectively and for those that are apprehensive about doing so we suggest initial consultations with a biologist experienced with the equipment being used.

16.4 Conclusions

Auditory monitoring of anuran populations, either by MCS or by ARS, is a useful tool for assessing the status and trends of populations, as well as the responses of populations and communities to environmental and anthropogenic change. The use of auditory monitoring has increased tremendously since the realization that amphibian populations are undergoing global population declines, and has become a standard approach to monitoring in many state, national, and international programs. Auditory monitoring has limitations in that (1) its utility in estimating abundance is restricted, (2) this method provides little information about the complete structure of a population (i.e. the status of non-calling adult females and subadults) or about whether a given population has experienced successful reproduction, and (3) this method cannot be used on non-calling amphibians, such as salamanders and caecilians. The two means by which auditory monitoring may be conducted (manual or automated) differ in their relative costs and benefits, and the appropriate approach for a given study ultimately depends on the study objectives and resources available. Nevertheless, MCS can provide vital information on changes in occupancy states of various sites, which can be overlaid with environmental data to assess potential causes of change in occupancy for a given species or community. Data from ARS can be used to optimize the effectiveness of MCS or interpret data based on MCS. As such, auditory monitoring has great utility in assessing the extent of declines of anuran amphibians, a topic of heightened concern in ecology and conservation biology.

16.5 Acknowledgments

We thank P. Stephen Corn, Erin Muths, and Charlotte K. Steelman for comments that improved the manuscript. Manuscript preparation was partially supported by the Department of Biology at Davidson College, Duke Power, and National Science Foundation grant (DEB-0347326) to M.E.D and the U.S. Geological Survey's Amphibian Research and Monitoring Initiative to SCW and WJB. The use of trade or product names does not imply endorsement by US government agencies.

16.6 References

Acevedo, M. A. and Villanueva-Rivera, L. J. (2006). Using automated digital recording systems as effective tools for the monitoring of birds and amphibians. *Wildlife Society Bulletin*, **34**, 211–14.

Agranat, I. D. (2007). *Automatic Detection of Cerulean Warblers Using Autonomous Recording Units and Song Scope Bioacoustics Software*. www.wildlifeacoustics.com/

Anthony, B. P. (2002). Results of the first batrachian survey in Europe using road call counts. *Alytes*, **20**, 55–66.

Barichivich, W. J. (2003). Appendix IV: Guidelines for building and operating remote field recorders. In C. K. Dodd, Jr, *Monitoring Amphibians in Great Smoky Mountains National Park*, pp. 87–96. U.S. Geological Survey Circular no. 1258. U.S. Geological Survey, Tallahassee, FL.

Blair, W. F. (1961). Calling and spawning seasons in a mixed population of anurans. *Ecology*, **42**, 99–110.

Bowers, D. G., Anderson, D. E., and Euliss, Jr, N. H. (1998). Anurans as indicator of wetland condition in the Prairie Pothole Region of North Dakota: an environmental monitoring and assessment program pilot project. In M. Lannoo (ed.), *Status and Conservation of Midwestern Amphibians*, pp. 369–78. University of Iowa Press, Iowa City, IO.

Bridges, A. S. and Dorcas, M. E. (2000). Temporal variation in anuran calling behavior: implications for surveys and monitoring programs. *Copeia*, **2000**, 587–92.

Corn, P. S. and Muths, E. (2002). Variable breeding phenology affects the exposure of amphibian embryos to ultraviolet radiation. *Ecology*, **83**, 2958–63.

Corn, P. S., Muths, E., and Iko, W. M. (2000). A comparison in Colorado of three methods to monitor breeding amphibians. *Northwestern Naturalist*, **81**, 22–30.

de Solla, S. R., Shirose, L. J., Fernie, K. J., Barrett, G. C., Brousseau, C. S., and Bishop, C. A. (2005). Effect of sampling effort and species detectability on volunteer based anuran monitoring programs. *Biological Conservation*, **121**, 585–94.

Dorazio, R. M. (2007). On the choice of statistical models for estimating occurrence and extinction from animal surveys. *Ecology*, **88**, 2773–82.

Dorcas, M. E. and Foltz, K. D. (1991). Environmental effects on anuran advertisement calling. *American Zoologist*, **31**, 3111A.

Droege, S. and Eagle, P. (2005). Evaluating calling surveys. In M. Lannoo (ed.), *Amphibian Declines: The Conservation Status of United States Species*, pp. 314–19. University of California Press, Berkeley, CA.

Duellman, W. E. and Trueb, L. (1986). *Biology of Amphibians*. McGraw-Hill Publishing Company, New York.

Genet, K. S. and Sargent, L. G. (2003). Evaluation of methods and data quality from a volunteer-based amphibian call survey. *Wildlife Society Bulletin*, **31**, 703–14.

Gooch, M. M., Heupel, A. H., Price, S. J., and Dorcas, M. E. (2006). The effects of survey protocol on detection probabilities and site occupancy estimates of summer breeding anurans. *Applied Herpetology*, **3**, 129–42.

Inkley, D. B. (2006). Final report assessment of utility of Frogwatch USA data 1998–2005. www.nwf.org/frogwatchUSA/FinalRptFW-ScienceAssess.pdf. National Wildlife Federation for U.S. Geological Survey, Washington DC.

Johnson, D. H. and Batie, R. D. (2001). Surveys of calling amphibians in North Dakota. *The Prairie Naturalist*, **33**, 227–47.

Kaiser, K. (2008). Evaluation of a long-term amphibian monitoring protocol in Central America. *Journal of Herpetology*, **42**, 104–10.

Kirlin, M., Gooch, M. M., Price, S. J., and Dorcas, M. E. (2006). Predictors of winter anuran calling activity in the North Carolina Piedmont. *Journal of the North Carolina Academy of Science*, **122**, 10–18.

Knutson, M. G., Sauer, J. R., Olsen, D. A., Mossman, M. J., Hemesath, L. M., and Lannoo, M. J. (1999). Effects of landscape composition and wetland fragmentation on frog and toad abundance and species richness in Iowa and Wisconsin, U.S.A. *Conservation Biology*, **13**, 1437–46.

Littlejohn, M. J. and Martin, M. M. (1969). Acoustic interaction between two species of leptodactylid frog. *Animal Behaviour*, **17**, 785–91.

Lotz, A. and Allen, C. R. (2007). Observer bias in anuran call surveys. *Journal of Wildlife Management*, **71**, 675–9.

MacKenzie, D. L., Nichols, J. D., Lachman, G. B., Droege, S., Royle, J. A., and Langtimm, C. A. (2002). Estimating site occupancy rates when detection probabilities are less than one. *Ecology*, **83**, 2248–55.

MacKenzie, D. I., Nichols, J. D., Royle, J. A., Pollock, K. H., Bailey, L. L., and Hines, J. E. (2006). *Occupancy Estimation and Modeling: Inferring Patterns and Dynamics of Species Occurrence*. Academic Press, San Diego, CA.

Marsh, D. M. and Trenham, P. C. (2008). Current trends in plant and animal population monitoring. *Conservation Biology*, **22**, 647–55.

Martof, B. S. (1953). Territoriality in the green frog, *Rana clamitans*. *Ecology*, **34**, 165–74.

Mohr, J. R. and Dorcas, M. E. (1999). A comparison of anuran calling patterns at two Carolina bays in South Carolina. *Journal of the Elisha Mitchell Scientific Society*, **115**, 63–70.

Nelson, G. L. and Graves, B. M. (2004). Anuran population monitoring: comparison of the North American Amphibian Monitoring Program's calling index with Mark-Recapture estimates for *Rana clamitans*. *Journal of Herpetology*, **38**, 355–9.

Oseen, K. L. and Wassersug, R. J. (2002). Environmental factors influencing calling in sympatric anurans. *Oecologia*, **133**, 616–25.

Pellet, J. and Schmidt, B. R. (2005). Monitoring distributions using call surveys: estimating site occupancy, detection probabilities and inferring absence. *Biological Conservation*, **123**, 27–35.

Penman, T. D., Lemckert, F. L., and Mahony, M. J. (2005). A cost-benefit analysis of automated call recorders. *Applied Herpetology*, **2**, 389–400.

Peterson, C. R. and Dorcas, M. E. (1992). The use of automated data acquisition techniques in monitoring amphibian and reptile populations. In D. McCullough and R. Barrett (eds), *Wildlife 2001: Populations*, pp. 369–78. Elsevier Applied Science, London.

Peterson, C. R. and Dorcas, M. E. (1994). Automated data acquisition. In W. Heyer, R. McDairmid, M. Donnelly, and L. Hayek (eds), *Measuring and Monitoring Biological Diversity. Standard Methods for Amphibians*, pp. 47–57. Smithsonian Institution Press, Washington DC.

Pierce, B. A. and Gutzwiller, K. J. (2004). Auditory sampling of frogs: detection efficiency in relation to survey duration. *Journal of Herpetology*, **38**, 495–500.

Pierce, B. A. and Gutzwiller, K. J. (2007). Interobserver variation in frog call surveys. *Journal of Herpetology*, **41**, 424–9.

Platz, J. E. (1993). *Rana subaquavocalis*, a remarkable new species of Leopard Frog (*Rana pipiens* complex) from southeastern Arizona that calls under water. *Journal of Herpetology*, **27**, 154–62.

Royle, J. A. and Nichols, J. D. (2003). Estimating abundance from repeated presence-absence data or point counts. *Ecology*, **84**, 777–90.

Saenz, D., Fitzgerald, L. A., Baum, K. A., and Conner, R. N. (2006). Abiotic correlates of anuran calling phenology: the importance of rain, temperature and season. *Herpetological Monographs*, **20**, 64–82.

Schmidt, B. R. (2005). Monitoring the distribution of pond-breeding amphibians when species are detected imperfectly. *Aquatic Conservation: Marine and Freshwater Ecosystems*, **15**, 681–92.

Scott, W. A., Pithart, D., and Adamson, J. K. (2008). Long-term United Kingdom trends in the breeding phenology of the common frog, *Rana temporaria*. *Journal of Herpetology*, **42**, 89–96.

Shirose, L. J., Bishop, C. A., Green, D. M., MacDonald, C. J., Brooks, R. J., and Helferty, N. J. (1997). Validation tests of an amphibian call count survey technique in Ontario, Canada. *Herpetologica*, **53**, 312–20.

Steelman, C. K. and Dorcas, M. E. (in press). Anuran calling survey optimization: developing and testing predictive models of anuran calling activity. *Journal of Herpetology*.

Stevens, C. E. and Paszakowski, C. A. (2004). Using chorus-size ranks from call surveys to estimate reproductive activity of the wood frog (*Rana sylvatica*). *Journal of Herpetology*, **38**, 404–10.

Sun, J. W. C. and Narins, P. M. (2005). Anthropogenic sounds differentially affect amphibian call rate. *Biological Conservation*, **121**, 419–27.

Todd, M. J., Cocklin, R. R., and Dorcas, M. E. (2003). Temporal and spatial variation in anuran calling activity in the western Piedmont of North Carolina. *Journal of the North Carolina Academy of Science*, **119**, 103–10.

Villanueva-Rivera, L. J. (2007). Digital recorders increase detection of *Eleutherodactylus* frogs. *Herpetological Review*, **38**, 59–63.

Walker, S. J. (2002). *Frog Census 2001: Community monitoring of water quality and habitat condition in South Australia using frogs as indicators*. Environment Protection Authority, Adelaide.

Weeber, R. C. and Vallitanios, M. (2000). *The Marsh Monitoring Program 1995–1999: Monitoring Great Lakes wetlands and their amphibian and bird inhabitants*. www.bsc-eoc.org/mmpreport.html. Bird Studies Canada with the U.S. Environmental Protection Agency, Ontario.

Weir, L. A. and Mossman, M. J. (2005). North American amphibian monitoring program (NAAMP). In M. Lannoo (ed.), *Amphibian Declines: The Conservation Status of United States Species*, pp. 307–13. University of California Press, Berkeley, CA.

Weir, L. A., Royle, J. A., Nanjappa, P., and Jung, R. E. (2005). Modeling anuran detection and site occupancy on North American Amphibian Monitoring Program (NAAMP) routes in Maryland. *Journal of Herpetology*, **39**, 627–39.

Woolbright, L. L. (1985). Patterns of nocturnal movement and calling by the tropical frog, *Eleutherodactylus coqui*. *Herpetologica*, **41**, 1–9.

Wright, A. H. and Wright, A. A. (1949). *Handbook of Frogs and Toads of the United States and Canada*. 3rd edn. Comstock Publishing Associates, Ithaca, NY.

17

Measuring habitat

Kimberly J. Babbitt, Jessica S. Veysey, and George W. Tanner

17.1 Introduction

Understanding wildlife–habitat relations is fundamental to sound management and conservation and provides important basic information on species' ecology. Obtaining accurate information on amphibian habitat associations can be challenging as many species have complex life cycles requiring very different habitat types during different life-history stages, and many species are highly cryptic during much of the year. However, numerous amphibian species are experiencing significant threats from habitat loss and degradation. Better information on habitat requirements and habitat use is necessary to understand how these changes affect habitat suitability for different species and to improve management targeted at enhancing amphibian habitat.

17.2 Habitat selection

Although this chapter is mainly concerned with measuring habitat, it is important to keep in mind that how an animal "selects" habitat is a complicated process involving multiple ecological and behavioral factors. Habitat selection also occurs at a variety of scales (Johnson 1980). The broadest scale (termed first-order selection) involves selection that determines the geographic distribution of a species. Second-order selection occurs at the level of general features in the landscape and dictates home range. Third-order selection is finer-scale selection for specific sites within a home range. Most habitat assessments focus on second- and third-order selection.

It is also important to understand differences between habitat selection, preference, and use. Habitat selection is defined as the process whereby an individual chooses a habitat among available alternatives. Habitat preference occurs when habitats are selected independent of availability. However, information on preference is difficult to obtain as certain conditions, such as equal access to resources provided on an equal basis, are a prerequisite for determining preference. Thus, most field research involving habitat assessments measure use or selection. It is assumed that animals select habitats that confer fitness advantages (i.e. increased survival or reproduction); however, only rarely are measures of fitness actually conducted along with habitat assessments. It is tempting, but not correct, to assume that habitat use implies selection. Garshelis (2000) provides a review of the problems inherent in making such an interpretation. Over-interpretation of habitat-use information can lead to erroneous assumptions about habitat selection, and thus improper recommendations regarding habitat management.

17.3 Spatial and temporal scale

The scale at which one examines habitat use can greatly influence the inferences drawn from the data. Features that have a significant relationship with habitat use at one scale may be of little importance at another scale (Wiens 1989; Levin 1992). It is important to collect data at a scale appropriate for the research question being asked. Although certain research questions may address only one scale, increased insight into habitat relations can be gained from examination of multiple scales (e.g. Welsh and Lind 2002). However, multi-scale approaches may be prohibitively costly. Spatial and temporal patterns of resources availability and abundance also influence habitat use and suitability and should be incorporated into study design (Southwood 1977). Year-to-year or seasonal changes in habitat use can reflect the availability and distribution of resources as modified by biotic (e.g. predators) and abiotic (e.g. weather) factors. Behaviors inherent in many species (e.g. seasonal breeding migration, over-wintering, estivation) and/or highly variable population dynamics (Pechman *et al.* 1991) also influence habitat use and detectability of individuals. Knowledge of natural history helps to guide appropriate timing for sampling.

17.4 Approaches for examining habitat selection

In this section we provide a brief introduction into the most common approaches for examining habitat selection. A detailed review of habitat selection theory, design, and analysis is beyond the scope of this chapter. More detailed information, reviews of study designs and their inherent strengths and weaknesses, and suggested statistical approaches can be found in Thomas and Taylor (1990, 2006), Garshelis (2000), Manly *et al.* (2002), and Morrison *et al.* (2006).

Use-availability designs are the most common approaches for examining habitat selection (Thomas and Taylor 1990, 2006; Garshelis 2000). These approaches compare the proportional use of each habitat type by an organism to the relative area of each habitat. Use-availability studies can be classified into four general designs (Thomas and Taylor 1990, 2006). In design 1, use is determined for all known individuals in a population but the individuals are not identified. Habitat is considered equally available to all individuals. This design is an assessment of habitat use at the population level and does not allow examination of variation in habitat selection among individuals. This design would be appropriate for studies where individuals are documented using visual encounter surveys or area-constrained searches but where no individuals are marked.

Design 2 is used when individuals are uniquely marked and, as in design 1, habitat availability is considered equivalent for all individuals. In design 3, use is determined for individually marked organisms but, unlike designs 1 and 2, available habitat is estimated for each individual animal. This design is appropriate when, for example, individual home ranges have been estimated and only areas within each home range are deemed available. Design 4 examines use of habitat by uniquely marked individuals at multiple time periods (e.g. each time an animal is located) and pairs measurements of used habitat with measurements of available habitat taken at each interval. This design is particularly useful for examination of small-scale habitat (i.e. microhabitat).

Use-availability studies generally focus on habitat selection based on broad habitat types, or multiple habitat parameters that are analyzed individually (Garshelis 2000). An alternative approach to use-availability studies is the site-attribute approach. This approach differs from use-availability studies in that the focus is on measuring habitat features that potentially influence use in areas where an individual has been documented and comparing those values to measurements made on the same features in areas that are not used. Differences between used and non-used areas are then assumed to reveal which features are responsible for selection for (or against) habitat use. Thus, the focus is not on the amount of use in one habitat compared to another, but rather whether the site was used or not and what specific features appear to determine use. Often, site-attribute studies are used to examine habitat variables at biologically important sites (e.g. breeding sites).

17.5 Determining availability

One of the most difficult aspects of examining habitat selection is determining habitat availability. Because we do not share the same perspective as the organisms we study, it is certain that some working assumptions will go into any determination of availability. At the broadest level, areas beyond the study area are not examined, and so by default are unavailable. If we delineate the home range of an organism we can define unavailable habitat as that outside the home-range boundary. However, within a home range, intraspecific interactions may mean that a portion of the home range is not just unused but unavailable. In contrast, some habitat may go unused during much of the year but yet be both available and critical (e.g. over-wintering sites, breeding sites). Knowledge of the organism's natural history can help to address some questions about what is unavailable and what is unused, but cautious interpretation is always warranted because improper designation of habitat availability can bias results.

17.6 What to measure

The list of variables one could measure is potentially quite long, but sampling is usually limited by time, personnel, and fiscal constraints. Previously published studies can provide initial guidance as to key parameters that should be measured. However, the specific research question may lead to selection of parameters that are not often measured. Further, because the importance of a habitat parameter may vary with scale, measurement of variables that have not previously appeared important may need to be measured when working at a different scale. The same can be argued when working in different areas to allow for regional comparisons. A reasonable compromise between measuring everything one can think of versus a small number of variables that appear most critical is to conduct a pilot study using a broader set of variables and then to reduce the number of variables by eliminating those that appear least important, or those that are correlated with one another. Because resource availability, abundance, and importance can change temporally, it is important to keep in mind that a pilot study conducted in one season may not provide useful guidance for selecting parameters in a different season.

17.7 Weather variables

As ectothermic organisms with semipermeable skin, amphibians are very sensitive to changes in abiotic conditions. This fact has two important influences on how habitat assessments should be conducted. First, abiotic habitat measures are critically important for understanding habitat use and selection because these features are often the key factors influencing habitat–amphibian associations. Second, study timing and location must be carefully selected because the abiotic environment can greatly influence levels of amphibian activity, thereby affecting our interpretation of habitat use. For example, movement patterns of pond-breeding amphibians are often tied to seasonal temperature and rainfall conditions. Thus, for many species assessment of breeding habitat use must be timed to match seasonal weather conditions rather than the calendar *per se*. For fossorial species, rainfall can trigger movement or surface activity, and interpretation of habitat use may vary greatly with precipitation patterns (Taub 1961).

It would be hard to imagine a research project that would not require measurement of basic weather data such as temperature and rainfall. Ambient air temperature can be measured with a standard handheld thermometer. This low-tech approach provides only a spot check of temperature but should be measured as it provides a record of specific conditions at the time and point of sampling. Temperature probes are available for water and soil temperature measurements. Most water-based meters measure temperature along with other parameters. Probes are useful for specific measures in space or time, but are not the best approach for providing integrative or frequent measurements. To provide a comprehensive assessment of abiotic and climate conditions, most projects will require additional temperature observations. Min/max thermometers, which can be used on land or in water, provide a relatively inexpensive way to obtain the range of temperature conditions an organism may experience, and measurements can be taken as frequently as one wants (e.g. over a 24 h period, over a week). If more detailed temperature conditions are desired, dataloggers should be used as these devices can be programmed to measure temperature at specific intervals throughout the day.

Rainfall data are easily collected with rainfall gauges. Gauge types range from simple plastic devices that are manually emptied at specified times intervals to electronic devices such as tipping buckets that record rainfall amounts on data sheets, on data drums, or directly to a computer and then empty themselves. Both collection frequency (i.e. hourly, daily, weekly) and gauge placement influence the data collected. For simple measurements of rainfall amounts, gauges should be located in open areas. However, to measure the amount of precipitation that reaches the forest floor (i.e. rain throughfall), rain gauges placed under the canopy provide more relevant data.

Relative humidity (and temperature) influences water loss rates and therefore affects both amphibian activity and habitat suitability. Similarly, wind currents across the skin surface affect water loss. Differences in habitat structure and aspect among sites, even at a local level, can influence these parameters making them potentially important to measure. Relative humidity can be measured using a sling psychrometer or a hygrometer. An anemometer is used to measure wind speed, and handheld versions are very inexpensive. Wind speed can also be estimated using a Beaufort scale of winds, which provides descriptive measures of wind effects that are given a number from 1 to 12. Wind direction can be determined at the same time as wind velocity by taking a compass reading in the direction of wind flow.

All the weather parameters mentioned can be measured using relatively inexpensive equipment or electronic data devices that require more money but less personnel time. Further, meteorological stations are maintained throughout many areas of the world and may provide free access to data. Availability may differ with geographic location, but researchers should check into availability before purchasing expensive equipment. These stations are an excellent source of climate information, but the utility of station data depends on the scale of the research project. If the purpose of an investigation is to understand the role of local habitat relations then weather-station data will likely be insufficient for examining among-site differences.

17.8 Aquatic habitat

Aquatic amphibian habitat includes lentic (i.e. pond or wetland) and lotic (i.e. stream) environments. Spatial and temporal variability, and systemic connections between aquatic habitat and the surrounding terrestrial landscape, are key features of aquatic habitats that structure amphibian communities. Aquatic sites constitute obligate habitat for some species, and are critical breeding habitat for species with complex life cycles involving aquatic egg or larval development. Because many amphibian species undergo ontogenetic shifts in habitat, generally from aquatic to terrestrial habitat, different foci and approaches are needed to understand habitat use for each life-history stage (i.e. eggs, larvae, juveniles, adults).

17.9 Physical habitat variables

Physical features of aquatic habitats provide a dynamic structure within which amphibians interact with their biotic and abiotic environments. Those physical habitat parameters that prescribe broadly whether aquatic habitats are suitable for amphibians are discussed in detail below. Many other physical variables influence the finer-scale suitability of aquatic habitat. Some of the more commonly measured of these are presented in Table 17.1. Welsh *et al.* (1997) and Welsh and Ollivier (1998) provide an extensive list of relevant stream habitat variables.

Table 17.1 *Additional aquatic variables for assessing amphibian habitat.*

Habitat variable	Relevance to amphibians	Measurement technique
Ultraviolet B radiation	Associated with altered growth, survival, and behavior	Polysulfone plastic dosimeter
		Pyranometer+datalogger
	Interacts synergistically with other variables	Spectrometer
Sun exposure	Through effects on water temperature, evaporation, and plant productivity; influences amphibian development rate; and oxygen, food, and water availability	*Light at surface*
		Pyranometer
		Pyroheliometer
		Quantum sensor
		Light at depth
		Submersible spectroradiometer or comparison between underwater and surface photocells
		See sections 17.11.1–17.11.3
Turbidity	Alters water transparency, with impacts on predator and amphibian behavior	Secchi disk
	Related to sedimentation and pollution loads, which can affect fitness	Nephelometer
		Turbidity meter
Stream order	Related to current intensity, water temperature, and microhabitat suitability	Determined via examination of maps or GIS
	See section 17.9	
Stream reach type (e.g. riffle, pool, run)	Flow regime and substrate interact to create particular reach types	Classify and map reach types (see Welsh *et al.* 1997 for review of reach classification systems)
	Amphibian species may be closely associated with particular reach types	
Water volume (e.g. of bromeliads, tree holes and other microhabitats)	Related to reproductive potential within a microhabitat	Remove water from microhabitat and quantify (e.g. in syringe or graduated cylinder), return water to microhabitat
Substrate depth	Related to the refuge-potential of the substrate	Ruler; meter stick; or long, graduated pole
Habitat age and type (i.e. natural, artificial, or restored) (also applies to terrestrial habitats)	Related to vegetative and substrate development, microhabitat quality, and the likelihood of amphibian use, immigration, and survival	Examine chronological series of topographic maps or aerial photos
		Examine historical records or management reports
Degree of human disturbance (in and adjacent to the habitat)	Associated with sedimentation, pollution, invasive species, and predators	Categorize according to land-use type and intensity

GIS, Geographic Information Systems.

Morphometry, the habitat's basic shape and structure, limits the community types possible in a habitat, through its influence on water temperature, current strength, sediment and vegetation patterns, water-holding capacity, substrate, and weathering rates. Morphometric variables include contours, perimeter, length, width, area, depth, slope, and surficial geology. For streams, morphometry can be assessed for the wetted or dry channel, or for individual habitat reaches. Channel slope is particularly important for streams because slope affects dispersal potential and flow velocity.

Many morphometric variables can be mapped using global positioning system (GPS) or professional survey equipment. Perimeter, length, and width can also be assessed with pacing or a measuring tape. Wet contours and water depth can be charted with an echo sounder, weighted line, or meter stick. Area can be calculated from length and width; or from maps, aerial photographs, or GIS layers. The reliability of map-derived area measurements increases with basin size (Skidds *et al.* 2007). Slope can be measured with a clinometer. Surficial geology can be determined from geologic maps or field-verified from test or gravel pits.

Hydroperiod, or the timing and duration that a habitat holds water during a given year, strongly influences community composition (Wellborn *et al.* 1996; Babbitt *et al.* 2003). For example, in ephemeral waterbodies, invertebrates are often the dominant predators and amphibians must have a short larval stage to complete metamorphosis before the habitat dries. Conversely, in permanent waterbodies, amphibians can over-winter as larvae and fish are typically major predators. Hydroperiod can be categorized for lentic habitats as temporary, semi-permanent, or permanent; and for lotic habitats as intermittent or perennial. Such categorizations can be made by examining vegetation, or water level and substrate, towards the end of the dry season. Detailed data can be obtained by recording changes in the extent of inundation (Skidds *et al.* 2007), or by tracking water depth and noting the drying date. Hydroperiod can vary widely from year to year in seasonal waterbodies, so hydroperiod data should be collected over several years.

Hydrologic connectivity (i.e. the extent to which aquatic environments are linked) influences both hydroperiod and the inputs to, and outputs from, a habitat. Connectivity may be observed by inspection for inflows and outflows during the rainy season, or detected from contour maps or historic aerial photographs. Monitoring wells or a hydrologic analysis may be necessary to establish groundwater connectivity.

For streams, flow intensity (e.g. velocity, variability) also strongly influences habitat suitability. Flow affects dispersal, predator distribution, scouring, and sedimentation, and the ability of amphibians, and their sperm (i.e. for external fertilization) and food (i.e. invertebrates and algae), to maintain channel position. Velocity is measured with a flow meter, a pitot tube, or by calculating the time required for a buoyant object or dye to travel a specified distance. Discharge, or flow volume per unit time, is equal to the stream's cross-sectional area times the velocity. Variability can be described by flow variance per unit time, or by the return interval of specific events (e.g. 100-year floods). Flow timing determines desiccation and scouring risks, and can be understood by examining velocity and discharge records.

Water temperature impacts amphibians at multiple ecological scales. For instance, temperature affects decomposition and photosynthesis, which influence food availability and thus amphibian abundance. Amphibian development rates also vary with water temperature. Temperature measurement is discussed in section 17.7. Drainage area, elevation and aspect, and canopy closure all affect water temperature. Drainage area and elevation can be determined from topographic maps. Elevation can also be measured with an altimeter. Canopy closure is discussed in section 17.11.

Aquatic substrates can be organic or inorganic. Organic substrates provide over-wintering habitat and are integral to aquatic food webs. Inorganic substrates (e.g. cobbles) provide oviposition sites and refuges for amphibians and their invertebrate and algal food. Substrate particle size is a critical stream-habitat variable, as particle size affects size of interstitial hiding places which affects susceptibility to predation (Barr and Babbitt 2002). Small particles (i.e. < 32 mm) can be quantitatively described via sieve analysis. Visual estimates of percent cover of each substrate type and size class (ranging from leaf litter and muck, to fine sediment up to boulders) can also be used to classify substrate composition and particle size (Welsh and Ollivier 1998).

17.10 Chemical variables

The chemical makeup of aquatic sites can significantly affect amphibian growth, survival, and behavior. Pollutants can be toxic or can act as endocrine disrupters and alter many life processes (e.g. sexual development; Orton *et al.* 2006). Therefore, most aquatic habitat studies include measurements of some chemical parameters. Chemical potency varies with a wide variety of factors, ranging from chemical formulation (e.g. Howe *et al.* 2004) to amphibian life stage (e.g. Griffis-Kyle 2005). Exposure to some chemicals is greatest in the water column, while other substances accrue in the substrate (e.g. Jofre and Karasov 1999). Finally, synergistic effects between chemicals and other abiotic or biotic variables are common (e.g. Relyea and Mills 2001). Therefore, it is important to consider context (e.g. site history) when planning chemical sampling.

Given the inherent variability of aquatic environments, a composite sample, drawn from multiple points in a habitat, is necessary for the sample to be representative of general water conditions. To characterize specific sites (e.g. riffles), however, single samples should be collected from point locations. While some variables (e.g. pH, conductivity) can be measured in the field, many chemicals require laboratory analysis. Sampling protocols can vary significantly between chemical variants, so protocols should be researched or laboratory personnel contacted prior to sampling. In particular, it is important to know whether and when a sample should be filtered or cooled; sample bottle type; and how quickly following collection a sample must be analyzed. Sampling methods for some commonly measured parameters are discussed below.

pH is a measure of water acidity. pH can be assessed with a pH meter or paper pH strips. Strips are less accurate than pH meters, however. Alkalinity, a measure of the site's ability to neutralize acids, is also often sampled, since it reflects the habitat's resistance to sudden changes in pH. Alkalinity is measured via titration with a strong acid.

Dissolved oxygen is required for respiration by aquatic amphibian life stages. Dissolved oxygen can be a strongly limiting variable for cold-water-adapted species (Jackson 2007). Dissolved oxygen also affects the redox potential and solubility of some metals and nutrients. Dissolved oxygen is usually measured in the field, using either a variation of the Winkler titration method or an oxygen meter. Samples can be titrated in the laboratory, but only if properly preserved while still in the field.

Conductivity, a measure of the water's ionic strength, is an easily sampled variable that can be used as a proxy for parameters that are more difficult to assess. Ions occur naturally in aquatic habitats as a result of weathering and nutrient fluxes, but are also produced by human activities (e.g. winter road salting). Very high conductivity levels, or significant changes in conductivity through time, suggest poor water quality and are cause for further analysis of more-specific chemical variables. Conductivity can be measured with a conductivity meter.

Nutrients influence amphibians through two main pathways: nutrients are necessary for growth and survival of, and can be toxic to, most life forms. Amphibians seem particularly sensitive to nitrogen, especially in the forms of nitrate, nitrite, and ammonia. Commonly sampled nutrients include: NH_4-N, NH_3-N, NO_3-N, NO_2-N, total Kjeldahl N, dissolved organic N, total ortho-PO_4, total P, and dissolved P. Dissolved organic carbon, which can attenuate ultraviolet radiation, is also of interest (e.g. Brooks *et al.* 2005). Other humic substances can also strongly impact amphibian biochemistry and communities (e.g. reduced plant photosynthesis), and should be considered for analysis (Steinberg *et al.* 2006). Important cationic micronutrients (e.g. calcium, magnesium) may also be analyzed.

Aluminum, heavy metals, polycyclic aromatic hydrocarbons (PAHs), polycyclic chlorinated hydrocarbons (PCHs), polychlorinated biphenyls (PCBs), and a variety of pesticides are potentially important chemicals for measurement as they have demonstrated lethal and sublethal impacts to amphibians and can significantly alter habitat suitability. Nutrients, dissolved organic carbon, cations, metals, pesticides, hydrocarbons, and other chemical samples are usually analyzed in the laboratory. Use of nationally certified laboratories is recommended for many analyses.

17.11 Measuring vegetation

Vegetation is a major component of most terrestrial habitat. Vegetation strongly influences habitat suitability through its effects on microclimate. Vegetation can affect thermal gradients by providing shade, impacting relative humidity and moisture retention, adding litter (from leaves to large woody debris) to the ground stratum, and altering wind flows and snow accumulation. Vegetation, as living individuals, snags, and down logs, also contributes to habitat suitability by providing amphibians foraging substrates, protective cover from predators, and structures for egg (leaves) and larval (bromeliads) development.

In aquatic habitats, living and decaying plants affect the productivity, and structural and chemical suitability of amphibian habitat. Vegetation provides refuge, oviposition, calling, nesting, and over-wintering sites. Vegetation can change the abiotic environment (e.g. water temperature), with effects on amphibian development and fitness. Aquatic, shoreline, and proximate terrestrial vegetation can all influence different aspects of aquatic amphibian habitat.

The techniques discussed below apply to terrestrial and aquatic environments, though some techniques must be modified for aquatic sites. In particular, for aquatic assessments of mid- and understory vegetation, emergent and submergent plants are often distinguished and branch density, which reflects oviposition site potential, may be of interest. Further, any examined microhabitat parameters will be habitat-specific. For example, four-toed salamanders (*Hemidactylium scutatum*) often brood nests within wetland-plant hummocks. For this species, it might be useful to know not only the abundance, distribution, and species composition, but also the depth, moisture content, and height, of each plant hummock (Wahl *et al.* 2008).

In the following sections we discuss measurement of overstory, midstory, and ground-stratum parameters. Many techniques can be used to quantify more than one of these categories. In some ecosystems, it may also be important to measure characteristics of vinous and epiphytic vegetation (see Bandoni de Oliveira and Navas 2004).

Vegetative parameters can be measured using quadrats, line-intercept, or plotless techniques. Quadrat shapes vary from circular, to square, to rectangular. Circular plots have less edge and can be permanently marked with one point. Depending on research objectives, quadrats may be placed within or across a moisture or elevation gradient.

Plant distribution often is spatially structured or patchy (Levin 1992) leading to a condition known as spatial autocorrelation. This condition complicates the measurement and statistical descriptions of species diversity; however, field sampling designs and statistical analyses have been developed to address this condition (e.g. Goslee 2006).

17.11.1 Tree (overstory) measurements

Parameters typically quantified for the tree stratum include species composition (i.e. richness and frequency of occurrence), density, total height, crown depth, diameter, and basal area. Since species richness and frequency are a function of presence or absence within sampling units, these parameters often are partial components of an overall sampling protocol that may be directed at determining plant density and canopy cover. Species composition and absolute density (number of trees/area) can be measured using quadrat or distance (plotless) techniques.

Appropriate quadrat size will vary inversely with the natural abundance of individuals in the sampled community. Conversely, adequate quadrat size will vary directly with increasing species richness following species/area relationships. Species richness may be described in terms of number of species/unit area or the total number of species encountered/area sampled. Frequency of occurrence is estimated by the proportion of sampling units within which a given species was found and describes how common a species is within the plant community. Density is estimated as the number of trees (either summed over all species or by individual species)/unit area.

While several distance measurement techniques exist, the point-center-quarter (PCQ) technique is commonly used (Higgins *et al.* 2005). This technique works best when describing structural parameters of just one or two species (Cottam and Curtis 1956). The PCQ method provides acceptable estimates of density of randomly dispersed trees. Variants of this method, either the angle-order (Morista 1957) or the corrected-point-distance (Batcheler 1973), are recommended for use when non-random distributions of individuals occur.

Tree height, up to 10 m, can be measured using a telescoping stage stadia rod elevated from the base of the tree to the highest point of the canopy crown. A manual or automated clinometer can be used to measure the height of any tree for which the base of the stem and top of the canopy (apex) can be seen. Stadia rods and clinometers also can be used to measure height to base of canopy and thus compute canopy thickness.

Tree canopy cover is defined as the vertical projection of the total areal extent of the canopy upon the ground and is typically expressed as m^2/ha or more commonly as a percentage. Changes in canopy cover can alter suitability of aquatic sites (e.g. via effects on temperature and hydroperiod), resulting in changes in amphibian community composition (Skelly *et al.* 1999). Light interception and irradiance at a point on the ground, however, are influenced by both the vertical and angular aspects of the canopy (Nuttle 1987). Several ground-based methods have been developed to measure tree canopy cover including, but not limited to, line-interception, moosehorn, spherical densitometer, and hemispherical photography (Fiala *et al.* 2006). The line-intercept method employs multiple line transects (each 15–20 m long) distributed within a stand. Tree canopy cover is estimated as the percentage of each line lying beneath the vertical projection of the canopy disregarding any breaks within the outline of the canopy.

The moosehorn and the spherical densitometer are both sighting devices that employ mirrors with etched markings. The moosehorn has a flat mirror to reflect the canopy directly above the sampling location onto the reflective grid. Percent canopy cover is determined by counting the number of cross-hair intersections or points intercepting the reflected canopy divided by the total number of points on the grid. The spherical densitometer uses a convex or concave spherical mirror. Canopy measurements are taken while facing in all four cardinal directions from a single point. Canopy cover is estimated by averaging the proportion of the 24 squares etched on the reflective spherical surface that is contacted by canopy. A simpler approach to measuring canopy cover is to use a tube and ocularly estimate percent canopy in broader categories (e.g. 0, 5, 25, 50, 75, 100%).

Horizontal photographs, when taken at a designated height with a hemispherical wide-angle lens, record the extent of the overhead canopy dependent upon the angle of view of the lens. Photographs taken at bright midday conditions should be avoided (Fiala *et al.* 2006). Photogrammetric data obtained using hemispherical lenses are analyzed following Rich (1989). These data are an estimate of light penetration through angular openings in the canopy and are not direct measurements of canopy cover (Fiala *et al.* 2006). From above the canopy, remotely-sensed data using recently developed laser altimetry (lidar) provides three-dimensional geospatially-referenced data that is useful in modeling structural attributes of terrestrial ecosystem, especially at the landscape level (Vierling *et al.* 2008).

The standard height above ground to measure tree diameter is 1.37 m and is called diameter at breast height (dbh). If a tree has a fluted or buttressed base, the diameter is measured immediately above this area. Flexible diameter tapes (that transform circumference to diameter) or calipers are typically used to measure dbh. Forest basal area is the proportion of ground occupied by the bases of trees, expressed as m^2/ha. Clear plastic prisms with known degrees of light refraction, called Basal Area Fraction (BAF), can be used to view and count the number of individual trees within a complete circle around a sampling point whose stems, when viewed through the prism, are not completely separated (Salek and Zahradnik 2008). This count is then multiplied by the prism's BAF (units are in multiples of m^2/ha) to estimate basal area in the vicinity of that sampling point. Individual trees whose stem displacements just touch are counted as one half. Amber-colored prisms aid in sighting under low-light conditions.

17.11.2 Shrub (midstory) measurements

The shrub or midstory stratum is composed mostly of woody plants that may never enter the overstory or of juvenile individuals of species that will mature into the overstory. Midstory structural and compositional parameters that may interest those studying amphibians are similar to those of the overstory, namely species composition, density, and cover.

Species composition can be estimated using quadrat and line-intercept methods (see section 17.11.1). When using quadrats, density of midstory plants is determined by counting the number of individual stems within the quadrat. Many shrubs and juvenile trees will produce multiple stems per plant (i.e. basal and coppice sprouts). Therefore, midstory plant density is characterized as the number of stems per unit area without regard to clonal associations. Quadrat size for estimating midstory plant density may range from 4 to 10 m^2 (Irwin and Peak 1979).

Midstory canopy cover can be estimated using the line-intercept method described for overstory canopy cover. The proportion of a fixed-length line transect intersected by the vertical projection of midstory canopy provides a measure of percent canopy cover. Since the canopy of midstory (compared to overstory) plants is closer to the observer, canopy cover of individual species can be determined. Canopies of different species may overlap; therefore, total midstory canopy cover may be less than the sum of the individual species' covers and should be measured as a separate parameter. Measuring maximum height of midstory plants at systematic or randomly located points along each line transect will provide a description of midstory plant-height dynamics.

In environments where shrubs are low-growing (i.e. less than 1 m tall), canopy and basal cover can be sampled using the point-intercept method (Bonham 1989). Sharp metal pins are inserted vertically or on an incline through the vegetation at random or systematically-placed locations. Canopy and basal cover are calculated as the proportion of inserted pins that intercept, respectively, any piece of vegetation (leaf or stem) or the base of a plant at ground level (Higgins *et al.* 2005). Point intercepts are often recorded along a line transect; however, note that the line is the sample unit. Measuring fewer points along a greater number of lines improves statistical power (Bonham 1989).

17.11.3 Ground (understory) measurements

Understory components may be vegetative or non-vegetative (e.g. rocks, bare ground), and can provide critical refuges for ground-dwelling amphibians. Typically measured characteristics are plant density, frequency, and percent and type of cover. Standing-crop biomass and moisture content may also be of interest, especially in fire-prone ecosystems.

Density of understory plants can be estimated using direct counts within quadrats. Quadrats typically vary in area from 0.5 to 1.0 m^2 (Higgins *et al.* 2005). Horizontal cover of vegetative and non-vegetative components can be assessed using ocular estimates within quadrats, intercepted distances along line transects, or vertical digital photographs (Luscier *et al.* 2006). Ocular estimates require observers to place estimated cover values into bracketed cover classes; the midpoints of those classes are used for statistical analyses. Line-intercept cover measurements are more precise and accurate but require more time to collect. Vertical digital photographs of the understory also can provide accurate estimates of object groups (e.g. grasses, forbs, litter, bare ground); however, this technology cannot easily differentiate among plant species (Luscier *et al.* 2006).

Coarse woody debris (CWD) consists of dead trees and tree parts that are in contact with the ground. Measurement of CWD is important in both terrestrial and aquatic environments. Typically measured structural attributes of CWD include areal cover and volume. CWD areal cover can be quantified using the line-intercept method while measuring other understory parameters. The proportion of lines intercepting the vertical projections of CWD can be presented as percent cover or translated to m²/ha (Baillie *et al.* 1999). Using quadrats, CWD volume (m³/ha) can be estimated from the density of downed logs and diameter measurements at the large and small ends of each log. CWD decomposition stage may be of interest and be classified according to Maser *et al.* (1979).

Litter depth and moisture content can greatly influence microhabitat suitability. Litter depth to mineral soil is easily measured using a ruler. To estimate biomass (g of dry weight/m²) and moisture content (%), litter can be collected in the field, weighed, and oven dried to a constant weight (Bonham 1989). Dry weight and moisture content of understory plants can be determined by clipping the vegetation at ground level within quadrats and weighing, drying, and re-weighing the vegetative samples.

17.12 Edaphic features

Many amphibian species spend a significant portion of their lives on or beneath the soil surface. Chemical and structural features of the edaphic environment can strongly influence habitat suitability. Soil pH, moisture, temperature, and texture are important parameters to measure. Soil pH outside the physiological tolerance of amphibians can be an important limiting factor (Wyman and Hawksley-Lescault 1987). Soil pH can be assessed by inserting a probe into an emulsion of sieved, dried soil in a 1:10 solution of distilled water (Davey and Conyers 1988). Soil moisture can be determined with a probe or gravimetrically, whereby the sample is weighed wet, dried to a constant weight, and reweighed (also known as weigh-dry-weigh). Soil thermometers allow measurement at specific depths. To characterize retreat temperatures, it may be useful to measure temperature at the soil surface and at greater depths. Texture can be assessed with a sieve analysis. Availability of underground retreats is an important habitat component for many fossorial amphibians. For example, mammal burrows provide important retreats for ambystomatid salamanders (Madison 1997), whereas spider burrows provide refuges for desert anurans (Dayton and Fitzgerald 2006). Burrow measurements can include dimension (e.g. length and diameter), distribution, and whether burrows are aligned vertically or horizontally.

17.13 Conclusion

Characterization of amphibian habitat has lagged behind many other taxa. Increased research on habitat suitability will provide valuable data that will assist in management and conservation of amphibian species. Habitat suitability is influenced by both the features of the environment outlined in this chapter as well as changes that occur at the landscape level. Researchers are encouraged to design studies that include approaches outlined in this chapter as well as those detailed in Chapter 20.

17.14 References

Babbitt, K. J., Baber, M. J. and Tarr, T. L. (2003). Patterns of larval amphibian distribution along a wetlands hydroperiod gradient. *Canadian Journal of Zoology*, **81**, 1539–52.

Baillie, B. R., Cummins, T. L., and Kimberley, M. O. (1999). Measuring woody debris in the small streams of New Zealand's pine plantations. *New Zealand Journal of Marine and Freshwater Research*, **33**, 87–97.

Bandoni de Oliveira, F. and Navas, C. A. (2004). Plant selection and seasonal patterns of vocal activity in two populations of the bromeligen treefrog *Scinax perpusillus* (Anura, Hylidae). *Journal of Herpetology*, **38**, 331–9.

Barr, G. E. and Babbitt, K. J. (2002). Effects of biotic and abiotic factors on distribution and abundance of larval two-lined salamanders (*Eurycea bislineata*): changes across spatial scales. *Oecologia*, **133**, 176–85.

Batcheler, C. L. (1973). Estimating density and dispersion from truncated or unrestricted joint point-distance nearest-neighbor distances. *Proceeding of the New Zealand Ecological Society*, **20**, 131–47.

Bonham, C. D. (1989). *Measurements for Terrestrial Vegetation*. John Wiley and Sons, New York.

Brooks, P. D., O'Reilly, C. M., Diamond, S. A., Campbell, D. H., Knapp, R., Bradford, D., Corn, P. S., Hossack, B., and Tonnessen, K. (2005). Spatial and temporal variability in the amount and source of dissolved organic carbon: implications for ultraviolet exposure in amphibian habitats. *Ecosystems*, **8**, 478–87.

Cottam, G. and Curtis, J. T. (1956). The use of distance measures in phytosociological sampling. *Ecology*, **37**, 451–60.

Davey, B. G. and Conyers, M. K. (1988). Determining the pH of acid soils. *Soil Science*, **146**, 141–50.

Dayton, G. H. and Fitzgerald, L. E. (2006) Habitat suitability models for desert amphibians. *Biological Conservation*, **132**, 40–9.

Fiala, A. C. S., Garman, S. L., and Gray, A. N. (2006). Comparison of five cover estimation techniques in the western Oregon Cascades. *Forest Ecology and Management*, **232**, 188–97.

Garshelis, D. L. (2000). Delusions in habitat evaluation: measuring use, selection, and importance. In L. Boitani and T. K. Fuller (eds), *Research Techniques in Animal Ecology: Controversies and Consequences*, pp. 111–64. Columbia University Press, New York.

Goslee, S. C. (2006). Behavior of vegetation sampling methods in the presence of spatial autocorrelation. *Plant Ecology*, **187**, 203–12.

Griffis-Kyle, K. L. (2005). Ontogenic delays in effects of nitrite exposure on tiger salamanders (*Ambystoma tigrinum tigrinum*) and wood frogs (*Rana sylvatica*). *Environmental Toxicology and Chemistry*, **24**, 1523–7.

Higgins, K. F., Jenkins, K. J., Clambey, G. K., Uresk, D. W., Naugle, D. E., Norland, J. E., and Barker, W. T. (2005). Vegetation sampling and measurement. In C. E. Braun (ed.), *Techniques for Wildlife Investigations and Management*, 6th edn, pp. 524–54. The Wildlife Society, Bethesda, MD.

Howe, C. M., Berrill, M., Pauli, B. D., Helbing, C. C., Werry, K., and Veldhoen, N. (2004). Toxicity of glyphosate-based pesticides to four North American frog species. *Environmental Toxicology and Chemistry*, **23**, 1928–38.

Irwin, L. L. and Peek, J. M. (1979). Shrub production and biomass trends following five logging treatments within the cedar-hemlock zone of northern Idaho. *Forest Science*, **24**, 415–26.

Jackson, D. C. (2007). Temperature and hypoxia in ectothermic tetrapods. *Journal of Thermal Biology*, **32**, 125–33.

Jofre, M. B. and Karasov, W. H. (1999). Direct effect of ammonia on three species of North American anuran amphibians. *Environmental Toxicology and Chemistry*, **18**, 1806–12.

Johnson, D. H. (1980). Comparison of usage and availability measurements for evaluating resource preference. *Ecology*, **61**, 65–71.

Levin, S. A. (1992). The problem of pattern and scale in ecology. *Ecology*, **73**, 1943–67.

Luscier, J. D., Thompson, W. L., Wilson, J. M., Gorham, B. E., and Dragut, L. D. (2006). Using digital photographs and object-based image analysis to estimate percent ground cover in vegetation plots. *Frontiers in Ecology and Environment*, **4**, 408–13.

Madison, D. M. (1997). The emigration of radio-implanted spotted salamanders, *Ambystoma maculatum*. *Journal of Herpetology*, **31**, 542–51.

Manly, B. F. J., McDonald, L. L., Thomas, D. L., McDonald, T. L., and Erickson, W. P. (2002). *Resource Selection by animals: Statistical Design and Analysis for Field Studies*, 2nd edn. Kluwer, New York.

Maser, C., Anderson, R. G., Cromack, Jr, K., Williams, J. T., and Martin, R. E. (1979). Dead and down woody material. In J. W. Thomas (ed.), *Wildlife Habitats in Managed Forests—The Blue Mountains of Oregon and Washington*, pp. 78–95. Agricultural Handbook 553. US Department of Agriculture, Washington DC.

Morista, M. (1957). A new method for the estimation of density by the spacing method applicable to non-randomly distributed populations. *Physiological Ecology*, **7**, 134–44.

Morrison, M. L., Marcot, B. G., and Mannan, R. W. (2006). *Wildlife-habitat Relationships: Concepts and Applications*, 3rd edn. Island Press, Washington DC.

Nuttle, T. (1997). Densiometer bias? Are we measuring the forest or the trees? *Wildlife Society Bulletin*, **25**, 610–11.

Orton, F., Carr, J. A., and Handy, R. D. (2006). Effects of nitrate and atrazine on larval development and sexual differentiation in the northern leopard frog *Rana pipiens*. *Environmental Toxicology and Chemistry*, **25**, 65–71.

Pechmann, J. H. K., Scott, D. E., Semlitsch, R. D., Caldwell, J. P., Vitt, L. T., and Gibbons, J. W. (1991). Declining amphibian populations: the problem of separating human impacts from natural fluctuations. *Science*, **253**, 892–5.

Relyea, R. A. and Mills, N. (2001). Predator induced stress makes the pesticide carbaryl more deadly to gray treefrog tadpoles (*Hyla versicolor*). *Proceedings of the National Academy of Sciences USA*, **98**, 2491–6.

Rich, P. M. (1989). *A Manual for Analysis of Hemispherical Canopy Photography.* Publication LA-11733-M. Los Alamos National Laboratory, Los Alamos, NM.

Salek, L. and Zahradnik, D. (2008). Wedge prism as a tool for diameter and distance measurement. *Journal of Forest Science,* **54**, 121–4.

Skelly, D. K., Werner, E. E., and Cortwright, S. A. (1999). Long-term distributional dynamics of a Michigan amphibian assemblage. *Ecology,* **80**, 2326–37.

Skidds, D. E., Golet, F. C., Paton, P. W.C., and Mitchell, J. C. (2007). Habitat correlates of reproductive effort in wood frogs and spotted salamanders in an urbanizing watershed. *Journal of Herpetology,* **41**, 439–50.

Southwood, T. R. E. (1977). Habitat, the template for ecological strategies? *Journal of Animal Ecology,* **46**, 337–65.

Steinberg, C. E. W., Kamara, S., Prokhotskaya, V. Y., Manusadzianas, L., Karasyova, T. A., Timofeyev, M. A., Jie, Z., Paul, A., Meinelt, T., Farjalla, V. F. *et al.* (2006). Dissolved humic substances – ecological driving forces from the individual to the ecosystem level? *Freshwater Biology,* **51**, 1189–1210.

Taub, F. B. (1961). The distribution of red-backed salamanders, *Plethonon c. cinereus,* within the soil. *Ecology,* **42**, 681–8.

Thomas, D. L. and Taylor, E. J. (1990). Study designs and tests for comparing resource use and availability. *Journal of Wildlife Management,* **54**, 322–30.

Thomas, D. L. and Taylor, E. J. (2006). Study designs and tests for comparing resource use and availability II. *Journal of Wildlife Management,* **70**, 324–36.

Vierling, K. T., Vierling, L. A., Gould, W. A., Martinuzzi, S., and Clawges, R. M. (2008). Lidar: shedding new light on habitat characterization and modeling. *Frontiers in Ecology and Environment,* **6**, 90–8.

Wahl, III, G. W., Harris, R. N., and Nelms, T. (2008). Nest site selection and embryonic survival in four-toed salamanders, *Hemidactylium scutatum* (Caudata: Plethodontidae). *Herpetologica,* **64**, 12–19.

Wellborn, G. A., Skelly, D. K., and Werner, E. E. (1996). Mechanisms creating community structure across a freshwater habitat gradient. *Annual Review of Ecology and Systematics,* **27**, 337–63.

Welsh, Jr, H. H. and Ollivier, L. M. (1998). Stream amphibians as indicators of ecosystem stress: a case study from California's redwoods. *Ecological Applications,* **8**, 1118–32.

Welsh, Jr, H. H. and Lind, A. J. (2002). Multiscale habitat relationships of stream amphibians in the Klamath-Siskiyou Region of California and Oregon. *Journal of Wildlife Management,* **66**, 581–602.

Welsh, Jr, H. H., Ollivier, L. M., and Hankin, D. G. (1997). A habitat based design for sampling and monitoring stream amphibians with an illustration from Redwood National Park. *Northwestern Naturalist,* **78**, 1–16.

Wiens, J. A. (1989). *The Ecology of Bird Communities. Vol. 1, Foundations and Patterns.* Cambridge University Press, Cambridge.

Wyman, R. L. and Hawksely-Lescauly, D. S. (1987). Soil acidity affects distribution, behavior and physiology of the salamander *Plethodon cinereus. Ecology,* **68**, 1819–27.

Part 5
Amphibian communities

18

Diversity and similarity

C. Kenneth Dodd, Jr

18.1 Introduction

Determining how many species are present in a particular habitat, their abundance, their importance, and how communities are related are critical questions in ecology and conservation biology. Ecologists use measures of diversity and similarity to understand community structure and function, particularly in terms of habitat use, food webs, predator–prey relationships, estimating how many species can co-exist within a community, energy flow, and nutrient cycling. Conservation biologists need to estimate diversity to identify areas for protection and management (Scott *et al.* 1987; Snodgrass *et al.* 2000), and to assess the effects of habitat change through time. Knowledge of these variables also is very practical for consultants or others doing rapid assessments, such as in the preparation of environmental impact assessments. Species diversity (richness, heterogeneity, evenness) is fairly simple to estimate, yet the use of these indices in amphibian studies is generally less than in many other areas of field research, and most field amphibian researchers have yet to explore the utility of measures of similarity.

There are many indices used to express various aspects of diversity and similarity; some of the most common are listed in Table 18.1. Magurran (1988, 2003), Krebs (1999), and Clarke and Warnick (2001) provide good discussions of some diversity and similarity indices, and these references can serve as a starting point for understanding what the indices do and how they are calculated, and for where to find more information about them. Whatever index is chosen, it is important to understand the assumptions surrounding the computation, the biases of the index, and the way the index should be interpreted. It is very common for biologists to sacrifice rigor (by violating assumptions) for ease of computation or precedence of use by other biologists. Some of the easiest indices to use (for example, Margalef) are also some of the most informative, whereas

Table 18.1 *Indices and measures of species diversity commonly used in ecological studies.*

Measure of…	Test
Species richness	Rarefaction
	Menhinick (D_{Mn})
	Jacknife
	Richness (S)
	Bootstrap
	Margalef (D_{Mg})
	Species-area curve estimates
	McIntosh (U)
Heterogeneity	α (logarithmic)
	Brillouin's Index (HB)
	λ (log normal)
	Fisher's α
	Simpson's Index ($1-D$)
	Shannon-Wiener Function (H')
Evenness and dominance	Simpson's (D')
	Berger–Parker (d)
	Camargo (E')
	Hill's Ratio (N_1)
	Smith and Wilson (E_Q)
	Pielou's (E)
	Modified Nee
	McIntosh (D)

other indices (for example, Shannon) are popular but not very informative by themselves.

18.2 Data transformation

Measures of diversity and similarity give differing weight to abundance; that is, some indices are more sensitive to the presence of rare species, whereas others are more sensitive to the abundance of the more common species. Thus, using the same data set but analyzing it with different indices could result in very different profiles of community diversity (Bloom 1981). Inclusion of rare species in diversity and similarity estimates might indicate that a community is far richer than it normally is, especially if the species is migratory or an extremely unusual find. On the other hand, metamorphosis could heavily weight an amphibian community index toward a species that happened to transform during the sampling period.

Table 18.2 *The effect of transformation on perceptions of abundance. Transformation normalizes data, and allows for variable weighting in studies of amphibian diversity. As you move to the right in the table, rare species assume greater importance in the analysis of community similarity/dissimilarity.*

No transformation (N)	Square root ($\sqrt{}$)	Double square root ($\sqrt{}\sqrt{}$)	Log(1+x)[1]	Presence/absence
0	0	0	0	0
1	1	1	0.3	1
10	3.3	1.7	1	1
100	10	3.3	2	1
1000	33	5.7	3	1
10 000	100	10	4	1

[1] Because the $\log(0) = \infty$, 1 is added to the value inasmuch as the $\log(1+y) = 0$ when $y = 0$.

One way to change weighting would be to transform the data. Table 18.2 shows how data transformation can weight the 'importance' of the less abundant or rare species, and thus allow them to have more influence in perceptions of community patterns. Data transformation also helps in clustering and ordination methods. Thus, it should be apparent that the objectives of the researcher become very important in the selection of diversity and similarity indices. Knowing the limitations and assumptions of the indices helps avoid confusion and aids in the interpretive process.

A second reason for transforming data is that sampling usually involves large numbers of zero captures; that is, all species are not caught on each day of sampling. Zeros present complications in data analysis, especially when using parametric measures which assume normal distributions. Transformation offers a means to validate the statistical assumptions of parametric techniques prior to using them. One must always remember to verify that the data transformation actually corrects the original issue with the data set. If it does not, transformation is not warranted. Of course, an alternative approach would be to avoid parametric statistics, and many of the diversity and similarity indices commonly used are based on non-parametric techniques.

18.3 Species diversity

18.3.1 Sampling considerations

There are a number of considerations to remember if one of the goals of sampling is to develop an estimate of species diversity within an area. First, estimates of diversity are only as accurate as the detection probabilities of finding individual species during the sampling period. This is especially important inasmuch as most species have heterogeneous detection probabilities due to a variety of both natural and observer-related causes (Boulinier *et al.* 1998). Second, all sampling techniques have biases, as numerous chapters in this volume have discussed. Estimates of species diversity will be biased to the extent that sampling techniques are biased. Examples of such biases include the following.

- Methods. If the techniques used will not detect the species, estimates of diversity within an area will be inaccurate. If the goal is to estimate amphibian community diversity, then multiple and appropriate techniques must be employed, especially if rare species are suspected to be in the study area.
- Timing. Sampling must extend throughout the activity period of all the species in question, or at least it must include representative subsamples throughout the activity period. For example, the timing of sampling has been shown to alter estimates of species richness by changing the shape of the species accumulation curve. If continuous sampling is not possible, biologists might consider randomization (for example, by randomizing the start of weekly sampling periods throughout an activity season).
- Location. Estimates of community diversity will only be accurate if all habitats (arboreal, aquatic, terrestrial, fossorial) are sampled. Species diversity estimates should not be extrapolated beyond the area sampled (that is, the statistical area of inference).
- Stochastic events. Weather conditions (heat, cold, drought, precipitation) and unexpected disturbances (catastrophic storms, earthquakes) all have the potential to influence sampling and thus estimates of diversity. Where necessary, biologists need to apply caveats to their estimates to acknowledge the limitations of their data.

18.3.2 Species richness

The number of species within a community, species richness, is the simplest way to describe local and regional diversity (Magurran 1988). Although species richness is a natural measure of diversity, it is also an elusive quantity to measure accurately (May 1988; Gotelli and Colwell 2001), especially if species have different detection probabilities during the sampling period (de Solla *et al.* 2005). The number of species observed is frequently an underestimate of actual species richness if based on counts conducted over limited periods of time. Estimates of species richness are very sensitive to the sample size and methods of sampling on which they are based (Smith and van Belle 1984).

Species richness does not imply anything about abundance, although the term frequently is used inaccurately as a synonym of diversity (see below). In some situations, species richness can be estimated in advance using field guides or museum specimens, coupled with knowledge of habitat requirements; this method is termed *interpolation*. For example, the number of species found within a forested area of a national park might be predicted from published literature and museum records. Interpolation can be very inaccurate, depending on scale. Knowledge of species richness in one watershed might be useful for predicting species occupancy in an adjacent watershed, but not on a regional scale. Thus, interpolation does not supplant the necessity of field sampling for an accurate understanding of richness since myriad factors may influence occupancy at a particular location.

The number of species within an area results from a complex interaction of resource availability, habitat complexity, biogeography, land-use history, and phylogenetic history. Habitats with large numbers of amphibian species include tropical rainforests and areas with diverse topography and, as might be expected, amphibian species richness is inversely correlated with latitude and, usually, elevation (Duellman 1999). Even within lowland rainforests, however, amphibian species richness changes spatially, so sampling considerations become paramount when estimating how many species are present. An estimate of species richness is only as accurate as the reliability of the sampling methods on which the estimate is based.

18.3.3 Species accumulation curves

One measure of species richness in a region is the rate at which new species are added to an inventory (Soberón and Llorente 1993). If repetitive samples are taken using standardized techniques, the rate at which species are detected and the point at which detection of new species levels off should give an indication of the number of species within an area. Such information is useful when little or nothing is known *a priori* concerning species richness.

Of course, sampling every species within a habitat may not be temporally or logistically possible, and rarely would all species be captured with 100% confidence within a limited sampling period. Thus, it usually becomes necessary to estimate statistically the number of species within the area of interest. This is done by plotting the increasing number of species captured (on the *y* axis, termed the *ordinate*) by either effort (assuming sampling is regularly spaced through time) or by the cumulative number of individuals observed or captured (on the *x* axis, termed the *abscissa*). Effort is defined quantitatively, such as by trap-days, hours searched, or number of pitfalls checked. Thus, an estimate of species richness is derived through cumulative sampling, preferably over a substantial period of time.

Species accumulation curves can be used in two primary ways. First, estimates of the number of amphibian species within the sampling area during the time of the survey can be derived (Thompson *et al.* 2003; Dodd *et al.* 2007), especially when that number is imprecisely known, such as during an inventory of tropical forest habitats. Second, estimates can be derived of the amount of effort needed to sample a particular area with a preset degree of confidence (usually within 80–90% of the asymptote) (Thompson and Thompson 2007; Thompson *et al.* 2007). During the initial phases of sampling, a species accumulation curve will rise steeply, especially in species-rich communities (Cam *et al.* 2002). When the curve levels off, it is likely that additional sampling will not detect many more species, and thus sampling could be curtailed. This technique helps avoid unnecessary sampling, especially when it is logistically difficult or expensive.

Species-richness estimates are of two types: *extrapolative* (inferring species richness based on subsamples of a data set) and *interpolative* (as above, inferring species richness based on comparisons with other areas or data sets). The former method is widely used, with three distinct classes of statistical approaches (Chazdon *et al.* 1998): (1) extrapolation of either species accumulation curves or species-area curves to an asymptotic value; (2) fitting data on the relative abundance of species in a single sample to a parametric distribution (e.g. log-series, log-normal, Poisson log-normal); and (3) non-parametric estimators.

The failure to detect rare species can dramatically underestimate the true local species richness. If, however, a limited fraction of a specific taxonomic group is sampled quantitatively, sampling bias can be reduced by using statistical extrapolation to estimate species richness (Colwell and Coddington 1994). Several non-parametric estimators either have been developed specifically for estimating species richness from samples, adapted to do so from mark–recapture applications, or were developed for the general class-estimation problem (Smith and van Belle 1984; Colwell and Coddington 1994). These non-parametric estimators only require the number of samples in which each species is found, rather than any parametric information about their abundance (Brose *et al.* 2003). Some of them can be reduced to a very simple form: $S_{estimated} = S_{observed} + R$, where R is an estimate based on whether rare species are present or undetected in the samples. Overall, non-parametric estimators appear to be less biased and more precise than the other two approaches.

The shape of species-accumulation curves are influenced by both abundance and diversity (Thompson and Withers 2003). If rare species are present, or if there are few species with high abundance, accumulation curves have low shoulders and long trajectories to the asymptote. Conversely, areas with large numbers of abundant species have steep trajectories and reach asymptotes quickly. Species diversity is positively correlated with the initial slope of the trajectory of the accumulation curve.

Both incidence-based (that is, occupancy) and abundance-based (that is, incorporating diversity) species accumulation curves can be generated. A computer program (such as the Mao Tau of EstimateS, see below) might use 1000 iterations of a data set to predict the expected range of shapes of the accumulation curve. This allows biologists to incorporate confidence intervals (95%) around the accumulation curve, which provides a better estimate of the actual numbers of species likely within the area than a simple curve. An example of species-accumulation curves based on intensive pitfall and other sampling techniques for amphibians is shown in Figure 18.1.

There are few applications of these methods dealing with amphibians. Pineda and Halffter (2004) used them to verify the completeness of inventories at both local and regional scales and determined the sampling effort needed for reaching a plateau. Heyer *et al.* (1999) tested the utility of museum collections for conservation decisions, focusing on frogs of the genus *Leptodactylus* from Amazonia. The results indicated that the data set was adequate in terms of sampling effort and useful for conservation decisions, at least for amphibians. Dodd *et al.* (2007) used species accumulation curves to compare changes in species presence through time as a consequence of habitat changes. They showed that long-term changes in habitat management resulted in decreases in species richness within the amphibian community.

18.3.4 Heterogeneity

The concept of heterogeneity (also sometimes termed 'species diversity') provides a way of expressing both the number of species and a measure of counts or abundance into a single index. What is more diverse: a community with many rare species or one with fewer but much more abundant species? What if two communities have the same number of species, but at different abundances? In computations, actual (rarely measured) or relative (based on counts) abundances are combined with species richness to measure the heterogeneity of the community. The result is an index which allows the observer to compare the heterogeneity of one or more communities (for example, Dodd 1992). An index by itself tells little; it gains value when it is used comparatively.

Amphibians are sampled as one might for other measures of diversity; that is, via a variety of techniques that provide capture histories through time.

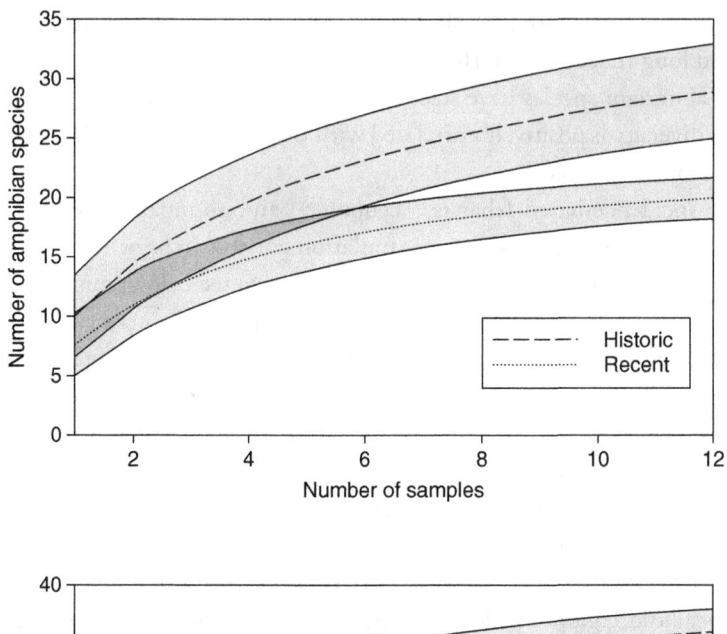

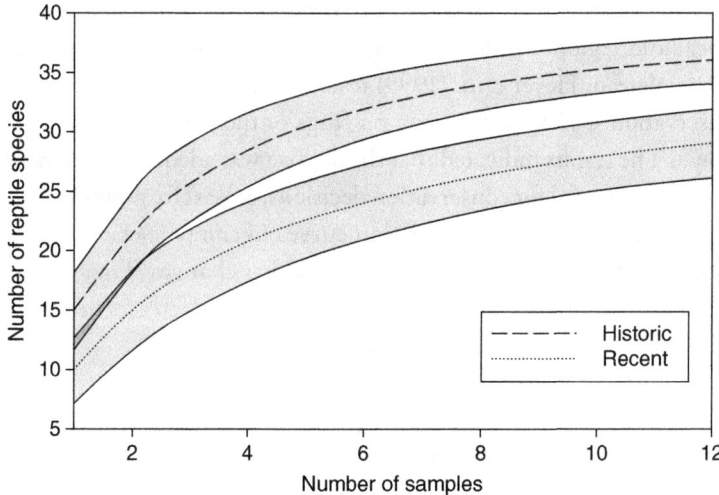

Fig 18.1 Species accumulation curves, with 95% confidence intervals, for amphibians (top) and reptiles (bottom) sampled at 12 study sites at St. Marks National Wildlife Refuge, Florida, USA. The historic curve shows relationships in 1977–9, whereas the recent curve shows the relationships in 2002–5. These species accumulation curves suggested that fewer species of amphibians were expected in the 2000s than in the 1970s. In the 1970s, the curves predicted that 29 species of amphibians might be present throughout the sampling sites, but only 19 in the 2000s. The actual values were 29 (1970s) and 24 (2000s). Reprinted from Dodd *et al.* (2007).

Abundance is usually and somewhat erroneously expressed in terms of counts taken during the course of a sampling period. Counts and abundance are not the same thing. A count is simply that: the number of individuals captured or observed during the course of a sampling period. On the other hand, true abundance is estimated by the equation:

$$\hat{N} = C/\hat{p}$$

where $\hat{N}$ is abundance, C is the number of animals counted, and $\hat{p}$ is the probability of detecting an individual. Thus, counts may or may not be a relative index of abundance and, hence, statistical indices based on counts may or may not be accurate (see Chapter 23). Given the unlikelihood of obtaining true estimates of abundance for all members of a community, counts will have to suffice if diversity indices are used.

18.3.5 Evenness and dominance

Evenness and dominance refer to the influence a species has in numbers on the community, particularly the most abundant species. Evenness is an important variable to measure, since high evenness is generally considered synonymous with high diversity (Magurran 1988); the most diverse community is thought of as one that has a large number of species and a large number of individuals uniformly abundant among samples. Like other measures of the proportional abundance of species, dominance and evenness indices incorporate estimates of richness and abundance. The result is a number that can be used to interpret whether a community is dominated by one or more species, or if species are more equitably distributed within the community. The biologist then uses this information to examine the data and identify the species responsible for dominance. Representative evenness/dominance indices are listed in Table 18.1.

18.4 Similarity

Measures of similarity are used to examine data from a number of sampling areas in order to compare community similarity, or more correctly, dissimilarity among those sampling areas. The result is a matrix of numbers representing paired comparisons which can be difficult to interpret without a visual context. However, these matrix-based comparisons can be fitted into a graphic depiction of similarity, aligning those communities or sampling areas most similar to one another into a cluster dendrogram. Cluster dendrograms then can be compared to see how communities change through time and how variables change from one community to another. One of the advantages of using similarity measures is that, depending on the measure, various types of data can be compared, such as species richness, abundance, biomass, or ecological parameters. Thus, these measures can be used to compare species diversity as well as some of the factors that might influence community structure or function.

Distance coefficients are often used to calculate the indices of similarity for some of the most commonly used approaches (for example, Bray–Curtis, Canberra), and computationally these are actually measures of dissimilarity. Thus, the measure of similarity is the reciprocal of the calculated value. When communities are exactly equal in the variables being compared, $d=0$; d approaches 1 as the communities become more dissimilar. Distance coefficients are computed in two ways, termed the *Euclidean* and *Manhattan* metrics. Think of walking between two points a few blocks apart in a city. Because of buildings and road structure, walking a direct line (Euclidean) is not possible, so it becomes necessary to go up and down streets in a somewhat non-direct pathway (Manhattan). Using a similar logic, popular metrics such as the Bray–Curtis Similarity Index use the Manhattan metric to estimate dissimilarity among all the possible comparisons within the data matrix, although other indices employ a somewhat less informative Euclidean metric.

Of course, it helps to know which species are responsible for perceptions of similarity or dissimilarity. There is a routine (SIMPER) in the statistical program Primer-E which can be used to compute the overall percent contribution that each species (or variable) makes to the dissimilarity between groups in a dendrogram cluster analysis (see below). The outcome is a list of species (or variables) in a decreasing order of importance in discriminating between all possible pairs of dissimilarity coefficients within the cluster (Clarke and Gorley 2006).

Similarity indices are just like any other indices as they have no real significance except comparatively. The indices will not tell the observer why certain communities are more similar or dissimilar to one another, or tell him or her why communities have changed in diversity through time. Instead, they help direct the observer toward the reasons for change, and allow for the development of hypotheses which can be tested using other data, future experimentation, or habitat manipulation. Some of the common similarity indices (see Magurran 1988, 2003; Krebs 1999; Clarke and Warnick 2001) include the following.

- Jaccard (*C*) uses binary data in a simplistic formula ($C_{jk} = p/p + m$) where p is the number of species in both communities j and k, and m is the number of species in one community but not the other. This is not a distance measure.
- Morisita (C_λ) is a true similarity index formulated for counts of individuals only.

- Horn is a modification of the Morisita C_λ that uses proportions instead of counts. It is thus not affected by sample size, and can be used for variables other than count or abundance data. However, it is not as robust as Morisita.

- Bray–Curtis (B) is a distance measure of dissimilarity, whereby coefficients are weighted toward abundant species, with rare species adding very little to the value. In addition to counts, matrices of catch-per-unit-effort or environmental variables can be analyzed for dissimilarity. See Beals (1984) and Clarke *et al.* (2006).

- Canberra (C) is another distance measure of dissimilarity, but one that is not as influenced by abundant species. This index gives more weight to rare species than the Bray–Curtis Index.

- Other similarity/dissimilarity indices include the Squared Euclidean Distance, Percent Similarity, Mountford, Dice, Kendall, Simplified Morisita, Sørensen, Product Moment Correlation Coefficient, Raup–Crick, and Whittaker.

In an example of the way similarity measures can be used, my colleagues and I conducted extensive field sampling for amphibians at St. Marks National Wildlife Refuge, Florida, USA, 30 years after similar sampling had occurred there. We attempted to use the same techniques at exactly the same sites. Based on capture results, we computed Bray–Curtis Similarity Indices for each community and sampling period, square-root-transformed the Bray–Curtis indices, and examined cluster dendrograms to visually compare community similarity. Changes in the dendrograms indicated significant changes in community similarity through time (Figure 18.2). We then re-examined weather conditions, habitat types, succession stages, and management practices during the two sampling periods. This led us to hypothesize four possible causes, acting interactively, that led to changes in species diversity within the amphibian community (Dodd *et al.* 2007).

18.5 Software

The program EstimateS does most computations required for species accumulation curves and non-parametric analyses of species richness, diversity, and dominance. A detailed description of the estimators computed can be found in Colwell and Coddington (1994), Chazdon *et al.* (1998), and Colwell (2006). EstimateS (version 8.0) computes randomized species accumulation curves, statistical estimators of true species richness (S), and a statistical estimator of

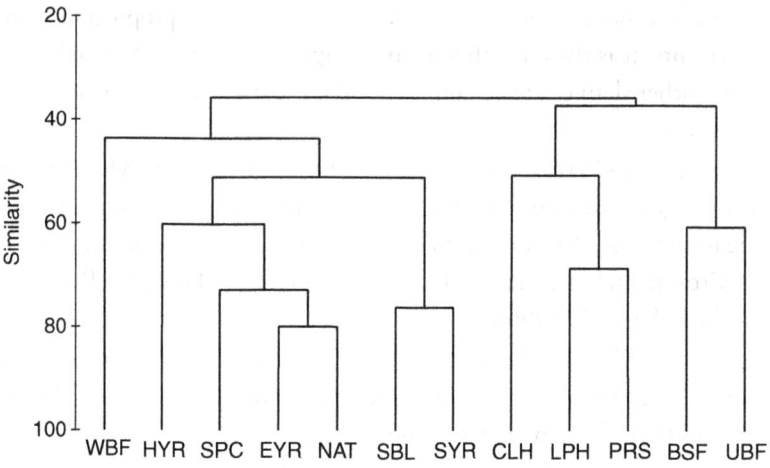

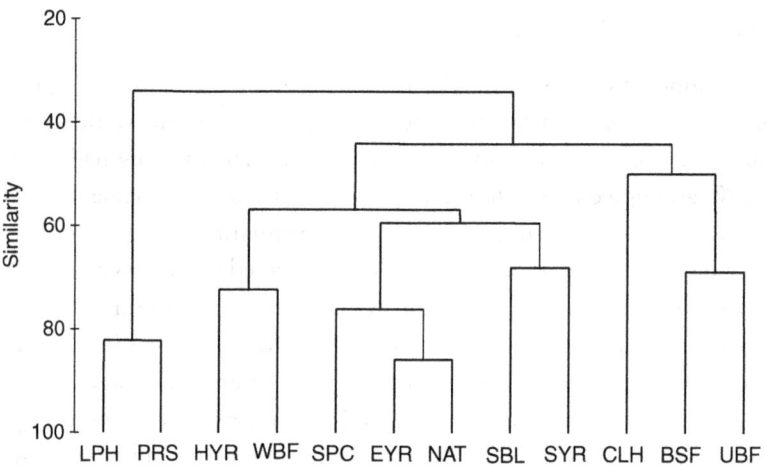

Fig 18.2 Cluster dendrograms based on square-root-transformed Bray–Curtis Similarity Index values showing the relationship among habitat types and the amphibian community at 12 study sites (coded on the *x* axis) at St. Marks National Wildlife Refuge, Florida, USA. The top cluster shows relationships in 1977–9, whereas the cluster at the bottom shows the relationships in 2002–5. The amphibian communities in the 1970s were most similar to one another within management units, regardless of habitat type or ongoing management practice. The pattern changed somewhat in the 2000 sampling period. Instead of a similarity based on east-to-west location, amphibian community similarities appeared to be based on dominant community type. Reprinted from Dodd *et al.* (2007).

the true number of species shared between pairs of samples, based on species-by-sample (or sample-by-species) incidence or abundance matrices. It can be used to compute an expected species accumulation curve, the Mao Tau (with 95% confidence limits). The Mao Tau is a sample-based rarefaction curve which provides a graphic estimate of expected species accumulation (Colwell *et al.* 2004; Figure 18.1). The program also can be used to compute both incidence-based coverage estimates (ICE) and abundance-based coverage estimates (ACE) of species richness among sampling sites. The derivation and use of these estimators is discussed by Chazdon *et al.* (1998) and Colwell (2006).

EstimateS further allows computation of Fisher's α, Shannon, and Simpson diversity indices; the Chao, Jacknife, ICE, ACE and other species richness estimators for abundance and incidence data; modified versions of the Sørensen and Jaccard similarity indices based on abundances, including the effects of unseen shared species (Chao *et al.* 2005); and classic Jaccard, Bray–Curtis, and Morisita–Horn (both incidence-based and abundance-based) similarity estimators. EstimateS can be downloaded free of charge from http://viceroy.eeb.uconn.edu/estimates.

Ecological software to accompany Kreb's *Ecological Methodology* is available from Exeter Software (Setauket, New York, USA; www.exetersoftware.com/cat/ecometh/ecomethodology.html). Version 6.1.3 sells for US$150 (as of May 2008). The software covers only the topics and analyses discussed in the book, which include various measures of richness, heterogeneity, and equitability, in addition to a wide range of other ecological analyses. These include binary coefficients, Euclidean distance coefficients, Bray–Curtis metric, Canberra metric, percentage similarity, Morisita's Index of Similarity, Horn's Index, Species Richness (Rarefaction method, Jackknife method for counts), logarithmic series, log-normal distribution, Simpson's Index of Diversity, Shannon–Wiener measure, Brillouin's Index of Diversity, and evenness measures. The program comes with a manual describing the analyses, and includes some information on the use and theory behind the various indices.

Primer 6 for Windows is another powerful tool for analysis of ecological data, including both diversity and dominance indices. Information on the program can be obtained from the website (www.primer-e.com/). Primer 6 allows univariate, graphical, and multivariate analyses of species, abundance, biomass, and physio-chemical data. The program facilitates grouping data into clusters, allows identification of species that are responsible for discrimination among two sample clusters, and graphs species abundance distributions through ordination and multidimensional scaling plots. As of May 2008 the cost is US$500 for a single-user license for academic research and is available from Primer-E

(Ivybridge, Devon, UK). Primer-E holds advanced workshops, the cost of which includes the manual (Clarke and Gorley 2006) and a book demonstrating the use of Primer 6 programs on data from the Plymouth Marine Laboratory (Clarke and Warnick 2001). Information on forthcoming workshops may be obtained from the website.

18.6 Summary

Indices of diversity and similarity offer a means of critically examining current ecological patterns and changes in community composition, especially when estimates of site occupation across a sufficient number of habitats are not available. In particular, the Bray–Curtis Similarity Index has proven useful in assessing the effects of habitat changes on herpetofauna and other taxa during monitoring programs (Pawar *et al.* 2004; Pieterson *et al.* 2006; Dodd *et al.* 2007), comparing stream and forest faunas (Parris and McCarthy 1999; Huang and Hou 2004), measuring the success of restoration efforts (Ruiz-Jean and Aide 2005), prioritizing areas for conservation (Seymour *et al.* 2001), assessing dietary differences (Whitfield and Donnelly 2006), and in analyses of geographic differences in diversity patterns (Urbina-Cardona and Londroño-M 2003; Menegon and Salvidio 2005; Smith-Vaniz *et al.* 2006).

An index-based approach offers insights into potential, if not definitive causes of community change. Once potential causes are identified, research can be designed to test hypotheses related to changes in species composition and relative abundance. Community ecology, conservation, and monitoring programs should incorporate an evaluation of habitat variables into data-collection protocols, and researchers must be aware of the potential importance of stochastic or periodic environmental disturbances, such as storms and flooding, when interpreting species presence/not detected and abundance data. Counting individual animals or determining the percentage of site occupancy neglects much information needed to understand changes in community composition through time. The use of diversity and similarity indices offers further insight into the comparative structure of amphibian communities.

18.7 References

Beals, E. W. (1984). Bray-Curtis ordination: an effective strategy for analysis of multi-variate ecological data. *Advances in Ecological Research*, **14**, 1–56.

Bloom, S. A. (1981). Similarity indices in community studies: potential pitfalls. *Marine Ecology Progress Series*, **5**, 125–8.

Boulinier, T., Nichols, J. D., Sauer, J. R., Hines, J. E., and Pollock, K. H. (1998). Estimating species richness: the importance of heterogeneity in species detectability. *Ecology*, **79**, 1018–28.

Brose, U., Martinez, N. D., and Williams, R. J. (2003). Estimating species richness: sensitivity to sample coverage and insensitivity to spatial patterns. *Ecology*, **84**, 2364–77.

Cam, E., Nichols, J. D., Sauer, J. R., and Hines, J. E. (2002). On the estimation of species richness based on the accumulation of previously unrecorded species. *Ecography*, **25**, 102–8.

Chao, A., Chazdon, R. L., Colwell, R. K., and Shen, T.-J. (2005). A new statistical approach for assessing similarity of species composition with incidence and abundance data. *Ecology Letters*, **8**, 148–59.

Chazdon, R. L., Colwell, R. K., Denslow, J. S., and Guariguata, M. R. (1998). Statistical methods for estimating species richness of woody regeneration in primary and secondary rainforests of northeastern Costa Rica. In F. Dallmeier and J. A. Comisky (eds), *Forest Biodiversity Research, Monitoring and Modelling: Conceptual Background and Old World Case Studies*, pp. 285–309. Smithsonian Institution/Man and Biosphere No. 20. Smithsonian Institution, Washington DC.

Clarke, K. R. and Warnick, R. M. (2001). *Change in Marine Communities: an approach to statistical analysis and interpretation*, 2nd edn. Primer-E, Plymouth.

Clarke, K. R. and Gorley, R. N. (2006). *PRIMER v6: user manual/tutorial*. Primer-E, Plymouth.

Clarke, K. R., Somerfield, P. J., and Chapman, M. G. (2006). On resemblance measures for ecological studies, including taxonomic dissimilarities and a zero-adjusted Bray-Curtis coefficient for denuded assemblages. *Journal of Experimental Marine Biology and Ecology*, **330**, 55–80.

Colwell, R. K. (2006). *EstimateS 8.0 user's guide.* http://viceroy.eeb.uconn.edu/EstimateSPages/EstSUsersGuide/EstimateSUsersGuide.htm.

Colwell, R. K. and Coddington, J. A. (1994). Estimating terrestrial biodiversity through extrapolation. *Philosophical Transactions of the Royal Society London Series B Biological Sciences* **345**, 101–18.

Colwell, R. K., Mao, C. X., and Chang, J. (2004). Interpolating, extrapolating, and comparing incidence-based species accumulation curves. *Ecology*, **85**, 2717–27.

de Solla, S. R., Shirose, L. J., Fernie, K. J., Barrett, G. C., Brousseau, C. S., and Bishop, C. A. (2005). Effect of sampling effort and species detectability on volunteer based anuran monitoring programs. *Biological Conservation*, **121**, 585–94.

Dodd, Jr, C. K. (1992). Biological diversity of a temporary pond herpetofauna in north Florida sandhills. *Biodiversity and Conservation*, **1**, 125–42.

Dodd, Jr, C. K., Barichivich, W. J., Johnson, S. A., and Staiger, J. R. (2007). Changes in a northwestern Florida Gulf Coast herpetofaunal community over a 28-y period. *American Midland Naturalist*, **158**, 29–48.

Duellman, W. E. (1999). Global distribution of amphibians: patterns, conservation, and future challenges. In W. E. Duellman (ed), *Patterns of Distribution of Amphibians. A Global Perspective*, pp. 1–30. Johns Hopkins University Press, Baltimore, MD.

Gotelli, N. J. and Colwell, R. K. (2001). Quantifying biodiversity: procedures and pitfalls in the measurement and comparison of species richness. *Ecology Letters*, **4**, 379–91.

Heyer, R. W., Coddington, J., Kress, J. W., Acevede, P., Cole, D., Erwin,T. L., Meggers, B. J., Pogue, M. G., Thorington, R. W., Vari, R. P., Weitzman, M. J., and Weitzman, S. H. (1999). Amazonic biotic data and conservation decisions. *Ciencia e Cultura, Journal of the Brazilian Association for the Advancement of Science*, **51**, 372–85.

Huang, C.-Y. and Hou, P.-C. L. (2004). Density and diversity of litter amphibians in a monsoon forest of southern Taiwan. *Zoological Studies*, **43**, 795–802.

Krebs, C. J. (1999). *Ecological Methodology*, 2nd edn. Benjamin-Cummings, Menlo Park, CA.

Magurran, A. E. (1988). *Ecological Diversity and its Measurement*. Princeton University Press, Princeton, NJ.

Magurran, A. E. (2003). *Measuring Biological Diversity*. Blackwell Publishing, Oxford.

May, R. M. (1988). How many species are there on Earth? *Science*, **241**, 1441–9.

Menegon, M. and Salvidio, S. (2005). Amphibian and reptile diversity in the southern Udzungwa Scarp Forest Reserve, south-eastern Tanzania. In B. A. Huber, B. J. Sinclair, and K.-H. Lampe (eds), *African Biodiversity. Molecules, Organisms, Ecosystems*, pp. 205–12. Springer, New York.

Parris, K. M. and McCarthy, M. A. (1999). What influences the structure of frog assemblages at forest streams? *Australian Journal of Ecology*, **24**, 495–502.

Pawar, S. S., Rawat, G. S., and Choudhury, B. C. (2004). Recovery of frog and lizard communities following primary habitat alteration in Mizoram, northeast India. *BMC Ecology*, www.biomedcentral.com/1472-6785/4/10.

Pieterson, E. C., Addison, L. M., Agobian, J. N., Brooks-Solveson, B., Cassani, J., and Everham III, E., M. (2006). Five years of the southwest Florida frog monitoring network: changes in frog communities as an indicator of landscape change. *Florida Scientist*, **69**, 117–26.

Pineda, E. and Halffter, G. (2004). Species diversity and habitat fragmentation: frogs in a tropical montane landscape in Mexico. *Biological Conservation*, **117**, 499–508.

Ruiz-Jean, M. C. and Aide, T. M. (2005). Restoration success: how is it measured? *Restoration Ecology*, **13**, 569–77.

Scott, J. M., Csuti, B., Jacobi, J. D., and Estes, J. E. (1987). Species richness. A geographic approach to protecting future biological diversity. *BioScience*, **37**, 782–8.

Seymour, C. L., De Klerk, H. M., Channing, A., and Crowe, T. M. (2001). The biogeography of the Anura of sub-equatorial Africa and the prioritisation of areas for their conservation. *Biodiversity and Conservation*, **10**, 2045–76.

Smith, E. P. and van Belle, G. (1984). Nonparametric estimation of species richness. *Biometrics*, **40**, 119–29.

Smith-Vaniz, W. F., Jelks, H. L., and Rocha, L. A. (2006). Relevance of cryptic fishes in biodiversity assessments: a case study at Buck Island Reef National Monument, St. Croix. *Bulletin of Marine Science*, **79**, 17–48.

Snodgrass, J. W., Komorosski, M. J., Bryan, Jr, L., and Burger, J. (2000). Relationships among isolated wetland size, hydroperiod, and amphibian species richness: implications for wetland regulation. *Conservation Biology*, **14**, 414–19.

Soberón, M. J. and Llorente, B. J. (1993). The use of species accumulation functions for the prediction of species richness. *Conservation Biology*, **7**, 480–8.

Thompson, G. G. and Withers, P. C. (2003). Effect of species richness and relative abundance on the shape of the species accumulation curve. *Austral Ecology*, **28**, 355–60.

Thompson, G. G. and Thompson, S. A. (2007). Using species accumulation curves to estimate trapping effort in fauna surveys and species richness. *Austral Ecology*, **32**, 564–9.

Thompson, G. G., Withers, P. C., Pianka, E. R., and Thompson, S. A. (2003). Assessing biodiversity with species accumulation curves; inventories of small reptiles by pit-trapping in Western Australia. *Austral Ecology*, **28**, 361–83.

Thompson, G. G., Thompson, S. A., Withers, P. C., and Fraser, J. (2007). Determining adequate trapping effort and species richness using species accumulation curves for environmental impact assessments. *Austral Ecology*, **32**, 570–80.

Urbina-Cardona, J. N. and Londoño-M, M. C. (2003). Distribución de la comunidad de herpetofauna asociada a cuatro áreas con diferente grado de perturbación en la Isla Gorgona, Pacífico Colombiano. *Revista de la Academia Colombiana de Cienias Exactas y Fisicas Naturales*, **27**, 105–13.

Whitfield, S. M. and Donnelly, M. A. (2006). Ontogenetic and seasonal variation in the diets of a Costa Rican leaf-litter herpetofauna. *Journal of Tropical Ecology*, **22**, 409–17.

19

Landscape ecology and GIS methods

Viorel D. Popescu and James P. Gibbs

19.1 Introduction

19.1.1 Relevance of landscape ecology to amphibian biology and conservation

The field of landscape ecology "deals with the effects of the spatial configuration of mosaics on a wide variety of ecological phenomena" (Wiens *et al.* 1993). The field has emerged from recognition of the need to link ecological processes to landscape configuration and composition (Turner *et al.* 2001). Another stimulus has been the revelation that population processes play out over much larger areas and time frames than previously assumed. Landscape ecology is a relatively young discipline that emerged in Central and Eastern Europe (Naveh and Lieberman 1994). The field has exploded over the past two decades largely because of technological advances (Turner *et al.* 2001). On the data side, a sudden wealth of spatial data from global positioning systems (GPS), satellite imagery, and high-resolution aerial photography has materialized, coupled with the Internet as a medium for its quick distribution. On the analysis side, computer processing speeds have increased exponentially (Moore 1975) and a large suite of software packages for manipulating and analyzing spatial data (that is, Geographic Information Systems or GIS) has evolved that capitalize on these vastly faster processing speeds. The result is a synergism between more data and faster computers for processing it, thereby creating opportunity for tackling increasingly complex questions in landscape ecology.

Technological advances have also occurred in a societal context in which environmental problems have become more and more pervasive. This, in turn, has forced recognition that conservation actions must be implemented over broad spatial scales that have not been addressed within the domain of classic ecological research; that is, conducted on a few hectares at most (Turner *et al.* 2001). As a result, the "landscape perspective" has become widespread in ecology and GIS has become an essential component of the ecologist's toolbox.

In the realm of amphibian biology, Douglas Gill's studies of the metapopulation ecology of red-spotted newts (Gill 1978a, 1978b) were among the first to catalyze this shift in scale in how we conceptualize population processes. These studies prompted recognition that the fates of individual, pond-based breeding populations of amphibians—the focus of nearly all prior research—are often tightly linked to the fates of other populations nearby as mediated by the flow of migrants among them. This landscape-as-population rather than pond-as-population perspective has been the subject of much subsequent research (Sjögren-Gulve 1994; Pope *et al.* 2000; Marsh and Trenham 2001; Hels 2002; Gamble *et al.* 2007; Richter-Boix *et al.* 2007; Werner *et al.* 2007) that has radically expanded the temporal and spatial scales at which we conceive of planning approaches for conserving amphibians (Semlitsch 2000; Trenham *et al.* 2003; Smith and Green 2005).

Landscape ecology is an increasing focus of amphibian research (Figure 19.1). The goal of this chapter is to assist amphibian biologists in

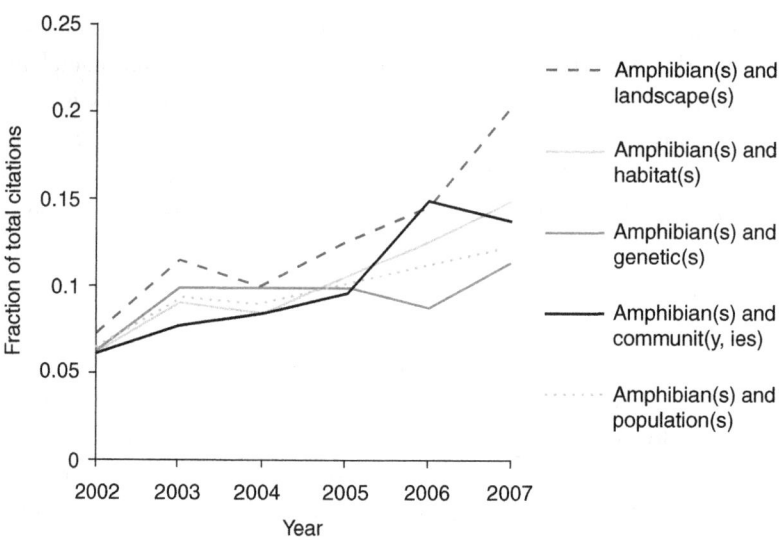

Fig 19.1 Fraction of amphibian studies published since 1985 featuring keywords that combine "amphibian(s)" with "landscape(s)," "habitat(s)," "genetic(s)," "communit(y, ies)," and "population(s)," indicating a recent surge in interest in landscape-level studies of amphibians relative to that on habitats, genetics, communities, or populations (Source: Thomson Scientific's *Web of Science*; data plotted for last 5 years only).

grasping the extraordinary opportunities (and significant limitations) for applying GIS to questions in amphibian biology. We emphasize GIS applications to research in the context of the broad spatial scales required to achieve amphibian conservation over the long term.

19.1.2 Defining a "landscape" from an amphibian's perspective

Landscape ecology is highly relevant for amphibian conservation because it deals with mosaics and patches, which very much define the biphasic life histories of the vast majority of amphibian species that require the use of different habitats/patches at different life stages. For landscape ecology to be informative, however, we must scale our analysis of landscape heterogeneity relative to the perception of the environment by the focal organism (Wiens 1989). For example, the "pond-as-patch" view of spatial organization of amphibian populations and metapopulations (Marsh and Trenham 2001) has as its advantage simplifying scientific investigations (i.e. binary landscapes of "good" versus "bad" habitat), but is not always appropriate. A growing body of evidence suggests that the habitat matrix surrounding ponds plays a critical role in shaping migration among pools and hence the resilience of amphibian metapopulations. What happens to an amphibian population when forest habitat used for migration is no longer contiguous but fragmented? What are the implications of replacing forest or wetland with agricultural or urban development? Answers to these questions cannot be provided unless traditional ecological research intertwines with description and modeling of spatial processes at broad spatial scales. Landscape ecology facilitates amphibian biologists to focus on generating a spatially explicit representation of real-world processes, such as metapopulation dynamics, gene flow, and spatial organization of amphibian populations (Figure 19.2), to help understand how actual populations operate in the landscape.

The concept of landscape connectivity is highly relevant to amphibian conservation because amphibian migration is so critical for maintaining viable amphibian populations. Most amphibians are generally regarded as highly philopatric and having poor dispersal abilities (Semlitsch 2000) despite evidence of long-distance movement in frogs and toads (Lemckert 2004). Landscape connectivity implies two complementary aspects: physical and functional. For example, reforesting an upland area between two breeding ponds restores physical connectivity for forest-dwelling amphibians, but is it enough to ensure the movement and hence the functional connectivity between the two? Moreover, the impact of habitat fragmentation on amphibian communities is difficult to detect because the negative effects of fragmentation occur with a time lag. Thus, a physical description of numbers and sizes of populations and the environment

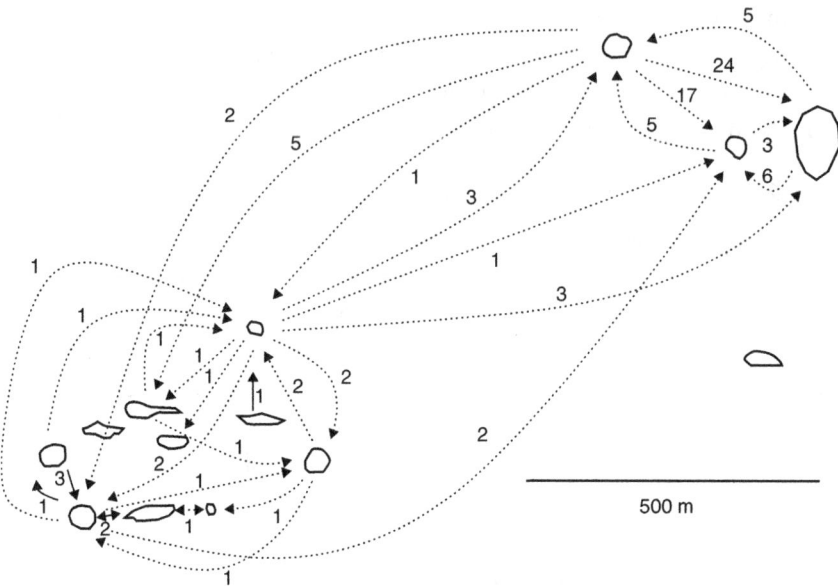

Fig 19.2 In spite of their comparatively low dispersal abilities, productivity of amphibians combined with movement patterns can result in large absolute numbers of amphibians moving across landscape. The figure depicts a network of ponds ($n = 14$) and numbers of dispersing first-time and experienced breeding marbled salamanders (*Ambystoma opacum*) during 1999 through 2005, in western Massachusetts, USA, displayed by origin and destination ponds. Two-ended arrows indicate an equal number of individuals dispersing in both directions (reprinted from Gamble *et al.* 2007, with permission from Elsevier).

they occupy might imply stability when in fact they are experiencing continuous decline and are functionally extinct (Löfvenhaft *et al.* 2004).

19.1.3 GIS

A GIS offers the tools to envision, analyze, process, and model spatial data. A GIS is essentially computer software capable of storing, manipulating, displaying, and analyzing geographically referenced data. Two-dimensional spatial data can be represented in vector or raster form. Vectors are points, lines, or polygons representing real-world features (i.e. sampling locations, roads, or land use, respectively) associated with individual data attributes. In contrast, rasters are continuous matrices of equally sized, rectangular cells, each usually containing a single value corresponding to the landscape feature they represent (e.g. 1 is deciduous forest, 2 is open water, 3 is urban). Adjacent cells of similar value represent homogenous spatial objects, such as a forest or a wetland. Raster resolution (or grain) refers to the actual dimension of the cells: high resolution is analogous to cells representing a small area on the ground (e.g. 1–100 m²) and captures a high degree of spatial complexity of the landscape, whereas low resolution refers to cells corresponding to large areas on the ground (e.g. > 1 km²) and provides a much more generalized representation of the landscape. Rasters do not reflect the exact boundaries of a spatial object in the way the vectors do (Figure 19.3), but their continuous nature is more conducive to the modeling of surfaces. At the same time, mathematical operations with rasters are less computationally intensive than with vectors. Nonetheless, using vector data is

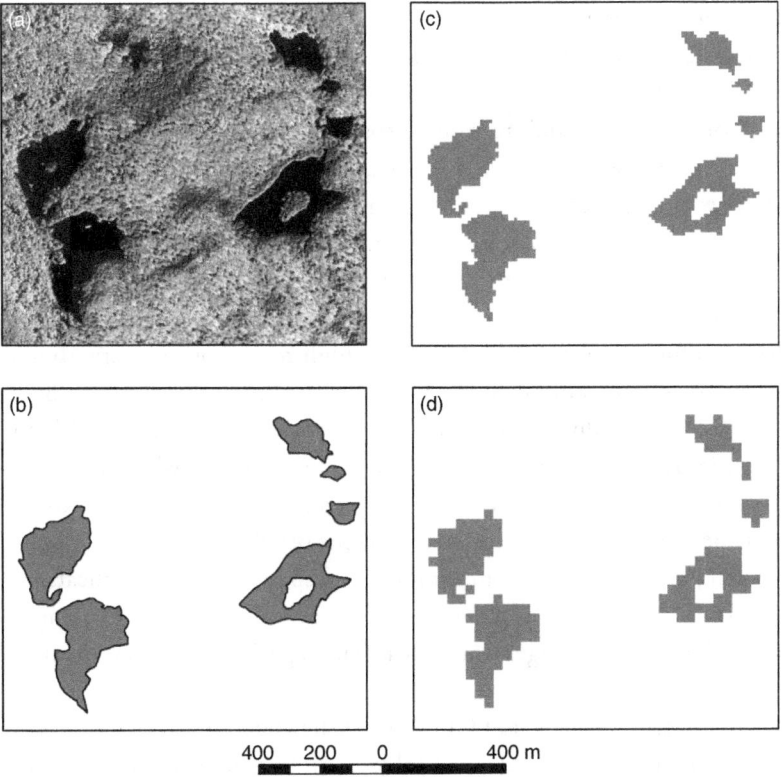

Fig 19.3 Spatial depiction of a set of ponds in a forested landscape using: (a) orthoimagery (1 m resolution aerial photo; the ponds correspond to the irregular shapes); (b) vector data (polygons); (c) raster data (cell size = 10 m × 10 m); and (d) raster data (cell size = 30 m × 30 m). Some spatial information is lost when increasing grain from 10 m × 10 m (c) to 30 m × 30 m (d).

necessary when topological relations are important, such as with analysis of networks.

GIS applications to amphibian research range widely. Minimally GIS can be used for simple assessment of the potential use of a region for study purposes. For example, a perfectly legitimate use of GIS is for storing, managing, and visualizing information on access, types, and distribution of habitats, their ownership, and location. GIS can also be used for more complicated analyses and modeling. For example, GIS tools and data can permit habitat-selection analyses (for example from telemetry studies), predict habitat suitability for amphibians in the face of climate change (amphibians of Europe; Araújo *et al.* 2006), and modeling invasive species' distributions (e.g. the global spread of bullfrogs, *Rana catesbeiana* (Ficetola *et al.* 2007) and the invasion of cane toads, *Bufo marinus*, in Australia (Urban *et al.* 2007)).

In recent years both raster and vector data have become readily available through Internet-based, primarily governmental data repositories. One of the most comprehensive and up-to-date spatial data portals we are aware of is Collins Software web-based Resource Management System (www.collinssoftware.com/freegis_by_region.htm). The recent increased availability of current spatial data is epitomized by online applications like Google Earth or Microsoft VirtualEarth; although these applications cannot be used for analysis purposes, they nevertheless represent GIS applications that capture a wealth of information contained, such as high-resolution imagery, digital elevation models, and transportation networks, as well as place names. Such tools are increasingly valuable aids in visualizing and rapidly assessing the landscape; for example for reconnaissance of field sites. Notably, what once was a special class of data—geospatial data—is increasingly being placed into the realm of generic information management by the proliferation of GPS applications (in cell phones and cars, for example), and spatial Internet applications. Thus, GIS may not persist as a stand-alone conceptual domain for long as conventional database management increases its capacity to integrate spatially explicit information.

In terms of software, a roster of GIS software bewildering to the uninitiated is available to manipulate spatial data (Table 19.1). In organizing the choices, a useful distinction can be made between commercial and free software. Primary in the commercial realm are the ESRI suite (ArcGIS and its predecessors ArcView and ArcInfo), IDRISI, MapInfo, or Manifold. In the realm of free and open-source applications, primary are GRASS GIS, MapServer, DIVA-GIS, ArcExplorer, JUMP, and Quantum GIS (for a list of all available free and open-source GIS software visit freegis.org and opensourcegis.org).

Table 19.1 *Popular GIS software packages.*

	GIS package	Developer	URL	Capabilities
Commercial software[1]	ArcGIS	ESRI	www.esri.com	Complete GIS for spatial analysis and modeling
	IDRISI	Clark Labs, Clark University	www.clarklabs.org/	Powerful raster-based GIS and image processing
	MapInfo	PB MapInfo Corporation	www.mapinfo.com/	Full range of spatial analysis and modeling
	Manifold	CDA International	www.manifold.net/	Database and mapping functionality
Free and open-source software	GRASS GIS	Open Source Geospatial Foundation	grass.itc.it/	Complete GIS similar to commercial software
	MapServer	University of Minnesota	mapserver.gis.umn. edu/	Spatially enabled Internet applications
	DIVA-GIS	R.J. Hijmans *et al.*	www.diva-gis.org/	Raster-based climate modeling
	JUMP GIS	The JUMP Project	www.jump-project. org/	Java-based vector GIS
	Quantum GIS	GNU Project	qgis.org/	Mostly a viewer with editing capabilities
	ArcExplorer	ESRI	www.esri.com/ software/ arcexplorer/ index1.html	Viewer with display, query, and data-retrieval applications

[1]Cost of the commercial GIS packages varies greatly depending on the type of license purchased (single versus multiple, educational versus business) as well as the array of features included.

Working with spatial data usually requires high computation requirements; generally speaking, the larger the extent of the area analyzed, the greater the necessity for a faster processor, larger random-access memory (RAM; usually > 1–2 GB), and increased hard-disk storage space. However, recent advances in computer hardware have made once crippling hardware constraints barely noticeable on even portable computers (for typical applications).

19.2 Applications of spatial data for amphibian conservation

19.2.1 Multiscale predictors of species occurrence and abundance

Many amphibian studies have reported that landscape-scale features influence both amphibian population and metapopulation dynamics (Sjögren-Gulve 1994; Pope *et al.* 2000). Studies of amphibian species richness, abundance, and occurrence invariably have relied on measures of landscape composition at different spatial scales surrounding central breeding pools. The most commonly applied approach uses GIS to extract variables from land cover or habitat raster data sets (for example, percent of forest, open water, wetland) within circular neighborhoods at distances between 30 m and 10 km from the pool edge. These "mini-landscapes" are usually concentric discs (i.e. overlapping neighborhoods 0–100 m, 0–200 m,..., 0–1000 m) or rings (i.e. non-overlapping neighborhoods 0–100 m, 100–200 m,..., 900–1000 m). Such were used in many studies predicting amphibian occurrence and abundance (Vos and Stumpel 1995; Hecnar and M'Closkey 1998; Joly *et al.* 2001; Pellet *et al.* 2004; Gibbs *et al.* 2005; Herrmann *et al.* 2005; Gagné and Fahrig 2007). Others have focused on potential barriers to movement such as highways or large rivers (Zanini *et al.* 2008) or specific habitat "resistances" (Ray *et al.* 2002) to delineate the neighborhoods used to extract the landscape composition variables, thus limiting the analysis to potential usable area only. Landscape-scale variables (e.g. density of roads and streams, distance to nearest wetland, number of occupied ponds, pond–forest adjacency) within the extracted neighborhoods along with site-specific variables (i.e. water pH, conductivity, wetland area, hydroperiod, presence of predators, etc.) were also used in predicting amphibian occurrence and abundance. Further, landscape and pond-scale variables have been used as variables to build predictive models for amphibian species richness or occurrence using ordinary linear models, generalized linear models (GLMs), generalized additive models (GAMs), or other modeling techniques (see Guisan and Zimmermann 2000 for a thorough review of predictive habitat distribution models used in ecological studies).

So-called extraction neighborhoods should reflect the dispersal ecology of the amphibian species of concern. Relying on the assumption that most amphibians have limited dispersal abilities, many studies extend their extraction of variables up to 1–1.5 km radii. However, during the design phase of the study, one must pay attention to the distance between the sampling points. If they are located too close to one another, they can be spatially correlated and the assumption of independence of observations is violated (i.e. the abundance of a species at one site is dependent on the abundance at a neighboring site); pseudo-replication then occurs (Guisan and Zimmerman 2000). At the same time, overlapping of the neighborhoods built around the sampling sites can also lead to lack of independence among extracted variables and pseudo-replication by re-measuring the same land unit over and over again (Figure 19.4). A further problem is inter-variable correlation (i.e. forest cover being a direct inverse linear function of open land). The primary consequence is that pseudo-replication biases model estimates and leads to erroneous predictions. Identification of appropriately spaced sampling locations and non-overlapping neighborhoods during the design phase of the study can be used as a cautionary method for avoiding issues related to non-independence. However, if the sampling situation does not allow such flexibility, statistical techniques such as GLM or GAM that are more relaxed to the assumption of independence among variables, and techniques for assessing and quantifying spatial autocorrelation have to be used (see section 19.3). Practical guidelines for addressing the issue of spatial autocorrelation for designing of field surveys and data analysis applicable to amphibian research are provided by Legendre *et al.* (2002).

19.2.2 Landscape thresholds

The relation between habitat fragmentation and the population response by amphibians is rarely linear. In other words, population persistence in a forest-dependent amphibian does not simply increase directly with more forest cover.

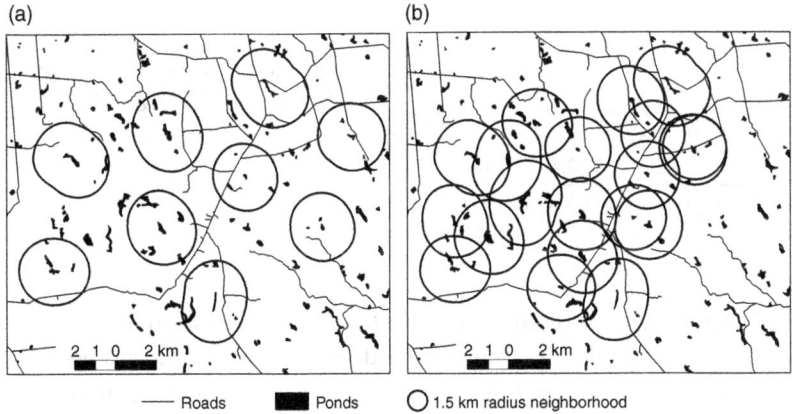

Fig 19.4 Circular neighborhoods used for extracting landscape scale variables around selected ponds: (a) non-overlapping (independent variables) and (b) highly overlapping (leading to non-independence of the predictors and pseudo-replication).

Often, small changes in the spatial pattern of resource distribution can produce sudden changes in the ecological response (Figure 19.5). These transitions are termed critical landscape thresholds (With and Crist 1995). Such critical thresholds have been identified in the response of amphibians to the habitat

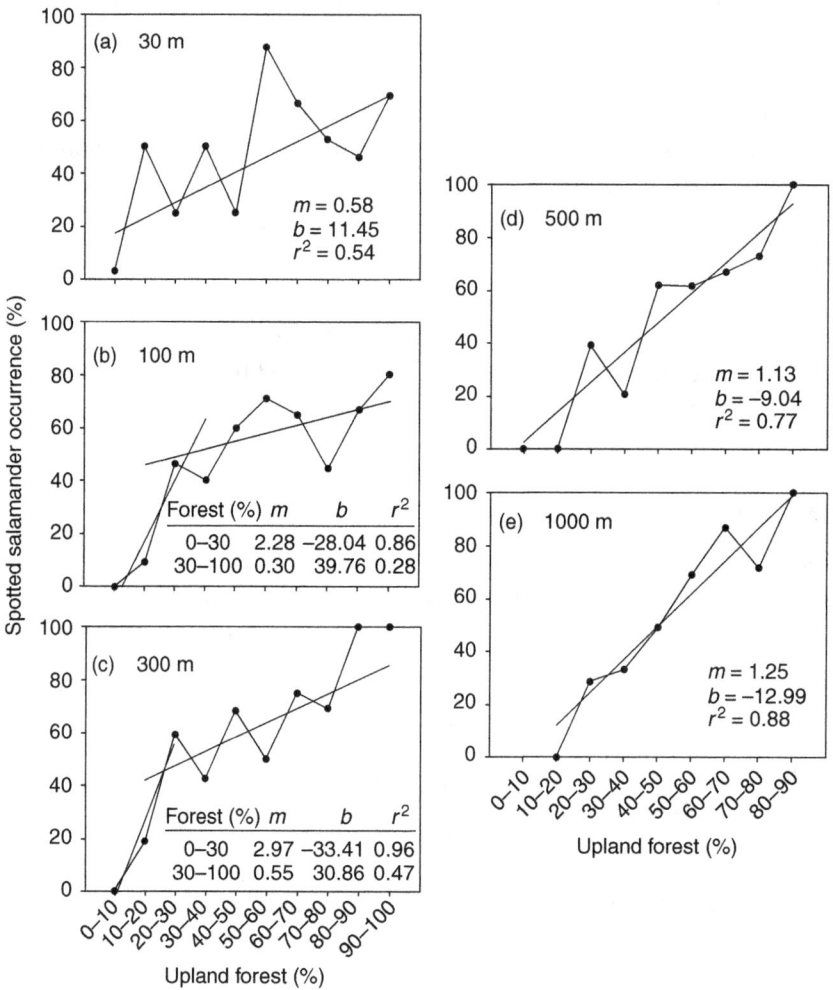

Fig 19.5 Occurrence of spotted salamanders (*Ambystoma maculatum*) plotted against the percentage of upland forest at five different spatial scales, measured from the edge of suitable breeding ponds. Where there is one line per graph, no significant threshold was identified. Figures with two regression lines indicate the location of a critical threshold. In panel (c), the lines represent the break identified *a priori* by visual inspection; there were three other statistically significant break points (reprinted from Homan *et al.* 2004, with permission from the Ecological Society of America).

alteration by human activities (Gibbs 1998a; Homan *et al.* 2004; Denoël and Ficetola 2007). Landscape-composition thresholds (e.g. the percentage of a land-use type within a certain neighborhood below which the long-term viability of the population is compromised), as well as landscape configuration (e.g. patch isolation as function of the distance between suitable breeding and foraging habitat) have been documented (Denoël and Ficetola 2007).

19.2.3 Connectivity/isolation in amphibian populations

Amphibian populations respond not only to the landscape composition but also to landscape configuration. In other words, what matters to a forest-dependent amphibian is not just how much forest is present. The number, size, and arrangement of the forest patches that comprise that amount may be more important. Of particular interest for amphibian conservation is assessment of the isolation of occupied patches (ponds in the case of pond-breeding amphibians) and measures of connectivity between patches (Tischendorf and Fahrig 2000). For amphibians, isolation can be simply defined as the lack of a permanent surface-water connection between ponds (Gibbons 2003). But biological and functional connectivity may occur in the absence of a permanent water connection if terrestrial corridors suitable for animal movement exist. Isolation and connectivity measures are highly relevant for the metapopulation level, but are difficult to approximate because they depend on species-specific dispersal abilities and physiology (Joly *et al.* 2003).

Generally speaking, straight-line (or Euclidian) distances are not suited for approximating patch isolation or connectivity. Depending on its composition and configuration, the habitat matrix may oppose particular resistance to animal movements so some landscapes are more permeable than others to these movements, even if the straight distances are the same. By extension, a well-watered stream bed 1 km in length may represent a vastly more functional linkage between two populations than a dry agricultural field representing just a few meters of separation between the same two ponds.

The significance of appropriately measured isolation/connectivity metrics has been recognized as an integral part of the efforts for managing amphibian populations (Semlitsch 2000). All connectivity measures rely on the assumption that the organisms use the habitat matrix for dispersal and the resistance opposed for movements is a function of quantifiable landscape features. Because different species perceive the environment uniquely and also differ in intrinsic vagility (as a function of leg length, crypsis, and desiccation resistance, for example), the spatial configuration and composition of the landscape affect their movements in non-intuitive ways. Consequently, researchers have developed a multitude of indices for patch isolation and connectivity that are species- and context-specific.

Typically, connectivity measures have not considered the habitat matrix between breeding sites and simply used "nearest neighbor" measures (Euclidian distances) (Knutson *et al.* 1999; Pope *et al.* 2000), but studies that account for habitat matrix complexity have started to emerge in recent years (Stevens *et al.* 2005; Compton *et al.* 2007; Figure 19.6).

In metapopulation dynamics theory, connectivity measures are critical for understanding and describing extinction and recolonization patterns and processes. Hanski *et al.* (1994) developed a connectivity index that was used for understanding amphibian metapopulation processes. This index requires specific knowledge of site occupancy and patch carrying capacity as well as a detailed map of the available patches:

$$\text{CONNECT} = \sum_{i=1}^{n} e^{-d_{ij}} A_j$$

where d_{ij} is the distance between patch i and patch j, and A_j is the carrying capacity of site j. This so-called Hanski index has provided a useful measure of pond connectivity for amphibian metapopulations (Schmidt and Pellet 2005).

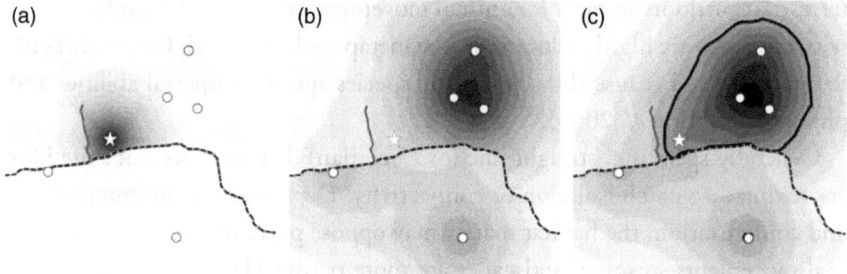

Fig 19.6 Compton *et al.* (2007) used a modified kernel estimator (commonly used in home-range studies) associated with a friction surface to model vernal pool connectivity for four ambystomatid salamanders at three scales in Massachusetts, USA. Their approach is different from a simple cost-distance calculation because they integrated the probability of pools exchanging individuals based on empirical salamander dispersal data (using the kernel estimator, $h = 339.6$ m) with a resistance map (expert-based habitat specific values). The figure depicts examples of the resistant-kernel estimator at three scales in a landscape with a focal pool (star), five neighboring pools (circles), and two roads: (a) local scale, showing connectivity to upland habitat from the focal pool; (b) neighborhood scale, showing the probability of the focal pool receiving dispersing animals from each neighboring pool; and (c) regional scale, with dark outline indicating pools that are interconnected by a specified level of dispersal. Darker shading indicates greater connectivity at each scale (reprinted with permission from Blackwell Publishing).

Many complementary indices for quantifying habitat fragmentation, habitat loss, connectivity, or other landscape characteristics have been developed. McGarigal *et al.* (2002) developed a computer software program (FRAGSTATS) capable of computing a multitude of landscape metrics for categorical maps, such as land cover or habitat maps. The software and documentation are available for free at www.umass.edu/landeco/research/fragstats/fragstats.html.

19.2.4 Landscape permeability

Landscape permeability refers to the resistance imposed by a habitat to movement. Permeability is usually associated to the costs incurred by an individual while using a particular type of habitat. At a population or metapopulation level dispersing individuals are essential for providing new colonists and maintaining typical levels of gene flow (Marsh and Trenham 2001). Any additional dispersal cost induced by a less permeable habitat matrix reduces the probability of pond recolonization in the case of human-induced or naturally occurring extinction.

Because most amphibians use terrestrial habitat for migration and dispersal, any change in the configuration and composition of the landscape surrounding the breeding pools potentially increases the costs for individual- and population-scale processes. At the level of the individual amphibian, low permeability can be reflected in longer exposure to inhospitable conditions. For example, low permeability may be associated with higher mortality risk due to desiccation in a clear-cut versus a closed forest, higher predation risk, or increased traveling time (and hence energy expended) associated with avoidance of inhospitable habitats.

Incorporating landscape permeability in the study of amphibian dispersal has been attempted using two different approaches: field experiments (Rothermel and Semlitsch 2002; Mazerolle and Desrochers 2005) and GIS modeling (Ray *et al.* 2002; Joly *et al.* 2003; Compton *et al.* 2007; Figure 19.6). Field experiments have explicitly measured the resistance opposed by low permeability habitat (agricultural land, barren land) relative to high permeability habitat (forest, wetland) to amphibian movement. Indices such as distance traveled, rate of movement, survival, probability of homing successfully, or habitat choice for movement were used for assessing habitat permeability empirically. This is valuable information that has the potential to greatly improve attempts at modeling amphibian dispersal. However, there are drawbacks of field experiments pertaining to the scale of the experiment, species-specific behaviour and vagility, and logistics. Measuring the permeability within 50 m dispersal arrays (Rothermel and Semlitsch 2002) or 70 m distance from a pond (Mazerolle and Desrochers 2005) is restrictive and does not reflect the extensive movements that some amphibians perform. The methods for obtaining the data (i.e. pitfall traps) may impede continuous movements and consequently not reflect habitat permeability best. Dissimilarities in vagility, habitat selection, and physiology between species, as well as age-specific differences, demand that such experiments be conducted across multiple species. Also, such experiments are labor-intensive and time-consuming, and the results cannot easily be translated to another geographical locality. So-called friction or resistance maps provide a potential GIS-based solution to modeling permeability realistically. The process includes assigning land-use or land-cover classes various permeability values based on expert opinion or empirical data, and using a GIS to create a least cost surface expressing connectivity between populations or breeding sites. Ray *et al.* (2002) used a land-use map to estimate potential migration zones for common toads (*Bufo bufo*) and alpine newts (*Triturus alpestris*) in Switzerland. Similarly, Joly *et al.* (2003) estimated biological connectivity of common toad (*B. bufo*) populations in the Rhone floodplain, France, while accounting for potential road mortality. Compton *et al.* (2007) used a statewide land-cover data set and a combination of expert-based and empirical permeability data to model connectivity for ambystomatid salamanders at three different scales in Massachusetts, USA (Figure 19.6).

19.2.5 Landscape genetics

Individual amphibians move genetic material around the landscape as they disperse to new areas and integrate themselves into new populations. Dispersal is the "glue" that causes adjacent populations to coalesce genetically when they would otherwise evolve in isolation, and steadily differentiate from one another through random events (drift) or natural selection and adaptation to their local environments. By disrupting amphibian movement, habitat fragmentation can alter gene flow and hence determine how genetic variation is distributed among populations in a landscape. Traditionally, population genetics has focused on estimating the extent to which genetic diversity was distributed within versus among populations; the spatially explicit forces shaping these patterns were largely ignored. In response, the field of landscape genetics arose recently as a means of relating genetic discontinuities to landscape features (Manel *et al.* 2003; Storfer *et al.* 2007).

Landscape genetics capitalizes on an emergence of spatial data and spatial statistical techniques along with the capacity to assay genetic variation cheaply and quickly, and with great resolution among of many populations (e.g. fine-scale spatial structure of spotted salamanders, *Ambystoma maculatum*; Zamudio and Wieczorek 2007). Advances in our ability to accumulate vast amounts of fine-scale genetic data are largely permitted by the development of polymerase chain reaction (PCR) methods that readily generate sufficient quantities of DNA from field samples for detailed analysis of patterns of genetic variation among individuals and populations (Chapter 22). Another attribute of landscape genetics is that genetic data are in some cases permitted to self-aggregate through statistical patterning, thus no longer requiring geneticists to arbitrarily define populations for sampling (Manel *et al.* 2003). GIS tools underpin the entire process of linking genetic data to spatial environmental data that essentially represents landscape genetics.

GIS-based landscape genetics studies for evaluating the functional connectivity of the landscape have been successfully applied in the case of the redback salamander (*Plethodon cinereus*) in Connecticut (Gibbs 1998b) and Virginia (Cabe *et al.* 2007), natterjack toad (*Bufo calamita*) in southern Belgium (Stevens *et al.* 2006), tiger salamander (*Ambystoma tigrinum*) in the western USA (Spear *et al.* 2005), dwarf squeaker (*Schoutedenella xenodactyloides*) in Taita Hills, Kenya (Measey *et al.* 2007), and spotted salamander (*Ambystoma maculatum*) in central New York (Zamudio and Wieczorek 2007). The procedure generally involves sampling distinct sites or populations and using high-resolution genetic markers (mainly microsatellite loci) to analyze population genetic structure (Chapter 22). Measures of genetic similarity between adjacent populations are then linked in a GIS to spatial variables such as topographical distance between sites/populations, elevation, and road/stream crossings to identify the environmental correlates of genetic similarity (and hence gene flow) between populations. Including habitat permeability (described above) has improved our understanding, for example, when friction-based distances outperform straight (Euclidian) distances in explaining the genetic differentiation (Stevens *et al.* 2006). Gibbs and Reed (2007) provide a review of genetics and landscape connectivity for vernal pool-breeding amphibians.

19.3 Spatial statistics

Tobler's First Law of geography is that "everything is related to everything else, but near things are more related than distant things" (Tobler 1970). In other words, data collected at any location in the landscape will have a greater similarity to, or influence on, those locations within its immediate neighborhood. Consequently, data tied to any point on the landscape is not independent but rather spatially autocorrelated with the identity of its neighbors. Spatial autocorrelation may lead to biased estimations of parameters and it needs to be accounted for in the modeling process or spatial statistical techniques need to be used. Legendre and Fortin (1989), Perry *et al.* (2002), and Dormann *et al.* (2007) review the statistical tools for testing for the presence of spatial autocorrelation in biological data and provide guidance for using the proper techniques. Practical guidelines for addressing the issue of spatial autocorrelation for designing of field surveys and data analysis are also provided by Legendre *et al.* (2002).

Common means for analyzing spatial patterns of the data include indices for global spatial autocorrelation (Moran's *I*, Geary's *c*), indices for spatial clustering of group-level data (Getis–Ord Local *G*), and spatial autocorrelation tests (Mantel and partial Mantel tests). Calculation of these indices and tests can be conducted either using spatial statistics software (GSLIB, Gstat, GS+, VARIOWIN) or as packages implemented in a general statistical framework with software (R, SAS, SYSTAT). Some GIS packages, such as ArcGIS (ESRI, Redlands, CA, USA) and IDRISI (Clark Labs, Worcester, MA, USA), also contain geostatistical tools to make these adjustments.

New modeling techniques that account for spatial autocorrelation in the data, such as autologistic regression (Augustin *et al.* 1998) and geographic weighted regression (Fotheringham *et al.* 2002) have been developed. For example, Knapp *et al.* (2003) used autologistic regression to model probability of pond occupancy in yellow-legged frogs (*Rana muscosa*) by incorporating an autocovariate term describing the degree of isolation of ponds. As a result, the autocovariate isolation term was the most important predictor of occupancy in yellow-legged frogs. A similar approach was taken by Davidson (2004) for explaining the relation between amphibian decline and historical pesticide use in California. Dormann (2007) brings forth further evidence on the importance of accounting for spatial autocorrelation: when modeling the distribution of organisms including amphibians, reptiles, mammals, birds, plants, insects, or mites, spatial autocorrelation biased the coefficient estimates, under- or overestimating by approximately 25%. Incorporating spatial autocorrelation also improved model fit.

19.4 Limitations and future directions

One primary limitation of GIS applications to amphibian biology and conservation is the chronic lag between need and availability of data with which to populate a GIS. Landscapes are changing at rates faster than is feasible to update of spatial information. Typically, there is a time lag of years from the acquisition of raw satellite data to the availability of synthesized data products that support landscape analyses. At a global scale, high-resolution spatial information may not be available or updated for large continental areas for many years. Most importantly, particularly data-deficient regions, such as tropical latitudes, with the least amphibian research attention, are nevertheless of critical importance for amphibian conservation.

Lack of a stable "taxonomy" for ecosystem classification also plagues GIS-based amphibian research. Researchers seeking the latest spatial information available typically analyze remotely sensed imagery (such as from Landsat, SPOT, or IKONOS satellite sensors) and implement their own classification of habitats or land use, often designed to suit the focus organism. This serves a given study well but reduces the comparability among different studies with respect to habitat use, availability, and selection by the same organism.

Inconsistent data resolution also creates problems. For example, habitat changes critical for amphibian population persistence might occur at a finer scale than satellite imagery can capture. Aerial photos may, however, be sufficient for the task. Whatever the case, misinterpretation due to an inability to resolve such patterns in some cases but not others may occur because of variation in coarseness of the spatial data available across studies. Not only spatial, but also temporal and informational resolution add to the problems of GIS use. The time lag between the publishing date of spatial data and the study period imposes obvious restrictions on the ability to model habitat relationships.

"GIS-philia" can also blind some biologists to the inadequacy of their efforts. GIS represents a thrilling mix of technologies for many amphibian researchers, and resulting maps often have a visually stunning effect that can distract from the maps' limitations. Moreover, no matter how accurate and up-to-date the spatial data, the animal–habitat correlations that are at the core of GIS analysis for many amphibian studies are always only part of the puzzle. Sophisticated analysis and modeling of spatial data require empirical data for model parameterization and calibration. Such data are also required for model validation, a critical step increasingly neglected in many modeling studies. Gathering sufficient field data to confront each model and thereby assess its validity should be regarded as a requirement for publication of any modeling studies or dissemination of any guidelines extending from such models. This said, model validation can be equally or more expensive than the cost of developing the original model, a cost most researchers (and their funders) are rarely willing to cover.

Landscape thresholds have the potential to become a powerful tool in managing amphibian populations because they offer specific management objectives (e.g. "retain more than 70% forest within a 2 km radius pond neighborhood to maintain a healthy population"). However, threshold research is still in its infancy (Huggett 2005), and the temporal and spatial behaviour of thresholds remains largely unstudied. Whether or not it is possible to translate thresholds identified within a landscape at one spatio-temporal scale to another needs further testing.

So much of what we can interpret from landscape-level data for amphibians depends on the understanding of amphibian movement ecology, a topic that remains poorly elaborated (Semlitsch 2008). We are only beginning to understand the processes distinguishing migration from dispersal, let alone understanding how and why amphibians move about the landscape, particularly juvenile forms, which typically represent the bulk of moving animals given the short-life spans that characterize amphibian demographics. There is a clear need for integrating basic amphibian ecology into modeling, especially more physiological data explaining movement preferences, before we can predict with confidence the implications of various landscape configurations and compositions.

In conclusion, habitat loss and degradation are now considered among the greatest threats to amphibians worldwide (Cushman 2006). As once naturally grading, heterogeneous landscapes are converted to fragmented mosaics with sharp, "hard" boundaries that impede movements, landscape ecology and GIS techniques are contributing significantly to understanding amphibian population dynamics, including the extinction and colonization processes so critical to sustaining amphibian populations. This understanding is slowly being parlayed into useful guidelines for managing amphibian populations effectively at the landscape scales at which most populations operate over the long term. Our ability to master GIS software to harness increasingly rich, abundant, and well-resolved spatial data will determine how well we realize the potentials of the field of landscape ecology and thereby inform us about broad-scale spatial processes that play such an important role in amphibian population persistence everywhere.

19.5 References

Araújo, M. B., Thuiller, W., and Pearson, R. G. (2006). Climate warming and the decline of amphibians and reptiles in Europe. *Journal of Biogeography*, **33**, 1712–28.

Augustin, N. H., Mugglestone, M. A., and Buckland, S. T. (1998). The role of simulation in spatially correlated data. *Environmetrics*, **9**, 175–96.

Cabe, P. R., Page, R. B., Hanlon, T. J., Aldrich, M. E., Connors, L., and Marsh, D. M. (2007). Fine-scale population differentiation and gene flow in a terrestrial salamander (*Plethodon cinereus*) living in continuous habitat. *Heredity*, **98**, 53–60.

Compton, B. W., McGarigal, K., Cushman, S. A., and Gamble, L. R. (2007). A resistant-kernel model of connectivity for amphibians that breed in vernal pools. *Conservation Biology*, **21**, 788–99.

Cushman, S. A. (2006). Effects of habitat loss and fragmentation on amphibians: a review and prospectus. *Biological Conservation*, **128**, 231–40.

Davidson, C. (2004). Declining downwind: amphibian population declines in California and historical pesticide use. *Ecological Applications*, **14**, 1892–1902.

Denoël, M. and Ficetola, G. F. (2007). Landscape-level thresholds and newt conservation. *Ecological Applications*, **17**, 302–9.

Dormann, C. F. (2007). Effects of incorporating spatial autocorrelation into the analysis of species distribution data. *Global Ecology and Biogeography*, **16**, 129–38.

Dormann, C. F., McPherson, J. M., Araujo, M. B., Bivand, R., Bolliger, J., Carl, G., Davies, R. G., Hirzel, A., Jetz, W., Kissling, W. D. (2007). Methods to account for spatial autocorrelation in the analysis of species distributional data: a review. *Ecography*, **30**, 609–28.

Ficetola, G. F., Thuiller, W., and Miaud, C. (2007). Prediction and validation of the potential global distribution of a problematic alien invasive species - the American bullfrog. *Diversity and Distributions*, **13**, 476–85.

Fotheringham, A. S., Brunsdon, C., and Charlton, M. (2002). *Geographically Weighted Regression—the Analysis of Spatially Varying Relationships*. Wiley, Chichester.

Gagné, S. and Fahrig, L. (2007). Effect of landscape context on anuran communities in breeding ponds in the National Capital Region, Canada. *Landscape Ecology*, **22**, 205–15.

Gamble, L. R., McGarigal, K., and Compton, B. W. (2007). Fidelity and dispersal in the pond-breeding amphibian, *Ambystoma opacum*: Implications for spatio-temporal population dynamics and conservation. *Biological Conservation*, **139**, 247–57.

Gibbons, J. W. (2003). Terrestrial habitat: a vital component for herpetofauna of isolated wetlands. *Wetlands*, **23**, 630–5.

Gibbs, J. P. (1998a). Distribution of woodland amphibians along a forest fragmentation gradient. *Landscape Ecology*, **13**, 263–8.

Gibbs, J. P. (1998b). Genetic structure of redback salamander *Plethodon cinereus* populations in continuous and fragmented forests. *Biological Conservation*, **86**, 77–81.

Gibbs, J. P. and Reed, J. M. (2007). Population and genetic linkages of vernal pool-associated amphibians. In A. J. K. Calhoun and P. G. DeMaynadier (eds), *Science and Conservation of Vernal Pools*, pp. 149–68. CRC Press, Boca Raton, FL.

Gibbs, J. P., Whiteleather, K. K., and Schueler, F. W. (2005). Changes in frog and toad populations over 30 years in New York State. *Ecological Applications*, **15**, 1148–57.

Gill, D. E. (1978a). Effective population size and interdemic migration rates in a metapopulation of the red-spotted newt, *Notophthalmus viridescens* (Rafinesque). *Evolution*, **32**, 839–49.

Gill, D. E. (1978b). The metapopulation ecology of the red-spotted newt, *Notophthalmus viridescens* (Rafinesque). *Ecological Monographs*, **48**, 145–66.

Guisan, A. and Zimmermann, N. E. (2000). Predictive habitat distribution models in ecology. *Ecological Modeling*, **135**, 147–86.

Hanski, I., Kuussaari, M., and Nieminen, M. (1994). Metapopulation structure and migration in the butterfly *Melitaea cinxia*. *Ecology*, **75**, 747–62.

Hecnar, S. J. and M'Closkey, R. T. (1998). Species richness patterns of amphibians in southwestern Ontario ponds. *Journal of Biogeography*, **25**, 763–72.

Hels, T. (2002). Population dynamics in a Danish metapopulation of spadefoot toads *Pelobates fuscus*. *Ecography*, **25**, 303–13.

Herrmann, H. L., Babbitt, K. J., Baber, M. J., and Congalton, R. G. (2005). Effects of landscape characteristics on amphibian distribution in a forest-dominated landscape. *Biological Conservation*, **123**, 139–49.

Homan, R. N., Windmiller, B. S., and Reed, J. M. (2004). Critical thresholds associated with habitat loss for two vernal pool-breeding amphibians. *Ecological Applications*, **14**, 1547–53.

Huggett, A. J. (2005). The concept and utility of 'ecological thresholds' in biodiversity conservation. *Biological Conservation*, **124**, 301–10.

Joly, P., Miaud, C., Lehmann, A., and Grolet, O. (2001). Habitat matrix effects on pond occupancy in newts. *Conservation Biology*, **15**, 239–48.

Joly, P., Morand, C., and Cohas, A. (2003). Habitat fragmentation and amphibian conservation: building a tool for assessing landscape matrix connectivity. *Comptes Rendus Biologies*, **326**, 132–9.

Knapp, R. A., Matthews, K. R., Preisler, H. K., and Jellison, R. (2003). Developing probabilistic models to predict amphibian site occupancy in a patchy landscape. *Ecological Applications*, **13**, 1069–82.

Knutson, M. G., Sauer, J. R., Olsen, D. A., Mossman, M. J., Hemesath, L. M., and Lannoo, M. J. (1999). Effects of landscape composition and wetland fragmentation on frog and toad abundance and species richness in Iowa and Wisconsin, U.S.A. *Conservation Biology*, **13**, 1437–46.

Legendre, P. and Fortin, M.-J. (1989). Spatial pattern and ecological analysis. *Plant Ecology*, **80**, 107–38.

Legendre, P., Dale, M. R. T., Fortin, M.-J., Gurevitch, J., Hohn, M., and Myers, D. (2002). The consequences of spatial structure for the design and analysis of ecological field surveys. *Ecography*, **25**, 601–15.

Lemckert, F. (2004). Variations in anuran movements and habitat use: implications for conservation. *Applied Herpetology*, **1**, 165–81.

Löfvenhaft, K., Runborg, S., and Sjögren-Gulve, P. (2004). Biotope patterns and amphibian distribution as assessment tools in urban landscape planning. *Landscape and Urban Planning*, **68**, 403–27.

Manel, S., Schwartz, M. K., Luikart, G., and Taberlet, P. (2003). Landscape genetics: combining landscape ecology and population genetics. *Trends in Ecology and Evolution*, **18**, 189–97.

Marsh, D. M. and Trenham, P. C. (2001). Metapopulation dynamics and amphibian conservation. *Conservation Biology*, **15**, 40–9.

Mazerolle, M. J. and Desrochers, A. (2005). Landscape resistance to frog movements. *Canadian Journal of Zoology*, **83**, 455–64.

McGarigal, K., Cushman, S. A., Neel, M. C., and Ene, E. (2002). *FRAGSTATS: Spatial pattern analysis program for categorical maps*. University of Massachusetts, Amherst, MA.

Measey, G. J., Galbusera, P., Breyne, P., and Matthysen, E. (2007). Gene flow in a direct-developing, leaf litter frog between isolated mountains in the Taita Hills, Kenya. *Conservation Genetics*, **8**, 1177–88.

Moore, G. E. (1975). Progress in digital integrated electronics. *Electron Devices Meeting 1975 International*, **21**, 11–13.

Naveh, Z. and Lieberman, A. (1994). *Landscape Ecology. Theory and Application*, 2nd edn. Springer-Verlag, New York.

Pellet, J., Hoehn, S., and Perrin, N. (2004). Multiscale determinants of tree frog (*Hyla arborea* L.) calling ponds in western Switzerland. *Biodiversity and Conservation*, **13**, 2227–35.

Perry, J. N., Liebhold, A. M., Rosenberg, M. S., Dungan, J., Miriti, M., Jakomulska, A., and Citron-Pousty, S. (2002). Illustrations and guidelines for selecting statistical methods for quantifying spatial pattern in ecological data. *Ecography*, **25**, 578–600.

Pope, S. E., Fahrig, L., and Merriam, H. G. (2000). Landscape complementation and metapopulation effects on leopard frog populations. *Ecology*, **81**, 2498–2508.

Ray, N., Lehmann, A., and Joly, P. (2002). Modeling spatial distribution of amphibian populations: a GIS approach based on habitat matrix permeability. *Biodiversity and Conservation*, **11**, 2143–65.

Richter-Boix, A., Llorente, G. A., and Montori, A. (2007). Structure and dynamics of an amphibian metacommunity in two regions. *Journal of Animal Ecology*, **76**, 607–18.

Rothermel, B. B. and Semlitsch, R. D. (2002). An experimental investigation of landscape resistance of forest versus old-field habitats to emigrating juvenile amphibians. *Conservation Biology*, **16**, 1324–32.

Schmidt, B. R. and Pellet, J. (2005). Relative importance of population processes and habitat characteristics in determining site occupancy of two anurans. *Journal of Wildlife Management*, **69**, 884–93.

Semlitsch, R. D. (2000). Principles for management of aquatic-breeding amphibians. *Journal of Wildlife Management*, **64**, 615–31.

Semlitsch, R. D. (2008). Differentiating migration and dispersal processes for pond-breeding amphibians. *Journal of Wildlife Management*, **72**, 260–7.

Sjögren-Gulve, P. (1994). Distribution and extinction patterns within a northern metapopulation of the pool frog, *Rana lessonae*. *Ecology*, **75**, 1357–67.

Smith, A. M. and Green, D. M. (2005). Dispersal and the metapopulation paradigm in amphibian ecology and conservation: are all amphibian populations metapopulations? *Ecography*, **28**, 110–28.

Spear, S. F., Peterson, C. R., Matocq, M. D., and Storfer, A. (2005). Landscape genetics of the blotched tiger salamander (*Ambystoma tigrinum melanostictum*). *Molecular Ecology*, **14**, 2553–64.

Stevens, V. M., Polus, E., Wesselingh, R., Schtickzelle, N., and Baguette, M. (2005). Quantifying functional connectivity: experimental evidence for patch-specific resistance in the Natterjack toad (*Bufo calamita*). *Landscape Ecology*, **19**, 829–42.

Stevens, V. M., Verkenne, C., Vandewoestijne, S., Wesselingh, R. A., and Baguette, M. (2006). Gene flow and functional connectivity in the natterjack toad. *Molecular Ecology*, **15**, 2333–44.

Storfer, A., Murphy, M. A., Evans, J. S., Goldberg, C. S., Robinson, S., Spear, S. F., Dezzani, R., Delmelle, E., Vierling, L., and Waits, L. P. (2007). Putting the "landscape" in landscape genetics. *Heredity*, **98**, 128–42.

Tischendorf, L. and Fahrig, L. (2000). How should we measure landscape connectivity? *Landscape Ecology*, **15**, 633–41.

Tobler, W. R. (1970). A computer movie simulating urban growth in the Detroit region. *Economic Geography*, **46**, 234–40.

Trenham, P. C., Koenig, W. D., Mossman, M. J., Stark, S. L., and Jagger, L. A. (2003). Regional dynamics of wetland-breeding frogs and toads: turnover and synchrony. *Ecological Applications*, **13**, 1522–32.

Turner, M. G., Gardner, R. H., and O'Neill, R. V. (2001). *Landscape Ecology in Theory and Practice: Pattern and Process*. Springer, New York.

Urban, M. C., Phillips, B. L., Skelly, D. K., and Shine, R. (2007). The cane toad's (*Chaunus [Bufo] marinus*) increasing ability to invade Australia is revealed by a dynamically updated range model. *Proceedings of the Royal Society of London Series B Biological Sciences*, **274**, 1413–19.

Vos, C. C. and Stumpel, A. H. P. (1995). Comparison of habitat-isolation parameters in relation to fragmented distribution patterns in the tree frog (*Hyla arborea*). *Landscape Ecology*, **11**, 203–14.

Werner, E. E., Yurewicz, K. L., Skelly, D. K., and Relyea, R. A. (2007). Turnover in an amphibian metacommunity: the role of local and regional factors. *Oikos*, **116**, 1713–25.

Wiens, J. A. (1989). Spatial scaling in ecology. *Functional Ecology*, **3**, 385–97.

Wiens, J. A., Stenseth, N. C., Horne, B. V., and Ims, R. A. (1993). Ecological mechanisms and landscape ecology. *Oikos*, **66**, 369–80.

With, K. A. and Crist, T. O. (1995). Critical thresholds in species' responses to landscape structure. *Ecology*, **76**, 2446–59.

Zamudio, K. R. and Wieczorek, A. M. (2007). Fine-scale spatial genetic structure and dispersal among spotted salamander (*Ambystoma maculatum*) breeding populations. *Molecular Ecology*, **16**, 257–74.

Zanini, F., Klingemann, A, Schlaepfer, R., and Schmidt, B. R. (2008). Landscape effects on anuran pond occupancy in an agricultural countryside: barrier-based buffers predict distributions better than circular buffers. *Canadian Journal of Zoology*, **88**, 692–9.

Part 6
Physiological ecology and genetics

20
Physiological ecology: field methods and perspective

Harvey B. Lillywhite

20.1 Introduction

Both public and scientific awareness of the phenomenon of "amphibian declines" (Blaustein and Wake 1995) have focused the importance of understanding the physiological underpinnings of amphibian diversity, distribution, and survival in varied and challenging environments. Contributions to understanding of ecology have been important goals of physiologists and integrative biologists interested in the structure, function, and behavior of amphibians (Feder and Burggren 1992). Such studies are rich in history and possibilities considering the biological diversity and evolutionary responses of amphibians to varied and extensive environments. These range from caves and subterranean burrows to high tree canopies, from forests to scrub and deserts, and from tropical lowlands to mountain tops. For many species complex life cycles connect developmental stages with both aquatic and terrestrial environments.

Scientific interest in the functional attributes of frogs began at least as early as 1671 when Malpighi used frogs as a model for investigating kidney function. This was followed by research in which amphibians were used for investigating various aspects of physiology from neuromuscular function to development and endocrinology. Frogs became familiar animals in college and high-school anatomy and physiology classes. However, in both scientific and education endeavors, the "amphibian" was simply a means to understanding physiological principles. Publications by G. Kingsley Noble (1931) and John Moore (1964) began to focus attention on the physiology of amphibians as a means to understanding their biology. The development of *physiological ecology* as an established discipline in the 1960s evoked parallel emphasis on the biology of amphibians specifically in contexts of adaptation to environment (e.g. Ray 1958; Gordon *et al.* 1961; Hutchison 1961; Brattstrom 1963; Thorson 1964; Warburg 1965; Whitford and Hutchison 1965; Bentley 1966; Lee 1968; Lillywhite 1970).

There are more than 6400 species of extant amphibians, represented by three orders (Anura, the frogs; Caudata, the salamanders; and Gymnophiona, the caecilians) and member species that are morphologically and physiologically different from other aquatic or terrestrial vertebrates. Field methods have been centrally important in helping us to understand how amphibians live and reproduce in varied habitats while meeting the challenges of dehydration, thermal extremes, energy balance, metabolic requirements, and life-history transitions from aquatic to terrestrial environments. Much remains to be learned about these important vertebrates, especially with respect to phylogenetic variation and challenges of environmental change. Important advances will be tied to key technological innovations related to miniaturization, remote sensing, and isotopic, genetic, and chemical analysis.

20.1.1 Important features of amphibians

Certain features of amphibians are characteristic and generally set them apart from most other tetrapods. These include:

- lack of a cleidoic egg, with early developmental stages limited to aquatic or moist environments;
- integument with comparatively few layers of keratin in the stratum corneum, and usually high permeability to water and respiratory gases;
- a complex life cycle in many species;
- strict ectothermy, with little evidence for incipient endothermy related to large mass, activity, or other muscle activation.

All of these features help to define what is "amphibian" and are centrally important to the ecology and conservation of amphibian diversity. The dynamic structure and function of skin are central to forcing functions that influence the abundance and distribution of individuals or populations. Attributes of skin are important not only with respect to temperature regulation, water relations, and gas exchange, but are also relevant to current evaluations of amphibian declines in the context of disease susceptibility and transmission, impacts of xenobiotics, and sensitivity of responsiveness to climate change. Increasingly, these will become the subjects of future field and laboratory research by herpetologists who are interested in amphibian diversity and conservation.

20.2 Heat exchange, body temperature, and thermoregulation

Research related to the ecophysiology of amphibians has tended to focus on energetics, temperature, and water balance, but traditional single-focused studies have evolved to include increasing integration with other facets of biology. Present understanding of the important role that temperature plays in the lives of amphibians began largely with observations of behavior, which in turn led to inferences and hypotheses of significance that could be tested by measurement and experiments, both in the laboratory and in the field. Amphibians are not as well studied as are fishes and reptiles. However, the data available should be viewed by researchers in context of a voluminous, phylogenetically broader literature on thermal biology. This rich literature stems partly from the ease of temperature measurement. Knowledge of the pervasive importance of temperature as a principal ecological factor that influences almost all biochemical and physiological processes, including behavior, is an important starting point for nearly all field studies of amphibian biology.

Anyone interested in the physiological ecology of amphibians must begin with appreciation for three things. First, the skin of amphibians couples them to their physical environment by the exchanges of energy, water, and respiratory gases. One must understand the physics of heat and water transfer between an amphibian and its environment to understand the thermal relations and ecology of these organisms. The implications of the interaction between heat and mass transfer for the environmental physiology of amphibians have been discussed by Tracy (1976) and Spotila (Spotila 1972; Spotila *et al.* 1992). Second, phenotypic responses to temperature can be plastic, such that attributes of morphology, physiology, and behavior are subject to reversible adjustments to temperature (Rome *et al.* 1992). Thus, the role of acclimatization is essential to understanding the thermal relationships of amphibians. Third, amphibians are a very diverse group of animals (Duellman and Trueb 1986), and one must appreciate that attributes and responses to variable microhabitats cannot be expected to be uniform across species. One should approach studies of amphibians with an open mind and without a narrow bias or expectation of stereotypic responses.

20.2.1 Thermal acclimation of physiological function

Some amphibians may undergo temperature changes of 30–35°C in relation to daily or seasonal changes, respectively (Rome *et al.* 1992). Temperature influences all chemical, physical, and physiological processes, and it is recommended that persons conducting field studies of amphibians be very familiar with what is known concerning the principles and patterns of temperature effects on amphibians. In particular, temperature affects energetics, muscle performance, behavioral activities, digestion, water exchange, reproduction, immune function, and susceptibility to disease and predation. To variable extents, amphibians can adjust by means of acclimation the sensitivity and responses of processes to temperature (Rome *et al.* 1992; Hutchison and Dupré 1992). The extent of developmental plasticity and adult phenotypic adjustments induced by environment, in addition to genetic adaptation, are both important to amelioration of stressful changes in environmental conditions, and hence distribution, population structure, and conservation.

20.2.2 Early methods in field studies of amphibian thermal relations

Insight and understanding of the thermal ecology of amphibians arose initially from observations of behavioral activities such as basking (adult anurans; e.g. Brattstrom 1963; Lillywhite 1970) and thermal aggregations (anuran tadpoles; e.g. Brattstrom 1962), in addition to measurements of body temperatures of animals in the field. Early summaries of detailed temperatures of free-living amphibians and their environments were published by Fitch (1956) and Brattstrom (1963). Published data on random measurements of amphibian body temperatures typically show a unimodal distribution, but interpretation of such data requires considerable caution unless other detailed information is available. Importantly, conclusions about thermoregulatory abilities require information about microenvironmental conditions that are available to animals and can be limited by time of day, season, and other factors. Moreover, because of the high evaporative rates characteristic of many amphibians, requirements for hydroregulation can compete with those for thermoregulation. And, as pointed out by Tracy (1976), the relative stability of body temperature need not represent or imply thermoregulation. Conclusions from the earlier literature were that some amphibians exercised control over body temperature, but active thermoregulation was not as well developed in amphibians generally as in other ectotherms such as lizards.

Interest in the body temperatures of field-active ecothermic animals, stimulated by the classic work of Cowles and Bogert (1944), created a need for simple means of measuring body temperatures that were not cumbersome. Hence, the Schultheis cloacal quick-reading thermometer was developed largely for field use and has been used extensively by biologists for more than 50 years. The Schultheis thermometer is a comparatively small (18 cm long) mercury thermometer with a slim, reduced bulb that registers changes of temperature quickly.

The thermometer is calibrated in 0.2°C divisions within the range of 0–50°C. It was manufactured originally by E.W. Schultheis and Sons of Brooklyn, New York, USA. The thermometer is now available from Miller and Weber of Queens, New York. In spite of easy breakage, this device made possible the rich data that now exist on field temperatures of amphibians and reptiles. With due caution against breakage, the thermometer has also been used to record temperatures in the microhabitats of animals, but the upper limit of measurement imposes a serious constraint on recording substrate temperatures in hot environments.

Cowles and Bogert (1944) used thermocouples to measure body temperatures of desert reptiles in outdoor enclosures, but the equipment was cumbersome. Subsequent development of small and easily portable instruments that reduced the size of batteries and electronic components of thermistor and thermocouple thermometers opened the way to a broader use of electronic thermometry in the field (Table 20.1). In particular, multiple probes enable one to record temperatures from a variety of microenvironment locations simultaneously without the serial measurements required of hand-held thermometers. Moreover, the quality of measurements have been further improved by rapid response time, accuracy, miniaturization, and readability of sensing probes constructed of thermistor or thermocouple sensors. These and infrared thermometers can also be used to measure surface temperatures of amphibians (Rowley and Alford 2007). Infrared probes are non-invasive and do not require manipulation of animals. However, accurate ones are expensive and require placing the sensor close to the animal, which can pose difficulties, especially for small species.

20.2.3 Precautions for temperature measurements

To ensure that spot measurements of body temperatures are meaningful, the investigator must have knowledge of the microenvironment, behavior, and preferably prior immediate history of the animal. This also necessitates knowledge and understanding of the biophysics of heat exchange as it applies to amphibians (Tracy 1976; Spotila *et al.* 1992). Even so, there are precautions that must be taken to validate the accuracy and precision of such body-temperature measurements.

First and foremost, the investigator needs to guard against transfer of heat between the animal and the investigator's hand, and between the animal and substrate or object on which it is placed during measurement, especially if the animal is small. Additionally, escape movements and attendant stress can alter the body temperature, and the skin may cool quickly if the animal is moved into a convective environment during measurement. Manipulation of smaller amphibians that causes only a few seconds delay can alter body temperatures

Table 20.1 Transducers and monitoring and logging devices that are useful for deployment in physiological field research with amphibians.

Device	Useful properties
Thermocouples	Electronic measurement of temperatures with fine wire probes and sensors; available with sensors <1 mm diameter. Battery power is available for field portable thermocouple thermometers. See also infrared thermometers.
Thermistors	Similar uses and advantages as thermocouples, but with non-linear thermal characteristics that can be unstable over time.
Infrared thermometers	Non-contact temperature measurement device that detects emitted infrared energy and converts this into a temperature reading. Critical considerations include field of view (target size and distance), spectral response, response time, signal processing, portability, hand-held options, and temperature range. Most infrared thermometers cover environmental and body temperatures that might be encountered in field studies of amphibians. Infrared thermocouples are small, low-cost infrared sensors that are self-powered and produce an output that mimics a thermocouple sensor. Infrared imagery may have varied roles in research with amphibians.
Transmitters	Smallest and lightest, 0.35–0.52 g total encapsulated transmitter with antenna and battery, for tracking movement. Crystal controlled two-stage design, pulsed by a multivibrator. Optional design available with pulse rate determined by temperature. Several configurations available intended for glue-on applications, but could be modified for implantation depending on ingenuity of the researcher. Larger units (0.39–3.8 g) can be supplied with helical or subcutaneous antenna. Medical grade epoxy or wax can be used to protect implanted units.
Passive integrated transponders (PITs)	Many types of implantable transponders are available for individual identification of animals from an implanted chip; location of animals over short range (1–10 m) by use of a diode tag and harmonic radar; or transmission of a temperature signal by passive acquisition of energy from an external electromagnetic field. No battery is required, and the small size permits implantation in the body cavity or just beneath the skin. However, usefulness for identification and temperature measurement is limited to very short distances (generally a few centimeters).

Dataloggers	Available with multiple channels for logging multiple inputs in the field, or as mini- or micro-units that can be used in multiple locations. Measurements include temperature, humidity, pressure, salinity, depth, and voltage. Microloggers can be used to generate thermal or humidity maps of the environment. Underwater dataloggers measure and record water temperatures to depths >305 m. Programmable with capability to record maximum/minimum or averaging. Sampling intervals 0.5 s–9 h. Up to 5-year battery life. Numerous units available from multiple sources with advantages of low-cost, real-time operation, programmable start time, reusable, and miniature size. Some units may be subject to calibration drift and should be recalibrated at intervals or before and after use.
iButtons™	Measurement of temperature and/or humidity, recorded from built in sensors into internal memory with user-defined time interval. Small, light-weight with stainless steel casing. A durable stainless steel package is resistant to dirt, dust, contaminants, moisture, and shock. Reusable and relatively inexpensive, with lithium battery operation for at least 10 years without maintenance. Units should be checked for calibration drift and should be recalibrated at intervals or before and after use.
Evaporimeters	Various products are available to measure area-specific transepidermal evaporative water loss and related variables. Many of these instruments are cumbersome, but some are portable with battery operation for field applications. The VapoMeter®, manufactured by Delfin Technologies, Kuopio, Finland, is a hand-held easily portable device developed for use with human skin and can also be used with small animals. Various other products known as dynamic diffusion porometers, leaf porometers, and leaf hygrometers are sold for use with plants and can measure conductance or resistance directly. Soil hygrometers and ceramic plates are used to measure water potential of soils.

Note: various products are available from multiple vendors, which can be accessed via the Internet. Further specifications are available online, providing details that are too numerous to be listed here.

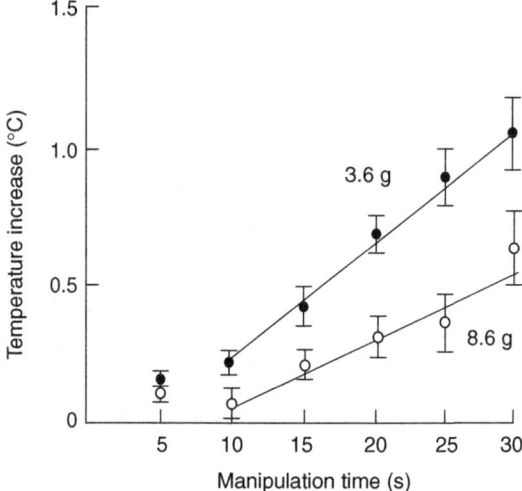

Fig 20.1 Increases in temperature of agar replicas of frogs ($N = 10$) following manipulation wherein two sizes of models were held in one hand during a given time lapse to simulate the expected change of body temperature in small anurans if they were handled similarly. Temperatures of the agar models were measured with thermocouples inserted into their centers and ranged initially from 22 to 23°C. Redrawn from Navas and Araujo (2000) with permission.

significantly (Figure 20.1) (Navas and Araujo 2000). Ideally, the measurement should be taken rapidly (within a matter of seconds) upon capture of the animal while it is held loosely but firmly by means of an insulating material such as dry cloth, cotton, or glove. Heat transfer can also be minimized if anurans are held by a hind limb while resting upon the insulation. Typically, investigators have inserted thermal sensors into the hindgut via entry at the cloaca, with the result that the core temperature is represented by the rear mass of the body. I have found it easier and quicker to insert the probe into the stomach via the mouth. If the tip of the sensor or thermometer is applied gently to the mouth, anurans usually open the mouth briefly when the probe can be inserted. This method typically measures the core temperature at a more central location of the body where the mass and thermal inertia is greatest. Insertion of the probe also is quicker compared with the time used in locating and probing the much smaller orifice of the cloaca. Finally, cloacal insertion often causes release of urine from the urinary bladder, which in turn increases evaporative cooling near the site of measurement.

Tolerance of broad thermal regimes may be characteristic of many species of amphibians which might be subjects of investigation. Whatever instruments of temperature measurement are utilized, they should be capable of measuring temperatures within the thermal range potentially encountered, including microhabitat temperatures. In consideration of reproduction, development, digestion, and activity, temperature extremes are important environmental factors and should be incorporated in ecological studies. With appropriate modifications, standard thermometers as well as electronic thermometers can be used to measure dry- and wet-bulb air temperatures, blackbody temperature (as an index of solar and thermal radiation), substrate, subterranean, and water temperatures. Whatever the method that is chosen for determining body temperatures of amphibians, quantitative indices have been described for evaluating the "effectiveness" of temperature regulation in these or other field-active ectotherms (Hertz *et al.* 1993).

20.2.4 Telemetry

Radiotelemetry offers many advantages over the use of thermometers. Body temperatures can be logged or recorded continuously or at chosen intervals without disturbance to the animal that is being monitored. The first use of radiotelemetry in field studies of amphibian body temperature was that of Lillywhite (1970) who used temperature-sensitive transmitters to record internal body temperatures of bullfrogs at shorelines and shallow water of ponds. Currently, a variety of transmitter designs, sizes, and other features are available commercially, but the options are generally not useful for studies involving smaller amphibian species (< 5 g). Where miniaturization is a requirement, dataloggers and iButtons™ have proven to be useful (Table 20.1). Micro-dataloggers can be used effectively to continuously monitor temperatures from both animals and their microhabitats. They can generate a thermal map of the environment, useful for understanding thermal selection behavior and constraints on physiology, and they can monitor temperatures of real animals or physical models in similar contexts. Available temperatures can have significant effects on amphibian activity, metabolism, muscle performance, and developmental rates. In extreme environments the available microhabitats can quickly reach lethal temperatures, and species must evolve strategies for avoiding thermal incapacitation or death.

For short-term measurements the unit can be simply fed to the animal (or physically lodged in the stomach), but longer-term studies require surgical implantation of the unit within the body cavity. Care must be taken to avoid infection and to allow adequate recovery from the surgery before an animal is released into the environment. Surgical implantation also avoids loss of the transmitter, interference with feeding, and postprandial temperature elevation that are possibly associated with units that are placed within the gut. Implantation procedures should incorporate due caution to avoid cutting the intestine and accidentally implanting units within the gut from which they are subsequently passed. The user must consider trade-offs between range, life of operation, and size, where the latter is determined largely by the mass of the batteries. It is recommended that transmitters or other implanted devices not exceed 3–5% of the animal's body mass. The use of passive integrated transponders (PITs) allows gathering of temperature signals from even small amphibians, but the range is very short (Table 20.1).

20.3 Water relations

Adequate levels of body water are essential to living organisms, and the maintenance of adequate hydration is especially challenging for amphibians (Warburg 1965; Shoemaker 1988; Tracy *et al.* 1993; Lillywhite and Mittal 1999; Bartelt 2000; Lillywhite and Navas 2006). Many species are subject to rapid dehydration and require water for the completion of complex life cycles; therefore, these vertebrates are especially sensitive to environmental changes that impact the amount and distribution of free water in time and in space.

20.3.1 Cutaneous water exchange: terrestrial environments

One of the more uniquely important features of amphibian morphology is the highly permeable integument that characterizes numerous species. The features of skin that are relevant to this discussion include a stratum corneum with only one or two layers of keratinized cells and the absence of lamellar bodies, lipogenic organelles which in amniotes give rise to a highly structured "permeability barrier" for water (Lillywhite 2006). Lacking these protective features, many species of amphibians lose water across the skin at rates similar to, or approaching, that of a free water surface. This is the principal underpinning of the well-known fact that dehydration is potentially a serious problem for many amphibian species living in terrestrial environments. The skin may offer little resistance to evaporative water loss, and consequently many species survive only by means of evolved behaviors and geographic distributions that allow them to remain near water or in relatively moist microenvironments (Child *et al.* 2008).

Early studies pioneered by Overton (1904), Adolph (1933), Thorson (1955), and others suggested that "typical" anurans evaporate water at rates similar to that of a free water surface. This initial impression, now perpetuated within a considerable literature, was based largely on studies of ranid and bufonid anurans. Authors of various publications and textbooks still tend to suggest or state that "most" amphibians have little or no skin resistance to cutaneous evaporative water loss. However, research during the past 40 years has shown that not all amphibians evaporate water in this manner (reviews in Toledo and Jared 1993; Lillywhite 2006).

Since the early 1970s a substantial number of anuran species have been found to occupy very arid habitats and to withstand xeric conditions due to plastic responses that modulate skin resistance (Shoemaker 1988; Navas *et al.* 2004). Arboreal species, in particular, exhibit specializations that enable them to withstand drying conditions including exposure to full sun and low ambient humidity. Key among these features are uricotelism and low rates of evaporation attributable to superficial lipids that are secreted from specialized cutaneous glands and wiped over the outer skin surfaces of the body. The evolution of uricotelism in xerophilic frogs only makes sense in the context of abilities of animals to reduce their rates of cutaneous water loss simultaneously (Shoemaker *et al.* 1972). Thus, a key adaptation is the ability of anurans to mitigate losses of body water by means of creating an external lipid barrier to transepidermal water loss (TEWL) (Lillywhite and Mittal 1999; Lillywhite 2006). Species with a skin resistance to evaporative water loss exceeding about 10 s/cm have been referred to as "waterproof" following the early publications of Loveridge (1970), Shoemaker *et al.* (1972), and Blaylock *et al.* (1976). This designation, however, is not literally accurate.

An alternative strategy for reducing losses of body water is the production of "cocoons" involving multiple layers of shed epidermis when fossorial species are subjected to dehydrating conditions while buried in soil (Lillywhite 2006). In the case of both secreted cocoons and the supraepidermal lipids that are secreted and wiped by arboreal frogs, subsequent function of the barrier depends on immobility of the animal for otherwise the structural integrity of the barrier is disturbed.

Understanding amphibian water balance in these animals requires careful observations in the field combined with appropriate mechanistic investigations of structure and function in the laboratory. Most physiological ecologists working with amphibians appreciate that only a small fraction of extant amphibian species—clearly less than 2%—have actually been studied in these or related contexts. Understanding the adaptive variation of skin resistance and its plasticity is central to the role of water economy in limiting the distribution of amphibians and their behavioral activities in time and in space. Increasingly, these studies will become more integrative. Most animals require access to fresh water or a moist substrate to replenish evaporative losses of body water, and specializations of ventral skin in pelvic and abdominal regions facilitate this process behaviorally and physiologically (Nagai *et al.* 1999). Water permeability generally varies directly with gaseous permeability (Lillywhite and Mittal 1999), and the skin exchanges respiratory gases and ions as well as water. Some amphibians rely exclusively on the skin for respiratory gas exchange. Most studies have focused on these aspects singly, but more integrative approaches are necessary to understand the multiple processes important to growth, distribution, and persistence of populations (Lillywhite *et al.* 1973; Bartelt 2000; Child *et al.* 2008).

20.3.2 Cutaneous water exchange: aquatic environments

Numerous amphibian species are aquatic or spend considerable time in water including aquatic larval life stages. Because the skin of amphibians is quite permeable in many species, the net movement of water across the skin reverses direction when animals move between land and water. The skin surfaces of submerged amphibians exchange dissolved gases, water, and ions, but the regulation of these exchanges is complex owing to the numerous sites at which such transfers can occur. Periodic losses of ions and water occur in urination, and periodic exchanges of gases occur whenever aquatic amphibians surface to ventilate lungs with air. The skin is not passive in these contexts, for the permeability and exchange processes can be altered by changes in blood flow, respiratory properties of blood, hormonal influence, extent of mucus covering the external surfaces, behavioral disruption of external boundary layer, ionic transporters, density of aquaporins, and other factors (Boutilier *et al.* 1992). Understanding the physiological ecology of aquatic amphibians will continue to rely on integration of field and laboratory studies, but advances will depend to a large degree on creative and diligent application of new methodologies and techniques in field methods. Such methods and integration will also characterize studies of estivation and hibernation (Pinder *et al.* 1992). It will be important to examine phylogenetically diverse taxa and diverse environmental conditions, as well as the many aspects of physiology that change during development and might differ between larval stages and adults (Ultsch *et al.* 1999). Finally, the aquatic environment of the eggs of amphibians is essential to understanding the distribution and persistence of species. Discussion of some of the interesting problems connecting physiology and ecology of eggs can be found in Seymour and Bradford (1995) and in Kam *et al.* (1998).

20.4 Measuring water exchange

The water exchanges of amphibians in the field can be measured directly by weighing frogs to quantify mass changes that include both cutaneous and pulmonary components. The difference between mass and weight is negligible, and such measurements can be easily converted to volume flux if changes are measured over time. If such data are part of experiments, animals are usually fasted and measured with the bladder empty ('standard' mass: Ruibal 1962) so that defecation and urination do not influence such measurements. Due to the low rate of metabolism and highly permeable skin characteristic of most amphibians, changes in mass over relatively short intervals (without ingestion or defecation) are assumed to result from exchanges of water insofar as the component due to metabolic losses of carbon will be negligibly small. Such measurements can be used to estimate daily water turnover rates and water budgets for amphibians. Due to the generally high water turnover rates of amphibians, isotopic labeling procedures such as the use of tritiated water (Nagy 1975) cannot be used without error. Urine flow rates have been used to estimate water turnover in the aquatic *Xenopus laevis* (Henderson *et al.* 1972).

Regional measurements of cutaneous evaporative water loss can be measured by use of a VapoMeter®, which is a portable, hand-held evaporimeter manufactured by Delfin Technologies, Kuopio, Finland (Table 20.1). These measurements all disturb the animal, but can be carried out in the field. To date, most information about the evaporative properties of amphibian skin is derived from laboratory studies (Toledo and Jared 1993; Lillywhite 2006).

20.5 Energetics

Chemical energy is important to amphibians, as the ability to maintain energy balance is crucial to both survival and reproduction. Energy budgets can be approached from perspectives of mass balance and are most meaningful when viewed over broad time scales such as annual cycles. Feeding rates are related to temperature, water balance, and seasonal and hormonal influences (e.g. growth hormone). Energy assimilation efficiencies can vary greatly, but most values average around 80–90%. Rates of metabolism (aerobic energy expenditure) are commonly estimated by manometry (pressure/volume changes) or respirometry (concentration changes of O_2 or CO_2), but cannot be measured in free-ranging amphibians by means of doubly labeled water technique due to characteristically high rates of water turnover (Nagy 1975). Therefore, energy data incorporated into amphibian energy budgets are based on laboratory estimates of metabolic energy expenditure coupled with time allocations for different activities (e.g. Jorgensen 1988). During periods of restricted resources, metabolic depression facilitates survival of amphibians by lowering levels of energy expenditure below normal resting values (generally 20–30% of standard metabolic rate; Guppy and Withers 1999). Brumation, dormancy, or inactivity are common responses of amphibians to cold or dehydrating conditions, especially in desert environments.

20.6 Modeling amphibian–environment interactions

Tracking variations in body temperature, energy expenditure, and water balance is critical to understanding the ecology of amphibians and their responses to environmental change. Many studies of ectotherms have emphasized thermoregulation and heat exchange, but we now appreciate that hydroregulation is equally important in amphibians (Tracy *et al.* 1993; Spotila *et al.* 1992). Moreover, the permeable skin of most amphibians influences body temperature, which is coupled to exchange of energy and water. Temperature and water balance interactively influence behavior, activity cycles, and habitat use. Such biophysical interactions potentially contribute to amphibian population declines linked to microclimate directly (Davidson *et al.* 2001) or to epidemic disease mediated by fungal parasites (Pounds *et al.* 1999, 2006).

20.6.1 Mathematical models

The transfer of heat and water between an amphibian and its environment occurs by means of several processes and can be described by energy and water budget equations based on physical principles. It is beyond the scope of this chapter to review detailed descriptions of heat energy and water budgets, or the processes involved, and the reader is referred to the reviews of Tracy and Spotila cited above, and the references therein. Further discussion of modeling with respect to amphibians can also be found in Chapter 21. Note that energy balance can be modeled at thermal equilibrium or in dynamic states of so-called "transient energy balance." Construction of a steady-state *climate space* describes the tolerable conditions and limits of thermal environment within which equilibrium body temperature is compatible with the physiological well-being of an animal (Spotila *et al.* 1992). Climate spaces can illuminate the biophysical constraints of the environment and are used to define the boundaries of microclimate in which an animal must remain to function and survive.

The status of heat exchange at any given moment determines the body temperature of an amphibian, which in many circumstances tends to be relatively stable (Tracy 1976). Aquatic amphibians are usually close to the temperature of surrounding water due to its high heat capacity and relatively rapid conduction to or from the animal. Heat exchange of fossorial amphibians is predominantly by conduction, and body temperatures of animals such as burrowing toads or caecilians are usually in equilibrium with surrounding soil. Terrestrial amphibians are sensitive to radiation, convection, and conductive exchange with the substrate. At high temperatures and low humidities, water evaporates at high rates and the skin as well as body is subject to rapid dehydration. Hence, the activities of amphibians often depend importantly on water storage and whether losses of body water exceed the possible uptake of water from soil, wet surfaces, ponds, or other sources. In relation to studies of population and community ecology, both thermal and hydric environments can be considered to represent ecological resources (Magnuson *et al.* 1979).

The significance of mathematical models is threefold. First, they illuminate the sensitivities of various processes in context of their relative importance in determining the magnitude of heat and mass exchange and resulting equilibrium body temperatures. Second, they can be used to generate hypotheses about amphibian interactions with their environment. Third, they have predictive value to enhance understanding of behaviors, energetics related to population growth, distributional limitations, and responses to climate change. On the other hand, modeling efforts have limitations and can perform no better than the assumptions and data that provide the inputs to the model. As example, substrate temperature can vary greatly between shaded and sunlit areas, providing thermoregulatory opportunities for amphibians that might shuttle between these two patches of microenvironment (Lillywhite 1970; Lillywhite *et al.* 1973). However, thermoregulation might appear to be much more constrained if one does not consider substrate temperature as one of the parameters in heat-balance models (Tracy 1976).

20.6.2 Physical models

Physical models are useful for simulating the thermal properties, and sometimes evaporative properties, of an animal in steady state. Like mathematical models, physical models integrate the physical attributes of a simulated animal with heat exchange processes related to the microenvironment. Thus, models can be placed at multiple locations and indicate the range of possible steady-state temperatures available to an animal. When compared with actual body temperatures of free-ranging animals, the models enable researchers to form conclusions related to thermoregulatory behavior, selection of retreat sites, timing of activity, and the range of possible environments occupied. Unlike mathematical models, physical models can indicate the dynamics of heat-exchange processes in local environments in real time.

Various objects have been used to represent amphibians in different thermal environments. One difficulty in the use of model amphibians is deciding if the animal should be represented as "wet-skinned" and therefore evaporates water as would the species represented. Various investigators have used models of frogs cast as agar (from alginate molds; Navas and Araujo 2000), plaster (O'Connor 1989), sponges (Hasegawa *et al.* 2005), and copper casts or tubes covered with water-saturated cotton or cloth (Bradford 1984; Bartelt and Peterson 2005). Care should be taken to ensure that the reflectance of the cloth or model surface is similar to that of the animals being compared, and that evaporation rates are not compromised by depletion of water associated with the model. Bartelt and Peterson (2005) have developed a physical model with a datalogger and water reservoir, which maintains a wet cloth sleeve over a copper model (Figure 20.2). To date, physical models have represented either wet- or dry-skinned animals. Agar replicas of anurans have been used as null models in studies of amphibians having near-zero resistance to evaporative water loss, and these models have been shown to exhibit rates of water evaporation and temperatures identical to those of living frogs which they are intended to mimic (Navas and Araujo 2000). The use of models for investigating amphibians having an intermediate resistance, or which vary the resistance of skin by wiping of lipids, still present special challenges.

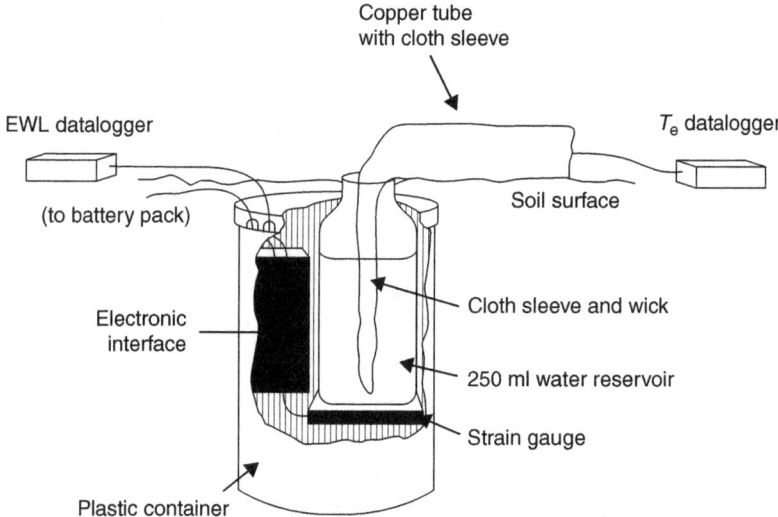

Fig 20.2 A physical model used to simulate the thermal and evaporative properties of toads. The model records operative (equilibrial) temperature (T_e) and evaporative water loss (EWL), both tested for accuracy, limitations, and applicability. The model assumes there is no physiological control over EWL and is designed specifically for western toads, *Anaxyrus* (= *Bufo*) *boreas*, but can be used or modified to measure T_e and EWL in other species of amphibians. Modified after Bartelt and Peterson (2005) with permission.

To understand the requirements for effective conservation or management of amphibian populations and communities, biologists need to understand how habitat structure and conditions of microenvironment affect thermal and hydric states of amphibians simultaneously. Recently, Bartelt (2000) compared operative temperatures and evaporative water losses of physical models with actual body temperatures and evaporative rates of toads determined by their natural behaviors in different habitats. The results suggest that habitats allowing these animals to conserve body water as well as to achieve warm body temperatures provide the more favorable conditions for the population. Future studies that integrate physiology with environmental interactions will be increasingly important to understanding the habitat requirements and conservation of amphibian populations and communities.

In other studies, thermal equilibria and evaporative properties of skin have been investigated using anurans themselves, either as live frogs with movement restricted by caging (Tracy 1975) or as dead or paralyzed frogs that are placed in locations of choice (Lillywhite *et al.* 1997, 1998). In these situations, care must be taken to avoid excess dehydration of animals or their exposed skin surfaces, for the evaporative properties of skin and the health of the animal changes if the skin itself is allowed to dehydrate (Lillywhite 1975). Moreover, the use of muscle-relaxing agents in field studies may not be allowed by animal-care-and-use committees, although there is no harm or excessive stress to the animal if the experiments are designed properly and executed carefully with due caution.

20.7 Other issues and future research directions

The subjects of temperature, water exchange, and energetics will continue to be centrally important subjects of future investigations into the physiological ecology of amphibians, but these will become increasingly integrated with other aspects of biology especially in the context of conservation and biodiversity concerns. Space does not permit detailed commentary on these other subjects, but here I will mention some directions that will likely be part of future advances in understanding amphibian biology based in field methods.

With respect to technology, we may expect to see increasing miniaturization of telemetry systems and creative employment of their usage in a variety of field contexts. Largely due to size limitations, there has been very little use of datalogger packages that can record physiological variables related to blood chemistry, gas exchange, cardiovascular function, muscle activity, and other parameters in addition to temperature and location. These will be particularly useful in contexts of understanding brumation (*sensu* Mayhew 1965), aquatic life, and survival in extreme or changing environments (Lillywhite and Navas 2006). Use might also be made of osmotic pumps, which provide accurate and reliable delivery of drugs or hormones in chronic infusion studies. Some newer pumps are suitable for implantation in mice or small rats and could be adopted for uses in amphibians related to nutrition, physiology, immunology, pharmacokinetic analysis, and other subjects of interest. Compact osmotic pumps can be used in conjunction with telemetry devices or catheters, and the versatility in drug delivery allows for behavioral and physiological analysis without disrupting study parameters.

The decline of amphibian populations in many parts of the globe beckons increasing attention to physiological studies related to threats such as chemical contaminants, emerging infectious diseases, overexploitation, habitat destruction, exotic species introductions, and climate change (Lips *et al.* 2008). Future biophysical studies related to climate change should increase attention to climate-linked hypotheses concerning reproduction, invasive disease outbreaks and their spread across landscapes, transmission processes, and amphibian population declines. Population dynamics and distribution of amphibians might be linked with features of habitat that are subject to temporal variation and behavioral choices related to physical interactions and physiological resources. Microbial communities associated with amphibian skin might also respond to aspects of biophysical exchange, pesticides, or other contaminants across gradients of habitats or in relation to habitat disturbance and climate change. Recently, anti-fungal skin bacteria have been implicated to form mutualistic relationships with amphibian species by assisting protection from pathogenic fungi (Lauer *et al.* 2007). These may well turn out to be very complex areas of research, and future investigations that might benefit from integration of physiology with amphibian ecology will likely involve teams of trained researchers working in different but complementary disciplines.

Stable isotopes are being used increasingly in ecological studies as indicators of water and nutrient metabolism, trophic positions of consumers in food webs, and potential energy and mass flow through ecological communities. Stable isotopes can be measured from very small tissue samples and provide a measure of energy and nutrients assimilated through trophic interactions by an organism (Lajtha and Michener 1994). The many applications and uses of stable isotope analysis have increased explosively during recent years, and their continued use in ecological studies will undoubtedly increase. Stable isotope techniques have not been widely used in studies of amphibians, but the methodology has much to offer in the way of advancing ecological knowledge of this important group of vertebrates.

The use of drugs and chemicals that are introduced into animals or the environment by investigators must be considered as a potential problem in future herpetological research in context of unwanted and unknown or inadequately studied effects on amphibians. Chemicals used as anesthetics, sedatives, chemical markers such as dyes, hormones, and various pharmaceutical agents all have potential cumulative, long-term effects on populations if used improperly or without knowledge of potential problems. As an example, the anesthetic tricaine methanesulfonate, popularly known as MS-222, has been used widely as an anesthetic in field studies of amphibians. A review of the literature indicates that MS-222 was used for decades before field biologists understood its mechanisms of action, and such uses potentially impact individual animals or populations by means of increasing stress, impairing sensory perception, and interfering with abilities of researchers to quantify parasite loads or diagnose bacterial infection (Byram and Nickerson 2009). During studies of declining populations of hellbenders (*Cryptobranchus alleganiensis*), thousands of these animals have been exposed to MS-222, many of them multiple times. As a consequence, Byram and Nickerson (2009) recommend that usage of MS-222 should be avoided when possible. When the anesthetic is used, it should be properly buffered, and the treated animals should be allowed adequate time and facilities for recovery. Field biologists need to recognize that the use of this or other anesthetics for immobilizing animals for marking, etc., can alter the animal's physiology and behavior as well as that of associated parasites and microbes. The authors also suggest that herpetologists and field biologists work as a community to develop humane and informed field anesthesia protocols. Such a recommendation would apply also to use of any other chemical in field research to ensure that researchers do not negatively impact the declining populations they might study.

Finally, it is important to comment briefly on the topics of reproduction and development. The evolutionary success of amphibians ultimately depends on their capabilities to reproduce, grow, and in many cases transition between different environments associated with metamorphosis. Amphibian growth and reproduction is characterized by diversity of reproductive modes, plasticity, and interplay between internal physiological cycles and external conditions of environment. The identification of factors in the environment that influence growth and development will continue to be important, especially with respect to climate, environmental chemistry, and the quality of food and water. Methods for measurement of egg properties, body mass, composition, and nutritional and hormonal status in the field (e.g. Licht *et al.* 1983) will be important to validate conclusions based strictly in laboratory studies, which means that access, timing, sampling effort, and other aspects of field investigation will need to be planned carefully by investigators. The important interplay between field and laboratory investigation will require careful integration as the role of endocrine system and other physiological controls are still not well understood with respect to reproduction and development in many species of amphibians. Insofar as most amphibians undergo "aquatic" embryonic and/or larval development before transitioning to more terrestrial (hence aerial) environments, survival depends on adaptation of structure and function—especially feeding, locomotory, osmoregulatory, cardiorespiratory, and sensory systems—in both environments during the lifetime of individuals (Burggren and Just 1992).

As with the many other topics related to physiological ecology of amphibians, aspects of reproduction and development are complex, diverse, and challenging with respect to the varied and intense selective pressures facing extant species. Integration of approaches to questions, methods, and data analysis will be an important theme in current and future studies having sufficient quality and novelty to advance understanding. Knowledge of the physiological ecology of amphibians has much to offer persons who are interested in the conservation and management of amphibian populations. To quote a postscript from Burggren and Just (1992): "Until we understand all the physiological responses to the environment in all developmental stages, we cannot hope to understand why a particular species thrives or disappears from the biosphere."

20.8 References

Adolph, E. F. (1933). Exchanges of water in the frog. *Biological Reviews*, **8**, 224–40.

Bartelt, P. E. (2000). *A Biophysical Analysis of Habitat Selection in Western Toads (Bufo boreas) in Southeastern Idaho*. PhD dissertation, Idaho State University, Pocatello, ID.

Bartelt, P. E. and Peterson, C. R. (2005). Physically modeling operative temperatures and evaporation rates in amphibians. *Journal of Thermal Biology*, **30**, 93–102.

Bentley, P. J. (1966). Adaptations of amphibian to arid environments. *Science*, **152**, 619–23.

Blaustein, A. R. and Wake, D. B. (1995). The puzzle of declining amphibian populations. *Scientific American*, **272**, 52–7.

Blaylock, L. A., Ruibal, R., and Plat-Aloia, K. (1976). Skin structure and wiping behavior of phyllomedusine frogs. *Copeia*, **1976**, 283–95.

Boutilier, R. G., Stiffler, D. F., and Toewes, D. P. (1992). Exchange of respiratory gases, ions, and water in amphibious and aquatic amphibians. In M. E. Feder and W. W. Burggren (eds), *Environmental Physiology of the Amphibians*, pp. 81–124. University of Chicago Press, Chicago, IL.

Bradford, D. F. (1984). Temperature modulation in a high-elevation amphibian, *Rana muscosa*. *Copeia*, **1984**, 966–76.

Brattstrom, B. H. (1962). Thermal control of aggregation behavior in tadpoles. *Herpetologica*, **18**, 38–46.

Brattstrom, B. H. (1963). A preliminary review of the thermal requirements of amphibians. *Ecology*, **44**, 238–55.

Burggren, W. W. and Just, J. J. (1992). Developmental changes in physiological systems. In M. E. Feder and W. W. Burggren (eds), *Environmental Physiology of the Amphibians*, pp. 467–530. University of Chicago Press, Chicago, IL.

Byram, J. K. and Nickerson, M. A. (2009). The use of tricaine (MS-222) in amphibian conservation. *Reptile and Amphibian Conservation Corps, Occasional Papers in Reptile and Amphibian Conservation* no .1, 1–15.

Child, T., Phillips, B. L., and Shine, R. (2008). Abiotic and biotic influences on the dispersal behaviour of metamorph cane toads (*Bufo marinus*) in tropical Australia. *Journal of Experimental Zoology*, **309A**, 215–24.

Cowles, R. B. and Bogert, C. M. (1944). A preliminary study of the thermal requirements of desert reptiles. *Bulletin of the American Museum of Natural History*, **83**, 261–96.

Davidson, C. H., Shaffer, B., and Jennings, M. R. (2001). Declines of the California redlegged frog: climate, UV-B, habitat, and pesticide hypotheses. *Ecological Applications*, **11**, 464–79.

Duellman, W. E. and Trueb, L. (1986). *Biology of Amphibians*. McGraw-Hill Book Co., New York.

Feder, M. E. and Burggren, W. W. (eds) (1992). *Environmental Physiology of the Amphibians*. University of Chicago Press, Chicago, IL.

Fitch, H. S. (1956). Temperature responses in free-living amphibians and reptiles of northeastern Kansas. *University of Kansas Publications of the Museum of Natural History*, **8**, 417–76.

Gordon, M. S., Schmidt-Nielsen, K., and Kelly, H. M. (1961). Osmotic regulation in the crab-eating frog (*Rana cancrivora*). *Journal of Experimental Biology*, **38**, 659–78.

Guppy, M. and Withers, P. C. (1999). Metabolic depression in animals: physiological perspectives and biochemical generalizations. *Biological Review*, **74**, 1–40.

Hasegawa, M., Suzuki, Y., and Wada, S. (2005). Design and performance of a wet sponge model for amphibian thermal biology. *Current Herpetology*, **24**, 27–32.

Henderson, I. W., Edwards, B. R., Garland, H. O. and Chester Jones, I. (1972). Renal function in two toads, *Xenopus laevis* and *Bufo marinus*. *General and Comparative Endocrinology Supplement* **3**, 350–9.

Hertz, P. E., Huey, R. B., and Stevenson, R. D. (1993). Evaluating temperature regulation by field-active ectotherms: the fallacy of the inappropriate question. *American Naturalist*, **142**, 796–818.

Hutchison, V. H. (1961). Critical thermal maxima in salamanders. *Physiological Zoology*, **34**, 92–125.

Hutchison, V. H. and Dupré, R. K. (1992). Thermoregulation. In M. E. Feder and W. W. Burggren (eds), *Environmental Physiology of the Amphibians*, pp. 206–49. University of Chicago Press, Chicago, IL.

Jorgensen, C. B. (1988). Metabolic costs of growth and maintenance in the toad, *Bufo bufo*. *Journal of Experimental Biology*, **138**, 319–31.

Kam, Y.-C., Yen, C.-F., and Hsu, J.-L. (1998). Water balance, growth, development, and survival of arboreal frog eggs (*Chirixalus eiffingeri*, Rhacophoridae): importance of egg distribution in bamboo stumps. *Physiological Zoology*, **71**, 534–42.

Lajtha, K. and Michener, R. H. (eds) (1994). *Stable Isotopes in Ecology and Environmental Science.* Blackwell Scientific Publications, Oxford.

Lauer, A., Simon, M. A., Banning, J. L., Andre, E., Duncan, K., and Harris, R. N. (2007). Common cutaneous bacterial from the eastern red-backed salamander can inhibit pathogenic fungi. *Copeia*, **2007**, 630–40.

Lee. A. K. (1968). Water economy of the burrowing frog, *Heleiporus eyrei* (Grey). *Copeia*, **1968**, 741–5.

Licht, P., McCreery, B. R., Barnes, R., and Pang, R. (1983). Seasonal and stress related changes in plasma gonadotropins, sex steroids, and corticosterone in the bullfrog, *Rana catesbeiana. General and Comparative Endocrinology*, **50**, 124–45.

Lillywhite, H. B. (1970). Behavioral temperature regulation in the bullfrog. *Copeia*, **1970**, 158–68.

Lillywhite, H. B. (1975). Physiological correlates of basking in amphibians. *Comparative Biochemistry and Physiology*, **52A**, 323–30.

Lillywhite, H. B. (2006). Water relations of tetrapod integument. *Journal of Experimental Biology*, **209**, 202–26.

Lillywhite, H. B. and Mittal, A. K. (1999). Amphibian skin and the aquatic-terrestrial transition: constraints and compromises. In A. K. Mittal, F. B. Eddy, and J. S. Datta Munshi (eds), *Water/Air Transitions in Biology*, pp. 131–44. Oxford and IBH Publishing, New Delhi.

Lillywhite, H. B. and Navas, C. A. (2006). Animals, energy, and water in extreme environments: perspectives from Ithala 2004. *Physiological and Biochemical Zoology*, **79**, 265–73.

Lillywhite, H. B., Licht, P., and Chelgren, P. (1973). The role of behavioral thermoregulation in the growth energetics of the toad, *Bufo boreas. Ecology*, **54**, 375–83.

Lillywhite, H. B., Mittal, A. K., Garg, T. K., and Agrawal, N. (1997). Wiping behavior and its ecophysiological significance in the Indian tree frog *Polypedates maculatus. Copeia*, **1997**, 88–100.

Lillywhite, H. B., Mittal, A. K., Garg, T. K., and Das, I. (1998). Basking behavior, sweating and thermal ecology of the Indian tree frog, *Polypedates maculatus. Journal of Herpetology*, **32**, 169–75.

Lips, K. R., Diffendorfer, J., Mendelson, III, J. R., and Sears, M. W. (2008). Riding the wave: Reconciling the roles of disease and climate change in amphibian declines. *PLOS Biology*, **6**, 441–54.

Loveridge, J. P. (1970). Observations on nitrogenous excretion and water relationships of *Chiromantis xerampelina* (Amphibia, Anura). *Arnoldia (Rhodesia)*, **5**, 1–6.

Magnuson, J. J., Crowder, L. B., and Medvick, P. A. (1979). Temperature as an ecological resource. *American Zoologist*, **19**, 331–43.

Mayhew, W. (1965). Hibernation in the horned lizard, *Phrynosoma m'calli. Comparative Biochemistry and Physiology*, **16**, 103–19.

Moore, J. A. (1964). *Physiology of the Amphibia.* Academic Press, New York.

Nagai, T., Koyama, H., Hoff, K. V. S., and Hillyard, S. D. (1999). Desert toads discriminate salt taste with chemosensory function of the ventral skin. *Journal of Comparative Neurology*, **408**, 125–36.

Nagy, K. A. (1975). Water and energy budgets of free-living animals: measurement using isotopically labeled water. In N. F. Hadley (ed.), *Environmental Physiology of Desert Organisms*, pp. 227–45. Dowden, Hutchinson and Ross, Stroudsburg, PA.

Navas, C. A. and Araujo, C. (2000). The use of agar models to study amphibian thermal ecology. *Journal of Herpetology*, **32**, 330–4.

Navas, C. A., Antoniazzi, M. M., and Jared, C. (2004). A preliminary assessment of anuran physiological and morphological adaptation to the Caatinga, a Brazilian semi-arid environment. In S. Morris and A. Vosloo (eds), International Congress Series, vol. 1275, *Animals and Environments*, pp. 298–305. Proceedings of the Third International Conference of Comparative Physiology and Biochemistry. Elsevier, Cambridge.

Noble, G. K. (1931). *The Biology of the Amphibia.* McGraw-Hill, New York.

O'Connor, M. P. (1989). *Thermoregulation in Anuran Amphibians: Physiology, Biophysics, and Ecology.* PhD dissertation, Colorado State University, Fort Collins, CO.

Overton, E. (1904). Neununddreissig Thesen über die Wasserökonomie der Amphibien und osmotischen Eigenschaften der Amphibienhaut. *Verhandlungen der Physikalisch-medizineschen Gesselschaft zu Würzburg*, **36**, 277–95.

Pinder, A. W., Storey, K. B., and Ultsch, G. R. (1992). Estivation and hibernation. In M. E. Feder and W. W. Burggren (eds), *Environmental Physiology of the Amphibians*, pp. 250–76. University of Chicago Press, Chicago, IL.

Pounds, J. A., Fogden, M. P. L., and Campbell, J. H. (1999). Biological response to climate change on a tropical mountain. *Nature*, **398**, 611–15.

Pounds, J. A., Bustamante, M. R., Coloma, L. A., Consuegra, J. A., Fogden, M. P.L., Foster, P. N., La Marca, E., Masters, K. L., Merino-Viteri, A., Puschendorf, R. *et al.* (2006). Widespread amphibian extinctions from epidemic disease driven by global warming. *Nature*, **439**, 161–7.

Ray, C. (1958). Vital limits and rates of desiccation in salamanders. *Ecology*, **39**, 75–83.

Rome, L. C., Stevens, E. D., and John-Alder, H. B. (1992). The influence of temperature and thermal acclimation on physiological function. In M. E. Feder and W. W. Burggren (eds), *Environmental Physiology of the Amphibians*, pp. 183–205. University of Chicago Press, Chicago, IL.

Rowley, J. J. and Alford, A. R. (2007). Non-contact infrared thermometers can accurately measure amphibian body temperatures. *Herpetological Review*, **38**, 308–11.

Ruibal, R. (1962). The adaptive value of bladder water in the toad, *Bufo cognatus*. *Physiological Zoology*, **35**, 218–23.

Seymour, R. S. and Bradford, D. F. (1995). Respiration of amphibian eggs. *Physiological Zoology*, **68**, 1–25.

Shoemaker, V. H. (1988). Physiological ecology of amphibians in arid environments. *Journal of Arid Environments*, **14**, 145–53.

Shoemaker, V., Balding, D., Ruibal, R., and McClanahan, Jr, L. (1972). Uricotelism and low evaporative water loss in a South American frog. *Science*, **175**, 1018–20.

Spotila, J. R. (1972). The role of temperature and water in the ecology and distribution of some lungless salamanders. *Ecological Monographs*, **42**, 95–125.

Spotila, J. R., O'Connor, M. P., and Bakken, G. S. (1992). Biophysics of heat and mass transfer. In M. E. Feder and W. W. Burggren (eds), *Environmental Physiology of the Amphibians*, pp. 59–80. University of Chicago Press, Chicago, IL.

Thorson, T. B. (1955). The relationship of water economy to terrestrialism in amphibians. *Ecology*, **36**, 100–16.

Thorson, T. B. (1964). The partitioning of body water in Amphibia. *Physiological Zoology*, **37**, 395–9.

Toledo, R. C. and Jared, C. (1993). Cutaneous adaptations to water balance in amphibians. *Comparative Biochemistry and Physiology*, **105A**, 593–608.

Tracy, C. R. (1975). Water and energy relations of terrestrial amphibians: insights from mechanistic modeling. In D. M. Gates and R. M. Schmerl (eds), *Perspectives in Biophysical Ecology*, pp. 325–46. Springer-Verlag, New York.

Tracy, C. R. (1976). A model of the dynamic exchange of water and energy between a terrestrial amphibian and its environment. *Ecological Monographs*, **46**, 293–326.

Tracy, C. R., Christian, K. A., O'connor M. P., and Tracy, C. R. (1993). Behavioral thermoregulation by *Bufo americanus*: the importance of the hydric environment. *Herpetologica*, **49**, 375–82.

Ultsch, G. R., Bradford, D. F., and Freda, J. (1999). Physiology: coping with the environment. In R. W. McDiarmid and R. Altig (eds), *Tadpoles: The Biology of Anuran Larvae*, pp. 189–214. University of Chicago Press, Chicago, IL.

Warburg, M. R. (1965). Studies on the water economy of some Australian frogs. *Australian Journal of Zoology*, **13**, 317–30.

Whitford, W. G. and Hutchison, V. H. (1965). Gas exchange in salamanders. *Physiological Zoology*, **38**, 228–42.

21

Models in field studies of temperature and moisture

Jodi J. L. Rowley and Ross A. Alford

21.1 Introduction

21.1.1 The importance of understanding the temperature and moisture environments of amphibians

In ectothermic organisms, variation in body temperature directly affects factors such as rates of energy acquisition, growth, and reproduction (Shoemaker and McClanahan 1975; McClanahan 1978). Temperature also exerts a strong influence on ecological interactions such as predator–prey and host–pathogen interactions, and changes in temperature can completely reverse the outcome of such interactions at both the individual and population levels (Elliot *et al.* 2002; Woodhams *et al.* 2003). Although the body temperature of terrestrial ectotherms is broadly correlated with environmental temperature, the actual body temperatures of many species can differ considerably from macroenvironmental temperatures due to species-specific behavior, physiology, morphology, and microenvironment use. As a result, ectotherms exposed to identical macroenvironmental conditions can experience very different body temperatures (Kennedy 1997).

The physiology and ecology of amphibians, and their thermal relations, are also strongly influenced by moisture. The degree of evaporative water loss (EWL) varies considerably within and among species (Shoemaker and Nagy 1977; Wygoda 1984; Buttemer *et al.* 1996; Young *et al.* 2005), and individuals of many species are able to adjust their rates of water loss over relatively short periods of time (Withers *et al.* 1982; Wygoda 1989a; Withers 1995; Tracy *et al.* 2008). The moisture and thermal environments, and interactions between them, can constrain the performance of amphibians (Snyder and Hammerson 1993;

Tracy *et al.* 1993). However, the actual temperature and humidity experienced by an amphibian can differ dramatically within a range of available macroenvironmental conditions, depending on microenvironment use (Schwarzkopf and Alford 1996; Seebacher and Alford 2002; Rowley and Alford 2007a). To understand how the environment is experienced by amphibians in the field, both the thermal and moisture environments must be characterized as they are experienced by amphibians.

Understanding the thermal and water relations of amphibians in the field is important not only for understanding their biology, but also for conservation. Recent declines in amphibian populations around the world have been attributed in part to the epidemic disease amphibian chytridiomycosis (Lips *et al.* 2006). Changes in thermoregulatory opportunities available to rainforest frogs may have contributed to their widespread declines in association with the disease (Pounds *et al.* 2006). Laboratory experiments have shown that elevated body temperatures, as can be produced by basking, can cure amphibians of the disease (Woodhams *et al.* 2003).

Global warming is also likely to have a large impact on many species, with macroenvironmental modeling suggesting that the distributions of many ectothermic species will be dramatically altered (Thomas *et al.* 2004). This approach is limited by the fact that the present distributions of many species may not reflect their fundamental climatic requirements (Parmesan *et al.* 2005). Understanding those requirements may aid in refining such predictions (Kennedy 1997; Parmesan *et al.* 2005).

There are two complementary approaches to collecting field data on temperature and water relations. The first is to sample living animals. This can involve single measurements from individuals (e.g. Snyder and Hammerson 1993; Navas 1996), or intensive monitoring via mark–recapture or tracking (e.g. Schwarzkopf and Alford 1996; Seebacher and Alford 2002). Sampling living animals typically results in a relatively small number of measurements per unit of investigator effort. In addition, it often requires attaching tags to animals or repeatedly handling them, both of which may alter normal behavior. The second approach is to place physical models that mimic the thermal and hydric properties of animals in the field. Because the thermal and hydric states of many models can be monitored simultaneously, a large number of microenvironments can be sampled intensively. The best approach is probably a combination of techniques. Living animals provide information on behavior and microhabitat use, which is used to inform the design of studies using models and to compare with their results. Models are used to collect large volumes of data that can provide an integrated picture of how the environment should affect animals across space and time. Sampling of living animals is covered elsewhere in this volume; here we focus on the collection and interpretation of data using physical models.

21.1.2 Physical models of amphibians

Physical thermal models mimic the thermal properties of a particular animal in steady state, thereby producing an approximation of body temperature, which is often referred to as the operative environmental temperature, although the original definition of operative environmental temperature requires that it be measured using a model with no thermal inertia (Bakken and Gates 1975). Relatively simple physical models made of substances such as hollow or water-filled copper casts or even tubes can be used to mimic the thermal properties of many reptiles (reviewed by Dzialowski 2005), and their use has led to a detailed understanding of many aspects of reptilian thermal biology (Dzialowski 2005).

Although such models are suitable for species with low and constant rates of EWL, the higher and often variable rates of EWL that occur in many amphibians complicate the design of physical models for these animals and the interpretation of data collected using them. Many amphibians have cutaneous resistances to EWL of zero or near zero, although it is increasingly clear that many anurans, particularly arboreal frogs, can have substantially higher levels of resistance (Wygoda 1984; Buttemer et al. 1996; Young et al. 2005). This is caused by mucus or lipid secretions (Shoemaker and McClanahan 1975; McClanahan 1978; Christian and Parry 1997) or possibly by physical barriers of dermal iridophores (Schmuck et al. 1988; Kobelt and Linsenmair 1992). In such species, cutaneous resistance to EWL may at times be more than 100 times greater than that of 'typical' ranid and bufonid frogs (Buttemer 1990; Buttemer and Thomas 2003), with the skin of some frogs being as resistant to water loss as that of terrestrial reptiles (McClanahan and Shoemaker 1987; Shoemaker et al. 1987). In many species, levels of skin resistance are variable, depending on the current physiological state and behavior of the individual (Toledo and Jared 1993; Lillywhite 2006; Tracy et al. 2008). The difficulty of accurately incorporating the effects of EWL has hindered the use of physical models in amphibian thermal biology (Bartelt and Peterson 2005). In species with variable resistance to EWL, measurements of thermal and water relations need to encompass the range of states possible for a given set of environmental conditions. In this chapter, we describe the use of models to measure this range of states directly.

21.2 Field models for investigating temperature and moisture

21.2.1 Models with zero resistance to EWL

To allow EWL to occur, physical models of amphibians for use in the field have been constructed as hollow copper tubes covered in wet fabric (Bradford 1984; Bartelt and Peterson 2005), periodically or continuously re-wetted plaster (O'Connor and Tracy 1987; Tracy *et al.* 2007), dead or immobilized amphibians (Wygoda 1989a, 1989b; Seebacher and Alford 2002), wet sponges (Hasegawa *et al.* 2005), or agar casts of animals (Schwarzkopf and Alford 1996; Navas and Araujo 2000). Most of these models lose water as free water surfaces, and can therefore only be used to model the body temperatures of amphibians with zero cutaneous resistance to EWL. The rates of EWL of sponge models can be adjusted by controlling the moisture content of the sponges (Hasegawa *et al.* 2005), which may make them preferable for species with constant skin resistances greater than zero. Modified agar models (Navas *et al.* 2002) may also be used to mimic constant higher skin resistance. Some types of models can also provide data on rates of evaporative water loss, and therefore of the potential for water stress, in the locations where they are deployed (Schwarzkopf and Alford 1996; Bartelt and Peterson 2005; Tracy *et al.* 2007). However, many species have variable skin resistance to EWL (Toledo and Jared 1993; Lillywhite 2006; Tracy *et al.* 2008). Models with zero resistance will significantly underestimate body temperature for such amphibians at many times, while models with fixed resistances above zero will overestimate them. No single model can therefore be used alone to accurately characterize the ranges of body temperatures available to many amphibians in the field.

21.2.2 A system of models that allows for variable EWL

21.2.2.1 Rationale

Instead of a single body temperature for any combination of environmental temperature, relative humidity, wind speed, and solar irradiance, the body temperatures available to an amphibian with variable cutaneous resistance, or one with unknown resistance, can be characterized as falling within an envelope the upper and lower boundaries of which are formed by the temperatures that would be experienced by animals with zero and very high skin resistance.

To define the upper and lower boundaries of body temperatures available to amphibians occupying any microenvironment, it is possible to use pairs of models that are identical except that one has zero resistance to evaporative water loss (a permeable model) and the other (an impermeable model) has either near-complete resistance or resistance equivalent to the highest known for the species of interest, if good data on this exist. Although we developed and have implemented this technique using agar models, it could be adapted to use with models of other types. If the upper boundary of skin resistance for a species is known, the technique we suggest could be modified by designing models similar to those we use, but with resistance that matches the species of interest. One way to do this has been suggested by Hasegawa *et al.* (2005); manipulating the water content of sponge models can vary their rates of EWL. Another technique would be to apply an impermeable coating in patches distributed over the body or concentrated in particular regions with lower EWL, and calibrate the resulting models against living animals (Navas *et al.* 2002).

A pair of permeable and impermeable models should always have equilibrium temperatures that indicate the lower and upper limits, respectively, of body temperature available to a frog at thermal equilibrium in any location. Placing pairs of models in a wide range of activity and retreat sites known to be used by a species will provide data that set boundaries on the body temperatures that can be experienced by that species in the habitat being sampled. If models with zero and near-complete resistance to EWL are used, this is true even if the range of skin resistance exhibited by the species is unknown, since the models span the possible extremes. Using pairs of models ensures that the measured range encompasses the range available to any individual of a species with variable skin resistance, regardless of its present state.

21.2.2.2 Model design and construction

Models of frogs in the water-conserving posture are constructed of 3% agar. Most frogs remain in the water-conserving posture during daylight hours, when temperature relationships are of most interest. Agar loses water as a free surface (equivalent to a cutaneous resistance of 0; Spotila and Berman 1976). The agar models are produced in latex molds, which themselves are made from plaster casts produced in alginate impressions taken from anesthetized or preserved frogs. To prevent water exchange with the substrate, the ventral surfaces of all models are coated with an impermeable plastic (PLASTI DIP®, clear, PLASTI DIP International, Blaine, MN, USA), which is intended for coating tool handles and other surfaces. In this condition, models should simulate the thermal performance of frogs with zero resistance to EWL in their dorsal skin (Spotila and Berman 1976), and with their ventral surfaces in contact with a dry, impermeable substrate. Leaving the ventral surface uncoated would allow water exchange with the substrate, but it is unlikely that this would occur in a manner similar to that experienced by living frogs due to the highly dynamic role of frog ventral skin in water exchange. Coating the ventral surface simulates frogs in a known, realistic situation, and eliminates what would otherwise be an uncontrolled source of variation in measurements. To create impermeable models, the remainder of the surface of half of the models is also coated with PLASTI DIP. These models match the permeable models in every respect, including spectral absorbence, which spectrophotometric measurements have shown is not measurably affected by clear PLASTI DIP between 330 and 800 nm. Coated and uncoated models also cannot be distinguished in infrared images taken through a B + W 094 infrared-pass filter, indicating that their reflectance does not differ substantially in the range from 800 to 1000 nm. Small thermal dataloggers (Thermochron iButtons, Dallas Semicondictor, Dallas, TX, USA; diameter 15 mm, height 6 mm) are embedded in each model to allow regular, repeated measurements of the models' core temperature.

21.2.2.3 Example of model construction and validation in the laboratory

Physical models (Figure 21.1) were made from 3% agar, in the shape of *Litoria caerulea* in the water-conserving posture. *L. caerulea* has a variable and moderate cutaneous resistance to EWL (maximum approx 10 s/cm; Buttemer 1990; Christian and Parry 1997).

The snout–vent length and weight of the models were approximately 75 mm and 44 g. The models were coloured green using four drops per 100 ml of green food colouring, made by mixing yellow, blue, and rose pink food colouring (Queen Fine Foods, Alderley, Queensland, Australia) at a ratio of 10:4:1. This produces spectral absorbence that falls within the range we have measured for *L. caerulea* between 330 and 800 nm, and qualitatively similar reflectance in the infrared between 800 and approximately 1000 nm as revealed by simultaneous photographs of models and living frogs taken through a B + W 094 infrared-pass filter.

The iButton embedded in each model was programmed to record the temperature every 30 min. Each datalogger was tied into a knot in a piece of

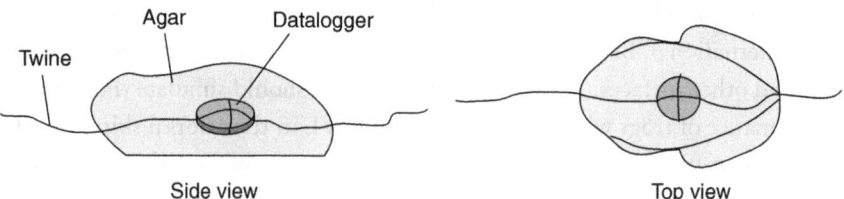

Fig. 21.1 Details of physical models.

twine, which was pegged at each end of the mold as the agar set, so that the datalogger was suspended in the centre of the model (Figure 21.1). The free twine remaining at each end of the models could be used to attach them to surfaces either by using tape over the twine or by tying the twine around features such as branches and roots. We set up three opaque, white, plastic containers (60 cm × 40 cm × 40 cm), each with a small water bowl in the center and a metal fly-screen lid. The containers were housed in a constant temperature room, which maintained ambient temperature between 19.5 and 21.5°C. Relative humidity fluctuated between 64 and 96% (mean 74%). A 150 W heat lamp was provided at one end of each container and was illuminated between 0930 and 2130 h to create a temperature gradient simulating the normal range of temperatures available to this species. We ran four temporal replicates of the experiment, creating a total of 12 sets of measurements of frog and model temperatures for comparison. Each replicate ran for 3 days.

At the start of each replicate, models were placed in each container in pairs, with one permeable and one impermeable model. The pairs were located so they spanned the range of thermal and light environments available in the containers. Twelve adult *L. caerulea* (11 males and one female) were captured near Townsville, Queensland, Australia. They ranged from 74.1 to 91.8 mm snout–vent length and 26 to 65 g body mass. Prior to experiments, they were maintained in smaller containers at ambient temperature in the constant-temperature room in which the experiments were carried out. Each frog was used in a single run of the experiment.

We recorded the body temperatures of frogs five times per day (0900, 1100, 1300, 1500, and 1700 h) over the 3 days of each temporal replicate, producing 15 measurements for each individual. The first time (0900 h) was chosen because at that time the frogs had not had access to any source of extra heat for almost 12 h, and their body temperatures should have been similar to those that would be measured during nocturnal readings in the field. Each temporal replicate was set up at least 60 min before the first temperature reading was taken, allowing models and frogs to reach a thermal steady state. At each reading, cloacal temperatures were measured using a small, chromel-alumel K-type thermocouple (diameter ≈1 mm) with the tip coated in plastic, attached to a digital thermometer (Small Pocket Thermometer model 90000, Industrial Automation, Joondalup, Western Australia). During temperature measurement, each frog was held by a single leg only. When temperatures were taken, the posture and location of the individual was noted; this information was later used to determine whether individuals had changed location, and to select the pair of models that should represent the upper and lower boundaries of the thermal envelope available to each frog at each measurement. All dataloggers and the thermo-couple were calibrated before the experiment.

All frog body temperatures fell within the broad envelope defined by the maximum and minimum temperatures of all impermeable and permeable models, respectively (Figure 21.2a). Temperatures of frogs fell within the narrower enve-lope defined by the temperatures measured in the same time interval in the permeable and impermeable models nearest to the frog on 144 of the 180 (80%)

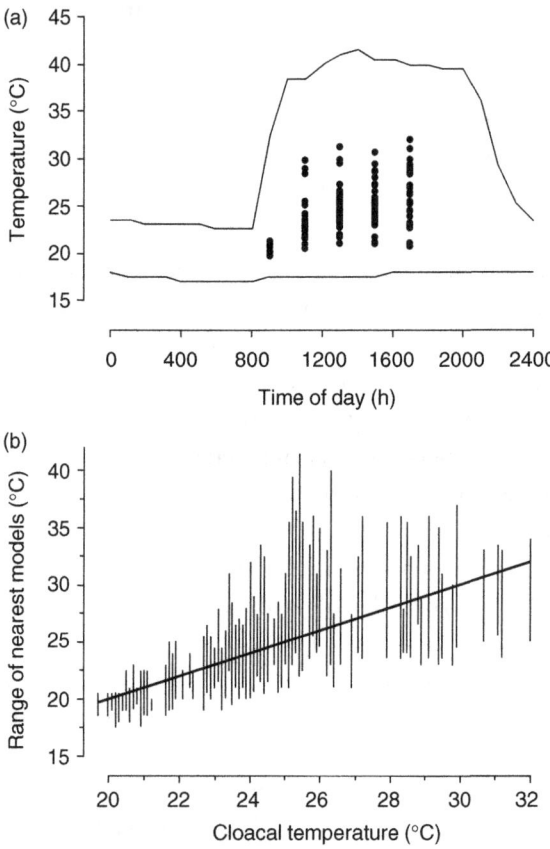

Fig. 21.2 (a) Range of temperatures available to *Litoria caerulea* in laboratory thermal gradients, as indicated by maximum and minimum temperatures of impermeable and permeable agar models (solid lines) and cloacal temperatures measured for individual frogs (points). (b) Lines indicating the range of temperatures between the permeable and impermeable models closest to each individual *L. caerulea* at each time its cloacal temperature was taken in laboratory thermal gradients. The line of equality is included to ease visualization of where cloacal temperature fell within each pair of model temperatures.

occasions on which we measured frog temperatures (Figure 21.2b). As expected, permeable models were cooler than frog body temperatures, while impermeable models were warmer than frog body temperatures. For 111 of our observations, frogs had changed their location since the last observation, or were being observed for the first time. On 28 of the 36 (77.8%) occasions when frog cloacal temperatures were outside the envelope defined by the nearest models, frogs had changed their location since the time of the last observation or were being observed for the first time, and therefore might not have equilibrated thermally. The association between changing location and being outside the thermal envelope was statistically significant (Fisher's Exact Test, two-tailed, $P = 0.034$). No cloacal temperature was more than 2.6°C below or 1.2°C above the boundaries of the envelope as defined by the nearest pair of models, and all were well within the extremes defined by all models.

21.2.2.4 Field validation of use of models to characterize available temperatures

We tracked 27 *Litoria lesueuri* using radiotelemetry and harmonic radar in the warm/wet season (19–21 March 2005) and the cool/dry season (5–13 August 2005) in Frenchman Creek, near Babinda, Queensland, Australia (17°20′S, 145°55′E; 20–100 m above sea level; Rowley and Alford 2007a). Weather conditions during both tracking periods varied considerably with respect to ambient temperature, humidity, solar radiation, wind speed, and rainfall. Body temperatures were recorded once during the day (between 1200 and 1600 h) and once at night (between 1900 and 0700 h) by holding a Raytek ST80 Pro-Plus Non-contact thermometer approximately 5 cm away from the frog and aiming at the lower dorsal area, between the thigh and point of transmitter attachment. We have demonstrated that the temperatures we measure in this manner accurately reflect the skin and cloacal temperatures of frogs (Rowley and Alford 2007b). During the tracking period, we placed pairs of permeable and impermeable models, constructed as previously described, in sites previously used as diurnal retreat sites by the tracked *L. lesueuri* (on bare ground, leaf litter, in vegetation, and on gravel, in exposed and sheltered positions; Rowley and Alford 2007c; Figure 21.3). Fifteen pairs of models were deployed during the warm/wet season, and eight pairs during the cool/dry season. The embedded iButtons recorded temperatures of the models every 30 min.

Models were left in the field and weighed every 24 h until at least one permeable model had lost 50% of its original weight. Although these models were reduced in size, they were within the range of body sizes of *L. lesueuri*, retained their basic shape, and were still 94% water, and thus continued to lose water at a

Fig. 21.3 Two of pairs of models placed in retreat sites used by frogs in the field

very similar rate per unit surface area. We would not recommend allowing models to fall below 50% of their original mass, as they begin to lose shape, and their proportional water content, and thus equivalent resistance to EWL, increases at an accelerating rate. Model size affects rates of heating and cooling, and could possibly alter equilibrium temperatures by altering the ratio of surface area to volume. However, our field data (Figure 21.4) suggest that any effects of changes as permeable models dehydrated on equilibrium temperature were very small. The distributions of temperature difference between the permeable and impermeable models were similar for all levels of desiccation, and the distribution in the 50–60% desiccated class was more similar to that in the 90–100% class than it was to those in the 60–70% and 70–80% classes, probably because of weather conditions at the time of measurement rather than any effect of drying on model temperature. The outlying extreme differences shown in Figure 21.4

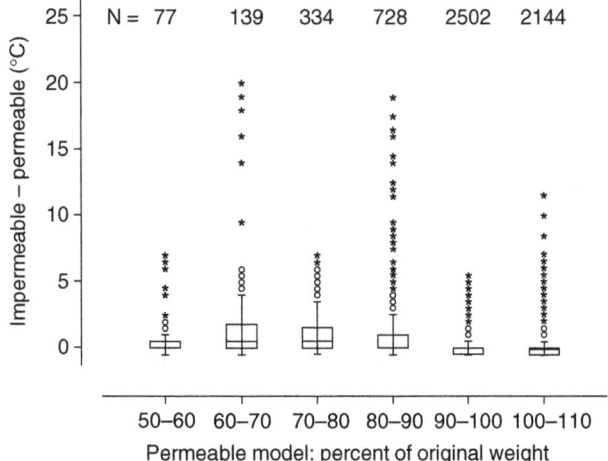

Fig. 21.4 Differences recorded between permeable and impermeable members of pairs of agar models of *Litoria lesueuri* placed in frog diurnal retreat sites in rainforest for up to 8 days near Babinda, Queensland, Australia. Box plots illustrate the median (heavy line within boxes), upper and lower quartile boundaries (box boundaries), range (lines), and outlying (circles) and extreme (asterisks) data points for differences, grouped by the weight of the permeable model as a percentage of its original weight. A small number of points at which the temperatures of permeable models exceeded those of impermeable models because only the permeable model was in sunlight have been omitted.

usually occurred at the same time each day and reflected occasions on which only the impermeable model was in the sun.

To determine the accuracy with which the models delineated the envelope of body temperatures available to frogs, we compared each measured frog body temperature with the minimum and maximum temperatures attained by models at the nearest 30 min mark (< 15 min time difference). We removed 16 frog body temperatures taken when frogs were found in microhabitats in which we had not placed models (i.e. under logs, on roads, in mown pastures, or in exposed sites high in terrestrial vegetation). Of the 145 field body temperatures remaining (30 taken in the warm/wet season and 115 in the cool/dry season), 121 (83.4%) were within the limits defined by the hottest impermeable model and the coldest permeable model, and 19 of the remaining 24 were within ±0.5°C of those limits. Only one body temperature was more than 0.9°C outside them (1.5°C above; Figure 21.5). Two-thirds of the temperatures that fell outside the limits were taken during nocturnal activity; in most cases these frogs were warmer than the models (Figure 21.5), which were all in diurnal retreat sites. This suggests that the nocturnal activity sites of frogs are warmer

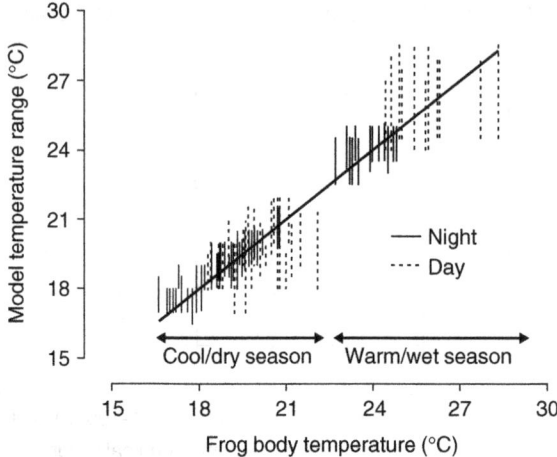

Fig. 21.5 Lines indicating the range of temperatures delineated by the maximum and minimum temperatures of all permeable and impermeable models at rainforest field site near Babinda, Queensland, Australia at the times body temperatures were measured for tracked *Litoria lesueuri* during the cool/dry and warm/wet seasons. Models were placed at known diurnal retreat sites of *L. lesueuri*. The line of equality is included to ease visualization of where cloacal temperature fell within each pair of model temperatures.

at night than their diurnal retreat sites are, and that models should be placed in nocturnal activity sites to more accurately delineate the environment experienced by frogs.

The physical models produced a relatively accurate outline of the temperature range into which the body temperatures of thermally equilibrated frogs fell, and the ways in which the model temperature envelope differed from the frog data were informative in themselves. This should hold true when this technique is used for any species of amphibian.

Our field validation is more rigorous than most that have been undertaken with reptiles (Dzialowski 2005). Many reptile studies evaluate the range of available temperatures by deliberately placing models to cover the maximum possible range of environmental temperatures (Dzialowski 2005). Had we done this, all of our field temperatures would have fallen within the range measured by models, as our laboratory temperatures did. This is not the main objective of our technique, which is intended to use models as a means of gaining more detailed information than is otherwise possible on the range of body temperatures a species is likely to experience, given its known behavior and microhabitat use. Measured body temperatures that fall outside the range experienced by models are not a failure of the models. Rather, they indicate that the species being studied uses a broader range of microenvironments than was provided by the locations in which models were placed.

21.2.2.5 The importance of matching the spectral properties of models to living animals

There has been substantial debate in the literature on the degree to which the spectral properties of models should match those of the species being modeled (reviewed by Dzialowski 2005). The color and therefore absorbence of many amphibians can alter when basking, or with thermal stress (Shoemaker *et al.* 1987; Snyder and Hammerson 1993; Withers 1995). For example, the total reflectance of *Litoria rubella* can range between approximately 19 and 32% (Withers 1995). This variation among and within individuals means that even a particular individual of many species cannot be precisely mimicked by a static model. We performed a short-term experiment to examine how color affected the thermal behavior of permeable and impermeable models of *L. caerulea*. We used green agar models with reflectance spectra that fell within the range of living individuals, constructed as described in section 21.2.2.3, and also constructed agar models with no added coloring and with 3 g/100 ml of carbon black added. Three replicate models of each type were placed, interspersed by color, in an open area during late summer, and their temperatures were measured over 24 h. Weather was largely clear with scattered clouds and no rain during the trial. We found that the relatively large difference in absorbance between green and clear agar models had only small effects on the daytime temperatures of impermeable models (Figure 21.6); green impermeable models were only slightly, and never significantly, warmer than colorless models. Even extreme differences in color had little effect on midday temperatures of permeable models (Figure 21.6). Although it is clearly best to produce models that approximate the overall absorbance of the species being examined, our results suggest that thermal envelopes estimated using models are likely to be reasonable approximations of the range of body temperatures available to amphibians, as long as obviously extreme colors such as carbon black are avoided.

21.2.2.6 Models as indicators of relative moisture availability of microenvironments

Although physical models have often been used to study rates of evaporative water loss under laboratory conditions, very few field studies have used them to

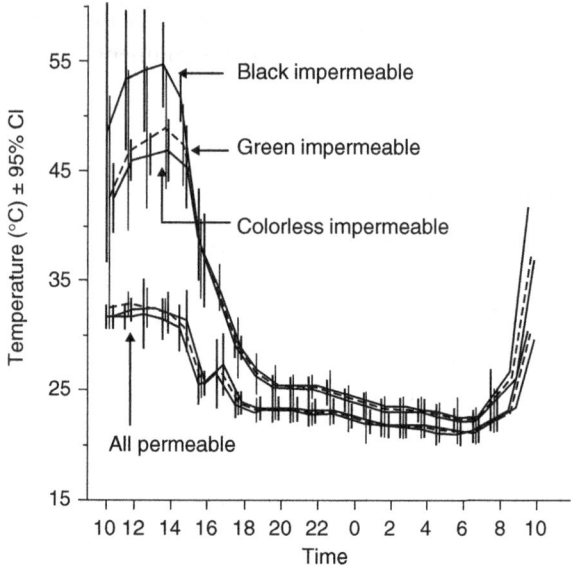

Fig. 21.6 Temperatures at hourly intervals for permeable and impermeable agar models of three colors exposed outdoors to ambient summer conditions. Vertical bars indicate 95% confidence intervals (CI) for the three replicate models of each color and permeability, lines connect the means for each color and type to aid in visualization of patterns. Because all permeable models performed almost identically while well hydrated, we have not attempted to make it possible to distinguish their lines, although the black models were slightly hotter. 95% Confidence interval bars were omitted for the last hour because steep slopes of curves made them unreadable. All were of magnitudes similar to those at hours 10–12 on the previous day.

assess the relative degree of water stress in a range of habitats or microhabitats, or across seasons. O'Connor and Tracy (1987) compared the water loss of wetted plaster models in different microhabitats to estimate water loss that would be experienced by *Rana pipiens*. Schwarzkopf and Alford (1996) placed agar models in retreat sites used by *Bufo marinus* to characterize the relative degree of water stress experienced by this introduced species in the Australian tropics. Their data strongly supported the hypothesis that seasonal shifts in retreat site use by *B. marinus* are related to water conservation. Seebacher and Alford (2002) obtained similar results using formalin-fixed *Bufo marinus* as models. Navas *et al.* (2002) compared water loss in frogs and in agar models under simulated field conditions, and Tracy *et al.* (2007) performed short-term comparisons of temperature and water loss between plaster models of *B. marinus* and living toads. It appears that no other published studies have used models extensively to examine the hydric conditions available to amphibians in the field.

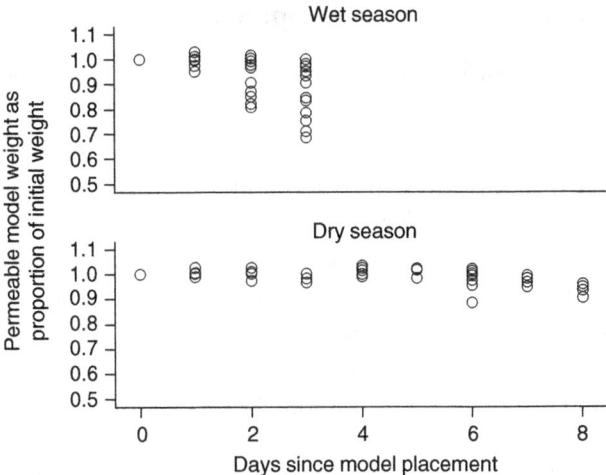

Fig. 21.7 Relative water loss by permeable agar models placed in diurnal retreat sites of *Litoria lesueuri* in rainforest near Babinda, Queensland, Australia. There was substantial cloud cover and rainfall during the winter, nominally dry season, sampling period (top), while the summer, nominally wet season, sampling period (bottom) was unusually dry. In both seasons, some retreat sites provided very low rates of evaporative water loss while some led to higher rates; in the summer, one of the models reached 50% water loss on day 4, causing data collection to cease.

Data collected using models to examine the envelope of thermal conditions available to amphibians can also be used to delineate an envelope of relative degrees of water stress. When agar models are used, the simplest approach is to weigh models daily to document cumulative change in water content. If automatically rehydrated models of other types, such as those developed by Bartelt and Peterson (2005) or Tracy *et al.* (2007) are used, cumulative water lost from reservoirs can be recorded manually or monitored automatically. Figure 21.7 presents data collected from the models of *L. lesueuri* placed in the field near Babinda, Queensland, as described in section 21.2.2.4. These data demonstrate that although the rainforest where the data were collected experiences annual wet and dry seasons, short-term weather patterns can result in greater opportunity for water stress in the wet season than in the dry season; this is probably compounded by the fact that the wet season occurs in the summer months, in which diurnal temperatures can be up to 10°C higher than in the winter dry season. They also indicate that the diurnal retreat sites that were monitored presented a wide range of potential levels of water stress. This is likely to affect the frogs' behavior and metabolism, and is also likely to affect their interactions with the amphibian chytrid fungus, which may reproduce more slowly on hosts with relatively dry skin surfaces (Pounds *et al.* 2006).

21.3 Summary and future developments

Models are useful for delineating the range of body temperatures and degrees of moisture stress available to amphibians in the field. Because numerous models can be placed in known retreat and activity sites, and temperatures and water loss can be recorded in models at frequent intervals, they can rapidly provide a detailed picture of the envelope of temperatures and moisture available to a species, without requiring that individuals be handled or otherwise disturbed. If a species is known to have zero resistance to EWL, permeable models should be sufficient to characterize its thermal and hydric environment. However, for species that have higher and usually variable levels of resistance to EWL, pairs of models should be employed to delineate the boundaries of the thermal envelope available.

Pairs of models cannot predict the exact body temperatures of specific individual frogs. Rates of EWL can fluctuate considerably over even short periods in individual frogs (McClanahan 1978; Withers *et al.* 1982; Wygoda 1989a; Withers 1995; Barbeau and Lillywhite 2005; Tracy *et al.* 2008). Although many frogs alter EWL with respect to temperature or humidity in a moderately predictable manner (Withers *et al.* 1982; Wygoda 1989a; Buttemer and Thomas 2003), in others, EWL fluctuates unpredictably (Barbeau and Lillywhite 2005). Because rates of EWL are so variable, measuring the limits of the envelope of body temperatures available to a species in a given microenvironment is likely to be more ecologically relevant to population-level questions for most species than precise measurements for particular individuals can be.

The physical models we have described are easy to construct in large numbers, and are also relatively inexpensive to produce, features that are highly desirable, since to truly characterize the range of conditions available to animals in the field it is necessary to sample a large number of locations (Dzialowski 2005). Our models are also compact and self-contained, and can therefore be placed in the exact shelter sites used by the study species without altering their structure, which might prove difficult with models of other designs (e.g. Bartelt and Peterson 2005; Tracy *et al.* 2007).

Using data obtained from models in the field in conjunction with the detailed knowledge of temporal patterns of microhabitat use made possible by advances in tracking technology, it will be possible to develop pictures of temporal patterns of body temperature experienced by frogs that are as detailed as those that have been available for some time for many reptiles. Variation in relative rates of water loss among sites and through time can also be determined. The knowledge that can be gained using models is likely to be increasingly useful in understanding the effects of climate change on amphibians. The majority of predictive models forecast a near-term global temperature rise of 1.5–4.5°C (Thomas *et al.* 2004), and these estimates have been used in predicting species responses (Thomas *et al.* 2004) despite their questionable relevance to micro-environmental conditions or body temperature (Kennedy 1997). Detailed species-specific data, as can be obtained using models, will therefore be essential for accurately predicting responses and designing management plans. This approach is also likely to be of use in developing more sophisticated models of energy use and growth within individuals, and thus of population dynamics, and in understanding the factors affecting host-pathogen interactions in nature (e.g. Woodhams *et al.* 2003).

21.4 References

Bakken, G. S. and Gates, D. M. (1975). Heat-transfer analysis of animals: some implications for field ecology, physiology, and evolution. In D. M. Gates and R. B. Schmerl (eds), *Perspectives in Biophysical Ecology*, pp. 255–90. Springer, New York.

Barbeau, T. R. and Lillywhite, H. B. (2005). Body wiping behaviors associated with cutaneous lipids in hylid tree frogs of Florida. *Journal of Experimental Biology*, **208**, 2147–56.

Bartelt, P. E. and Peterson, C. R. (2005). Physically modeling operative temperatures and evaporation rates in amphibians. *Journal of Thermal Biology*, **30**, 93–102.

Bradford, D. F. (1984). Temperature modulation in a high-elevation amphibian, *Rana muscosa*. *Copeia*, **1984**, 966–76.

Buttemer, W. A. (1990). Effect of temperature on evaporative water loss of the Australian tree frogs *Litoria caerulea* and *Litoria chloris*. *Physiological Zoology*, **63**, 1043–57.

Buttemer, W. A. and Thomas, C. (2003). Influence of temperature on evaporative water loss and cutaneous resistance to water vapour diffusion in the orange-thighed frog (*Litoria xanthomera*). *Australian Journal of Zoology*, **51**, 111–18.

Buttemer, W. A., van der Wielen, M., Dain, S., and Christy, M. (1996). Cutaneous properties of the Green and Golden Bell Frog *Litoria aurea*. *Australian Zoologist*, **30**, 134–8.

Christian, K. A. and Parry, D. (1997). Reduced rates of water loss and chemical properties of skin secretions of the frogs *Litoria caerulea* and *Cyclorana australis*. *Australian Journal of Zoology*, **45**, 13–20.

Dzialowski, E. M. (2005). Use of operative temperature and standard operative temperature models in thermal biology. *Journal of Thermal Biology*, **30**, 317–34.

Elliot, S. L., Blanford, S., and Thomas, M. B. (2002). Host-pathogen interactions in a varying environment: temperature, behavioural fever and fitness. *Proceedings of the Royal Society of London Series B Biological Sciences*, **269**, 1599–1607.

Hasegawa, M., Suzuki, Y., and Wada, S. (2005). Design and performance of a wet sponge model for amphibian thermal biology. *Current Herpetology*, **24**, 27–32.

Kennedy, A.D. (1997). Bridging the gap between general circulation model (GCM) output and biological microenvironments. *International Journal of Biometeorology*, **40**, 119–22.

Kobelt, F. and Linsenmair, K. E. (1992). Adaptations of the reed frog *Hyperolius viridiflavus* (Amphibia: Anura: Hyperoliidae) to its arid environment. *Journal of Comparative Physiology B*, **162**, 314–26.

Lillywhite, H. B. (2006). Water relations of tetrapod integument. *Journal of Experimental Biology*, **209**, 202–26.

Lips, K. R., Brem, F., Brenes, R., Reeve, J. D., Alford, R. A., Voyles, J., Carey, C., Livo, L., Pessier, A. P., and Collins, J. P. (2006). Emerging infectious disease and the loss of biodiversity in a Neotropical amphibian community. *Proceedings of the National Academy of Sciences USA*, **103**, 3165–70.

McClanahan, L. L. (1978). Skin lipids, water loss, and energy metabolism in a South American tree frog (*Phyllomedusa sauvagei*). *Physiological Zoology*, **51**, 179–87.

McClanahan, L. L. and Shoemaker, V. H. (1987). Behavior and thermal relations of the arboreal frog *Phyllomedusa sauvagei*. *National Geographic Research*, **3**, 11–21.

Navas, C. A. (1996). Implications of microhabitat selection and patterns of activity on the thermal ecology of high elevation neotropical anurans. *Oecologia*, **108**, 617–26.

Navas, C. A. and Araujo, C. (2000). The use of agar models to study amphibian thermal ecology. *Journal of Herpetology*, **34**, 330–4.

Navas, C. A., Jared, C., and Antoniazzi, M. M. (2002). Water economy in the casque-headed tree-frog *Corythomantis greeningi* (Hylidae): role of behaviour, skin, and skull skin co-ossification. *Journal of Zoology*, **257**, 525–32.

O'Connor, M. P. and Tracy, C. R. (1987). Thermal and hydric relations of leopard frogs in the field. *American Zoologist*, **27**, 118A.

Parmesan, C., Gaines, S., Gonzalez, L., Kaufman, D. M., Kingsolver, J., Peterson, A. T., and Sagarin, R. (2005). Empirical perspectives on species borders: from traditional biogeography to global change. *Oikos*, **108**, 58–75.

Pounds, J.A. , Bustamante, M. R., Coloma, L. A., Consuegra, J. A., Fogden, M. P. L., Foster, P. N., La Marca, E., Masters, K. L., Merino-Viteri, A., Puschendorf, R. *et al.* (2006). Widespread amphibian extinctions from epidemic disease driven by global warming. *Nature*, **439**, 161–7.

Rowley, J. J. L. and Alford, R. A. (2007a). Movement patterns and habitat use of rainforest stream frogs in northern Queensland, Australia: implications for extinction vulnerability. *Wildlife Research*, **34**, 371–8.

Rowley, J. J. L. and Alford, R. A. (2007b). Non-contact infrared thermometers can accurately measure amphibian body temperatures. *Herpetological Review*, **38**, 308–11.

Rowley, J. J. L. and Alford, R. A. (2007c). Behaviour of Australian rainforest stream frogs may affect the transmission of chytridiomycosis. *Diseases of Aquatic Organisms*, **77**, 1–9.

Schmuck, R., Kobelt, F., and Linsenmair, K. E. (1988). Adaptations of the reed frog *Hyperolius viridiflavus* (Amphibia, Anura, Hyperoliidae) to its arid environment: V. Iridophores and nitrogen metabolism. *Journal of Comparative Physiology B*, **158**, 537–46.

Schwarzkopf, L. and Alford, R. A. (1996). Desiccation and shelter-site use in a tropical amphibian: comparing toads with physical models. *Functional Ecology*, **10**, 193–200.

Seebacher, F. and Alford, R. A. (2002). Shelter microhabitats determine body temperature and dehydration rates of a terrestrial amphibian (*Bufo marinus*). *Journal of Herpetology*, **36**, 69–75.

Shoemaker, V. H. and McClanahan, L. L. (1975). Evaporative water loss, nitrogen excretion and osmoregulation in phyllomedusine frogs. *Journal of Comparative Physiology*, **100**, 331–45.

Shoemaker, V. H. and Nagy, K. A. (1977). Osmoregulation in amphibians and reptiles. *Annual Review of Physiology*, **39**, 449–71.

Shoemaker, V. H., McClanahan, L. L., Withers, P. C., Hillman, S. S., and Drewes, R. C. (1987). Thermoregulatory response to heat in the waterproof frogs *Phyllomedusa* and *Chiromantis*. *Physiological Zoology*, **60**, 365–72.

Snyder, G. K. and Hammerson, G. A. (1993). Interrelationships between water economy and thermoregulation in the canyon tree frog *Hyla arenicolor*. *Journal of Thermal Biology*, **25**, 321–9.

Spotila, J. R. and Berman, E. N. (1976). Determination of skin resistance and the role of the skin in controlling water loss in amphibians and reptiles. *Comparative Biochemistry and Physiology*, **55A**, 407–11.

Thomas, C. D., Cameron, A., Green, R. E., Bakkenes, M., Beaumont, L. J., Collingham, Y. C., Erasmus, B. F. N., de Siquera, M. F., Grainger, A., Hannah, L. *et al.* (2004). Extinction risk from climate change. *Nature*, **427**, 145–8.

Toledo, R. C. and Jared, C. (1993). Cutaneous adaptations to water balance in amphibians. *Comparative Biochemistry and Physiology*, **105A**, 593–608.

Tracy, C. R., Christian, K. A., O'Connor, M. P., and Tracy, C. R. (1993). Behavioral thermoregulation by *Bufo americanus*: the importance of the hydric environment. *Herpetologica*, **49**, 375–82.

Tracy, C. R., Betts, G., Tracy, C. R., and Christian, K. A. (2007). Plaster models to measure operative temperatures and evaporative water loss of amphibians. *Journal of Herpetology*, **41**, 597–603.

Tracy, C. R., Christian, K. A., Betts, G., and Tracy, C. R. (2008). Body temperature and resistance to evaporative water loss in tropical Australian frogs. *Comparative Biochemistry and Physiology*, **150A**, 102–8.

Withers, P. C. (1995). Evaporative water loss and colour change in the Australian desert tree frog *Litoria rubella* (Amphibia: Hylidae). *Records of the Western Australian Museum*, **17**, 277–81.

Withers, P. C., Hillman, S. S., Drewes, R. C., and Sokol, O. M. (1982). Water loss and nitrogen excretion in Sharp-nosed Reed frogs (*Hyperolius nasutus*: Anura, Hyperoliidae). *Journal of Experimental Biology*, **97**, 335–43.

Woodhams, D. C., Alford, R. A., and Marantelli, G. (2003). Emerging disease of amphibians cured by elevated body temperature. *Diseases of Aquatic Organisms*, **55**, 65–7.

Wygoda, M. L. (1984). Low cutaneous evaporative water loss in arboreal frogs. *Physiological Zoology*, **57**, 329–37.

Wygoda, M. L. (1989a). Body temperature in arboreal and nonarboreal anuran amphibians: differential effects of a small change in water vapour density. *Journal of Thermal Biology*, **14**, 239–42.

Wygoda, M. L. (1989b). A comparative study of heating rates in arboreal and nonarboreal frogs. *Journal of Herpetology*, **23**, 141–5.

Young, J. E., Christian, K. A., Donnellan, S., and Tracy, C. R. (2005). Comparative analysis of cutaneous evaporative water loss in frogs demonstrates correlation with ecological habits. *Physiological and Biochemical Zoology*, **78**, 847–56.

Genetics in field ecology and conservation

Trevor J. C. Beebee

22.1 Background: the importance of genetics in ecology and conservation

The genetic analysis of wild populations has become increasingly popular with the advent of techniques based on large-scale and non-destructive sampling. Nevertheless, using molecular methods remains one of the most expensive and time-consuming aspects of ecological research. It is therefore important to consider carefully when it is useful to acquire genetic data. Certainly there are such circumstances for amphibian researchers (Beebee 2005). Areas in which genetics can be invaluable include the following.

1) Identification of cryptic species or life stages. There are many examples among the amphibia of difficulties with identification based on morphology alone. This can be true for adults, but is a more common problem with larvae. Molecular genetic markers permit species identification in virtually all situations.

2) Measuring genetic diversity, and detecting inbreeding and genetic load effects. Small and isolated populations are vulnerable to the accumulation of mildly deleterious alleles. These decrease the fitness of individuals, and eventually the viability of the population, contributing to a potential "extinction vortex" in combination with other factors such as demographic and environmental stochasticity. Genetic methods can provide useful indicators of this problem, potentially leading to genetic restoration. They can also estimate the "effective size" of a population (roughly the number of successful breeders, and usually smaller than the census size), which is what matters in risk assessment.

3) Identification of barriers to individual movement. Amphibians are, in general, poor dispersers but long-term population viability may often depend

on occasional movements of animals between distant breeding sites. Such connectivity maintains a large effective population size over wide areas, but can be difficult to demonstrate when individual movements are rare. Genetic techniques can track and quantify such movements indirectly.

4) Defining populations. Knowing the extent and size of individual populations is important for demographic studies, and for conservation management. Assemblages in ponds have commonly been used as surrogates for "populations" of pond-breeding amphibians but this is often too simplistic. Many species exist as metapopulations with individuals moving regularly between neighbouring ponds. Genetic analysis can refine population size and boundary information.

5) Historical issues. Genetic data can be used to reveal the sources of alien introductions, and to show whether a population of uncertain origin is recently introduced, perhaps warranting eradication, or is truly native and thus meriting conservation.

6) Behavior and sexual selection. High-resolution molecular markers now permit the determination of individual DNA profiles. This means that issues such as parentage, and thus individual reproductive success, can be determined and correlated with features such as body size and condition.

In the following sections, I describe how genetic data can be obtained from wild amphibian populations, and how they can be analysed to address the questions outlined above. More detailed accounts can be found in Plötner *et al.* (2007) and Vences and Wake (2007). The overwhelming majority of genetic studies on wildlife populations thus far have used "neutral" markers; that is, changes in DNA sequences that are invisible to natural selection. These types of marker perform well for addressing most of the issues listed above, although there is a serious question about their application to inbreeding depression and genetic load. In the final section I consider likely future developments, including the identification of genetic markers under the influence of selection for investigating population viability.

22.2 Molecular methods for investigating amphibian populations

22.2.1 Laboratory facilities

There is a substantial amount of core equipment necessary for any genetic study. This includes: weighing balances; pipettes; plasticware including disposable pipette tips, and microcentrifuge and polymerase chain reaction (PCR) tubes;

fridges and freezers; water baths; microcentrifuges; vortex mixers; a vacuum desiccator; PCR machines; gel electrophoresis rigs; and either an automated DNA sequencer or high-voltage electrophoresis equipment to run sequencing gels, together with X-ray development or phosphorimager facilities. Also very useful are ultraviolet (UV) spectrophotometers for quantifying DNA yields, UV photographic equipment for identifying DNA bands on gels, microwave ovens, and autoclaves (although a pressure cooker will often suffice!). For more sophisticated work, such as the generation of DNA libraries for isolating micro-satellite loci, microbiology facilities are also necessary, including sterile hoods and incubators. These facilities are often not readily available to ecologists, so there is as strong case for collaborations with suitably equipped laboratories when contemplating genetic studies. All laboratory work should be carried out using high standards of cleanliness, and especially with the use of disposable gloves (non-latex when handling live animals) for all procedures.

22.2.2 Sampling

All the genetic methods described below are based on PCR amplification of very small amounts of DNA. Thus tiny amounts of tissue suffice for analysis, and sacrifice of individuals is not necessary. There are several ways of obtaining tissue samples for DNA extraction. Small (2–3 mm^2) segments of larval tailfins or tail tips can be excised with a sharp scalpel; terminal digits can be removed with sharp scissors or buccal swabs can be taken from immature or adult amphibians (Pidancier *et al.* 2003). None of these procedures are likely to cause significant mortality. It is important to ensure that sampling is representative of a population. Taking all the samples from one aggregation of tadpoles, for example, may bias in favour of a few sibships.

Each tissue sample should be preserved immediately in a small volume ($\approx$1 ml) of at least 70% ethanol, in sealed tubes, and returned to the laboratory. DNA is stable in alcohol for many months at field or room temperature. Buccal swabs (e.g. from Epicentre Biotechnologies, Madison, WI, USA) are an exception, and should be treated as per the manufacturer's instructions. With future statistical analysis in mind, it is desirable to sample at least 20 individuals per population.

22.2.3 DNA extraction

Several methods are available for rapid extraction of DNA from multiple small tissue samples. This initial step is important, because the PCR requires high-quality DNA with minimal contamination. Buccal swabs come with extraction solutions and method details. Four other techniques are outlined below, followed by a summary of their relative merits.

22.2.3.1 Proteinase K/phenol/chloroform method

Tissue samples are digested overnight at high temperature (> 50°C) in the presence of the enzyme proteinase K, conditions that leave DNA intact. The mixture is then extracted with organic solvents (phenol and chloroform) that precipitate residual proteins and other cell debris, leaving DNA in solution above the organic layer after a brief centrifugation. The DNA is then concentrated by precipitation using ethanol, and obtained as a pellet following centrifugation. After drying, the DNA is redissolved in less than 50 μl of sterile distilled water and stored (always at −20°C).

22.2.3.2 Kit-based DNA extraction

Many companies (e.g. Qiagen, Crawley, West Sussex, UK) produce kits designed for efficient DNA extraction from multiple samples. These kits include all the buffer solutions required, and mini-columns with a silica-based plug that selectively binds DNA. Detailed protocols are provided, and DNA is recovered from the columns in a final elution step.

22.2.3.3 Chelex method

Chelex-100 is an ion-exchange resin marketed by several chemical companies. It forms a suspension in distilled water, and when tissue is added and incubated overnight (at > 50°C) DNA is released into solution. A final boiling step is necessary to complete the process, followed by a brief centrifugation. The supernatant above the resin pellet is stored as a source of DNA (Walsh *et al.* 1991).

22.2.3.4 FTA cards

A recent development is the use of FTA cards (Whatman, Maidstone, Kent, UK), a substrate onto which tissue extracts (including buccal swabs) can be blotted directly and which can then store DNA without degradation for months or years. Small sections of the card are removed using a punch, and after brief washes can be used directly in PCR amplifications. Several hundred amplifications can be made from half an FTA card, and subsequent genotyping is very reliable (Livia *et al.* 2006).

22.2.3.5 Summary of DNA-extraction methods

The first two DNA extraction methods discussed above generate high-quality native (double-stranded) DNA suitable for virtually all types of genetic analysis. However, care must be taken with the use of the organic solvents, which are potentially toxic. Kits are more expensive than the phenol/chloroform method, and tend to produce lower yields. However, extraction time is shorter with a kit and there is less risk of contamination with substances that can interfere with subsequent PCR amplifications. Some amphibians have high concentrations of tissue pigments, and these are not always removed by the phenol/chloroform approach. The Chelex method is attractively cheap and simple, and contamination problems are rare. However, Chelex extraction produces denatured (single-stranded), not native, DNA. This makes it unsuitable for techniques such as amplified fragment length polymorphism (AFLP; see below) that absolutely require native DNA. FTA cards may also be unsuitable for AFLP analysis, where DNA needs to be digested with restriction enzymes prior to PCR amplification.

22.2.4 Quantifying DNA recoveries

It is important to estimate DNA recoveries from the tissue samples to ensure that yields are both sufficient and consistent. This is best done using a UV spectrophotometer because DNA absorbs light strongly at 260 nm. Native DNA has an absorbance of 1.0 at 50 μg/ml, whereas denatured DNA and oligonucleotides at this concentration have an absorbance of about 1.25.

22.2.5 The basis of PCR analysis

The PCR is fundamental to the molecular genetic methods relevant to ecology and conservation. It is based on the exponential increase of a particular DNA sequence that can be obtained in just a few hours using a series of denaturation, priming, and polymerization cycles in a programmable laboratory machine. In principle, a single DNA molecule can give rise to 1 billion copies (more than enough for any analysis) in just 30 cycles of amplification. In each cycle, a denaturation step separates the two DNA strands, at over 90°C; a priming step allows specially designed oligonucleotide primers, typically 18–25 nucleotides long and complementary in base sequence to flanking sections of the piece of DNA, to anneal to the separated strands, usually at 50–60°C; and a polymerization step uses a heat-stable (*Taq*) DNA polymerase and nucleotide substrates to synthesise new strands of DNA between the priming sites, starting at the inward-facing ends of each primer, at 70–72°C. The ability to amplify such tiny amounts of starting material permits nondestructive sampling, but also renders the process vulnerable to traces of DNA contamination from unexpected sources (such as shed human skin cells), a problem that must always be tested for by using blanks with no added DNA. PCR machines can process up to 100 samples simultaneously, and are thus well suited to population studies.

22.2.6 Choice of analytical methods

Given a good batch of DNA samples, what are the options for genetic analysis? A lot depends on the problem under consideration but there are nevertheless only a small number of common marker types from which to choose (Beebee and Rowe 2008).

22.2.6.1 Mitochondrial DNA (mtDNA)

Mitochondria contain multiple copies of a single, circular DNA chromosome. The structure of this chromosome is highly conserved in vertebrates, and all have the same set of 13 protein-encoding genes, together with genes coding for ribosomal and transfer RNAs, and a control region (or CR) or D-loop. The control region is concerned with the regulation of RNA and DNA synthesis. mtDNA has a mutation rate of around tenfold higher than nuclear protein-coding genes, with the result that genetic diversity accrues relatively rapidly. It does not undergo meiosis, so sequence changes can be tracked over time as distinct lineages unscrambled by recombination. Inheritance is purely maternal because although mtDNA occurs in sperm, it does not survive fertilization. Finally, because most cells have many mitochondria, each with many copies of mtDNA, there are typically hundreds or thousands of mtDNA molecules per diploid cell compared with just two copies of each nuclear chromosome. This means that mtDNA is relatively easy to detect in old or partly degraded tissue samples such as museum specimens, despite constituting less than 1% of a cell's total DNA. In a typical mtDNA analysis, oligonucleotide primers complementary to parts of a mtDNA gene are used in PCRs to generate partial gene fragments. These are then purified (commercial kits are available to clean up PCR products) and sequenced. Sequencing can be done in-house or sent to one of the many companies that provide this service at low cost. The sequences are then compared for polymorphisms among many different individuals, each of which will carry only a single mtDNA sequence, referred to as its mtDNA haplotype.

mtDNA is particularly useful for phylogeographic studies, tracing the origin of populations over periods of up to a few million years. mtDNA has, for example, provided a mass of evidence concerning postglacial colonization routes of amphibians in the northern hemisphere over the past 20 000 years. However, it is usually less powerful than the nuclear DNA markers now in widespread use for fine-scale studies of population structure. Although high by comparison with nuclear protein-coding genes, the mtDNA mutation rate is often still too low to generate enough diversity for inter-population analysis. Exceptions are places where species have persisted for a long time, such as glacial refugia in southern Europe, or in the tropics. A particular advantage in the use of mtDNA is the existence of conserved (or universal) primers for PCR reactions that function with many different species, particularly for the cytochrome b (*cyt*b) gene. This means that study of a new species can be initiated with minimal lead-in time for the development of genetic markers. A disadvantage is that due to its lack of recombination mtDNA can only be considered as a single locus despite the multiplicity of genes that it bears. Inferences about population structure from a single locus, particularly one that relies solely on maternal inheritance, are questionable because they may not accurately reflect events across the genome as a whole.

22.2.6.2 *Random amplification of polymorphic DNA (RAPD) and amplified fragment length polymorphism (AFLP) analyses*

Both of these techniques aim to analyse up to several hundred nuclear DNA loci simultaneously (a locus is any specific sequence of DNA, whether a functional gene or not). Both RAPD and AFLP are "anonymous" because the genome is sampled randomly with no knowledge of the sequences that ultimately turn up. Both methods have the advantage that no DNA sequence information is required from the species under study before starting the investigation. The techniques use oligonucleotide primers that can be purchased cheaply from biotechnology companies.

22.2.6.2.1 *RAPD analysis*

Each PCR reaction includes a single, random-sequence primer just 10 nucleotides in length. Any 10-nucleotide sequence should occur by chance, on average, once every million base pairs along a DNA molecule. RAPD PCR produces a product when two such sequences happen to occur in facing orientation and relatively close to each other. These PCR products are then visualized as bands on agarose gels following electrophoresis, after staining with ethidium bromide (caution is required, as this is a mutagenic chemical), followed by photography using a UV light source. More sophisticated analytical procedures are also possible (see AFLP section below). A primer may generate anything from no bands to more than 20, and most of them will be the same for every individual, and thus uninformative. It is normal practice to screen initially with many different primers (maybe 50 or more) to identify those generating a useful number of polymorphisms (i.e. those that produce bands in some individuals but not others).

Despite the potentially tedious initial screening of multiple primers, this is the simplest and cheapest type of DNA-based genetic analysis. However, for consistent results it is vital to use the same reagents (such as the same manufacturer's *Taq* polymerase), the same PCR machine, and especially the same amount of DNA (≈25 ng) in all analyses. Because of reproducibility problems, some high-profile journals such as *Molecular Ecology* usually reject population studies based solely on RAPD analyses. However, RAPDs remain an excellent approach to interspecific problems such as the identification of cryptic species and life stages.

22.2.6.2.2 AFLP analysis

In this method, DNA samples are first digested with two restriction endonucleases that cut at different, specific sequences (Figure 22.1). Short, double-stranded DNA linkers of known sequence are then ligated onto the ends of the DNA fragments, and after that the DNA is amplified by PCR using primers partly complementary to the linkers, but with one to three random nucleotide additions at the internal ends (i.e. overlapping the DNA fragments between the linkers). This means that only a subset of the DNA fragments, those where the sequences happen to complement the primer ends, will be amplified. Like RAPDs, this procedure generates a large number of fragments. Polyacrylamide gels (with higher resolving power than agarose) are usually used to separate them, and the PCR products are labelled (e.g. with a radionuclide containing the ^{33}P isotope for detection by autoradiography, or fluorescently for analysis using an automated DNA sequencer). All this is more complicated than RAPDs and, because of the need for restriction endonuclease digestion, more DNA is required. However, a major advantage of the technique, relative to RAPDs, is its reproducibility. AFLPs are widely used for intraspecific population studies and have been employed successfully with amphibians. Some investigations including one with the alpine salamander *Salamandra atra* have indicated that they may not reveal high-resolution population structure (Riberon *et al.* 2004), but it is commonplace for AFLP analysis to generate several hundred polymorphic markers, and thus be broadly representative of the genome.

22.2.6.2.3 Microsatellite analysis

Microsatellites consist of di-, tri-, or tetranucleotides repeated multiple times at a particular locus (e.g. GTGTGTGTGT, GTCGTCGTC, GTACGTACGTAC). Most vertebrate genomes have hundreds or thousands of microsatellite loci spread across their nuclear chromosomes, and almost all are in non-coding (i.e. neutral) regions. Microsatellites have even higher mutation rates than mtDNA, and largely for this reason have become the marker of choice for most population studies (Jehle and Arntzen 2002). Microsatellite loci are amplified by the PCR using primers complementary to their flanking regions. Alleles differ in the number of repeats, and thus in size, and can be separated and identified by

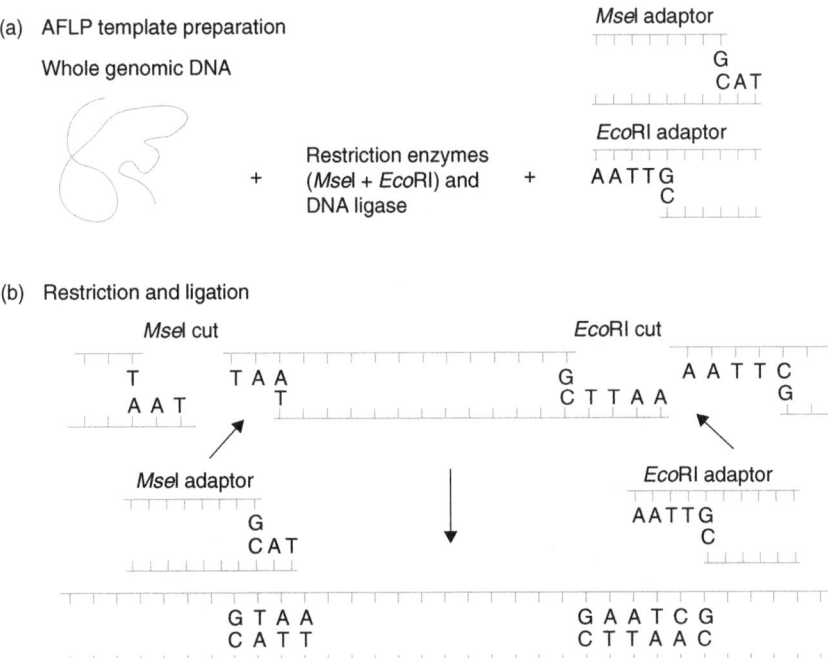

(a) AFLP template preparation

Whole genomic DNA

Restriction enzymes
+ (*Mse*I + *Eco*RI) and +
DNA ligase

*Mse*I adaptor

*Eco*RI adaptor

(b) Restriction and ligation

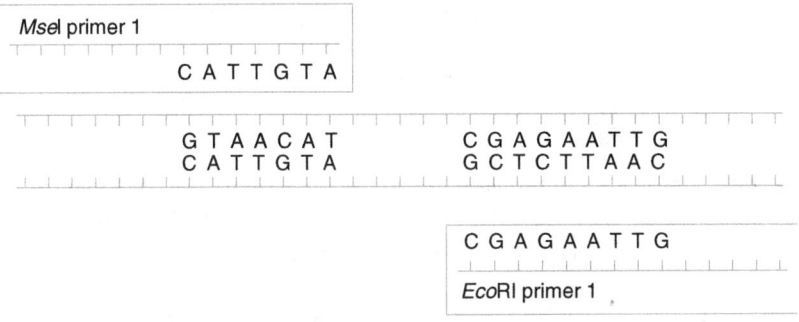

(c) Selective amplification (one of many primer combinations shown)

Fig. 22.1 Basis of AFLP analysis using the restriction enzymes *Mse*I and *Eco*RI.

electrophoresis (Figure 22.2, natterjack toads) or on automated DNA sequencers. It is often possible to "multiplex" microsatellite loci, meaning to amplify several at a time in the same reaction tube, using primers with different fluorescent labels.

The main problem with microsatellites is characterizing them for a new species. To design primers it is necessary to know the flanking sequences, and particularly in the amphibia these seem to be poorly conserved even among species

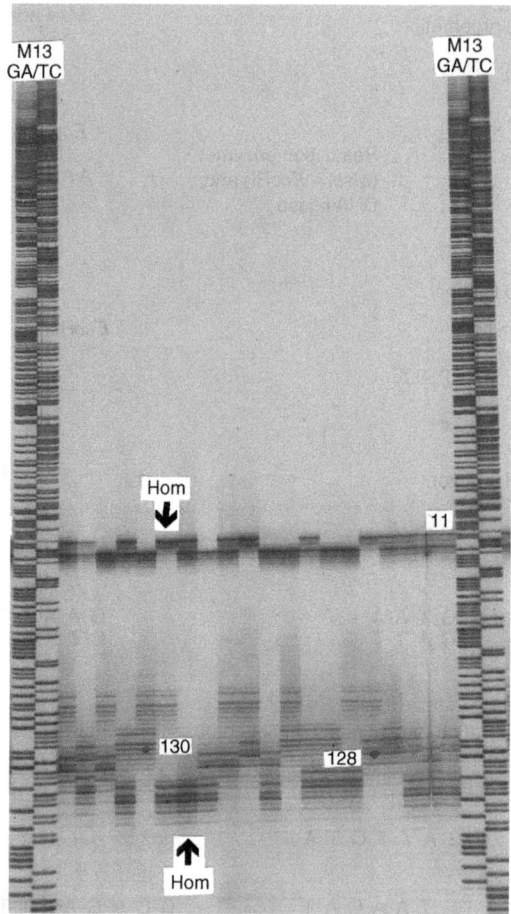

Fig. 22.2 Gel analysis of two microsatellite loci in 20 individuals. M13 GA/TC are size markers, Hom show examples of homozygotes. Lanes with two band sets are heterozygotes. Other numbers are reference samples.

in the same genus (Primmer and Merila 2002). It is nevertheless always worth trying primers from a previously characterized close relative if these are available (*Molecular Ecology Resources* and *Conservation Genetics* publish details for dozens of new species every year), but often it is necessary to start from scratch. This means creating a genomic library enriched for microsatellite sequences by selective hybridization, screening clones for microsatellites, sequencing positive clones, and then designing primers from the flanking sequence information (Zane *et al.* 2002). All of this is expensive and time-consuming. The rewards, however, can be great. Both RAPD and AFLP analyses generate "dominant" markers, in which heterozygotes cannot be distinguished from homozygotes.

Microsatellites, however, are "codominant" markers because, due to differing allele sizes, homozygotes and heterozygotes are readily distinguished (Figure 22.2). This provides very valuable extra information. Typically, seven or eight polymorphic microsatellite loci are sufficient for basic fine-scale analysis of population structure, and to identify individual animals with negligible ($< 10^{-10}$) probabilities of mistakes.

22.3 Analysis of genetic data

The results of genotype analyses are usually in one of three forms: DNA sequences (e.g. mtDNA haplotypes), binary (presence or absence of a band for dominant RAPD or AFLP data), or allele composition (such as 120.160, the sizes of two alleles in a microsatellite heterozygote). Many computer programs are available as freeware for analyses of all these data types.

22.3.1 Cryptic species or life-stage identification

This usually just requires visual inspection of gel photographs following RAPD analyses. The banding patterns (RAPD "phenotypes") should be distinct and reproducible for each species, and consistent across multiple widely separated populations. There have been numerous RAPD identification successes with amphibians (e.g. Bardsley *et al.* 1998; Mikulicek and Pialek 2003).

22.3.2 Genetic diversity, inbreeding, and bottlenecks

Analysis of mtDNA sequences involves establishing the extent of change between haplotypes, as the numbers of mutational events that have led (usually) to single nucleotide polymorphisms (SNPs). Software packages such as DNASTAR and T-COFFEE align multiple DNA sequences and highlight SNPs. ARLEQUIN will use the full sets of haplotypes from different population samples and quantify genetic diversity in each population, in terms of both haplotype diversity and nucleotide diversity. The two measures are not identical, because haplotypes can vary by more than one SNP. For dominant data such as RAPDs and AFLPs, genetic diversity is commonly estimated as gene diversity, defined as $1 - \Sigma(P_i)^2$, where P_i is the frequency of the ith allele, assuming a biallelic state for each locus and a state of Hardy–Weinberg equilibrium in the population. Most commonly, however, genetic diversity is estimated from codominant data such as those provided by microsatellites. Many programs, such as GENEPOP and FSTAT, calculate heterozygosities, allelic richness (mean numbers of alleles per locus), and percentage of loci that are polymorphic in a population. It is important to test (the software provides methods) for concordance with Hardy–Weinberg equilibrium, and to ensure that the loci are in linkage equilibrium. Microsatellites often have tens of alleles, and mean heterozygosity estimates across loci in large populations are commonly more than 0.6.

Low levels of genetic diversity are possible indicators of inbreeding depression or high genetic load. This was true in a study of natterjack toads *Bufo calamita* (Rowe and Beebee 2003). However, there are important caveats on this interpretation. Firstly, the estimates must be low relative to other populations, *for exactly the same suite of loci*. Second, low genetic diversity at neutral loci can arise for various reasons apart from inbreeding, such as proximity to a range edge after postglacial colonization, as with Italian agile frogs *Rana latastei* (Ficoleta *et al.* 2007). Such populations were fully viable, whereas isolated populations with equally low genetic diversity showed fitness reductions (Figure 22.3). It is therefore important to test directly for low fitness in suspect populations (for example by measuring embryonic or larval survival and growth rates) rather than relying solely on genetic diversity measures. However, another test can be applied to any population with data from multiple codominant loci. When population size declines rapidly, allelic richness is lost much faster than average heterozygosity. BOTTLENECK quantifies this "heterozygote excess," which is transient but extends over several generations during and after a decline (Luikart *et al.* 1998). This approach accurately identified known recent population crashes in natterjack toads, *B. calamita* (Beebee and Rowe 2001).

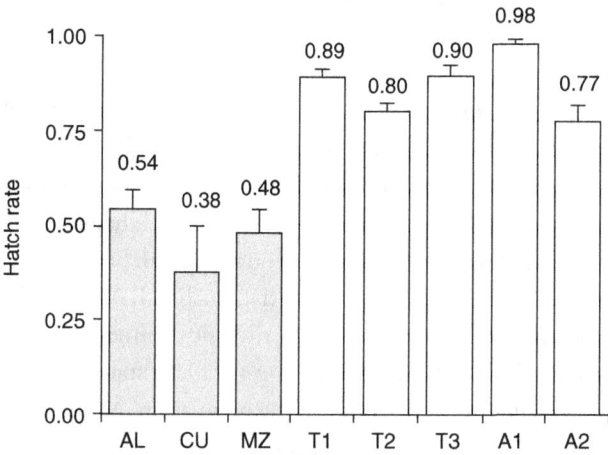

Fig. 22.3 Low fitness of isolated *Rana latastei* populations in Italy. Three isolated populations (> 3 km from the nearest neighbor) are shown as gray bars, and five non-isolated populations as white bars. Standard errors are shown, based on more than 25 egg clutches per population.

22.3.3 Identification of barriers to movement

A powerful use of genetic data is to identify barriers limiting, or corridors facilitating, the migration of animals. There are two main ways of doing this. Programs such as ARLEQUIN, GENEPOP, and FSTAT estimate F statistics from any type of genetic data, especially F_{ST}, which is a measure of genetic divergence between two populations. F_{ST} varies from 0 (no differentiation, the two "populations" are really just one) to 1 (complete distinction). F_{ST} values over 0.2 imply such low gene flow that random genetic drift will dominate over migration, and the populations will become increasingly divergent over time. More precise interpretations of migrant numbers are now shunned because they make unverifiable assumptions (Whitlock and McCauley 1999). Nevertheless, F statistics are still useful, especially in multiple pairwise comparisons between populations. Isolation by distance (or IBD), a significant correlation between geographical distance and F_{ST}, implies regular historical gene flow (i.e. migration) between populations, and unusually high F_{ST} suggests the existence of barriers to movement. Many programs such as GENEPOP have algorithms for assessing isolation by distance. Gene flow and isolation by distance in amphibians based on F statistics have been widely reported, for example with mtDNA in the case of the Yosemite toad *Bufo canorus* in California (Shaffer *et al.* 2000) and with microsatellites in the Cascades frog *Rana cascadae* (Monsen and Blouin 2004).

A problem with F_{ST} is that it assesses movement as an average over some unknown past time period. Assignment methods based on multilocus genetic data give a more precise and contemporary indication of gene flow. These cluster genotypes into populations, and then highlight genotypes which were sampled in the "wrong" place (i.e. which genetically belonged to one population, but were found in another). Such methods that can identify individual migrants are computationally intensive but very powerful. STRUCTURE and BAPS are perhaps the best known. It is even possible to detect, using BAYESASSNM, the first-generation progeny of migrants by genotyping amphibian larvae, as shown with crested and marbled newts *Triturus cristatus* and *Triturus marmoratus* (Jehle *et al.* 2005a). There are many published studies of amphibian population structures, mostly based on microsatellite analyses, from all over the world. Examples include North American wood frogs, *Rana sylvatica* (Newman and Squire 2001), European crested newts (Jehle *et al.* 2005b), and Australian tree-frogs, *Litoria aurea* (Burns *et al.* 2004).

Landscape genetics defines the spatial arrangement of populations using genetic data and identifies both impermeable habitats (barriers) and permeable ones (corridors) between population clusters. A study with the Columbia spotted frog

Rana luteiventris exemplifies this approach (Funk *et al.* 2005). Programs such as GENELAND explicitly combine both genetic and geographical information for landscape studies. Figure 22.4 shows an example of population differentiation in the long-toed salamander *Ambystoma macrodactylum*, determined using seven microsatellite loci, mediated primarily by altitude in Montana, USA (Giordano *et al.* 2007). Relevance to conservation is exemplified by a study of eastern red-backed salamander (*Plethodon cinereus*) populations, which were much more fragmented in semi-urban than in pristine habitats in Canada (Noel *et al.* 2007). This approach can therefore reveal the significance of urban development as disrupters of population connectivity.

22.3.4 Defining populations and population size

Software such as STRUCTURE, BAPS, and GENELAND can test whether genetic clustering corresponds to individual aggregations, or to groups of them. STRUCTURE, for example, can take a full genotype data set from multiple sampling sites and make no initial assumptions about how many populations the data represent. It can then estimate the number of "true" (genetically distinct) populations and list the individuals that constitute each, as with natterjack toads in Britain (Rowe and Beebee 2007). This requires a lot of genetic information; preferably from more than 10 highly polymorphic loci, and more than 30 individuals per sampling site. Genetic data can also be

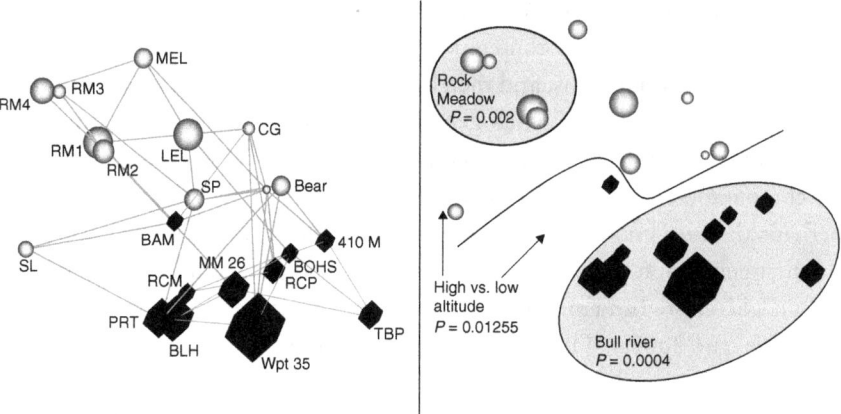

Fig. 22.4 Population genetic structure in long-toed salamanders. Spheres are high-altitude sites, cubes are low-altitude sites, and sizes are proportional to intrapopulation microsatellite variation. Connection lengths are proportional to site similarity, and the probabilities of genetic distinction between low- and high-altitude groups, as well as the distinctiveness of two main clusters, are also shown.

used to estimate effective population size, N_e, once populations are defined. N_e is virtually always lower than census size N, on average by a factor of 10 in wildlife populations. N_e can be estimated most accurately by two samplings of a population, preferably separated by an interval of several generations (so >5 years for most amphibians), followed by genotyping each sample set across multiple loci. The larger the change in allele frequencies, the greater has been the effect of genetic drift, and thus the smaller is the N_e. Programs such as NeESTIMATOR and TM3 are available for these calculations. An alternative is to estimate N_b, the numbers of successfully breeding adults in a single generation, by collecting genetic data from parents and offspring in the same year. This is more convenient than waiting for several generations, but has large error limits especially when the parental sample includes more than one age cohort. Most estimates of N_e in amphibian populations have been low, often in the tens or hundreds (Beebee and Griffiths 2005).

22.3.5 Historical issues

To show the relationships between population clusters, tree-building programs such as PHYLIP or PAUP are required. Statistical rigour is provided by bootstrapping, which involve random resampling of the primary data set to put percentage confidence estimates on each tree branch. Trees are central to *phylogeography* (the application of intraspecific phylogenetics), and provide valuable insights into population origins and histories. They can be generated in a variety of ways using any type of genetic data. However, a further level of analysis is possible with mtDNA sequences. Haplotype sequences can be connected as networks, showing in detail their likely evolutionary histories. These in turn can be subject to nested clade analysis to infer historical events such as range or population expansion (although the reliability of nested clade analysis is controversial), using software packages such as TCS and GEODIS. The Miocene and Pleistocene histories of several amphibian species have been unravelled using these techniques, including (for example) *Rana arvalis* in Europe (Babik *et al.* 2004), *Bufo fowleri* in North America (Smith and Green 2004), and the origins of tropical amphibian distributions such as *Dendrobates tinctorius* (Noonan and Gaucher 2006) in South America and *Xenopus* species in Africa (Evans *et al.* 2004). The tailed frog *Ascaphus* provides an example (Figure 22.5) of how phylogeography has improved understanding of Pacific northwest biogeography. Inland and coastal populations were separated several million years ago into two distinct species (*Ascaphus montanus* and *Ascaphus truei*), whereas internal differences among populations in each region were probably generated in refugia during the Pleistocene glaciations (Nielson *et al.* 2001).

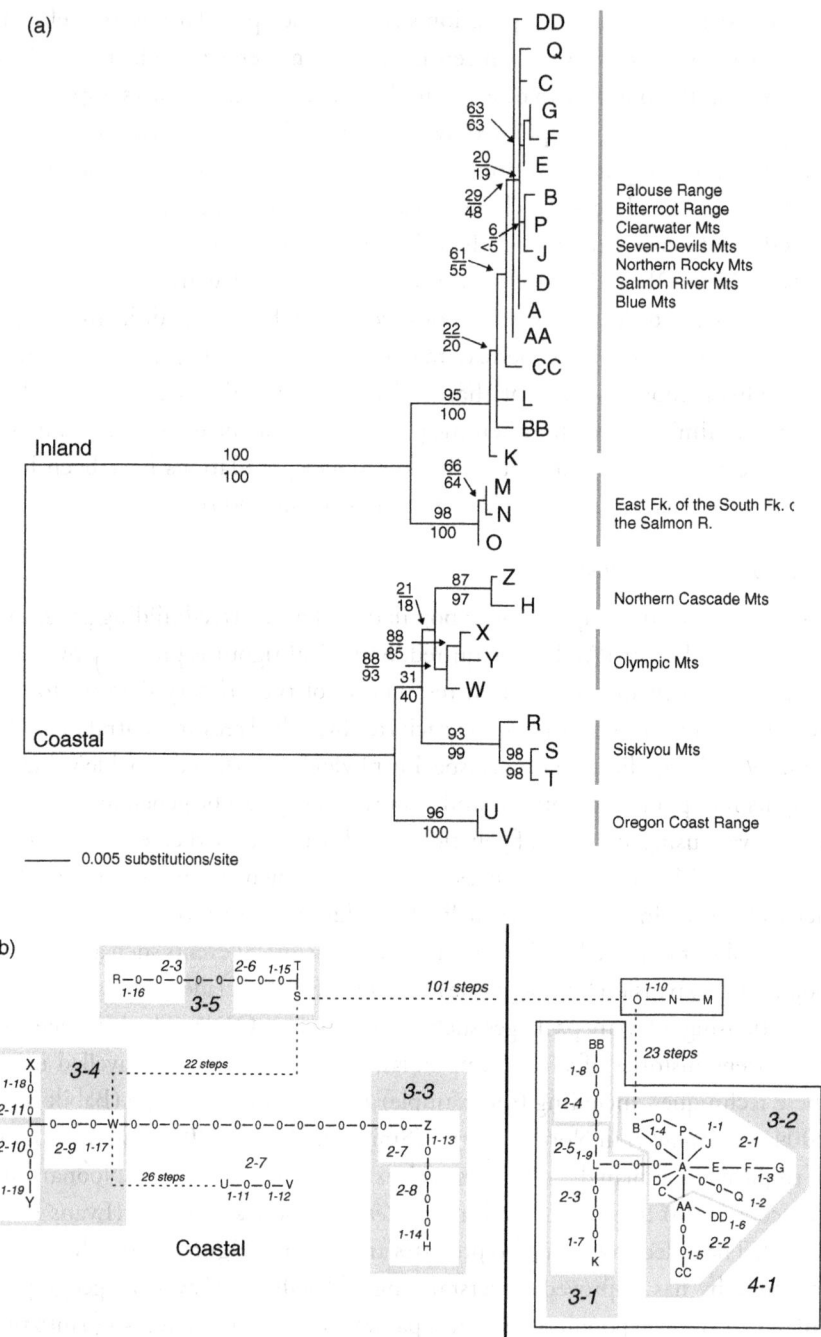

Fig. 22.5 Relationships among tailed frog populations, based on mtDNA. (a) Phylogenetic tree. (b) Haplotype network in nested clades.

Phylogeography has also proved useful when investigating the origins of introduced species, such as marine toads *Bufo marinus* in Australia (Estoup *et al.* 2001). This approach also demonstrated that the pool frog *Rana lessonae* was a longstanding native of Britain rather than a nineteenth century introduction as previously believed (Zeisset and Beebee 2001).

22.3.6 Behavior and sexual selection

Work in this area requires genotype information that can identify individuals with minimal error rates. Suites of highly polymorphic microsatellite loci are the most commonly used to generate unique DNA fingerprints (or profiles). It is then possible to investigate multiple paternity and individual reproductive success. Molecular markers demonstrated, for example, multiple paternity as a result of "clutch piracy" in the European common frog *Rana temporaria* (Vieites *et al.* 2004). In the salamander *Plethodon cinereus* multiple paternity occurs even in species that seems socially monogamous (Liebgold *et al.* 2006). Computer programs such as GENECLASS, KINSHIP, and CERVUS carry out such inferences.

22.4 Future developments

The range of neutral molecular markers available for genetic studies on wild populations has increased dramatically in recent decades. What will happen next? Molecular ecology is expensive, and perhaps the greatest hope is that costs will continue to decline as the necessary technology and consumables become more widely available. Nevertheless, collaboration between ecologists, conservation biologists, and molecular biologists will remain essential for the foreseeable future. Although microsatellites seem likely to dominate the neutral-marker field for some time, other possibilities are on the horizon. SNP analysis of nuclear DNA sequences, for example, has some promise. For adequate power this requires 100 loci or more, but high-throughput genotyping makes nuclear SNP studies increasingly feasible (Morin *et al.* 2004). The main advantage is that the mutation mechanism for SNPs is simpler than that for microsatellites, permitting more rigorous statistical approaches to data analysis.

However, it is in the field of adaptive rather than neutral variation that we can expect to see the most exciting new developments. It is increasingly important to assess genetic loci directly related to fitness, rather than relying on the surrogacy of neutral loci. There are already several ways of doing this, some of which are non-molecular. It is possible to carry out controlled crosses in the laboratory and, with half-sib experimental designs, to relate the fitness of offspring to parents from different source populations. The genetic and environmental contributions to critical attributes such as larval growth rates and size at metamorphosis can be determined in this way. There are also statistical approaches for estimating population differentiation based on such quantitative traits, using Q_{ST} as an analogue of F_{ST}. Comparisons of these two statistics can be particularly informative. Where $Q_{ST} > F_{ST}$, the implication is that selection rather than random drift dominates any genetic differentiation (and vice versa). Work along these lines has been very informative with *R. temporaria* (Laugen *et al.* 2005)

Breeding experiments are time-consuming and, like molecular methods, require specialized laboratory facilities. It can also be problematic to work on a large scale, and sacrifice of animals (especially males to provide sperm) is often necessary. There are, however, also non-destructive molecular approaches to the study of adaptive variation. The use of highly polymorphic microsatellites can, after analysis with programs such as KINSHIP, determine kin relations retrospectively in samples from wild populations and thus reduce the need for controlled crosses (although this cannot be applied to interpopulation comparisons). Fortunately it is now possible to identify loci involved in adaptive variation and investigate them directly. One such approach is genome scanning. In this method, AFLPs are used initially to generate multiple polymorphic bands. Each band (locus) is then analysed separately, in cross population comparisons, to estimate pairwise F_{ST} values. Most loci will be neutral, but those with F_{ST} estimates significantly greater than predicted under neutrality can be identified. These (or sequences in linkage disequilibrium with them) may be under directional selection, between (for example) different habitat types. Once identified, diversity at these loci can be quantified to generate a "population adaptive index". These anonymous loci represent a valuable comparison with definitely neutral loci and are thus a significant step forward. This method has been used successfully with *R. temporaria*, tentatively identifying loci associated with adaptation to different altitudes (Bonin *et al.* 2007).

Identifying loci definitely under selection is still problematic for most species, but an exception is the major histocompatibility complex (MHC) of proteins involved in the cellular immune response. These are the most polymorphic protein-encoding genes known in vertebrates, and are under strong selection to maintain resistance to pathogens. Understanding this defence mechanism is particularly relevant in an age of global amphibian declines at least partly mediated by chytrid fungi. PCR primers for amplification of the most variable part of MHC class II loci (exon 2) are now available for some species such as North American ambystomatid salamanders (Bos and DeWoody 2005), and population studies will certainly follow. There are technical difficulties with MHC, largely because most animals have multiple loci that can be hard to distinguish, but this gene family has great prospects for estimating adaptive variation in wild populations.

Genetic studies have already told us a lot about amphibians that we could not have discovered without them. Much more is undoubtedly in store, and this will remain an exciting and challenging research area for decades ahead.

22.5 References

Babik, W., Branicki, W., Sandera, M., Lutrinchuk, S., Borkin, L. J., Irwin, J. T., and Rafinski, J. (2004). Mitochondrial phylogeography of the moor frog, *Rana arvalis*. *Molecular Ecology*, **13**, 1469–80.

Bardsley, L., Smith, S., and Beebee, T. J. C. (1998). Identification of *Bufo* larvae by molecular methods. *Herpetological Journal*, **8**, 145–8.

Beebee, T. J. C. (2005). Amphibian conservation genetics. *Heredity*, **95**, 423–7.

Beebee, T. J. C. and Rowe, G. (2001). Application of genetic bottleneck testing to the investigation of amphibian declines: a case study with natterjack toads. *Conservation Biology*, **15**, 266–70.

Beebee, T. J. C. and Griffiths, R. A. (2005). The amphibian decline crisis: a watershed for conservation biology? *Biological Conservation*, **125**, 271–85.

Beebee, T. J. C. and Rowe, G. (2008). *An Introduction to Molecular Ecology*, 2nd edn. Oxford University Press, Oxford.

Bonin, A., Nicole, F., Pompanon, F., Miaud, C., and Taberlet, P. (2007). Population adaptive index: a new method to help measure intraspecific genetic diversity and prioritise populations for conservation. *Conservation Biology*, **21**, 697–708.

Bos, D. H. and DeWoody, J. A. (2005). Molecular characterisation of major histocompatibility complex class II alleles in wild tiger salamanders (*Ambystoma tigrinum*). *Immunogenetics*, **57**, 775–81.

Burns, E. L., Eldridge, M. D. B., and Houlden, B. A. (2004). Microsatellite variation and population structure in a declining Australian Hylid *Litoria aurea*. *Molecular Ecology*, **13**, 1745–57.

Estoup, A., Wilson, I. J., Sullivan, C., Cornuet, J. M., and Moritz, C. (2001). Inferring population history from microsatellite and enzyme data in serially introduced cane toads, *Bufo marinus*. *Genetics*, **159**, 1671–87.

Evans, B. J., Kelley, D. B., Tinsley, R. C., Melnick, D. J., and Cannatella, D. C. (2004). A mitochondrial DNA phylogeny of African clawed frogs: phylogeography and implications for polyploid evolution. *Molecular Phylogenetics & Evolution*, **33**, 197–213.

Ficoleta, G. F., Garner, T. W. J., and DeBernardi, F. (2007). Genetic diversity, but not hatching success, is jointly affected by postglacial colonisation in the threatened frog, *Rana latastei*. *Molecular Ecology*, **16**, 1787–97.

Funk, W. C., Blouin, M. S., Corn, P. S., Maxell, B. A., Pilliod, D. S., Anrish, S., and Allendorf, J. W. (2005). Population structure of Columbia spotted frogs (*Rana luteiventris*) is strongly affected by the landscape. *Molecular Ecology*, **14**, 483–96.

Giordano, A. R., Ridenhour, B. J., and Storfer, A. (2007). The influence of altitude and topography on genetic structure in the long-toed salamander (*Ambystoma macrodactylum*). *Molecular Ecology*, **16**, 1625–37.

Jehle, R. and Arntzen, J. W. (2002). Microsatellite markers in amphibian conservation genetics. *Herpetological Journal*, **12**, 1–9.

Jehle, R., Burke, T., and Arntzen, J. W. (2005a). Delineating fine-scale genetic units in amphibians: probing the primacy of ponds. *Conservation Genetics*, **6**, 227–34.

Jehle, R., Wilson, G. A., Arntzen, J. W., and Burke, T. (2005b). Contemporary gene flow and the spatio-temporal genetic structure of subdivided newt populations (*Triturus cristatus*, *T. marmoratus*). *Journal of Evolutionary Biology*, **18**, 619–28.

Laugen, A. T., Kruuk, L. E. B., Laurila, A., Rasanen, K., Store, J., and Merila, J. (2005). Quantitative genetics of larval life-history traits in *Rana temporaria* in different environmental conditions. *Genetical Research*, **86**, 161–70.

Liebgold, E. B., Cabe, P. R., Jaeger, R. G., and Leberg, P. L. (2006). Multiple paternity in a salamander with socially monogamous behaviour. *Molecular Ecology*, **15**, 4153–60.

Livia, L., Antonella, P., Hovirag, L., Mauro, N., and Panara, F. (2006). A non-destructive, rapid, reliable and inexpensive method to sample, store and extract high-quality DNA from fish body mucus and buccal cells. *Molecular Ecology Notes*, **6**, 257–60.

Luikart, G., Sherwin, W. B., Steele, B. M., and Allendorf, F. W. (1998). Usefulness of molecular markers for detecting population bottlenecks via monitoring genetic change. *Molecular Ecology*, **7**, 963–74.

Mikulicek, P. and Pialek, J. (2003). Molecular identification of three crested newt species (*Triturus cristatus* superspecies) by RAPD markers. *Amphibia-Reptilia*, **24**, 201–7.

Monsen, K. J. and Blouin, M. S. (2004). Extreme isolation by distance in a montane frog *Rana cascadae*. *Conservation Genetics*, **5**, 827–35.

Morin, P. A., Luikart, G., and Wayne, R. K. (2004). SNPs in ecology, evolution and conservation. *Trends in Ecology and Evolution*, **19**, 208–16.

Newman, R. A. and Squire, T. (2001). Microsatellite variation and fine-scale population structure in the wood frog (*Rana sylvatica*). *Molecular Ecology*, **10**, 1087–1100.

Nielson, M., Lohman K., and Sullivan, J. (2001). Phylogeography of the tailed frog (*Ascaphus truei*): implications for the biogeography of the Pacific Northwest. *Evolution*, **55**, 147–60.

Noel, S., Ouellet, M., Galois, P., and Lapointe, F. J. (2007). Impact of urban fragmentation on the genetic structure of the eastern red-backed salamander. *Conservation Genetics*, **8**, 599–606.

Noonan, B. P. and Gaucher, P. (2006). Refugial isolation and secondary contact in the dyeing poison frog *Dendrobates tinctorius*. *Molecular Ecology*, **15**, 4425–35.

Pidancier, N., Miquel, C., and Miaud, C. (2003). Buccal swabs as a non-destructive tissue sampling method for DNA analysis in amphibians. *Herpetological Journal*, **13**, 175–8.

Plötner, J., Kohler, F., Uzzell, T., and Beerli, P. (2007). Molecular systematics of amphibians. In H. Heatwole (ed.), *Amphibian Biology*, vol. 7, *Systematics*, pp. 2672–2756. Surrey Beatty & Sons, Chipping Norton, NSW.

Primmer, C. R. and Merila, J. (2002). A low rate of cross-species microsatellite amplification success in Ranid frogs. *Conservation Genetics*, **3**, 445–9.

Riberon, A., Miaud, C., Guyetant, R., and Taberlet, P. (2004). Genetic variation in an endemic salamander, *Salamandra atra*, using amplified fragment length polymorphism. *Molecular Phylogenetics & Evolution*, **31**, 910–14.

Rowe, G. and Beebee, T. J. C. (2003). Population on the verge of a mutational meltdown? Fitness costs of genetic load for an amphibian in the wild. *Evolution*, **57**, 177–81.

Rowe, G. and Beebee, T. J. C. (2007). Defining population boundaries: use of three Bayesian approaches with microsatellite data from British natterjack toads (*Bufo calamita*). *Molecular Ecology*, **16**, 785–96.

Shaffer, G., Fellers, G. M., Magee, A., and Voss, R. (2000). The genetics of amphibian declines: population substructure and molecular differentiation in the Yosemite Toad, *Bufo canorus* (Anura, Bufonidae) based on single-strand conformation polymorphism analysis (SSCP) and mitochondrial DNA sequence data. *Molecular Ecology*, **9**, 245–57.

Smith, M. A. and Green, D. M. (2004). Phylogeography of *Bufo fowleri* at its northern range limit. *Molecular Ecology*, **13**, 3723–33.

Vences, M. and Wake, D. B. (2007) Speciation, species boundaries and phylogeography of amphibians. In H. Heatwole (ed.), *Amphibian Biology*, vol. 7, *Systematics*, pp. 2613–71. Surrey Beatty & Sons, Chipping Norton, NSW.

Vieites, D. R., Nieto-Roman, S., Barluenga, M., Palanca, A., Vences, M., and Meyer, A. (2004). Post-mating clutch piracy in an amphibian. *Nature*, **431**, 305–8.

Walsh, P. S., Metzger, D. A., and Higuchi, R. (1991). Chelex R100 as a medium for simple extraction of DNA for PCR-based typing from forensic material. *Biotechniques*, **10**, 506–13.

Whitlock, M. C. and McCauley, D. E. (1999). Indirect measures of gene flow and migration: F-ST not equal 1/(4Nm+1). *Heredity*, **82**, 117–25.

Zane, L., Bargelloni, L., and Patarnello, T. (2002). Strategies for microsatellite isolation: a review. *Molecular Ecology*, **11**, 1–16.

Zeisset, I. and Beebee, T. J. C. (2001). Determination of biogeographical range: an application of molecular phylogeography to the European pool frog *Rana lessonae*. *Proceedings of the Royal Society of London Series B Biological Sciences*, **268**, 933–8.

Part 7

Monitoring, status, and trends

23

Selection of species and sampling areas: the importance to inference

Paul Stephen Corn

23.1 Introduction

Inductive inference, the process of drawing general conclusions from specific observations, is fundamental to the scientific method. Platt (1964) termed conclusions obtained through rigorous application of the scientific method as "strong inference" and noted the following basic steps: generating alternative hypotheses; devising experiments, the results of which will exclude one or more hypotheses; conducting the experiment to get a "clean result"; and repeating the process with revision based on the information obtained. Every student is exposed to these basics in introductory courses, and a considerable proportion of a modern graduate education in the sciences is devoted to acquiring the analytic (statistical) skills necessary to apply the scientific method. Not even considering the field of mathematical statistics or applied statistics in disciplines such as social sciences, library shelves groan under the weight of texts on applied statistics, ranging from introductory (Hayek and Buzas 1997) to advanced (Williams *et al.* 2002), for conducting research in ecology, and new works are published every year. Much effort is currently devoted to the mechanisms of analysis and the issues involved in choosing among statistical methods; specifically, traditional hypothesis testing versus information-theoretic or Bayesian approaches (Hobbs and Hilborn 2006). This chapter does not address these topics, but instead discusses some of the issues related to selection of study sites and species necessary to obtain a "clean result." Regardless of the method of analysis, successful inference relies on correctly designed data collection, meaning that the observations represent the population of interest. As Anderson (2008, p. 7) puts it, valid inference requires, "some type of probabilistic sampling of the well-defined population."

Experimental design and sampling are well-developed topics in applied statistics. There are numerous books available (see Anderson 2001 for a sample of representative works), and many universities offer their graduate students specific courses in these areas. Most biologists have had the requisite education and are familiar with and employ statistically valid designs. However, good study designs are not universal. Anderson (2008, p. 143) stated:

…there is no excuse for collecting data that are fundamentally flawed. Still, I see data collected from convenience sampling where any valid inductive inference is precluded. In some cases the population of interest is not even defined.

Convenience sampling is a broad term that describes non-random study-site selection (Hayek 1994; Anderson 2001). Inadequate design and sampling is often justified on logistical grounds, or, to put it more crudely, that it is unrealistic to satisfy the demands of a statistician, who is usually office-bound in an ivory tower. As a field biologist and not a statistician, I sometimes find myself sympathetic to such rationalizations. However, the solution is not to politely thank the statistician and then go ahead and sample as originally planned (see the first phrase in the Anderson quote above). Instead, more planning is necessary. Either a means must be devised for obtaining a valid sample, or the project's objectives should be revised so that a population can be defined that will allow a valid sample, or, in extreme cases, the project might need to be abandoned.

In this chapter, I provide a brief overview of sampling design, but this is a large, complex topic and it is impossible to do more than scratch the surface here. For more thorough instruction and discussion, including statistics specific to different designs, consult one of the many texts that have been written for that purpose (e.g. Cochran 1977; Thompson 1992; Hayek and Buzas 1997; Williams *et al.* 2002). The US Geological Survey (www.pwrc.usgs.gov/monmanual/) and National Park Service (http://science.nature.nps.gov/im/monitor/) also have extensive and very useful web-based resources to aid in study design for inventory and monitoring studies. Following this introduction to sampling, I will discuss issues related to selection of study sites and effects of estimating abundance on inference. This discussion will use a few selected cases to illustrate problems that can arise from either poor study design or interpretation of results that cannot be supported by the design. These examples include issues arising from both selection of study sites and collection and analysis of data from individual species, and describe some of the methods that can be used to minimize bias.

23.2 Sampling

In all but a tiny number of exceptional cases, it is impossible to survey every possible habitat or catch every individual in a population. Therefore, inference depends on statistics generated from a subset of individuals or habitats drawn from the population. Ideally, a sample has three qualities (Williams *et al.* 2002): it comprises separate individual units, and these share the same underlying distribution and are statistically independent. These criteria are more difficult to satisfy in inventory and monitoring studies than in a carefully designed experiment, but a good study design should try to come as close as possible.

Before a sample can be taken, the sampling units must be defined and be available for selection. Sample units can be individual animals, artificially bounded areas (quadrats), or natural habitat features, such as ponds, streams, or drainage basins. For population studies, where the individual is the sample unit, the availability for selection is often assumed, but this assumption is tested when the data are analyzed. For inventory or monitoring studies, site delineation and selection has become easier with the increasing availability of tools such as satellite imagery and Geographic Information Systems (GIS) software. A good example is the study by Kroll *et al.* (2008), which took advantage of detailed GIS data layers to select stream reaches for sampling. However, detailed GIS coverage exists mainly where there is an economic use for the data (e.g. the commercial forests surrounding the streams studied by Kroll *et al.* 2008), and even when GIS data are available the amphibian habitats may be poorly described. Temporary wetlands are notoriously underrepresented in aquatic data layers, and studies of lentic-breeding amphibians often will have more success in defining sampling units if areas are used instead of specific habitats, for example, 1 km^2 blocks (Johannson *et al.* 2006) or drainage basins (Corn *et al.* 2005).

The most basic form of probabilistic sampling is a simple random sample. All the available sample units are put into a metaphorical hat (the mechanics of choosing a sample now usually involve a computer) and the desired number of units are selected in random order. The number of sample units should be sufficient to estimate parameters of interest with sufficient precision, but samples that are too large should be avoided. Determining the desired sample size requires knowing the variability of the parameter to be estimated, the magnitude of the effect to detect (e.g. a 10% difference or a 5% annual trend), and the strength of the inference. The formula for determining sample size depends on the sampling scheme. Hayak and Buzas (1997) and Williams *et al.* (2002) describe the details for determining sample size.

If, as is usually the case, habitats are not uniform, species are patchily distributed, or the number of samples is small relative to the area of interest, simple random sampling may be inadequate to characterize the variability of the system being studied. There are several slightly more complex designs that can be used to achieve a more representative sample. In the case where the number of samples is relatively small, strictly random selection can result in sample sites clumped together instead of dispersed throughout the study area. If the study area is not uniform, this is not desirable, and a common modification is to employ a systematic random sample (Figure 23.1), in which, after a random start, every nth unit is selected, where the number of available units divided by n equals the desired sample size. Williams *et al.* (2002) cautioned that systematic sampling risks a biased result if the units are arranged in such a way that environmental gradients are correlated with the order of the sampling units.

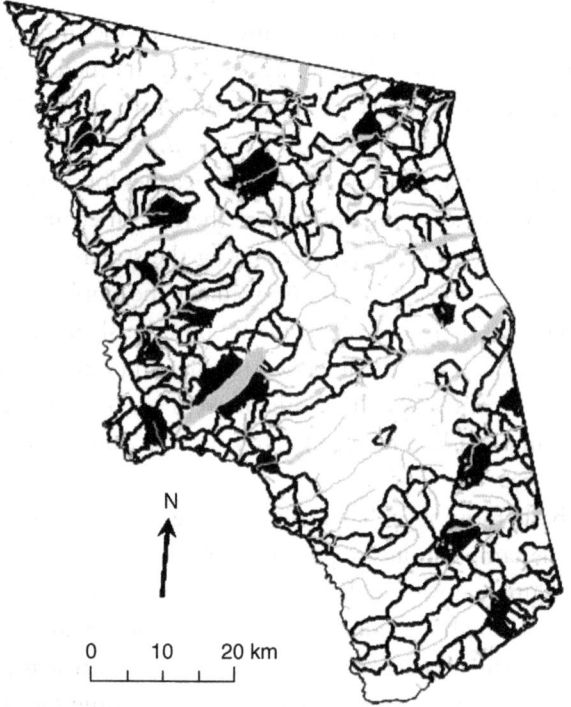

Fig. 23.1 Sample locations for monitoring amphibian populations in Glacier National Park, MT, USA. Twenty small drainage basins (shaded) composed a systematic random sample from a frame of 235 basins (outlined). Areas without sufficient surface water, where the elevation precluded amphibian occurrence, or where extremely rugged terrain prevented safe access were excluded from the frame. Inference about status of amphibians in the park is limited to the frame.

If a study area is not homogenous and differences within the area can be described (e.g. altitudinal gradients, different types of wetlands, differences in ability to access potential sample units) then a stratified design is usually preferable to a simple random or systematic random sample. Once the strata and sampling units within them are delineated, then simple or systematic random samples can be drawn from each stratum. The formulas for calculating various statistics vary among sampling designs. See Williams *et al.* (2002) for a concise description of these. For long-term monitoring studies, strata should be based on features that are not expected to change significantly during the course of the study. For example, strata based on geological differences would be preferable to strata based on vegetative land cover.

This section is not a comprehensive treatment of sampling design. Advanced schemes, such as adaptive sampling (Thompson 1992; Williams *et al.* 2002), are beyond the scope of this chapter, but may be necessary or desirable in many cases. Sources listed above should be consulted when beginning the design of any study.

Any valid sampling design must incorporate replication of sample units. A study that compared toad populations in only two ponds could only describe the differences in abundance between the two ponds. Tests of hypotheses of the causes of the differences are not appropriate because there is no replication (Underwood 1998). Most inventory or monitoring studies include numerous sample units, but care must be taken to ensure that external factors are dealt with in the design stage (e.g. through stratification). If variation among sample units can be attributed to something other than random processes, then the study would suffer from pseudoreplication (Hurlbert 1984). Large field studies are vulnerable to temporal pseudoreplication (sample units vary in some systematic fashion over time), because it is seldom possible to visit all sample units simultaneously. This is a particular problem for studies of amphibians. For example, surveys that use breeding activity are affected by the changing composition of breeding choruses over time, both among and within species, and external factors, such as weather, that influence behavior of individuals. Studies that focus on larval stages must deal with growth and development and changes in abundance that may influence detection. It is always a good practice to minimize the time span of field surveys where possible.

23.3 Study sites and consequences of convenience sampling

If study sites or sample units are not selected using a probalistic design, the result is a convenience sample. The motivations for convenience sampling are several, but common reasons are to choose study sites that are most accessible, or where the target species are most abundant. The latter case poses particular problems for analysis of trends in abundance, because there may be a built-in bias for detecting declining populations (Alford and Richards 1999; but see Green 2003).

The effects of convenience sampling on inference can be subtle. Results of ecological experiments are often interpreted to demonstrate the generality of effects or even causality, but experiments by themselves are insufficient to explain complex ecological phenomena; such an effort requires integrating observation and theory with experimentation (Werner 1998). Broad inference is limited when experiments, even those that are internally well designed, are conducted at locations that are convenient for the researcher. The results may be useful for investigating possible mechanisms, but cannot be generalized without making the unsupportable assumption that the study sites are an unbiased representation of the habitats in question. For example, Blaustein *et al.* (1994) found that ambient ultraviolet-b (UV-B) radiation caused higher mortality of amphibian embryos than when embryos were shielded from UV-B at four lakes in the Cascade Mountains in Oregon, USA. This and subsequent research formed the basis of a theory of how climate change, UV-B, and pathogens might play a significant role in global amphibian declines (Kiesecker *et al.* 2001; Pounds 2001; Blaustein and Kiesecker 2002). The issue of amphibian declines and UV-B has been controversial with a large literature that I will not delve into here. Relevant to this chapter, the sites used for the UV-B studies were not chosen with respect to the potential for exposure to UV-B, but were a convenience sample. The primary study site turned out to have water much more transparent to UV-B transmission than most amphibian breeding habitats in the Pacific Northwest (Palen *et al.* 2002), severely restricting the generality of the proposed hypothesis.

Convenience sampling is almost always done without any intention to produce a biased result. The water chemistry and UV-B transmission at the study sites in the Oregon UV-B studies were not known beforehand. These sites were known to the researchers to have suitable amphibian populations and included locations where long-term studies had been conducted. Similarly, Corn and Bury's (1989) study of the effects of logging on stream amphibians in western Oregon did not include conscious bias in selection of streams to be sampled, but it was none-the-less based on a convenience sample. In this study, Bruce Bury and I identified likely sample locations on topographic maps beforehand, but the decision to sample was made in the field after inspecting each stream. We attempted to select typical streams and sample reaches based on our knowledge of the characteristics of headwater streams in the region. However, we did not begin with a well-defined population of streams, and we did not apply any probabilistic sampling. As stated by Hayek and Buzas (1997, p. 113), "If the basis for inclusion in a sample is judgment, regardless of how expert, we will not have a reproducible measure of our field study's usefulness." Corn and Bury (1989) found strong differences in abundance and diversity of amphibians between streams in logged and unlogged forest. The paper was an early influence on what has become a spate of research on stream amphibians and forest management in the Pacific Northwest. Although the original results are largely supported by subsequent work (Olson *et al.* 2007), the convenience sampling design employed limits the scope of the conclusions and Corn and Bury (1989) should be viewed as hypothesis-generating work rather than a definitive demonstration of differences between streams in managed and unmanaged forests.

Cautions about convenience sampling apply equally to the sample frame, the pool from which the sample is drawn, as to individual study sites. Study areas, meaning the regions containing the sample frames, are almost never chosen at random. Study areas may be defined by a relevant management question (e.g. what is the status of amphibians in a National Park?), or they might be contain habitats or species of interest, yet be located conveniently near a researcher's institution. Probabilistic methods can be used properly to select study sites, but if the frame is defined by convenience, then inference is restricted to the study area and the generality of the results may be limited. Two recent studies, conducted less than 200 km apart in Switzerland, illustrate this point. Pellet *et al.* (2004) found that urbanization and roads had a strong negative influence on presence of European treefrogs (*Hyla arborea*). Conversely, Van Buskirk (2005) found only weak support for landscape variables (including urbanization) to explain occurrence and abundance of treefrogs. Both studies were well designed, and although different methods of analysis were used (logistic regression versus information theoretic analysis) the different conclusions likely resulted from intrinsic differences between the study areas.

The North American Amphibian Monitoring Program (NAAMP; Weir and Mossman 2005), patterned after the North American Breeding Bird Survey (BBS; Peterjohn *et al.* 1995), also suffers from a sampling frame defined by convenience. NAAMP conducts manual calling surveys of breeding amphibians on prescribed road routes. Species are identified by their breeding calls, and data are collected mainly by volunteers. Survey routes are generated through a random process, but the goal of the program is to monitor trends in amphibian populations throughout the region where routes are conducted (Weir and Mossman 2005). Reliance on roadside observations limits the inference to those areas accessed by roads, or requires investigators to make an additional, untested assumption that the roadside amphibian population experiences the same trends in abundance as those found in populations away from roads. The biases that potentially undermine this assumption can be related to both habitat and the observations themselves (Peterjohn et al. 1995). In Chapter 16 in this volume Dorcas et al. discuss assumptions about auditory observations. Habitat assumptions require that roadside wetlands reflect the same conditions found at wetlands away from roads, and that habitat condition changes over time in the same direction and at a similar pace alongside and away from roads. These assumptions are more likely to be violated than satisfied. For example, Keller and Scallan (1999) found that land cover types were similar near and away from BBS routes in Maryland and Ohio, but that in Maryland, urbanization was proceeding at a more rapid pace along roads. If urbanization is associated largely with existing road networks, then roadside habitats may diverge from distant habitats more rapidly in Maryland than in Ohio.

A more immediate and concrete difference between wetlands near and distant from roads (and also a difference between amphibians and birds) is that roads, especially paved roads with higher volumes of traffic, are a significant source of mortality of adult amphibians moving in and out of breeding ponds. Studies in North America and Europe have found negative relationships between traffic intensity and amphibian mortality and breeding activity (Fahrig et al. 1995), habitat occupancy (Pellet et al. 2004), and species richness and abundance (Eigenbrod et al. 2008). Similarly in New York, Karraker et al. (2008) found a twofold increase in density of egg masses of two amphibian species away from roads compared to roadside ponds. This may have been more an indirect road effect than from direct mortality. Demographic models showed significant negative effects on these species in roadside ponds resulting from the application of salt for de-icing (Karraker et al. 2008). Road de-icers epitomize a confounding variable contributing to differences that would be very difficult to model in analyses of NAAMP data. Application of de-icers varies among geographic regions, states, road types, and chemistry from year to year, and the data quantifying application are likely to be extremely difficult to obtain. The assumption that roadside habitats are equivalent to habitats away from roads is not supported by research to date.

The road studies illustrate the point that concerns about convenience sampling also apply to choosing variables for explaining patterns or trend. Data may be readily available in GIS data layers (e.g. road density), but less available data (e.g. traffic intensity, de-icer application) may be more important for generating well-supported models. Anderson (2008) emphasizes that considerable effort should be devoted to generating hypotheses before data are collected; this obviously applies to selection of the variables that make up the candidate models.

23.4 Abundance and inference

Issues of inference regarding species typically arise when deciding how to quantify abundance or the related estimate of habitat occupancy. These are important and somewhat controversial topics, and other chapters deal with estimating occupancy (Chapter 24) and abundance (Chapter 25) in detail. In this section, I focus on how uncertainty about numbers affects inference.

Studies often use an index, either simple counts of individuals observed or counts converted to relative abundance, to represent the abundance of a species (Hayek and Buzas 1997), and many standard field methods (e.g. in this volume or Heyer *et al.* 1994) are designed to obtain counts. However, inference about trends over time or differences among habitats requires that there is no trend in the detection probability (the relationship between the index and the true abundance). Bart *et al.* (2004) contended that this was a reasonable assumption for bird studies. Anderson (2001) regarded this as unlikely, and recently put it even more strongly (Anderson 2008, p. 20): "... the evidence is conclusive that they [index values] represent an amateur, unthinking approach and is not scientific" and "... index values are not data, they are just numbers." This viewpoint is not universal. Reliable estimation of detection probability may be difficult in many situations (Johnson 2008), and variation in detection probability, particularly among individuals, may result in unreliable estimates of abundance (Link 2003). Nevertheless, evidence from amphibian studies tends to support the idea that use of indices should be avoided, because estimates incorporating detection are more closely correlated to actual abundance than are counts (Schmidt 2004).

Welsh and Droege (2001) advocated use of count data from surveys of plethodontid salamanders to monitor forest condition and biological diversity. Terrestrial salamanders are associated with habitat features that are often disrupted by activities such as logging, and counts of salamanders in several studies sampled by a variety of techniques show relatively consistent numbers from year to year (Welsh and Droege 2001). Low inter-annual variation increases ability to detect trends. However, at any given time, the majority of *Plethodon* in a population are subterranean and unavailable to be captured (Bailey *et al.* 2004a), and counts of terrestrial plethodontids vary considerably among years, habitats, and sampling methods, violating many of the assumptions required for use of indices in monitoring (Hyde and Simons 2001; Dodd and Dorazio 2004). More critically, temporary immigration between surface and subterranean habitats varies among sampling occasions, so that the proportion of the population available to be sampled is not constant. Additionally, capture–recapture studies on plethodontids have found low detection probabilities, often less than 0.05 (Jung *et al.* 2000; Welsh and Droege 2001; Dodd and Dorazio 2004). Low capture probabilities are not a problem if they are relatively uniform among habitats and observers (Welsh and Droege 2001), but there is considerable reason to doubt that this is true. Detection probabilities of plethodontids varied from 0.06 to 0.41 among years (Dodd and Dorazio 2004), and.between 0.01 and 0.58 among sampling occasions in one model evaluated by Bailey *et al.* (2004a). Low detection magnifies any bias due to variation among observers, habitats, or years. For example, an increase in detection from 0.50 to 0.55 would result in a 10% increase in numbers of animals observed, but an increase from 0.06 to 0.38 would result in 6.5 times as many observations in the second count. This could produce a result similar to that illustrated in Schmidt (2004, figure 1), where counts were similar among years, but capture probability varied, with the result that counts did not reflect a large increase in actual abundance.

Concerns about use of count data apply more broadly than to just terrestrial salamanders. Johannson *et al.* (2006) used uncorrected counts of common frog (*Rana temporaria*) egg masses as an index to population size, and concluded that population size declined with increasing latitude and smaller populations had less genetic variability. Johannson *et al.* conceded that egg mass counts likely underestimated true abundance of breeding females but contended it was an unbiased index, because sampling was the same at all study sites. However, counts often fail to detect all egg masses present in a pond for a variety of reasons, such as differences in habitat complexity, weather conditions that might affect visibility, or variation in ability among observers. Grant *et al.* (2005) found detection probabilities of ranid egg masses to vary between 0.78 and 1.0. Variation in detection introduces uncertainty into conclusions about population size; this uncertainty is magnified if detection varies in a systematic manner across a study area.

Count data are incorporated into many indices of species diversity (Hayek and Buzas 1997), but calculation of these indices for amphibian assemblages are not appropriate, unless unbiased estimates of abundance are used instead of the raw counts. It has long been known that number of captures varies among sampling methods, so that a diversity index that included counts of species made using different techniques (for example, pitfall traps for one species and time-constrained searches for another) would not be valid (Corn 1994). Hyde and Simons (2001) demonstrated sampling efficiency varied among methods, but also among habitat types for some species of plethodontids when the same method was employed. Interpretations of diversity indices suffer from the same problems as interpretations about abundance.

The uncertainties about count data, the expense of obtaining unbiased estimates of population size, and high amount of year-to-year variation in abundance of many amphibian species prompted Green (1997) to suggest that tracking the changes in the occurrence of a species across the landscape—that is, whether or not a species occupied a given patch of habitat—was a more efficient way to monitor status and trend of amphibians. Such presence/absence (or, more accurately, detected/not detected) data used to be considered inferior to or less scientific than count data (MacKenzie et al. 2003). In part, this is because surveys to detect presence are typically at a reduced level of effort than surveys that include counts, and in most situations it is impossible to prove absence even with great effort. False negatives, or failures to detect species that are actually present, introduce bias that underestimates occupancy and can lead to errors in interpretation, such as incorrectly identifying the influence of habitat variables on occurrence (Mazerolle et al. 2005; Hossack and Corn 2007).

Occupancy estimation (MacKenzie et al. 2006; see also Chapter 24 in this volume) has been implemented in a variety of studies of amphibians, including effects of disturbance (e.g. Mazerolle et al. 2005; Hossack and Corn 2007), in addition to more general monitoring efforts (e.g. Corn et al. 2005; Schmidt 2005). Occupancy analysis has been recommended in the US Geological Survey's Amphibian Research and Monitoring Initiative (Muths et al. 2005), mainly through surveys for tadpoles or egg masses to indicate breeding populations. Tadpoles typically have high detection probabilities (Brown et al. 2007; Hossack and Corn 2007), which increases the precision of estimated occupancy. Occupancy analysis also shows promise as a means for monitoring terrestrial salamanders. Detection probabilities for occurrence of plethodontids on small plots were much higher than capture–recapture studies have found for detection of individual salamanders, and estimated occupancy was much less variable among years compared to estimated abundance (Bailey et al. 2004b).

Occupancy analysis may not be the best choice in all situations. It includes assumptions about the relationship between occurrence on the landscape and abundance of a species (Royle and Nichols 2003), although combining occupancy and count data is topic of active development (Royle et al. 2005; see also Chapter 24). Occupancy analysis can be difficult to implement in habitats that are not discrete, for example extensive wetlands. It also may not be possible to obtain reliable estimates of occupancy for rare species with low detection probabilities (Bailey et al. 2004b). A reservation about use of occupancy for detecting trends in salamanders is that local populations must go extinct or be recolonized for change to be observed. This may be a common occurrence for pond-breeding amphibians (Green 1997), but it is less likely that populations of terrestrial salamanders undergo the same dynamics.

One final note of caution: research can be well designed, but the resulting data and analysis can still be undermined if the investigators do not pay sufficient attention to the biology of the organisms they are studying. Kroll *et al.* (2008) recently conducted a study on stream amphibians that employed a rigorous statistically valid design for sample selection, estimated occupancy among streams for several species, and analyzed effects on detection and occupancy using information-theoretic methods. One of their findings was that detection of coastal tailed frogs (*Ascaphus truei*) declined as the field season progressed. This was likely a consequence of the timing of sampling, which was conducted from 19 July to 6 October. Although in parts of the Pacific Northwest, tadpoles of *A. truei* require 2 or more years to reach metamorphosis, many of the streams examined by Kroll *et al.* are in a region where tadpoles metamorphose in less than 1 year, beginning as early as late June (Bury and Adams 1999). During the time when Kroll *et al.* were sampling, only adult frogs or hatchling tadpoles (which tend to remain clustered for several weeks under rocks near the nest site) would have been present in many streams. Both of these life stages are less observable than older tadpoles. Reduced detection probability results in positive bias in the occupancy estimate. Studies that employ occupancy analysis should be designed so that detection does not vary among sample units in a systematic manner.

23.5 Conclusions

The path to strong inference leads through good study design that incorporates probabilistic sampling from a well-defined population. Inventory and, especially, monitoring studies stray from this path when scientific rigor is sacrificed to logistic constraints and convenience in data collection. Tension often exists between the field biologist and the consulting statistician regarding the requirements of good study design and the logistical realities of data collection. Having been on the field biologist's side of the argument, I can testify that the attitude summarized by "Yes, we realize valid sample selection is important, and it would be nice, but we have to collect data from the real world", is fairly common. Constraints in site selection can be incorporated into study design, such as by stratifying based on accessibility, and the resulting analysis can test hypotheses about whether populations that are easily accessible differ from those that are not.

The perils of convenience sampling also apply to choice of life stage to study or explanatory variables to incorporate in a model. The easiest life stage to study may not be the same one that is most sensitive to external factors, and variables should not be included in a model simply because the data are available. There is no "magic bullet" for sampling amphibians. No single technique encompasses the variety of life histories of amphibians or the habitats in which they can be found. Occupancy analysis provides a useful tool for avoiding the pitfalls of using simple count data or the logistic difficulties of obtaining unbiased estimates of abundance, but it is not a panacea. Ultimately, the design that allows the strongest inference will be one that avoids convenience sampling and minimizes untested assumptions when the data are analyzed.

23.6 Acknowledgments

I thank Larissa Bailey, Blake Hossack, Benedikt Schmidt, and Susan Walls for suggestions that greatly improved the manuscript.

23.7 References

Alford, R. A. and Richards, S. J. (1999). Global amphibian declines: a problem in applied ecology. *Annual Review of Ecology and Systematics*, **30**, 130–65.

Anderson, D. R. (2001). The need to get the basics right in wildlife field studies. *Wildlife Society Bulletin*, **29**, 1294–7.

Anderson, D. R. (2008). *Model Based Inference in the Life Sciences: a Primer on Evidence*. Springer, New York.

Bailey, L. L., Simons, T. R., and Pollock, K. H. (2004a). Estimating detection probability parameters for plethodon salamanders using the robust capture–recapture design. *Journal of Wildlife Management*, **68**, 1–13.

Bailey, L. L., Simons, T. R., and Pollock, K. H. (2004b). Estimating site occupancy and species detection probability parameters for terrestrial salamanders. *Ecological Applications*, **14**, 692–702.

Bart, J., Droege, S., Geissler, P., Peterjohn, B., and Ralph, C. J. (2004). Density estimation in wildlife surveys. *Wildlife Society Bulletin*, **32**, 1242–7.

Blaustein, A. R. and Kiesecker, J. M. (2002). Complexity in conservation: lessons from the global decline of amphibian populations. *Ecology Letters*, **5**, 597–608.

Blaustein, A. R., Hoffman, P. D., Hokit, D. G., Kiesecker, J. M., Walls, S. C., and Hays, J. B. (1994). UV repair and resistance to solar UV-B in amphibian eggs: a link to population declines? *Proceedings of the National Academy of Sciences USA*, **91**, 1791–5.

Brown, G. W., Scroggie, M. P., Smith, M. J., and Steane, D. (2007). An evaluation of methods for assessing the population status of the threatened alpine tree frog *Litoria verreauxii alpina* in southeastern Australia. *Copeia*, **2007**, 765–70.

Bury, R. B. and Adams, M. J. (1999). Variation in age at metamorphosis across a latitudinal gradient for the tailed frog, *Ascaphus truei*. *Herpetologica*, **55**, 283–91.

Cochran, W. G. (1977). *Sampling Techniques*, 3rd edn. John Wiley & Sons, New York.

Corn, P. S. (1994). Straight-line drift fences and pitfall traps. In W. R. Heyer, M. A. Donnelly, R. W. McDiarmid, L.-A. C. Hayek, and M. S. Foster (eds), *Measuring and Monitoring Biological Diversity: Standard Methods for Amphibians*, pp. 109–17. Smithsonian Institution Press, Washington DC.

Corn, P. S. and Bury, R. B. (1989). Logging in western Oregon: responses of headwater habitats and stream amphibians. *Forest Ecology and Management*, 29, 39–57.

Corn, P. S., Hossack, B. R., Muths, E., Patla, D. A., Peterson, C. R., and Gallant, A. L. (2005). Status of amphibians on the Continental Divide: surveys on a transect from Montana to Colorado, USA. *Alytes*, 22, 85–94.

Dodd, Jr, C. K. and Dorazio, R. M. (2004). Using counts to simultaneously estimate abundance and detection probabilities in a salamander community. *Herpetologica*, 60, 468–78.

Eigenbrod, F., Hecnar, S. J., and Fahrig, L. (2008). The relative effects of road traffic and forest cover on anuran populations. *Biological Conservation*, 141, 35–46.

Fahrig, L., Pedlar, J. H., Pope, S. E., Taylor, P. D., and Wegner, J. F. (1995). Effect of road traffic on amphibian density. *Biological Conservation*, 73, 177–82.

Grant, E. M., Jung, R. E., Nichols, J. D. and Hines, J. E. (2005). Double-observer approach to estimating egg mass abundance of pool-breeding amphibians. *Wetland Ecology and Management*, 13, 305–20.

Green, D. M. (1997). Perspectives on amphibian population declines: defining the problem and searching for answers. In D. M. Green (ed.), *Amphibians in Decline: Canadian Studies of a Global Problem. Herpetological Conservation*, vol. 1, pp. 291–308. Society for the Study of Amphibians and Reptiles, St. Louis, MO.

Green, D. M. (2003). The ecology of extinction: population fluctuation and decline in amphibians. *Biological Conservation*, 111, 331–43.

Hayek, L.-A. C. (1994). Research design for quantitative amphibian studies. In W. R. Heyer, M. A. Donnelly, R. W. McDiarmid, L.-A. C. Hayek, and M. S. Foster (eds), *Measuring and Monitoring Biological Diversity: Standard Methods for Amphibians*, pp. 21–39. Smithsonian Institution Press, Washington DC.

Hayek, L.-A. C. and Buzas, M. A. (1997). *Surveying Natural Populations*. Columbia University Press, New York.

Heyer, W. R., Donnelly, M. A., McDiarmid, R. W., Hayek, L.-A. C., and Foster, M. S. (eds), (1994). *Measuring and Monitoring Biological Diversity: Standard Methods for Amphibians*. Smithsonian Institution Press, Washington DC.

Hobbs, N. T. and Hilborn, R. (2006). Alternatives to statistical hypothesis testing in ecology: a guide to self teaching. *Ecological Applications*, 16, 5–19.

Hossack, B. R. and Corn, P. S. (2007). Responses of pond-breeding amphibians to wildfire: short-term patterns in occupancy and colonization. *Ecological Applications*, 17, 1403–10.

Hurlbert, S. N. (1984). Pseudoreplication and the design of ecological field experiments. *Ecological Monographs*, 54, 187–211.

Hyde, E. J. and Simons, T. R. (2001). Sampling plethodontid salamanders: sources of variability. *Journal of Wildlife Management*, 65, 624–32.

Johannson, M., Primmer, C. R., and Merilä, U. (2006). History vs. current demography: explaining the genetic population structure of the common frog (*Rana temporaria*). *Molecular Ecology*, 15, 975–83.

Johnson, D. H. (2008). In defense of indices: the case of bird surveys. *Journal of Wildlife Management*, **72**, 857–68.

Jung, R. E., Droege, S., Sauer, J. R., and Landy, R. B. (2000). Evaluation of terrestrial and streamside salamander monitoring techniques at Shenandoah National Park. *Environmental Monitoring and Assessment*, **63**, 65–79.

Karraker, N. E., Gibbs, J. P., and Vonesh, J. R. (2008). Impacts of road deicing salt on the demography of vernal pool-breeding amphibians. *Ecological Applications*, **18**, 724–34.

Keller, C. M. E. and Scallan, J. T. (1999). Potential roadside biases due to habitat changes along Breeding Bird Survey routes. *The Condor*, **101**, 50–7.

Kiesecker, J. M., Blaustein, A. R., and Belden, L. K. (2001). Complex causes of amphibian population declines. *Nature*, **410**, 681–4.

Kroll, A. J., Risenhoover, K., McBride T., Beach E., Kernohan B. J., Light, J., and Bach, J. (2008). Factors influencing stream occupancy and detection probability parameters of stream-associated amphibians in commercial forests of Oregon and Washington, USA. *Forest Ecology and Management*, **255**, 3726–35.

Link, W. A. (2003). Nonidentifiability of population size from capture-recapture data with heterogeneous detection probabilities. *Biometrica*, **59**, 1123–30.

MacKenzie, D. I., Nichols, J. D., Hines, J. E., Knutson, M. G., and Franklin, A. B. (2003). Estimating site occupancy, colonization, and local extinction when a species is detected imperfectly. *Ecology*, **84**, 2200–7.

MacKenzie, D. I., Nichols, J. D., Royle, J. A., Pollock, K. H., Bailey, L. L., and Hines, J. E. (2006). *Occupancy Estimation and Modeling: Inferring Patterns and Dynamics of Species Occurrence*. Academic Press, Burlington, MA.

Mazerolle, M. J., Desrochers, A., and Rochefort, L. (2005). Landscape characteristics influence pond occupancy by frogs after accounting for detectability. *Ecological Applications*, **15**, 824–34.

Muths E., Jung, R. E., Bailey, L. L., Adams, M. J., Corn, P. S., Dodd, Jr, C. K., Fellers, G. M., Sadinski, W. J., Schwalbe, C. R., Walls, S. C. *et al.* (2005). The U.S. Department of Interior's Amphibian Research and Monitoring Initiative: a successful start to a national program. *Applied Herpetology*, **2**, 355–71.

Olson, D. H., Anderson, P. D., Frissell, C. A., Welsh, Jr, H. H., and Bradford, D. F. (2007). Biodiversity management approaches for stream–riparian areas: perspectives for Pacific Northwest headwater forests, microclimates, and amphibians. *Forest Ecology and Management*, **246**, 81–107.

Palen, W. J., Schindler, D. E., Adams, M. J., Pearl, C. A., Bury, R. B., and Diamond, S. A. (2002). Optical characteristics of natural waters protect amphibians from UV-B in the U. S. Pacific Northwest. *Ecology*, **83**, 2951–7.

Pellet, J., Guisan, A., and Perrin, N. (2004). A concentric analysis of the impact of urbanization on the threatened European tree frog in an agricultural landscape. *Conservation Biology*, **18**, 1599–1606.

Peterjohn, B. G., Sauer, J. R., and Robbins, C. S. (1995). Population trends from the North American Breeding Bird Survey. In T. E. Martin and D. M. Finch (eds), *Ecology and Management of Neotropical Migrants*, pp. 3–39. Oxford University Press, New York.

Platt, J. R. (1964). Strong inference. *Science*, **146**, 347–53.

Pounds, J. A. (2001). Climate and amphibian declines. *Nature*, **410**, 639–40.

Royle, J. A. and Nichols, J. D. (2003). Estimating abundance from repeated presence–absence data or point counts. *Ecology*, **84**, 777–90.

Royle, J. A., Nichols, J. D., and Kéry, M. (2005). Modelling occurrence and abundance of species when detection is imperfect. *Oikos*, **110**, 353–9.

Schmidt, B. R. (2004). Declining amphibian populations: the pitfalls of count data in the study of diversity, distributions, dynamics, and demography. *Herpetological Journal*, **14**, 167–74.

Schmidt, B. R. (2005). Monitoring the distribution of pond-breeding amphibians when species are detected imperfectly. *Aquatic Conservation: Marine and Freshwater Ecosystems*, **15**, 681–92.

Thompson, S. K. (1992). *Sampling*. John Wiley & Sons, New York.

Underwood, A. J. (1998). Design, implementation, and analysis of ecological and environmental experiments. In W. J. Resetarits, Jr. and J. Bernardo (eds), *Experimental Ecology: Issues and Perspectives*, pp. 325–49. Oxford University Press, New York.

Van Buskirk, J. (2005). Local and landscape influence on amphibian occurrence and abundance. *Ecology*, **86**, 1936–47.

Weir, L. A. and Mossman, M. J. (2005). North American Amphibian Monitoring Program (NAAMP). In M. Lannoo (ed.), *Amphibian Declines: The Conservation Status of United States Species*, pp. 307–13. University of California Press, Berkeley, CA.

Welsh, Jr, H. H. and Droege, S. (2001). A case for using plethodontid salamanders for monitoring biodiversity and ecosystem integrity of North American forests. *Conservation Biology*, **15**, 558–69.

Werner, E. E. (1998). Ecological experiments and a research program in community ecology. In W. J. Resetarits, Jr and J. Bernardo (eds), *Experimental Ecology: Issues and Perspectives*, pp. 3–26. Oxford University Press, New York.

Williams, B. K., Nichols, J. D., and Conroy M. J. (2002). *Analysis and Management of Animal Populations*. Academic Press, San Diego, CA.

24

Capture–mark–recapture, removal sampling, and occupancy models

Larissa L. Bailey and James D. Nichols

24.1 Introduction

Understanding the distribution and abundance of organisms is frequently stated as the objective of ecological investigations (Elton 1927; Krebs 1972). Similarly, distribution and abundance are primary criteria used to classify the status of species (e.g. threatened, endangered) for conservation purposes (Gardenfors *et al.* 2001). Thus, both scientific and conservation perspectives lead us to select abundance and occupancy as reasonable state variables for investigation. We define abundance as the number of organisms present in some area of interest and occupancy as the proportion of sample units or, more generally, of area, that is occupied by a species of interest.

Estimation of these quantities is typically based on count statistics of some sort. For example, abundance estimation may be based on the number of animals caught in traps, detected by visual encounter surveys, or perhaps by auditory means. Estimation of occupancy is typically based on detection/non-detection data (frequently referred to as presence/absence data) representing counts of sample units at which the focal species is (or is not) detected. Use of count statistics to draw inferences about such quantities as abundance and occupancy requires attention to two important sources of variation in counts (Pollock *et al.* 2002; Williams *et al.* 2002; Lancia *et al.* 2005).

The first important source of variation is spatial, with counts, abundance, and occupancy of organisms varying over space. It is frequently not possible to count organisms over the entire area about which inference is sought, requiring selection of a subset of geographic sample units for survey. Selection of these units must be done in such a way that counts on these units permit inference about the units not selected, even in the face of geographic variation. Many such sampling designs have been developed (e.g. simple random sampling, stratified random sampling, adaptive cluster sampling), and we refer the reader to sources that describe these designs in some detail (Cochran 1977; Williams *et al.* 2002).

The second important source of variation concerns the recognition that counts, whether of organisms or occupied sample units, are nearly always incomplete, in that some organisms and species go undetected no matter how intensive the sampling effort. Detection probability associated with the count must be estimated to provide information needed to translate the count(s) into an inference about abundance or occupancy. Various means of estimating detection probabilities have been developed for both counts of organisms (Seber 1982; Williams *et al.* 2002; Lancia *et al.* 2005) and counts of sample units occupied by a focal species (MacKenzie *et al.* 2002, 2006; Royle and Dorazio 2008).

Both scientific and conservation interests frequently focus not only on the values of relevant state variables, but also on the estimation of the vital rates responsible for the dynamics of those state variables. Rates of survival, reproduction, and movement in and out of the population are responsible for all changes in abundance of organisms and are thus important topics of study. Similarly, rates of local extinction and colonization determine the dynamics of species occurrence across units of the landscape. Ecological investigations of population and occupancy dynamics frequently focus on the environmental and ecological variables that induce changes in the relevant vital rates. Conservation requires actions that influence vital rates in the desired manner. As with estimation of abundance, inference about vital rates requires models that explicitly incorporate detection probabilities that reflect the incompleteness of virtually all wildlife sampling.

In this chapter we describe methods that we believe should be useful for estimating abundance, occupancy, and their respective vital rates in studies of amphibian populations. Our intention is to provide the reader with information about how these methods work, as well as an entrée to the scientific literature that describes these methods in detail.

24.2 Estimating amphibian population size and vital rates

24.2.1 Marking: tag type and subsequent encounter

Robust estimation of amphibian abundance and the vital rates that cause changes in abundance (e.g. survival and movement) usually require capturing and marking individual animals from the population(s) of interest. Numerous methods have been used to 'mark' amphibians and most require physical handling of individuals during an initial capture and marking occasion (see Chapters 8 and 11 in this volume for some methods). 'Marking' can involve applying a mark, tag, or radiotransmitter, or simply photographing or otherwise recording individuals with unique natural patterns, or obtaining unique identification via genetic material. Collectively, we group these marking methods into two sets: those permitting unique, individual identification on subsequent encounters and those where the individual can only be identified as marked or unmarked (so-called batch marks). Batch marks are relatively easy to apply and may identify the time, location, or life-history stage when the individual was first caught, but batch marks are generally not as useful for estimating survival and movement probabilities. In the next section we will discuss capture–mark–recapture models with the understanding that marked individuals can be uniquely identified. We will discuss the use of batch-marked animals for abundance estimation in the subsequent section on removal sampling (section 24.2.3).

24.2.2 Capture–mark–recapture

Capture–mark–recapture studies typically involve one to a few study areas or populations. The duration of the sampling period, the intensity of sampling within the period, and the length of time between successive sampling periods can vary substantially depending on study objectives. These features (objectives and sampling design) are critical components that determine which capture–mark–recapture methods are most appropriate for a given study. Capturing, marking, and recapturing individuals within a single day have been used to estimate amphibian abundance at different life-history stages: egg masses (Grant *et al.* 2005) and tadpoles (Jung *et al.* 2002). Alternatively, many studies of amphibian populations involve capturing individuals during the species' breeding season over multiple years, where each sampling period may be 2–12 weeks (Schmidt and Anholt 1999; Fretey *et al.* 2004; Church *et al.* 2007).

Data resulting from capture–mark–recapture studies are summarized into individual capture histories, which are simply rows of ones and zeros indicating whether an individual was captured (1) or not (0) during each sampling occasion. For example, a study of terrestrial or stream-side salamanders might sample an area for four consecutive days yielding individuals with the following capture history: 0 1 1 0. These individuals were first captured and marked on day 2, recaptured on the third day, but not encountered on the fourth and final day of sampling. If we assume that the population exposed to sampling does not change over the 4-day period, then these data can be modeled using closed-population models (closed population implies no births, deaths, immigration, or emigration; Williams *et al.* 2002). Here, we define p_t as the capture probability for an unmarked individual in the population exposed to sampling on day t, and c_t as the probability of recapture for previously marked animals. Assuming the capture events are independent, we would model the above capture history, 0 1 1 0, as:

$$\Pr(0\ 1\ 1\ 0) = (1-p_1)p_2c_3(1-c_4)$$

Since all individuals in the exposed population are assumed to be present at the initial sampling occasion, $(1-p_1)p_2$ denotes the probability a salamander was missed on the first day and captured for the first time on day 2. The probability of encountering an individual may change after first capture: thus the remainder of the expression indicates the probabilities associated with the events of recapture on day 3, c_3, but not on day 4, $(1-c_4)$.

We would compile similar probability expressions for each captured salamander. The product of these probabilities over all captured salamanders would constitute a model (i.e. the likelihood) for the entire data set, and estimates of the capture and recapture probabilities could be obtained. In this model, the likelihood is conditioned on the total number of individuals captured during the entire study, M_{K+1}, where K indexes the final sample occasion. Abundance N can be estimated as a derived parameter, specifically as:

$$\hat{N} = \frac{M_{K+1}}{\left[1-(1-\hat{p}_1)(1-\hat{p}_2)(1-\hat{p}_3)(1-\hat{p}_4)\right]} \tag{24.1}$$

Abundance is simply estimated as the total number of animals caught, divided by the probability that a member of the population is caught at least once during the study. This latter probability (the denominator of eqn 24.1) is obtained as the complement of the probability that the animal is not captured at any of the four sample occasions. Modern closed-population models include conditional likelihood models (Huggins 1989, 1991) and full likelihood models where the abundance parameter, N, remains in the likelihood (Otis *et al.* 1978; see Williams *et al.* 2002 and Chao and Huggins 2005 for detailed reviews). Parameter estimates can be obtained for various models using common, free software packages such as programs MARK (White and Burnham 1999) and CAPTURE (Rexstad and Burnham 1991).

In most studies related to amphibian ecology and conservation, the point estimates of abundance and various capture probabilities are not of primary interest. Rather, biologists are more interested in the relationships between these parameters and relevant environmental, habitat, or biological variables (i.e. covariates).

Usually, biologists have competing hypotheses about the influence of specific covariates on model parameters, and these hypotheses are translated into models that are then fit to data using covariate modeling. For example, suppose capture and recapture probabilities were thought to be equivalent but to vary with temperature on the day of the sampling occasion, and both capture probabilities and local abundances were thought to differ between two groups of individuals (say males and females). Capture probability can be modeled as a linear-logistic function of both temperature and sex:

$$p_{tg} = c_{tg} = \frac{e^{\beta_0 + \beta_1 x_t + \beta_2 x_g}}{1 + e^{\beta_0 + \beta_1 x_t + \beta_2 x_g}} \tag{24.2}$$

where x_t is the temperature at sampling occasion t and x_g is an indicator variable denoting whether the individual is male ($x_g = 0$) or female ($x_g = 1$). This additive model contains three parameters—β_0, β_1 and β_2—with β_1 and β_2 defining the relationships between capture probability and temperature and sex, respectively. Abundance estimates for males and females can be calculated using eqn 24.1, where both the numerator and denominator are gender-specific. Inference about the relationship between capture probability and both temperature and sex can be based on confidence intervals of the estimates of β_1 and β_2, and on comparisons of the model incorporating the relationship of eqn 24.2 with other models that do not include these relationships. Such model comparisons could be based on model selection statistics (Burnham and Anderson 2002) or on likelihood ratio tests in the case of nested models (e.g. Williams *et al.* 2002).

Program MARK has numerous classes of closed-population models that include conditional and full likelihood models, as well as models that allow for individual heterogeneity in capture probabilities that cannot be modeled via covariates (Norris and Pollock 1996; Pledger 2000). Discrimination among competing biological hypotheses is accomplished by first fitting their corresponding models to a given set of capture histories and then computing model selection statistics (Burnham and Anderson 2002). This approach to model selection presupposes that at least one model in the model set fits the data adequately. Goodness-of-fit tests have been developed for closed-population models (Otis *et al.* 1978), but they do not always perform well. Thus, practitioners are encouraged to use model-averagee estimates of abundance based on supported models in the candidate set (Stanley and Burnham 1998).

The application of likelihood-based closed-population capture–mark–recapture models to amphibian populations is surprisingly rare (but see Bailey *et al.* 2004b; Jung *et al.* 2002, 2005). Most inference about factors contributing to variation in amphibian abundance is still based on relative abundance indices (counts). The lack of use of closed-population models may relate to suspected violations in the closure assumption and the belief that the population of amphibians available for capture is likely to change between sampling occasions. Tests of the closure assumption have been developed by Otis *et al.* (1978) and Stanley and Burnham (1999). This topic is discussed in detail in Chapter 25, and practitioners should consider carefully both the target population and associated components of capture probability when designing abundance-related studies.

Often, it is the processes that cause change in a population that may be the primary interest for amphibian research and conservation. Investigation of these processes requires the use of capture–mark–recapture models that are "open" to changes in population size and composition between sampling periods. There are four fundamental demographic parameters responsible for changes in amphibian abundance: reproduction, mortality, emigration, and immigration. The estimation of these parameters and their relationship to environmental and habitat variables is key to understanding the dynamics of amphibian populations and predicting possible responses to management or conservation actions. Methods for estimating amphibian reproduction are discussed in previous chapters in this volume (Chapters 4 and 9), and here we focus primarily on estimating survival and movement probabilities.

Consider our previous four-occasion study, but let's assume that the four occasions represent successive years. Now our observed capture history, 0 1 1 0, is interpreted as an individual that was first captured, marked, and released in year 2, recaptured in year 3, and not seen in year 4. The first capture–mark–recapture models for open populations were conditional on first release (year 2 in this case), in the sense that the initial capture is used as a starting point and is not itself modeled. The classic Cormack–Jolly–Seber (CJS) model (Cormack 1964; Jolly 1965; Seber 1965) contains two types of parameter: p_t is the probability that an amphibian present in the study area is captured during occasion t, and φ_t is the probability that an amphibian alive and present in the study area during occasion t is still alive and in the study area at occasion $t + 1$ (apparent survival probability). Thus, we would write a model for the observed capture history, 0 1 1 0, as:

$$\Pr (0\,1\,1\,0\,|\text{ released in year 2}) = \varphi_2\,p_3\left[\varphi_3(1 - p_4) + (1 - \varphi_3)\right] = \varphi_2\,p_3(1 - \varphi_3\,p_4)$$

The first two terms, $\varphi_2 p_3$, represent the probability the amphibian survives between periods 2 and 3 and is recaptured in year 3 (corresponding to the events 1 1 in the capture history). The remainder of the expression $[\varphi_3(1 - p_4) + (1 - \varphi_3)]$ represents two possibilities: (1) the amphibian survived and remained in the study area from year 3 to 4, but was not captured in year 4, or (2) the amphibian died or permanently emigrated from the study area between year 3 and 4. Probability statements such as this are compiled for each capture history (i.e. each individual) and combined into a likelihood function. Estimates can be obtained via maximum likelihood using computer software such as programs MARK (White and Burnham 1999) or M-SURGE (Choquet *et al.* 2005).

Many extensions are available for this sort of modeling. Parameters can be constrained to be constant over time, they can be modeled as functions of group (e.g. males and females) membership, and they can be modeled as functions of time-specific or individual-specific covariates (Lebreton *et al.* 1992). So-called unconditional modeling also permits the estimation of time-specific abundance, recruitment, and even population growth rate (see Williams *et al.* 2002). Program MARK (White and Burnham 1999) provides user-friendly software for such modeling.

These models for open populations are capable of dealing with population gains and losses between sampling periods, but they do not provide a means of distinguishing movement from other processes. Indeed, under the CJS model, the processes of emigration and immigration are confounded with survival and capture probabilities. For example, if individuals permanently emigrate from the study area, then the complement of 'apparent survival' probability will include both mortality and permanent emigration (e.g. Burnham 1993). In the case of random temporary emigration, CJS estimates of capture probability represent the product of the probability that the individual is not a temporary emigrant and the probability that the individual is caught, conditional on presence (Kendall *et al.* 1997; see also Chapter 25 in this volume). Temporary emigration is particularly important in amphibian ecology, as individuals are often unavailable for capture during specific life-history phases (i.e. post-metamorphic juveniles or adults that skip breeding opportunities). Movement processes can be separately estimated in some cases using multistate models (see below) or models that incorporate extra sources of information (Burnham 1993; Kendall *et al.* 1997; Fujiwara and Caswell 2002; Kendall and Nichols 2002; Williams *et al.* 2002; Bailey *et al.* 2004a; Schaub *et al.* 2004).

Multistate models are open-population capture–mark–recapture models that can be used when animals are categorized by some state variable that can be assessed each time an animal is encountered (Arnason 1972, 1973; Brownie *et al.* 1993; Schwarz *et al.* 1993). When location or study site is the state, direct inferences about movement among sampled locations are possible (e.g. Hestbeck *et al.* 1991; Brownie *et al.* 1993). In addition to location, state variables can include characteristics of the captured animals themselves. For example, size class is a reasonable state variable in some amphibian studies (Wood *et al.* 1998).

Multistate models can also be used when one or more of the states may be unobservable. Such uses focus on situations in which animals move back and forth between study sites, where capture and resighting efforts are made, and other locations at which no sampling effort is expended. For example, when amphibians are sampled at breeding sites, non-breeding animals are temporary emigrants and thus unobservable. In the simple two-state case, the model contains parameters that are state-specific, r, where $r = O$ denotes the observable or breeding state, and $r = U$ denotes the unobservable or non-breeding state. Parameters include: p_t^o, the aforementioned capture probability which is non-zero for breeders only ($p_t^U = 0$, by definition); S_t^r, the probability that an individual in state r at time t survives until time $t + 1$; and ψ_t^{rs}, the probability that an individual that is in state r at time t, and survives until time $t + 1$, is in state s at time $t + 1$. Using this parameterization we would write a model for the observed capture history, 0 1 1 0 1, as:

$$\Pr(0\,1\,1\,0\,1 \mid \text{released in year 2}) = S_2^O \psi_2^{OO} p_3^O S_3^O \left[(\psi_3^{OO}(1 - p_4^O) S_4^O \psi_4^{OO} + \psi_3^{OU} S_4^U \psi_4^{UO}) \right] p_5^O$$

The fact that the individual was not captured in year 4 could result from one of two processes: either the amphibian was in the study area but not captured, or the amphibian was not in the study (or breeding) area in year 4, but returned to the area in year 5. Note that the above parameterization assumes that survival between sampling periods t and $t + 1$ depends on state at occasion t, e.g. S_t^r. For situations in which this assumption is not appropriate, it is possible to combine the survival and transition parameters as $\phi_t^{rs} = S_t^r \psi_t^{rs}$ and to estimate this survival-transition parameter (ϕ_t^{rs}) directly.

Parameters of these multistate models can be estimated using maximum likelihood methods and modeled as a function of relevant covariates. Useful recent software includes programs MARK (White and Burnham 1999) and M-SURGE (Choquet *et al.* 2005). Practitioners using these models should verify that all parameters are uniquely identifiable as there are some cases where multiple parameter values will yield the same maximized likelihood values (Kendall and Nichols 2002; Gimenez *et al.* 2004; Schaub *et al.* 2004; Bailey *et al.* 2009; Hunter and Caswell 2009). Simplifying assumptions, such as equating parameters over time or state, may be necessary (Kendall and Nichols 2002; Schaub *et al.* 2004; Bailey *et al.* 2009; Hunter and Caswell 2009), but the models offer tremendous flexibility and the ability to estimate important demographic parameters that have long eluded researchers (Biek *et al.* 2002).

The original robust design (Pollock 1982; Kendall *et al.* 1997) combines open and closed models and involves sampling at two temporal scales: each primary period (e.g. year) consists of two or more secondary occasions that are closely spaced in time, such that the population may be considered demographically and geographically closed. For example, during the breeding season individuals at a pool may be sampled multiple times over the course of a few days. Each independent sample of the breeding population would be considered a secondary occasion, and this process would be repeated for several years (primary periods), over which time survival, breeding and/or movement probabilities could be estimated (see Kendall 2004 for review of robust design models). The robust design typically yields improved precision of vital rate parameter estimates, but can also be used to estimate period- (year-) specific abundance, while accounting for modeled or unmodeled heterogeneity in capture probabilities. The robust design also permits direct estimation of probabilities of temporary emigration (Kendall *et al.* 1997) and of the separate contributions of recruitment from immigration and *in situ* reproduction (Nichols and Pollock 1990; Nichols *et al.* 2000; see Schmidt *et al.* 2005 for an amphibian example).

24.2.3 Removal sampling

Removal sampling is simply a special case of a closed-population abundance model where recapture probability is zero (i.e. $c_t \neq p_t$, where $c_t = 0$; Otis *et al.* 1978; White *et al.* 1982). On each sampling occasion, captured individuals are removed from the population in some manner. Removal can be accomplished by either physically removing individuals from the site (e.g. temporarily retaining individuals until sampling is completed) or by batch marking individuals. The data then consist of the number of new individuals captured on each sampling occasion. As with closed models, one can directly model the effect of sampling effort (e.g. hours of search) or other time-specific covariates on the probability of initial capture. For applications of removal models to estimate amphibian abundances, see Bruce (1995), Jung *et al.* (2002), and Bailey *et al.* (2004b).

24.3 Estimating amphibian occupancy and vital rates

Macroecological studies focus on the distribution of amphibians across the landscape. Even in large-scale studies for which abundance is of interest, abundance estimation at many study locations may be prohibitively expensive. In these and other cases, biological hypotheses and conservation objectives may shift from individual populations to the number or proportion of patches occupied by a target amphibian species. In these cases it may be reasonable to use occupancy, or the proportion of sites that are occupied, as a state variable.

Quantitative occupancy modeling has exhibited rapid development recently, with much of this work motivated by amphibian systems (Muths *et al.* 2005). These methods can be used for single or multiple species over one to many seasons. All methods acknowledge the likely scenario that species are not always detected when present and emphasize that reliable inference can still be made from detection/non-detection information if detection and occupancy probabilities are simultaneously estimated. MacKenzie and his colleagues offer an initial review of these methods and apply them to several amphibian problems (MacKenzie *et al.* 2006). Here we consider methods that estimate occupancy and associated vital rates for a single species only.

24.3.1 Occupancy estimation

Occupancy studies typically involve large areas containing numerous sampling units, or sites. These sites may be naturally occurring patches of habitat (i.e. wetlands or stream reaches) or independent subunits, such as plots or transects, within different habitat types. A subset of sites is chosen using a probabilistic sampling method (e.g. simple or stratified random sample), and sites are sampled repeatedly over a period of time during which there is no change in the occupancy status at the sites (i.e. sites are either occupied or unoccupied by the target species during the sampling period). The possible duration of the sample period will depend on the species life history and behavioral characteristics.

During each site visit the target species is recorded as detected (1) or not (0), creating a detection history for each site reminiscent of the individual capture histories in capture–mark–recapture methods. The main difference between capture–mark–recapture and occupancy scenarios is that sites can have detection histories consisting only of zeros. Such histories never occur in capture–mark–recapture data, because the data are conditional on individuals that are captured/detected at least once. As with abundance estimation, a probabilistic model is used to describe the sequence of events that produced each detection history.

MacKenzie *et al.* (2002, 2006) defined two types of parameter for occupancy models: ψ_i represents the probability that site i is occupied by the target species, and p_{ij} is the probability of detecting the species at site i during the jth independent visit to the site. Dropping the i notation for each site, we would write a model for the observed detection history, 0 1 1 0 as:

$$\text{Pr } (0\ 1\ 1\ 0) = \psi(1 - p_1)p_2 p_3(1 - p_4)$$

The site is clearly occupied by the target species, detected during the second and third visit, but not detected during the first and fourth visit. A detection history consisting of all zeros, 0 0 0 0, would have two possible explanations: either the site was occupied, but the species was not detected during any visit, or the site was unoccupied. Written as a mathematical expression:

$$\Pr (0\ 0\ 0\ 0) = \psi(1-p_1)(1-p_2)(1-p_3)(1-p_4)+(1-\psi)$$

Again, the detection data and the corresponding probability model are combined to form a likelihood function, and estimates can be obtained via software such as program MARK (White and Burnham 1999) or program PRESENCE (MacKenzie et al. 2006; www.mbr-pwrc.usgs.gov/software/).

An alternative approach to estimation and modeling is to view the above models hierarchically. One model component concerns the true spatial process of species occurrence across the sampled sites. Conditional on this true spatial process, a sampling component models the detection process. These hierarchical components (process and sampling) are combined under a Bayesian framework, and statistical inference is achieved using Markov chain Monte Carlo (MCMC) methods (Royle and Dorazio 2008). Using either procedure, it is possible to model occupancy and detection probability as functions of measured covariates. Models incorporating different combinations of covariates represent competing hypotheses about factors believed to influence amphibian distribution or probability of detection. These occupancy models have gained popularity in amphibian studies around the globe (e.g. Bailey et al. 2004c; Royle 2004; Mazerolle et al. 2005; Muths et al. 2005; Pellet and Schmidt 2005; Royle and Link 2005) and represent an important improvement over logistic regression models of amphibian–habitat relationships that ignore imperfect detection (Gu and Swihart 2004; MacKenzie et al. 2006).

24.3.2 Estimation of occupancy vital rates: extinction and colonization

Amphibian systems are often viewed as collections of sites that are sometimes occupied by the target species, and sometimes not, depending on the dynamic processes of extinction and colonization. These vital rates are the processes by which occupancy status changes among sites throughout the landscape and can be the primary parameters of interest for many amphibian conservation and management programs (Muths et al. 2005). MacKenzie et al. (2003) extended the so-called single-season occupancy models discussed in the previous section to include these two dynamic parameters: ε_t is the probability that an occupied site in season t becomes unoccupied in season $t+1$ (local extinction) and γ_t is the probability that an unoccupied site in season t is occupied by the target species in season $t+1$ (colonization). These multi-season models still assume that all (or a subset of) sites are visited multiple times within a season, over a period where the occupancy state at each site is static (notice this design resembles the robust design in capture–mark–recapture studies). Probability models and likelihoods are developed in the usual fashion, and inference can be based on either maximum likelihood or MCMC implementation of hierarchical models. Covariates can be modeled and constraints can be imposed to address interesting biological hypotheses about variables believed to influence extinction and colonization probabilities (MacKenzie *et al.* 2003, 2006).

24.4 Summary and general recommendations

Understanding factors that influence amphibian abundances and distributions are fundamental to amphibian ecology and conservation. The estimation methods we discussed here should be useful to both scientific studies and conservation efforts, as they have the flexibility to accommodate both spatial and temporal variation while accounting for imperfect detection of individuals or species of interest. Scientific studies of amphibian population or occupancy dynamics typically involve efforts to discriminate among competing biological hypotheses about the processes that underlie such dynamics. These hypotheses can be translated into models that include both the ecological processes of interest and the sampling processes that give rise to the collected data. Conditional on selection of appropriate study design and field methods, the competing models can be fit to the resulting data and evaluated (e.g. via model selection criteria; Burnham and Anderson 2002).

Conservation decisions are also based on hypotheses, and their corresponding models, about how studied populations respond to potential management actions. The same process as described above for scientific studies is used to determine the degrees of faith that should be placed in the different models of population response to conservation or management action(s). In addition, conservation efforts require estimates of state variables (e.g. abundance or occupancy) for the purposes of making state-dependent decisions and assessing the degree to which conservation objectives are being met. The inference methods described in this chapter should be useful in the conduct of science and conservation for amphibian populations.

Sampling design should be carefully considered when designing capture–mark–recapture or occupancy studies. Pilot data or guesses about capture/detection probabilities can be used with recently-developed simulation- or approximation-based software to compare estimator precision for different sample sizes and sampling scenarios (see Devineau *et al.* 2006 for capture–mark–recapture studies and Bailey *et al.* 2007 for occupancy studies). Common capture–mark–recapture recommendations include trying to achieve high (> 0.20) capture probabilities, minimizing variation in p among individuals, and collecting important covariates likely to influence capture probabilities. These recommendations are echoed for the species detection probabilities used to estimate occupancy and associated dynamic parameters (see MacKenzie *et al.* 2002, 2006).

Ultimately, sampling design, including data collection and selection of inference methods, should be inherited from the nature and objectives of the larger scientific or conservation program. Conditional on the larger program, the methods discussed in this chapter can be tailored to meet program needs. The evolution of capture–mark–recapture and occupancy inference methods has been characterized by substantive interactions between statisticians and both ecologists and conservation biologists, with model development being motivated almost exclusively by the needs of ecologists. Many of the recent advances in occupancy methods were motivated by collaborations between amphibian researchers and statisticians (Muths *et al.* 2005; MacKenzie *et al.* 2006; Royle and Dorazio 2008), and we expect such interactions to continue and to lead to development of even more useful inference methods in the future.

24.5 Disclaimer

Any use of trade, product, or firm names is for descriptive purposes only and does not imply endorsement by the US Government.

24.6 References

Arnason, A. N. (1972). Parameter estimates from mark-recapture experiments on two populations subject to migration and death. *Researches on Population Ecology*, **13**, 97–113.

Arnason, A. N. (1973). The estimation of population size, migration rates, and survival in a stratified population. *Researches on Population Ecology*, **15**, 1–8.

Bailey, L. L., Kendall, W. L., Church, D. R., and Wilbur, H. M. (2004a). Estimating survival and breeding probability for pond-breeding amphibians: A modified robust design. *Ecology*, **84**, 2456–66.

Bailey, L. L., Simons, T. R., and Pollock, K. H. (2004b). Comparing population size estimators for plethodontid salamanders. *Journal of Herpetology*, **38**, 370–80.

Bailey, L. L., Simons, T. R., and Pollock, K. H. (2004c). Estimating site occupancy and species detection probability parameters for terrestrial salamanders. *Ecological Applications*, **14**, 692–702.

Bailey, L. L., Hines, J. E., Nichols, J. D., and MacKenzie, D. I. (2007). Sampling design trade-offs in occupancy studies with imperfect detection: examples and software. *Ecological Applications* **17**, 281–90.

Bailey, L. L., Kendall, W. L., and Church, D. R. (2009). Exploring extensions to multi-state models with multiple unobservable states. In D. L. Thomson, E. G. Cooch, and M. C. Conroy (eds), *Modeling Demographic Processes in Marked Populations*, pp. 693–710. Springer Science+Business Media, New York.

Biek, R., Funk, W. C., Maxell, B. A., and Mills, L. S. (2002). What is missing in amphibian decline research: Insights from ecological sensitivity analysis. *Conservation Biology*, **16**, 728–34.

Brownie, C., Hines, J. E., Nichols, J D., Pollock, K. H., and Hestbeck, J. B. (1993). Capture-recapture studies for multiple strata including non-Markovian transitions. *Biometrics*, **49**, 1173–87.

Bruce, R. C. (1995). The use of temporary removal sampling in a study of population dynamics of the salamander *Desmognathus monticola*. *Australian Journal of Ecology*, **20**, 403–12.

Burnham, K. P. (1993). A theory for combined analysis of ring recovery and recapture data. In J. D. Lebreton and P. M. North (eds), *Marked Individuals in the Study of Bird Populations*, pp. 199–213. Birkhauser-Verlag, Basel.

Burnham, K. P. and Anderson, D. R. (2002). *Model Selection and Multimodel Inference.* Springer-Verlag, New York.

Chao, A. and Huggins, R. M. (2005). Modern closed-population capture-recapture models. In S. C. Amstrup, T. L. McDonald, and B. F. J. Manly (eds), *Handbook of Capture-Recapture Analysis*, pp. 58–87. Princeton University Press, Princeton, NJ.

Choquet, R., Reboulet, A. M., Pradel R., Gimenez, O., and Lebreton, J. D. 2005. *M-SURGE 1–8 User's Manual.* CEFE, Montepellier. ftp.cefe.cnrs.fr/biom/soft-cr/.

Church, D. R., Bailey, L. L., Wilbur, H. M., Kendall, W. L., and Hines, J. E. (2007). Iteroparity in the variable environment of the salamander *Ambystoma tigrinum. Ecology*, **88**, 891–903.

Cochran, W. G. (1977). *Sampling Techniques.* Wiley, New York.

Cormack, R. M. (1964). Estimates of survival from the sighting of marked animals. *Biometrika*, **51**, 429–38.

Devineau, O., Choquet, R., and Lebreton, J. D. (2006). Planning capture-recapture studies: straightforward precision, bias, and power calculations. *Wildlife Society Bulletin*, **34**, 1028–35.

Elton, C. (1927). *Animal Ecology.* Sidgwick & Jackson, London.

Fretey, T., Cam, E., Le Garff, B., and Monnat, J. Y. (2004). Adult survival and temporary emigration in the common toad. *Canadian Journal of Zoology*, **82**, 859–72.

Fujiwara, M. and Caswell, H. (2002). A general approach to temporary emigration in mark-recapture analysis. *Ecology*, **83**, 3266–75.

Gardenfors, U., Hilton-Taylor, C., Mace, G. M., and Rodriguez, J. P. (2001). The application of IUCN red list criteria at regional levels. *Conservation Biology*, **15**, 1206–12.

Gimenez, O., Viallefont, A., Catchpole, E. A., Choquet, R., and Morgan, B. J. T. (2004). Methods for investigating parameter redundancy. *Animal Biodiversity and Conservation*, **27**, 561–72.

Grant, E. H. C., Jung, R. E., Nichols, J. D., and Hines, J. E. (2005). Double-observer approach to estimating egg mass abundance of pool-breeding amphibians. *Wetlands Ecology and Management*, **13**, 305–20.

Gu, W. and Swihart, R. K. (2004). Absent or undetected? Effects of non-detection of species occurrence on wildlife-habitat models. *Biological Conservation*, **116**, 195–203.

Hestbeck, J. B., Nichols, J. D., and Malecki, R A. (1991). Estimates of movement and site fidelity using mark resight data of wintering Canada geese. *Ecology*, **72**, 523–33.

Huggins, R. M. (1989). On the statistical analysis of capture experiments. *Biometrika*, **76**, 133–40.

Huggins, R. M. (1991). Some practical aspects of conditional likelihood approach to capture experiments. *Biometrics*, **47**, 725–32.

Hunter, C. M. and Caswell, H. (2009). Rank and redundancy of multistate mark-recapture models for seabird populations with unobservable states. In D. L. Thomson, M. C. Conroy, and E. G. Cooch (eds), *Modeling Demographic Processes in Marked Populations*, pp. 799–828. Springer Science+Business Media, New York.

Jolly, G. M. (1965). Explicit estimates from capture-recapture data with both death and immigration-Stochastic model. *Biometrika*, **52**, 225–47.

Jung, R. E., Dayton, G. H., Williamson, S. J., Sauer, J. R., and Droege, S. (2002). An evaluation of population index and estimation techniques for tadpoles in desert pools. *Journal of Herpetology*, **36**, 465–72.

Jung, R. E., Royle, J. A., Sauer, J. R., Addison, C., Rau, R. D., Shirk, J. L., and Whissel, J. C. (2005). Estimation of stream salamander (Plethodontidae, Desmognathinae and Plethodontinae) populations in Shenandoah National Park, Virginia, USA. *Alytes*, **22**, 72–84.

Kendall, W. L. (2004). Coping with unobservable and mis-classification state in capture-recapture studies. *Animal Biodiversity and Conservation*, **27**, 97–107.

Kendall, W. L. and Nichols, J. D. (2002). Estimating state-transition probabilities for unobservable states using capture-recapture/resighting data. *Ecology*, **83**, 3276–84.

Kendall, W. L., Nichols, J. D., and Hines, J. E. (1997). Estimating temporary emigration using capture-recapture data with Pollock's robust design. *Ecology*, **78**, 563–78.

Krebs, C. J. 1972. *Ecology*. Harper and Row, New York.

Lancia, R. A., Kendall, W. L., Pollock, K. H., and Nichols, J. D. (2005). Estimating the number of animals in wildlife populations. In C. E. Braun (ed.), *Techniques for Wildlife Investigations and Management*, 6th edn, pp. 106–53. The Wildlife Society, Bethesda, MD.

Lebreton, J. D., Burnham, K. P., Clobert, J., and Anderson, D. R. (1992). Modeling survival and testing biological hypotheses using marked animals—a unified approach with case-studies. *Ecological Monographs*, **62**, 67–118.

MacKenzie, D. I., Nichols, J. D., Lachman, G. B., Droege, S., Royle, J. A., and Langtimm, C. A. (2002). Estimating site occupancy rates when detection probabilities are less than one. *Ecology*, **83**, 2248–55.

MacKenzie, D. I., Nichols, J. D., Hines, J. E., Knutson, M. G., and Franklin, A. B. (2003). Estimating site occupancy, colonization and local extinction probabilities when species are not detected with certainty. *Ecology*, **84**, 2200–7.

MacKenzie, D. I., Nichols, J. D., Royle, J. A., Pollock, K. H., Bailey, L. L., and Hines, J. E. (2006). *Occupancy Estimation and Modeling: Inferring Patterns and Dynamics of Species Occurrence*. Academic Press, Boston, MA.

Mazerolle, M. J., Desrochers, A., and Rochefort, L. (2005). Landscape characteristics influence pond occupancy by frogs after accounting for detectability. *Ecological Applications*, **15**, 824–34.

Muths, E., Jung, R. E., Bailey, L. L., Adams, M. J., Corn, P. S., Dodd, Jr, C. K., Fellers, G. M., Sadinski, W. J., Schwalbe, C. R., Walls, S. C. *et al.* (2005). Amphibian Research and Monitoring Initiative (ARMI): a successful start to a national program in the United States. *Applied Herpetology*, **2**, 355–71.

Nichols, J. D. and Pollock, K. H. (1990). Estimation of recruitment from immigration versus in situ reproduction using Pollock's robust design. *Ecology*, **71**, 21–6.

Nichols, J. D., Hines, J. E., Lebreton, J.-D., and Pradel, R. (2000). Estimation of contributions to population growth: a reverse-time capture-recapture approach. *Ecology*, **81**, 3362–76.

Norris, J. L. and Pollock, K. H. (1996). Nonparametric MLE under two closed capture recapture models with heterogeneity. *Biometrics*, **52**, 639–49.

Otis, D. L., Burnham, K. P., White, G. C., and Anderson, D. R. (1978). Statistical-inference from capture data on closed animal populations. *Wildlife Monographs*, **62**, 1–135.

Pellet, J. and Schmidt, B. R. (2005). Monitoring distributions using call surveys: estimating site occupancy, detection probabilities and inferring absence. *Biological Conservation*, **123**, 27–35.

Pledger, S. (2000). Unified maximum likelihood estimates for closed capture- recapture models using mixtures. *Biometrics*, **56**, 434–42.

Pollock, K. H. (1982). A capture-recapture design robust to unequal probability of capture *Journal of Wildlife Management*, **46**, 757–60.

Pollock, K. H., Nichols, J. D., Simons, T. R., Farnsworth, G. L., Bailey, L. L., and Sauer, J. R. (2002). Large scale wildlife monitoring studies: statistical methods for design and analysis. *Environmetrics*, **13**, 105–19.

Rexstad, E. and Burnham, K. P. (1991). *User's Guide for Interactive Program CAPTURE*. www.mbr-pwrc.usgs.gov/software/. Colorado Cooperative Fish and Wildlife Unit, Fort Collins, CO.

Royle, J. A. (2004). Modeling abundance index data from anuran calling surveys. *Conservation Biology*, **18**, 1378–85.

Royle, J. A. and Link, W. A. (2005). A general class of multinomial mixture models for anuran calling survey data. *Ecology*, **86**, 2505–12.

Royle, J. A. and Dorazio, R. M. (2008). *Hierarchical Modelling and Inference in Ecology*. Academic Press, San Diego, CA.

Schaub, M., Gimenez, O., Schmidt, B. R., and Pradel, R. (2004). Estimating survival and temporary emigration in the multistate capture-recapture framework. *Ecology*, **85**, 2107–13.

Schmidt, B. R. and Anholt, B R. (1999). Analysis of survival probabilities of female common toads, *Bufo bufo*. *Amphibia-Reptilia*, **20**, 97–108.

Schmidt, B. R., Feldmann, R., and Schaub, M. (2005). Demographic processes under-lying population growth and decline in *Salamandra salamandra*. *Conservation Biology*, **19**, 1149–56.

Schwarz, C. J., Schweigert, J. F., and Arnason, A. N. (1993). Estimating migration rates using tag-recovery data. *Biometrics*, **49**, 177–93.

Seber, G. A. F. (1965). A note on the multiple-recapture census. *Biometrika*, **52**, 249–59.

Seber, G. A. F. (1982). *The Estimation of Animal Abundance and Related Parameters*. Macmillian, New York.

Stanley, T. R. and Burnham, K. P. (1998) Estimator selection for closed-population capture-recapture. *Journal of Agricultural Biological and Environmental Statistics*, **3**, 131–50.

Stanley, T. R. and Burnham, K. P. (1999). A closure test for time-specific capture-recapture data. *Environmental and Ecological Statistics*, **6**, 197–209.

White, G. C. and Burnham, K. P. (1999). Program MARK: survival estimation from populations of marked animals. *Bird Study*, **46**, 120–39.

White, G. C., Anderson, D. R., Burnham, K. P., and Otis, D. L. (1982). *Capture-recapture and Removal Methods for Sampling Closed Populations*. Los Alamos National Lab, Los Alamos, NM.

Williams, B. K., Nichols, J. D., and Conroy, M. J. (2002). *Analysis and Management of Animal Populations*. Academic Press, San Diego, CA.

Wood, K. V., Nichols, J. D., Percival, H. F., and Hines, J. E. (1998). Size-sex variation in survival rates and abundance of pig frogs, *Rana grylio*, in northern Florida wetlands. *Journal of Herpetology*, **32**, 527–35.

Quantifying abundance: counts, detection probabilities, and estimates

Benedikt R. Schmidt and Jérôme Pellet

25.1 Background: imperfect detection in amphibian ecology and conservation

Understanding temporal and spatial variation in distribution and abundance has been, and will remain, a central goal in amphibian ecology and conservation. Even though these two quantities, distribution and abundance, are so fundamental, we usually cannot observe them directly in the field. We rarely observe all individuals, all populations, or all species in an area of interest. Imperfect detection is the rule rather than the exception and is a characteristic that all field studies share.

Imperfect detection is a trivial fact and herpetologists are often aware of it. For example, Hairston and Wiley (1993) attributed all fluctuations in salamander counts to variation in weather conditions (salamanders tend to remain underground during cold weather) and motivation of students (a class of highly motivated students found an exceptionally high number of salamanders). This implies that salamander detection was imperfect and, equally important, variable among years (Hyde and Simons 2001).

Nevertheless, imperfect detection is all too often ignored when estimating abundance or when temporal and spatial trends are analysed. In this chapter we pinpoint patterns and the consequences of imperfect detection and show how to deal with it in general (Chapter 24 introduces many of the estimators required to deal with imperfect detection). We focus on the effects of imperfect detection on population abundance estimation but we call attention to the fact that imperfect detection also affects other ecological quantities, such as distribution and species richness (e.g. Schmidt 2004; Royle and Dorazio 2006; Mazerolle *et al.* 2007). We focus on how the issue of imperfect detection should be incorporated into the study of amphibians populations; that is, how to sample amphibian populations and how to estimate abundance. By doing so, we assume that clear study objectives have been formulated in advance (Yoccoz *et al.* 2001) and sites appropriately selected (see Chapter 23).

25.2 Imperfect detection

It is convenient to use a simple equation to conceptualize imperfect detection:

$$E(C) = Np$$

where N is the true value of the parameter of interest (i.e. number of individuals, density, number of populations in an area, or species richness) and p is the detection probability (Gill 1985; Yoccoz *et al.* 2001; Pollock *et al.* 2002; Schmidt 2004). $E()$ denotes a statistical expectation. The expectation $E(C)$ is the average of the count C over repeated realizations of the sampling process. This equation has three major implications that we discuss below.

Sampling a population should be viewed as a stochastic process because it involves a probability of detection. This is why there is a statistical expectation $E(C)$ in the above equation. Even under identical conditions we should not expect to obtain the same result if we sample the same population multiple times. We should therefore expect variability in the counts. Variability in counts (C) does not imply variation in abundance (N) or detection probability (p); it can simply be random variation. Technically, counts are random variables. This is illustrated in Figure 25.1. The figure shows that, as expected under

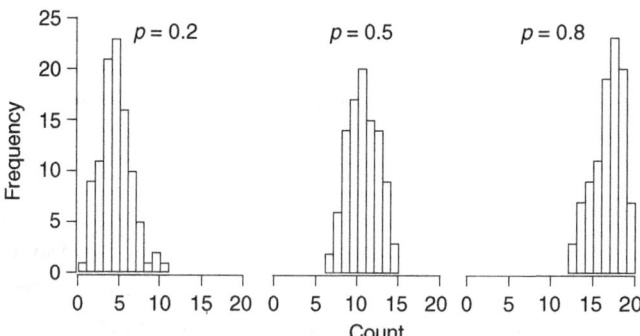

Fig. 25.1 Expected variation in counts of individuals under identical conditions for 100 repeated counts, a true population size $N = 20$ and detection probabilities $p = 0.2, 0.5,$ and 0.8. Data were simulated using the R code rbinom(n = 100, size = 20, p = x), where x had the values 0.2, 0.5, and 0.8, respectively.

binomial sampling, variability of the counts is greatest if $p = 0.5$. The range of likely counts is large no matter what the detection probability is.

25.2.1 Counts underestimate abundance

Because p can take any value between 0 and 1, counts almost always underestimate true abundance (or other biological parameters such as species richness or occupancy probabilities). Detection probabilities vary depending on the methods used to sample a population. In studies using drift fences, detection probabilities can be very close to 1 (Bailey *et al.* 2004a), meaning that very few individuals escape detection. When using other methods, such as hand capture, trapping, netting, or cover-board surveys, detection probabilities are usually far below 1 (see, for example, case studies and estimates of detection probabilities in Mazerolle *et al.* 2007). Consequently, the discrepancy between counts and true abundances increases with decreasing detection probability. Table 25.1 shows the correlations between counts and abundance where the true number of individuals was known. Counts and true abundances are somewhat correlated but the proportion of variance explained can be quite low.

Pellet *et al.* (2007) analysed the relationship between the chorus counts, number of captures and mark-recapture estimates of abundance in detail (Table 25.2). Their study of the European treefrog *Hyla arborea* showed that calling males represented a variable fraction of the total male population present at breeding ponds. In two sites studied over 3 years, the proportion of males calling varied between 0.32 and 0.65, suggesting that there is no solid link between chorus activity and (estimated) population size. Similarly, their capture effort (total number of males captured) did not truly reflect actual population size,

Table 25.1 *Results of linear regressions between counts (C), population size estimates ($\hat{N}$), and censuses for tadpoles of two anuran species. R^2 and F tests were calculated using PROC GLM in SAS. Asterisks indicate significance at $\alpha = 0.05$. The Lincoln–Peterson estimator was used to estimate abundance. Table adapted from Schmidt (2004). Data are from Jung et al. (2002).*

Species	Studied in	Intercept	Slope	R^2	F
Between count and census					
Hyla arenicolor	Natural ponds	−13.92	1.73	0.66	34.21*
Scaphiopus couchii	Mesocosms	12.44	1.17	0.96	283.1*
Between estimate and census					
Hyla arenicolor	Natural ponds	18.35	0.87	0.97	1005.0*
Scaphiopus couchii	Mesocosms	15.25	0.94	0.99	1509.0*

Table 25.2 *Counts and estimates of male European tree frogs* (Hyla arborea) *in two breeding aggregations. Maximum and mean chorus size, as well as number of males captured, correlate only weakly with actual (estimated) male population size, demonstrating the effects of imperfect detection on population size estimates. Adapted from Pellet* et al. *(2007).*

	Year	Maximum chorus size	Mean chorus size	Total males captured	Modeled male population size$\pm$SE	Proportion of calling males
Males in	2002	27	11.4	35	57.9$\pm$9	47%
population	2003	18	6.8	34	49.6$\pm$6.9	36%
Camp Romain	2004	20	12.5	75	62.1$\pm$15	32%
Males in	2002	25	6.9	29	38.5$\pm$5.7	65%
population Les	2003	15	7.2	30	30.8$\pm$1.3	49%
Mossières	2004	20	7.3	45	46.8$\pm$2.2	43%

thus demonstrating that population trends estimated from counts (callers or captures) are likely to reflect a mixture of population and detectability trends. In addition, Pellet *et al.* (2007) compared mean chorus counts and maximum chorus counts. Mean counts were only 25–63% of the maximum counts. Such variability probably arises from the fact that a maximum count is an extreme value that is likely to be highly variable spatially and temporally. We argue that the maximum count is not a suitable metric if the goal of a study is a comparison between years and/or sites. If anything, mean counts would be better comparable across sites and years than maxima (but see Schmidt and Pellet 2005 for a study where maximum counts predicted population persistence).

Funk *et al.* (2003) compared the efficiency of abundance estimates based on visual encounter surveys, distance sampling, and mark–recapture methods for monitoring population trends of forest-floor-dwelling *Eleutherodactylus* frogs. They found that mark–recapture methods were best at estimating abundance and the method had the greatest power to detect population declines. Like in the examples above, the message is clear: estimates clearly outperform counts in reflecting true population abundances.

25.2.2 Per-visit and cumulative detection probabilities

Detection probabilities come in two flavors: per visit and cumulative. Whereas per-visit detection probabilities may be low, cumulative detection probabilities are usually much higher. Cumulative detection probability, p_c, is given by

$$p_c = 1 - (1 - \bar{p})^n$$

where $\bar{p}$ is the average per-visit detection probability and n is the number of visits or capture events. Per-visit (i.e. per day) detection probabilities of the frog *Colostethus stepheni* in the study of Funk and Mills (2003) ranged from 0 to 1 with a mean of 0.58 (W.C. Funk, personal communication). Because Funk and Mills (2003) had six capture events during a short time period when the population did not change, the cumulative capture probabilities were greater than 0.987. This implies that Funk and Mills (2003) achieved an almost complete census of the frogs and suggests that multiple capture events are an obvious way to deal with imperfect detection because cumulative detection probabilities can be quite high.

25.2.3 Temporal and spatial variation in detection probabilities

Counts of animals are usually made to assess temporal or spatial variation in population size. Such comparisons are only valid if $E(p)$ remains constant in time or space; however, this is hardly ever the case (MacKenzie and Kendall 2002), even under strictly standardized methods. If p varies temporally and/or spatially then variation in p and N are confounded. In the form of an equation, this may be expressed as

$$E(T) = C_1/C_2 = N_1 p_1 / N_2 p_2$$

Spatial or temporal variation in p_i can lead to the detection of spurious temporal or spatial trends $E(T)$. The contrary may also be true: one may miss true patterns in abundance or (under- or) overestimate the effects of variables that are thought to explain variation in abundance (Mazerolle *et al.* 2005). Obviously, the absolute values of p matter. If p is high, then bias will be low. If p is low, however, then most variation in C is likely due to variation in p. For example, if p varies between 80 and 90%, then variation in C will be relatively small. However, if a low p varies by the same absolute amount, say between 10 and 20%, then variation in C will be relatively greater.

Temporal and spatial variation is the rule rather than the exception. Tacitly, Hairston and Wiley (1993) attributed variation in salamander counts entirely to variation in detection probability. In the study of Funk and Mills (2003), p varied from 0 to 1 depending on prevailing weather conditions (W.C. Funk, personal communication). Daily means (across sites) ranged from 0.43 to 0.75 and different sites had different means (across days) ranging from 0.43 to 0.90 (W.C. Funk, personal communication). Most variation in detection probabilities of the European treefrog *H. arborea* in the study of Pellet *et al.* (2007) was between sites whereas Schmidt *et al.* (2007) found that detection probabilities of *Salamandra salamandra* were low in autumn and high in spring. Bailey *et al.* (2004a) used a drift fence to capture salamanders (*Ambystoma tigrinum*) on their way to and from the pond. Detection probabilities were high in most years (greater than 90%) but inexplicably lower in one year (76%). Although detection probabilities of egg masses of *Rana sylvatica* and *Ambystoma maculatum* were generally high (usually > 80%), Grant *et al.* (2005) documented substantial spatial and temporal variation in detection probabilities that could bias estimates of population trends based on unadjusted counts.

25.3 Components of imperfect detection

There are many reasons why detection is imperfect and it makes sense to decompose detection probabilities into its components (Pollock *et al.* 2004; Nichols *et al.* 2008):

$$E(C) = Npe$$

where e is exposure to sampling (also referred to as availability for sampling). In this equation, p should best be called detection probability *given exposure*. Nichols *et al.* (2008) describe how detection probability and exposure to sampling can be further decomposed. Detection probability (p) and exposure to sampling (e) are easy to distinguish at the conceptual level. In practice, the distinction may not always be obvious.

Under different names, exposure to sampling is a well-known phenomenon of mark–recapture studies dealing with survival estimation. The two cases are transients and temporary emigrants (Williams *et al.* 2002). Transients are animals that show up only once in the study area and then leave. Temporary emigrants are animals that leave the study area temporarily during the sampling period and then return (e.g. animals that skip a breeding season). In other words, these animals are only partly exposed to sampling and this can cause negative bias in survival estimates.

Incomplete exposure to sampling also affects abundance estimation (Kendall 1999; Bailey *et al.* 2004b; Royle and Dorazio 2006). Non-exposure to sampling can result from both biological and methodological reasons. Biological reasons include brooding salamanders that are underground and not at the surface during the time of survey or frogs that do not breed in a particular year and hence do not migrate to the study site which may be a breeding site (e.g. pond). Another biological reason may be that only a proportion of all males present at a pond may be calling at a particular time. This would explain why there is a difference between mean and maximum chorus counts (Pellet *et al.* 2007). Methodological reasons may play a part when funnel or minnow traps are placed along the edge of a pond and some individuals spend all their time near the center of the pond. Non-exposure to sampling may also result from breeding phenology. Some individuals may breed early and some may breed later in the season (e.g. Sinsch 1988). If the population is sampled only early in the season, then many individuals will not be exposed to sampling. This might be a case where both biology and method cause non-exposure to sampling.

Non-exposure to sampling can have profound consequences for abundance estimation (Kendall 1999). Imagine that the goal of a study is to estimate abundance in a particular area. If amphibians do not move in and out of the study area, then an abundance estimator such as the Lincoln–Peterson estimator will provide an unbiased estimate the number of amphibians in the area. However, if some individuals move randomly in and out of the area (i.e. if they are not always exposed to sampling), then the very same estimator will estimate a different quantity. It will now estimate the size of the superpopulation where the superpopulation is defined as the total number of amphibians exposed to sampling at least once. This includes all amphibians that are residents in the study area, but also all individuals that move in and out or that move through the study area. Bailey *et al.* (2004b) encountered a related problem in their study of salamanders in the Appalachian Mountains, USA. They argued that short-term studies where salamander movement was negligible yielded estimates of the "surface" population (i.e. salamanders exposed to sampling only). Long-term studies, where salamanders had time to move from the surface to deeper ground and vice versa, gave estimates of the total number of salamanders in the sampled area. Kinkead and Otis (2007) describe a similar situation with breeding and non-breeding ambystomatid salamanders that were sampled at the breeding site.

If many individuals are not exposed to sampling (low *e*), then the mismatch between the spatial or temporal scale at which a population is sampled and the desired temporal or spatial scale of inference is likely to be great. In conclusion, study design (both the spatial and the temporal scales) and species behavior can jointly determine which biological entity is quantified.

25.4 How to deal with imperfect detection

25.4.1 Estimation of abundance

Evidently, an elegant way to deal with imperfect detection is to estimate detection probability ($\hat{p}$). An estimate of detection probability can then be used to correct counts (C) and estimate abundance ($\hat{N}$):

$$\hat{N} = C / \hat{p}$$

This equation is the conceptual basis for all kinds of abundance estimators, be they mark–recapture, distance sampling, point count, removal, or other methods (Williams *et al.* 2002; Mazerolle *et al.* 2007; see also Chapter 24 in this volume). When non-exposure to sampling is a problem, then $\hat{N}$ can only be estimated if one knows the fraction of the population that is exposed to sampling as well as the probability of detecting exposed individuals, populations, or species:

$$\hat{N} = C / \hat{p}\hat{e}$$

The best tool to estimate abundance in the presence of non-exposure to sampling is the robust design (see Williams *et al.* 2002) which has been successfully used with plethodontid and ambystomatid salamanders (Bailey *et al.* 2004b; Kinkead and Otis 2007). The robust design allows estimating both abundance and non-exposure (known as temporary emigration in this particular case). In the previously cited case of the European treefrog (*H. arborea*), the robust design also allowed demonstrating that annual male temporary emigration was negligible (i.e. that males rarely skipped breeding seasons; Pellet *et al.* 2007). In contrast, Bailey *et al.* (2004a) showed that temporary emigration in *A. tigrinum* was substantial.

Other approaches to dealing with non-exposure were described by Royle and Dorazio (2006) and Condit *et al.* (2007). Royle and Dorazio (2006) describe a method for point counts that allows dealing with non-exposure to sampling that arises from a mismatch between the scale at which data was collected and the desired scale of inference. They describe a case where quadrats have only been partially searched (in that study, only a fraction of the area of 1 km² quadrats was surveyed). Their method works if there is a suitable covariate (such as transect length) that can statistically link the exposed population to the true population.

Condit *et al.* (2007) developed a method for estimating the size of a population when individuals are asynchronously present. Their method may be particularly relevant for pond-breeding amphibians where the breeding season of the population is longer than the breeding season of an individual (e.g. in the natterjack toad *Bufo calamita* and the European treefrog *H. arborea*; Sinsch 1988; Friedl and Klump 2005).

Methods to estimate abundance are not without problems. Detection probabilities can be low. This will have the effect that confidence intervals can be very wide (Williams *et al.* 2002). It is possible to make confidence intervals shrink with more effort or better capture techniques. That is, researchers should either use better methods that increase per-visit detection probabilities or increase the number of repeat visits such that cumulative detection probabilities are increased. However, standard errors and confidence intervals are not a nuisance. Rather, they are an advantage of estimation methods. Standard errors and confidence intervals are a measure of uncertainty that allow an assessment of the estimates' reliability. Consequently, we view wide confidence intervals as an honest statement whether a particular estimate is or is not particularly reliable and useful. Wide confidence intervals are no reason to discard mark–recapture estimates and to prefer the simple counts (C; e.g. Alford and Richards 1999). There is also uncertainty associated with counts (because detection probability p is unknown) but it is not explicit and is, in fact, unknowable.

Heterogeneity in detection probabilities among individuals can be a problem in mark–recapture studies (Link 2003). Heterogeneity usually leads to negative bias in abundance estimates and in the worst case it may not be possible to identify a best model that should be used for inference. We believe that amphibian ecologists and conservationists should attempt to minimize detection probability heterogeneity among individuals by adopting methods that account for variation in detectability among individuals (i.e. grouping individuals into homogeneous sets by sexes, colour morphs, age classes).

25.4.2 Other approaches to dealing with imperfect detection

Amphibian ecologists have dealt with imperfect detection in many ways. Some authors simply did not analyse data from species where detection was uncertain and variable. For example, Pechmann *et al.* (1991) analysed data from ambystomatid salamanders that are unlikely to trespass a drift fence but they did not analyse data from treefrogs that could easily trespass a drift fence.

The most common objection to the use of methods that allow estimating population sizes and detection probabilities instead of counts is that they are demanding in terms of money, human resources, and statistical knowledge. This argument was not true in the detailed study of Funk *et al.* (2003) where different methods were compared. We discard this argument because every conservation action based on counts is likely to be biased to the point where they will be inefficient or, worse, counterproductive (Yoccoz *et al.* 2001). Moreover, it will be impossible to evaluate the success of actions taken because no data will be available to detect population trends accurately, and to adapt management actions accordingly.

Some authors have expressed the view that there is no need to estimate detection probabilities and to adjust counts accordingly (i.e. estimate abundance). One argument put forward in the context of long-term monitoring programs is that variability in detection probability does not matter as long as there is no temporal trend in detection probability (Bart *et al.* 2004; tacitly, this is also the reasoning of Hairston and Wiley 1993). In such a situation, variation in detection probability likely causes extra variability (sampling variability) in the counts (in comparison to variation in absolute abundance which is the phenomenon of biological interest). Detectability-induced extra variation in the counts means that a monitoring program loses power to detect temporal trends. However, one should keep in mind that detection probabilities likely show temporal trends. Reasons include, but are not limited to, habitat succession or changes through time in the ability of the observer to detect the study species (Link and Sauer 1997).

One commonly held view is that field methods can be standardized to the extent where detection probability is constant. If this is the case, then counts or other estimates of relative abundance should serve well as proxies for absolute abundance. Unfortunately, variation in detection probabilities is the rule rather than the exception (MacKenzie and Kendall 2002). Whenever detection probabilities of amphibians have been estimated, they were found to be variable both within and across seasons (Bailey *et al.* 2004a, Kinkead and Otis 2007; Mazerolle *et al.* 2007; Pellet *et al.* 2007; Schmidt *et al.* 2007). This was the case even when researchers used standard(ized) methods; even when drift fences were used—where the assumption is that detection probability is 1—there was variation in detection probability (Bailey *et al.* 2004a). Pellet *et al.* (2007) used the same methods at two sites yet detection probabilities differed between sites by a factor of approximately two. Hyde and Simons (2001) showed that counts obtained from applying four standard methods gave results that were only weakly correlated. That is, the use of standardized methods does not guarantee that detection probabilities are constant. We believe that standardization of field methods is important because it can help to keep variation in detection probabilities within bounds, but it should certainly not be viewed as a panacea.

Standardization of field methods is one solution to limit variation in detection probabilities. Another solution is to measure covariates that may affect detection probabilities and use these covariates at the analysis stage to adjust counts (Link and Sauer 1997, 1998). This approach may work well as long as the important covariates are known and has been successfully used in large-scale bird monitoring programs. However, it may be that the effect of a covariate on the counts varies from one site to the next. Lauber (2004) counted alpine salamanders (*Salamandra atra*) along fixed transects at four sites and tested whether weather covariates could be used to predict the salamander counts. An analysis of covariance found no main effect of air humidity on counts but there was

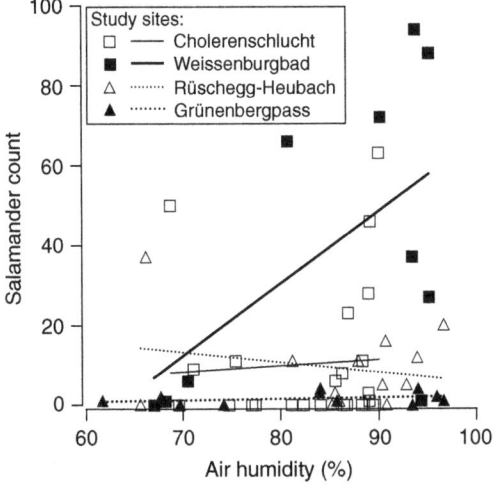

Fig. 25.2 Geographic variation in the relationship between salamander (*Salamandra atra*) counts and air humidity. Salamanders were counted multiple times at four sites along transects. Analysis of covariance showed no significant main effect of air humidity on counts but there was a significant interaction between air humidity and site. Data were taken with permission from Lauber (2004).

a significant site-by-air-humidity interaction (Figure 25.2). Air humidity can thus be used to adjust counts at some sites but not at others; a likely explanation is that sites differ in overall humidity. Grant *et al.* (2005) also found that no single explanatory variable or set of variables best explained variation in detection probabilities across sites of egg masses of *Rana sylvatica* and *Ambystoma maculatum*. This implies that herpetologists must be very cautious when extrapolating results from one study site to another. At the planning stage, it also implies that one should try to replicate all experiments spatially and temporally to find out whether biological patterns are universally applicable.

In conclusion, we believe that the use of standard methods is always valuable. Because it does not always avoid variation in detection probabilities, it is better to rely on adjusting counts than on the strong assumption that detection probabilities are not showing any trends. We argue that one should assume *a priori* that detection probabilities are less than one and that they are variable in space and time. Thus, amphibian ecologists and conservationists should provide evidence for their studies that the counts that they report are indeed reliable indices of abundance. Because detection probabilities can vary from one study to the next even under apparently highly similar conditions (see examples in Mazerolle *et al.* 2007), the proof has to be provided every time anew.

25.5 Designing a sampling protocol

The adequate design of a sampling protocol is a fundamental aspect of research, whether it be on individuals, populations, or species. Once the research or monitoring question addressed has been explicitly formalized (Yoccoz *et al.* 2001) and sites were appropriately selected (see Chapter 23 in this volume), it is essential that aspects of imperfect detection are incorporated into the design. To do so, the procedures of data collection and data analysis must be identified in advance. Having a sound knowledge of your study species' ecology will be instrumental in determining which biological quantity (e.g. above-ground population size, breeders, super-population) your estimator will represent.

Importantly, one must be fully aware that detection probability (p) is not only a species trait (see examples in Mazerolle *et al.* 2007); it also depends on methods, observers, year, site, and a myriad of other factors. Because p (and exposure to sampling, e) are both variable in space and time, one cannot apply values obtained in one study to another. For these reasons we recommend that detection probability is explicitly integrated in all amphibian study protocols. Non-exposure to sampling is often hard to deal with at the analysis stage (except when using the robust design). We therefore recommend that researchers carefully plan a study such that all animals are exposed to sampling.

25.6 Software

There are many computer programs freely available to estimate population abundance while incorporating detection probability. The most versatile and widely used of them is program MARK (White and Burnham 1999). This software, available at www.phidot.org/, allows the analysis of a wide range of capture–recapture-based data sets. Every new version incorporates the latest development in capture–recapture and thus allows the user to choose from a wide range of models the one that will fit its data best. As the name implies, DISTANCE (www.ruwpa.st and.ac.uk/ distance/) is the software tool that allows one to design and analyse distance sampling surveys. More recent developments have been integrated in statistical software such as R and WinBugs. These tools have the inconvenience of being less user-friendly than the previously listed programs, which benefit from a graphical user interface. There is software that can be used when planning a mark–recapture study (Devineau *et al.* 2006; Zucchini *et al.* 2007).

25.7 Outlook

Imperfect detection is the feature that the vast majority of all amphibian surveys have in common. Complete censuses where all amphibians that are present at the study site are captured are not impossible, but require a lot of work (drift fences: Bailey *et al.* 2004a; many capture events: Funk and Mills 2003; Pellet *et al.* 2007). We argue that detection probabilities can be low and highly variable among years and/or sites. While counts that are not adjusted for imperfect detection can certainly indicate negative population trends (Laurance *et al.* 1996), variability in imperfect detection can seriously bias inference from surveys. Amphibian ecologists and conservationists should therefore estimate detection probabilities as the best tool to calibrate a survey and use robust methods for abundance estimation (Williams *et al.* 2002; Chapter 24). Unfortunately, the use of such methods is not yet widespread (Alford and Richards 1999).

The number of methods available for estimation of abundance that account for imperfect detection has increased tremendously in the recent past. Existing methods are constantly being refined, while new methods are being developed (e.g. Royle 2004; Royle and Dorazio 2006). Still, all methods need to be used with care as sampling design, the behavior of the species, and the estimator used all determine which biological quantity is being estimated. Notwithstanding, the quality of inference from methods that adjust for imperfect detection will be stronger than inference from any other kind of method.

In the future, we ought to be able to estimate abundance with a precision and freedom from bias that was not achievable in the past. We should now be able to determine which factors influence abundance rather than study patterns of an inseparable combination of abundance, detectability, and exposure to sampling (such as counts). This will help us gain new insights into fundamental and applied aspects of amphibian ecology and conservation.

25.8 References

Alford, R. A. and Richards, S. J. (1999). Global amphibian declines: a problem in applied ecology. *Annual Reviews of Ecology and Systematics*, **30**, 133–65.

Bailey, L. L., Kendall, W. L., Church D. R., and Wilbur, H. M. (2004a). Estimating survival and breeding probability for pond-breeding amphibians: a modified robust design. *Ecology*, **85**, 2456–66.

Bailey, L. L., Simons, T. R., and Pollock, K. H. (2004b) Comparing population size estimators for plethodontid salamanders. *Journal of Herpetology*, **38**, 370–80.

Bart, J., Droege, S., Geissler, P., Peterjohn, B., and Ralph, C. J. (2004). Density estimation in wildlife surveys. *Wildlife Society Bulletin*, **32**, 1242–7.

Condit, R., Le Boeuf, B. J., Morris, P. A., and Sylvan, M. (2007). Estimating population size in asynchronous aggregations: A Bayesian approach and test with elephant seal censuses. *Marine Mammal Science*, **23**, 834–55.

Devineau, O., Choquet, R., and Lebreton, J. D. (2006). Planning capture-recapture studies: straightforward precision, bias, and power calculations. *Wildlife Society Bulletin*, **34**, 1028–35.

Friedl, T. W. P. and Klump, G. M. (2005). Sexual selection in the lek-breeding European treefrog: body size, chorus attendance, random mating and good genes. *Animal Behaviour*, **70**, 1141–54.

Funk, W. C. and Mills, L. S. (2003). Potential causes of population declines in forest fragments in an Amazonian frog. *Biological Conservation*, **111**, 205–14.

Funk, W. C., Almeida-Reinoso, D., Nogales-Sornosa, F., and Bustamante, M. R. (2003). Monitoring population trends of *Eleutherodactylus* frogs. *Journal of Herpetology*, **37**, 245–56.

Gill, D. E. (1985). Interpreting breeding patterns from census data: a solution to the Husting dilemma. *Ecology*, **66**, 344–54.

Grant, E. H. C., Jung, R. E., Nichols, J. D., and Hines, J. E. (2005). Double-observer approach to estimating egg mass abundance of pool-breeding amphibians. *Wetlands Ecology and Management*, **13**, 305–20.

Hairston, Sr, N. G. and Wiley, R. H. (1993). No decline in salamander (Amphibia: Caudata) populations: a twenty-year study in the southern Appalachians. *Brimleyana*, **18**, 59–64.

Hyde, E. J. and Simons, T. R. (2001). Sampling plethodontid salamanders: sources of variability. *Journal of Wildlife Management*, **65**, 624–32.

Jung, R. E., Dayton, G. H., Williamson, S. J., Sauer, J. R., and Droege, S. (2002). An evaluation of population index and estimation techniques for tadpoles in desert pools. *Journal of Herpetology*, **36**, 465–72.

Kendall W. L. (1999). Robustness of closed capture-recapture methods to violations of the closure assumption. *Ecology*, **80**, 2517–25.

Kinkead, K. E. and Otis, D. L. (2007). Estimating superpopulation size and annual probability of breeding for pond breeding salamanders. *Herpetologica*, **63**, 151–62.

Lauber, A. (2004). *Methodenevaluation zum Monitoring der Alpensalamanderpopulation*. Diploma thesis, Eidgenössische Technische Hochschule Zürich (ETHZ), Zürich.

Laurance, W. F., McDonald, K. R., and Speare, R. (1996). Epidemic disease and the catastrophic decline of Australian rain forest frogs. *Conservation Biology*, **10**, 406–13.

Link, W. A. (2003). Nonidentifiability of population size from capture-recapture data with heterogeneous detection probabilities. *Biometrics*, **59**, 1123–30.

Link, W. A. and Sauer, J. R. (1997). Estimation of population trajectories from count data. *Biometrics*, **53**, 488–97.

Link, W. A. and Sauer J. R. (1998). Estimating population change from count data: Application to the North American Breeding Bird Survey. *Ecological Applications*, **8**, 258–68.

MacKenzie, D. I. and Kendall, W. L. (2002). How should detection probability be incorporated into estimates of relative abundance? *Ecology*, **83**, 2387–93.

Mazerolle, M. J., Desrochers, A., and Rochefort, L. (2005). Landscape characteristics influence pond occupancy by frogs after accounting for detectability. *Ecological Applications*, **15**, 824–34.

Mazerolle, M. J., Bailey, L. L., Kendall, W. L., Royle, J. A., Converse, S. J., and Nichols, J. D. (2007). Making great leaps forward: accounting for detectability in herpetological field studies. *Journal of Herpetology*, **41**, 672–89.

Nichols, J. D., Thomas, L., and Conn, P. B. (2008). Inferences about landbird abundance from count data: recent advances and future directions. In D. L. Thomson, E. G. Cooch, G. Evan, and M. J. Conroy (eds), *Modeling Demographic Processes in Marked Populations*, pp. 203–38. Springer Science+Business Media, New York.

Pechmann, J. H. K., Scott, D. E., Semlitsch, R. D., Caldwell, J. P., Vitt, L. J., and Gibbons, J. W. (1991). Declining amphibian populations: the problem of separating human impacts from natural fluctuations. *Science*, **253**, 892–5.

Pellet, J., Helfer, V., and Yannic, G. (2007). Estimating population size in the European tree frog (*Hyla arborea*) using individual recognition and chorus counts. *Amphibia-Reptilia*, **28**, 287–94.

Pollock, K. H., Nichols, J. D., Simons, T. R., Farnsworth, G. L., Bailey, L. L., and Sauer, J. R. (2002). Large scale wildlife monitoring studies: statistical methods for design and analysis. *Environmetrics*, **13**, 105–19.

Pollock, K. H., Marsh, H., Bailey, L. L., Farnsworth, G. L., Simons, T. R., and Alldredge, M. W. (2004). Separating components of detection probability in abundance estimation: an overview with diverse examples. In W. L. Thompson (ed.), *Sampling Rare or Elusive Species*, pp. 43–58. Island Press, Washington DC.

Royle, J. A. (2004). *N*-mixture models for estimating population size from spatially replicated counts. *Biometrics*, **60**, 108–15.

Royle, J. A. and Dorazio, R. M. (2006). Hierarchical models of animal abundance and occurrence. *Journal of Agricultural, Biological, and Environmental Statistics*, **11**, 249–63.

Schmidt, B. R. (2004). Declining amphibian populations: the pitfalls of count data in the study of diversity, distributions, dynamics and demography. *Herpetological Journal*, **14**, 167–74.

Schmidt, B. R. and Pellet, J. (2005). Relative importance of population processes and habitat characteristics in determining site occupancy of two anurans. *Journal of Wildlife Management*, **69**, 884–93.

Schmidt, B. R., Schaub, M., and Steinfartz, S. (2007). Apparent survival of the salamander *Salamandra salamandra* is low because of high migratory activity. *Frontiers in Zoology*, **4**, e19.

Sinsch, U. (1988). Temporal spacing of breeding activity in the natterjack toad, *Bufo calamita*. *Oecologia*, **76**, 399–407.

White, G. C. and Burnham, K. P. (1999). Program MARK: survival estimation from populations of marked animals. *Bird Study*, **46**, 120–38.

Williams, B. K., Nichols, J. D., and Conroy, M. J. (2002). *Analysis and Management of Animal Populations*. Academic Press, San Diego, CA.

Yoccoz, N. G., Nichols, J. D., and Boulinier, T. (2001). Monitoring of biological diversity in space and time. *Trends in Ecology and Evolution*, **16**, 446–53.

Zucchini, W., Borchers, D. L., Erdelmeier, M., Rexstad, E., and Bishop, J. (2007). *WiSP 1.2.4*. Institut für Statistik und Ökonometrie, Georg-August-Universität Göttingen, Göttingen.

26

Disease monitoring and biosecurity

D. Earl Green, Matthew J. Gray, and Debra L. Miller

26.1 Introduction

Understanding and detecting diseases of amphibians has become vitally important in conservation and ecological studies in the twenty-first century. Disease is defined as the deviance from normal conditions in an organism. The etiologies (causes) of disease include infectious, toxic, traumatic, metabolic, and neoplastic agents. Thus, monitoring disease in nature can be complex. For amphibians, infectious, parasitic, and toxic etiologies have gained the most notoriety. Amphibian diseases have been linked to declining amphibian populations, are a constant threat to endangered species, and are frequently a hazard in captive breeding programs, translocations, and repatriations. For example, a group of viruses belonging to the genus *Ranavirus* and the fungus *Batrachochytrium dendrobatidis* are amphibian pathogens that are globally distributed and responsible for catastrophic population die-offs, with *B. dendrobatidis* causing known species extinctions (Daszak *et al.* 1999; Lips *et al.* 2006; Skerratt *et al.* 2007). Some infectious diseases of amphibians share similar pathological changes; thus, their detection, recognition, and correct diagnosis can be a challenge even by trained veterinary pathologists or experienced herpetologists.

This chapter will introduce readers to the most common amphibian diseases with an emphasis on those that are potentially or frequently lethal, and the techniques involved in disease monitoring. It will also outline methods of biosecurity to reduce the transmission of disease agents by humans. We start by covering infectious, parasitic, and toxic diseases. Next, surveillance methods are discussed, including methods for sample collection and techniques used in disease diagnosis. Finally, biosecurity issues for preventing disease transmission will be covered, and we provide protocols for disinfecting field equipment and footwear.

26.2 Amphibian diseases of concern

Amphibians are susceptible to a variety of pathogens, including internal and external parasites, viruses, bacteria, and fungi. Each of the three major life stages of amphibians (embryos, larvae, and adults) has a distinct suite of diseases, with some overlap between life stages. Aquatic amphibian embryos and larvae share many diseases with fish, whereas post-metamorphic stages often share few infectious diseases with earlier life stages. For detailed information on the amphibian diseases, we recommend that readers consult recent reviews (e.g. Converse and Green 2005a, 2005b; Green and Converse 2005a, 2005b) and the veterinary literature (e.g. Wright and Whitaker 2001).

26.2.1 Infectious diseases

Major infectious diseases for each amphibian life stage are summarized in Tables 26.1–26.3. Many viruses have been reported in amphibians, and include *Ranavirus*, herpesvirus, and adenovirus (Converse and Green 2005a, 2005b; Green and Converse 2005a, 2005b). Of these, *Ranavirus* has been the most significant contributor to population declines, resulting in significant morbidity and mass mortality (Daszak *et al.* 1999; Green *et al.* 2002; Cunningham *et al.* 2007). In North America, ranaviruses are responsible for the majority of catastrophic die-offs in ambystomid salamanders and late-stage anuran larvae, with the number of reported cases each year exceeding all other pathogens by three to four times (Green *et al.* 2002; Muths *et al.* 2006). Although many die-offs have been with common species, declines in several species of conservation concern (e.g. *Rana muscosa*, *Rana aurora*, *Bufo boreas*, and *Ambystoma tigrinum stebbinsi*) have been reported (Jancovich *et al.* 1997; Converse and Green 2005a). There is evidence that ranaviruses may function as a novel or endemic pathogen, with the former likely associated with the movement of infected amphibians by humans (Storfer *et al.* 2007). Anthropogenic stressors also may facilitate emergence (Forson and Storfer 2006; Gray *et al.* 2007a). Additionally, subclinically infected individuals (i.e. those that do not appear sick) may serve as reservoirs for more susceptible amphibian species (Brunner *et al.* 2004).

Likewise, numerous bacteria have been cultured from anurans (Mauel *et al.* 2002). Of these, *Mycobacterium liflandii*, a mycolactone-producing mycobacteria, is of concern because it is closely related to the human pathogen *Mycobacterium ulcerans* (Yip *et al.* 2007), which causes severe skin lesions in humans. Nevertheless, *Aeromonas hydrophila* remains the most recognized bacterial pathogen in amphibians because of its association with red-leg disease

Table 26.1 *Significant diseases of amphibian eggs and embryos.*

Disease agent	Common host species	Mortality rate	Organ of choice	Test methods
Lucke tumor herpesvirus	*Rana pipiens* only	0	Mesonephros or whole embryo	Culture on *Rana pipiens* cell line; PCR
Chlamydomonas (symbiotic alga)[1]	*Ambystoma* spp.	0	Egg capsule	Gross or microscopic exam
Watermolds[2]	*Bufo* spp., *Hyla* spp., *Pseudacris* spp., *Rana* spp., *Ambystoma* spp., *Taricha* spp.	Variable	Egg capsule	Culture; histology; DNA sequencing
Microsporidium schuetzi	*Rana pipiens*	<10%	Whole swollen eggs	Histology; electron microscopy
Tetrahymena/Glaucoma (ciliated protozoa)	*Ambystoma* spp.	15–25%	Egg capsule, brain and subcutis of embryos/ larvae	Submerged exam of eggs/embryos under dissecting microscope; histology; exam by protozoologist

[1] *Chlamydomonas* sp. is a symbiotic blue-green alga in the egg capsule of *Ambystoma maculatum* in eastern North America and *Ambystoma gracile* in western North America, and not considered a disease agent.
[2] Watermold infections (oomycetes of several genera) referred to as saprolegniasis.

(Green and Converse 2005a). However, it is important to note that red-leg disease is a gross descriptor of a specific lesion (i.e. swollen red legs) and not specific for a particular etiology. Many pathogens (e.g. *Ranavirus, A. hydrophila*, alveolates) can cause edema (i.e. swelling) and erythema (reddening) in amphibians (Figure 26.1a). This emphasizes the importance of diagnostic testing to determine the correct pathogen causing the disease.

Finally, numerous fungal and fungus-like organisms (Converse and Green 2005a, 2005b; Green and Converse 2005a, 2005b) and newly characterized pathogens (Davis *et al.* 2007) are known to cause catastrophic mortality of amphibian populations. *B. dendrobatidis* (Figure 26.1b) has resulted in global population declines and species extinctions (Wake and Vredenburg 2008). The newly discovered alveolate organism has only been diagnosed in a few

Table 26.2 Significant diseases of larval amphibians.

Disease agent	Common host species	Mortality rate	Organ of choice	Test methods
Ranaviruses	*Bufo* spp., *Hyla* spp., *Rana* spp., *Pseudacris* spp., *Ambystoma* spp., *Notophthalmus* spp.	50–99%	Liver, skin ulcers mesonephroi[1]	Culture at 20–25°C; PCR on liver, spleen, skin ulcers, mesonephroi
Batrachochytrium dendrobatidis	*Rana* spp., *Pseudacris* spp.	0%	Oral disc, toe tips	Histology; PCR; culture
Watermolds[1]	*Bufo* spp., *Hyla* spp., *P seudacris* spp., *Rana* spp., *Ambystoma* spp., *Taricha* spp.	Variable	Oral disc, skin	Culture; histology
Ichthyophonus sp.	*Pseudacris* spp., *Rana* spp., *Ambystoma* spp.	0–≈50%	Skeletal muscle	Histology
Perkinsus-like organism	*Rana* spp., rarely *Hyla* spp., rarely *Pseudacris* spp.	5–99%	Liver	Histology; PCR
Tetrahymena/glaucoma (ciliated protozoa)	*Ambystoma* spp.	15–25%	Egg capsule, brain and subcutis of embryos/larvae	Submerged exam of eggs/embryos under dissecting microscope; histology; examination by protozoologist
Ribeiroia ondatrae	*Bufo* spp., *Pseudacris* spp., *Rana* spp., *Ambystoma* spp.	Variable	Skin around vent and proximal hindlimbs	Examination by parasitologist; PCR
Other metacercariae	Most aquatic genera	Low	Parasite	Examination by parasitologist
Lernaea sp. ("anchorworm")	*Rana* spp.	Low	Parasite in skin	Examination by parasitologist
Leeches	*Rana* spp.	Low	Parasite	Examination by parasitologist

[1]Mesonephroi, "body kidneys" (versus pronephroi or "head kidneys" found only in larvae); "true" kidneys of reptiles, birds, and mammals are metanephroi.
[2]Watermold infections (oomycetes of several genera) referred to as saprolegniasis.

Table 26.3 *Significant diseases of post-metamorphic amphibians.*

Disease agent	Common host species	Mortality rate	Organ of choice	Test methods
Ranaviruses	*Bufo* spp., *Hyla* spp., *Pseudacris* spp., *Rana* spp., *Ambystoma* spp., *Notophthalmus* spp.	Low in adults; variable in recently metamorphosed amphibians	Liver, skin ulcers mesonephroi	Culture at 20–25°C; PCR on liver, spleen, skin ulcers, mesonephroi; electron microscopy
Lucke tumor herpesvirus	*Rana pipiens*	Variable in >2 yr old *Rana pipiens* only	Mesonephroi	Histology of tumors
Batrachochytrium dendrobatidis	Most anuran genera	Very high in many anurans especially at high elevations in tropical latitudes	Skin of pelvic patch, toe webs	Histology; PCR; culture; electron microscopy
Ichthyophonus sp.	*Rana* spp., *Ambystoma* spp., *Notophthalmus* spp.	Very low	Skeletal muscle	Histology
Amphibiothecum penneri[1]	*Bufo* spp.	0	Ventral skin nodules	Cytology of discharge; histology of nodule
Hepatozoon spp.	*Rana* spp.	Low	Blood smear; liver	Cytology; histology
Ribeiroia ondatrae	*Bufo* spp., *Pseudacris* spp., *Rana* spp., *Ambystoma* spp.	0	Skin around vent, proximal hindlimbs and at tip of urostyle	Examination of metacercaria by a parasitologist; PCR; radiographs of malformations
Rhabdias spp. (nematode lungworm)	*Bufo* spp. *Rana* spp.	Unknown	Lungs	Visible at dissection; histology; examination by a parasitologist

[1]*Amphibiothecum* (formerly *Dermosporidium*) *penneri*, referred to as dermosporidiosis.

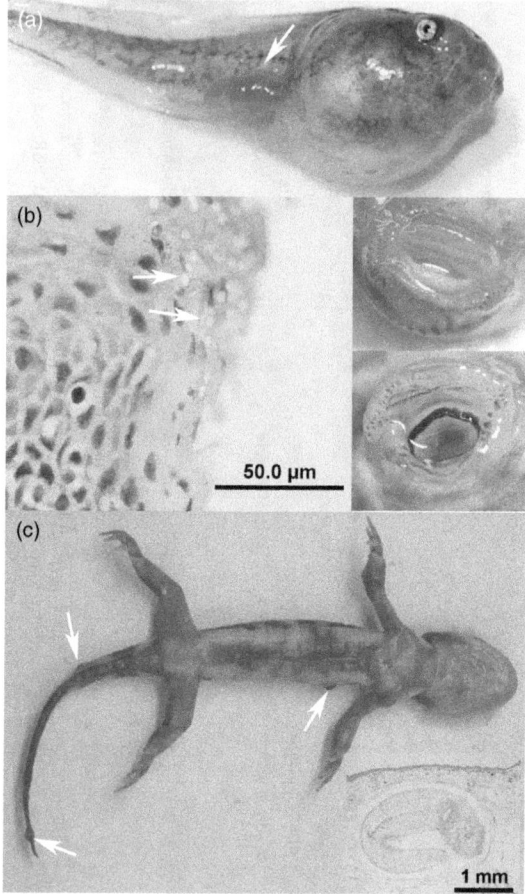

Fig. 26.1 (a) Tadpoles with swollen bodies and swollen red legs (arrow) are often diagnosed as red-leg disease but the etiology is varied and may include *Aeromonas hydrophila*, *Ranavirus*, and alveolates. (b) The amphibian fungus, *Batrachochytrium dendrobatidis* (arrows), infects the keratin-producing cells of amphibians. Tadpole skin is not keratinized; rather, only their 'teeth' contain keratin. Grossly, this is seen by loss of pigmentation (upper inset) of the tooth rows. Lower inset is of a normal tadpole for comparison. (c) Trematode cercaria encyst within the skin (arrows) and body cavities of amphibians serving as a secondary host and may be easily seen grossly. Histologically, the organisms are found in thin-walled cysts (inset).

isolated geographical areas so far (Davis *et al.* 2007). Still other organisms, such as the watermolds *Saprolegnia*, may be beneficial (e.g. by facilitating decomposition of dead eggs) but also have the potential to be opportunistic pathogens of amphibians at any life stage (Converse and Green 2005a, 2005b; Green and Converse 2005a, 2005b).

26.2.2 Parasitic diseases

As with any species, parasites are commonly found on and in amphibians (Figure 26.1c). External parasites include leaches, anchorworms, and mites, whereas internal parasites include various trematodes, cestodes, nematodes, and protozoa (Converse and Green 2005a, 2005b; Green and Converse 2005b; Wright and Whitaker 2001). Many species of helminthes (trematodes, cestodes, nematodes) have been documented in amphibians, and often they are considered incidental (Miller *et al.* 2004), but their presence may be an indicator of stress or aquatic food-web restructuring related to human land use (Johnson and Lunde 2005; Gray *et al.* 2007b). Likewise, many protozoans (e.g. myxozoa) are often considered incidental findings but their numbers may increase when amphibians are stressed, and they may potentially contribute to morbidity.

26.2.3 Toxins

Contaminants in the environment may kill larvae or post-metamorphs (Relyea 2005, 2009), and may have non-lethal impacts including reducing growth, impacting metamorphosis, disrupting gonadal development and secondary sex characteristics, or causing musculoskeletal, skin, and visceral malformations (Boone and Bridges 2003; Davidson *et al.* 2007; Ouellet *et al.* 1997; Storrs and Semlitsch 2008). Often these changes are not detected by external examinations until metamorphosis is complete or until the animals attain a size for reproduction. Amphibians are often considered sentinels or bio-indicators of environmental quality because they are sensitive to toxins and many species have the potential to be exposed to stressors in aquatic and terrestrial systems due to their typical biphasic life cycle (Blaustein and Wake 1995).

26.3 Disease monitoring: detection and diagnosis

26.3.1 Disease surveillance

Recently, the World Animal Health Organization (the OIE; www.oie.int/eng/en_index.htm) included two amphibian diseases (chytridiomycosis and ranaviral disease) on their listing of reportable diseases. The OIE listing provides the impetus for disease surveillance and required testing of amphibians prior to transport among states or between nations. The need for required testing of amphibians for pathogens has been expressed by several researchers (Gray *et al.* 2007a; Griffiths and Pavajeau 2008; Picco and Collins 2008). Historically, pre-transport pathogen testing and health certification for amphibians has been essentially non-existent, unlike for domestic livestock and pets and some wild mammals (e.g. cervids). The OIE has established guidelines for surveillance and requirements necessary for countries to declare *Ranavirus*-free status (www.oie.int/eng/norms/fcode/en_chapitre_2.4.2.htm#rubrique_ranavirus) and *B. dendrobatidis*-free status (www.oie.int/eng/normes/fcode/en_chapitre_2.4.1.htm#rubrique_batrachochytrium_dendrobatidis). The OIE-approved methods for conducting surveillance and diagnosis of infection are in development. In the meantime, guidelines from the 2006 OIE Manual of Diagnostic Tests for Aquatic Animals (www.oie.int/ eng/normes/fmanual/A_summry.htm) and from the fisheries industry (USFWS and AFS-FHS 2005) can be helpful for general monitoring of amphibian population health. In general, the criteria of a population health assessment should include (1) determination of the status and trends of amphibian pathogens, (2) determination of the risk of disease for threatened or endangered amphibians, (3) investigation of unexplained population declines, (4) evaluation of populations following a morbidity or mortality event, (5) detection of pathogens in non-indigenous species, (6) evaluation of a site or population prior to translocation, (7) evaluation of sympatric amphibians prior to release of captive-raised animals, and (8) the potential for amphibians and their diseases to "piggy back" with fish translocation.

Disease testing should not focus on one pathogen. For surveillance programs, we recommend that animals are tested for infection by at least the OIE pathogens: ranaviruses and *B. dendrobatidis*. For diagnosis of morbid or dead individuals, we encourage a full diagnostic work-up (i.e. necropsy, histology, bacterial culture, virus testing, and parasite testing) to attempt to identify all etiologic agents. It is important to note that simultaneous infection by multiple pathogens is possible. Further, histological examination of organs often is required to determine which of the pathogens identified are causing the changes responsible for the diseased state (Miller *et al.* 2008, 2009). Histological examination is also important in discovering introduced pathogens or pathogens that have not been described previously (Longcore *et al.* 1999; Davis *et al.* 2007).

Population health assessments can include non-lethal or lethal collection of tissue samples from individual amphibians (Greer and Collins 2007), and collection of environmental samples (e.g. water, soil; Walker *et al.* 2007). Ideally, we recommend that tissue samples are collected from all species in a community and from pre- and post-metamorphic life stages. Amphibian species differ in susceptibility to pathogens, and some age classes may serve as a reservoir (e.g. larval for *B. dendrobatidis* and adults for *Ranavirus*; Daszak *et al.* 1999; Brunner *et al.* 2004; Schock *et al.* 2008). Further, some infectious diseases become evident only after the post-metamorph has overwintered (e.g. Lucke tumor herpesvirus, *Amphibiothecum* (formerly *Dermosporidium*) *penneri*). The lack of gross signs of disease also does not imply healthy populations. We and others have found tadpoles with no signs of illness but that are infected with ranaviruses (Gray *et al.* 2007a; Harp and Petranka 2006; Miller *et al.* 2009). Laboratory studies have demonstrated that amphibian pathogen infection and mortality rates frequently track each other (e.g. Brunner *et al.* 2007); thus, high prevalence in a population could signal that a die-off is imminent.

In some cases, it may not be possible to collect sufficient tissue for disease testing. For example, small amphibians (e.g. *Bufo* larvae) may not have adequate tissue for tests, especially for toxicological analysis. Also, non-lethal testing may be required because a species is listed as a conservation concern. We found that testing for *Ranavirus* from tail clips results in about 20% false-negatives (D. L. Miller and M. J. Gray, unpublished results). In cases when a small amount of tissue is collected, multiple individuals within a species could be pooled to acquire sufficient tissue for testing. If contaminants are suspected as the cause of a die-off, we also recommend collecting and testing water and sediment at the amphibian breeding site.

Monitoring for malformations can be challenging, because typically malformed individuals have low survival. Although amphibian malformations have been documented for many years (Rostand, 1958), an increase in malformation rates occurred in the late twentieth century (Johnson and Lunde 2005). Generally, malformation studies have targeted recently metamorphosed amphibians (Meteyer *et al.* 2000), because metamorphs with prominent abnormalities are quickly removed from the population by predation or starvation. Additionally, the bony skeleton of metamorphosed amphibians is more conducive for radiographically visualizing deformities compared to the cartilaginous skeleton of larvae. However, monitoring of larval abnormalities is needed because it is likely that some abnormalities prevent metamorphosis, thus are not detected in post-metamorphic cohorts.

Finally, comprehensive disease surveillance should include captive amphibians in zoological and ranaculture facilities, because disease transmission can occur between captive and free-ranging populations. Maintenance of health in zoological facilities is especially important for rare species or in captive breeding populations intended for release. High densities in ranaculture facilities, pet shops, and stores that sell amphibians (e.g. *Ambystoma tigrinum*) for fishing bait can be cauldrons for disease transmission and pathogen evolution (Picco and Collins 2008). Ranaviruses isolated from ranaculture facilities and bait shops appear to be more virulent than wild strains (Majji *et al.* 2006; Storfer *et al.* 2007). This emphasizes the importance for disease monitoring at facilities with captive amphibians. In the event of a die-off in a captive facility, freshly dead animals should be submitted for diagnostic evaluation. Live animals that are infected should be euthanized or treated if a treatment exists (discussed in section 26.4.3), and facilities decontaminated with bleach or an equivalent disinfectant (discussed in section 26.4.2).

26.3.2 Sample size

Determination of statistically appropriate sample sizes for amphibian disease surveillance remains in its infancy. Although not established for amphibians, health assessment of fish is based on the minimum assumed pathogen prevalence level (APPL). The commonly used APPLs in aquatic health investigations are 2, 5, and 10% (Lavilla-Pitogo and de la Pena 2004; USFWS and AFS-FHS 2005). The APPL is used with an estimate of amphibian population size to determine the number of individuals that should be tested to have 95% confidence in pathogen detection. If it is assumed that APPL is 10%, required sample size ranges from 20 to 30 depending on the size of the amphibian population (Table 26.4). Required sample size increases with decreasing APPL (Table 26.4). Unpublished findings of the US Geological Survey National Wildlife Health Center suggest that APPL for ranaviruses, *B. dendrobatidis*, and alveolates is 10% or less (Table 26.5). In Tennessee, USA, health monitoring of two common anuran species inhabiting farm ponds revealed 29% prevalence for *Ranavirus*, 0% for *B. dendrobatidis* and alveolates, and 43% for parasites (Miller *et al.*, 2009). *Ranavirus* prevalence in plethodontid salamanders in the southern Appalachian Mountains can range from 3 to 81% depending on the watershed (M. J. Gray and D. L. Miller, unpublished results). Thus, we recommend that biologists determine required sample sizes for amphibian disease monitoring using either 5 or 10% APPL (Table 26.4).

26.3.3 Sample collection and shipment

Sample collection may include whole live animals, dead animals, sections of tissues, swabs of lesions or orifices, environmental samples, or sympatric species. It is important to wear disposable gloves when handling amphibians and to change gloves between animals. This is necessary to prevent disease transmission between amphibians and to protect biologists from zoonotic diseases (discussed in section 26.4). Gutleb *et al.* (2001) and Cashins *et al.* (2008) reported that disposable gloves (especially latex gloves) may be toxic to amphibian larvae. Therefore, when handling amphibians, biologists and researchers should use disposable vinyl gloves that have been rinsed with distilled or sterilized water (Cashins *et al.* 2008).

Mortality events involving all amphibian species should be investigated, even if it is not part of a disease surveillance program. There is a paucity of information

Table 26.4 *Sample size (i.e. number of amphibians) to assure 95% confidence in detection of pathogens in a population (modified from USFWS and AFS-FHS 2005).*

Estimated population size	Number of amphibians for	
	5% APPL	10% APPL
100	45	23
500	55	26
2000	60	27
>10 000	60	30

APPL, assumed pathogen prevalence level.

Table 26.5 *Previously unreported low disease prevalence in free-ranging amphibian populations in the USA (US Geological Survey National Wildlife Health Center).*

Pathogen	Host species	Life stage	US state	Sample size	Prevalence (number positive)	Test method	Case number
Ranavirus	R. catesbeiana	L	OR	15	7% (1)	Culture	44276
	Pseudacris maculata	L	WY	11	18% (2)	Histology	4779
Perkinsus-like organism	R. catesbeiana	L	OR	12	8% (1)	Histology	44276
	R. sphenocephala	L	FL	15	27% (4)	Histology	4864
	R. sphenocephala	L	LA	12	8% (1)	Histology	18626
	R. sphenocephala	RM	MD	14	21% (3)	Histology	18761
	R. sphenocephala	L	MS	11	9% (1)	Histology	18642
Ichthyophonus	R. sphenocephala	L	FL	19	11% (2)	Histology	4864
	R. grylio	L	FL	16	19% (3)	Histology	4864
	R. sphenocephala	L	LA	12	8% (1)	Histology	18626
	R. clamitans	L	ME	16	31% (5)	Histology	4824

L, larvae; RM, recently metamorphosed.

on the occurrence of pathogen-related die-offs in amphibian populations. The majority of samples submitted to diagnostic laboratories are from biologists that encountered a dead or morbid amphibian during other work activities. Morbid or freshly dead amphibians are preferred, because amphibians decompose rapidly. Decomposed carcasses are not suitable for cultures, histology, and parasitological examinations, but may have limited diagnostic usefulness for molecular tests that detect pathogens and for toxicological analyses. In general, we recommend that amphibians be collected live or within 24 h of death. Mummified (i.e. desiccated) carcasses with dry, leathery, and stiff digits or limbs usually have limited diagnostic usefulness.

Dead amphibians should be collected, put individually in plastic bags (e.g. Nasco Whirl-Pak® bags), and placed on ice for transport. Live amphibians can be placed in separate plastic containers and humanely euthanized (Baer 2006) via transdermal exposure for 10 min to tricaine methanesulfonate (100–250 mg/L) or benzocaine hydrochloride (> 250 mg/L or 20% benzocaine over-the-counter gel; Oragel, Del Paharmaceuticals, Uniondale, New York, USA) after returning from the field. It is important that amphibians are bagged separately to prevent cross-contamination of samples. Biologists that are experienced in blood collection may collect blood antemortem from the ventral vein in adult anurans or tail vein in salamanders, or collect blood antemortem or immediately postmortem from the heart of larvae or adults (Wright and Whitaker 2001). Blood can be tested for various biochemical parameters and examined for cellular composition, blood parasites, and viral inclusions (discussed in section 26.3.4).

We recommend that half of the individuals collected are frozen immediately for cultures and molecular tests. Samples can be frozen in a standard −20°C freezer if stored for short duration (< 1 month); otherwise, samples should be stored in a −80°C freezer. The other half of samples should be promptly fixed in 75% ethanol or 10% neutral buffered formalin for histology. For the first 2–4 days of fixation, the volume of fixative should be 10 times the volume of the animals. After this initial fixation, carcasses can be stored in a smaller volume of fixative that is sufficient to cover the tissues. The body cavity of amphibians that are more than 1 g in body mass should be cut along the ventral midline prior to immersion in fixative to assure rapid fixation of internal organs. Body cavities of frozen individuals should not be opened.

Special processing is required for amphibians with skin, digital, limb, head, or vertebral abnormalities. Whenever possible, amphibians with suspected malformations should be submitted alive for examinations. Dead individuals should be promptly frozen until time exists to properly fix individuals. Fixation can be done with ethanol or formalin but should be done in a pan so that carcasses can be positioned on a flat surface with limbs and digits extended from the body during fixation. Positioning amphibians in the standard museum configuration is ideal. Digits and limbs may be taped in position prior to fixation. Amphibians should be covered with fixative and additional fixative added if a significant amount evaporates. Placing a cover over the pan will help reduce evaporation. After 2–4 days of fixation, the carcass and limbs will be hardened in position and may be stored in a smaller volume of fixative. The positioning described above is necessary for radiographic examination of the malformation.

Given that it often is not possible to obtain collection permits for threatened amphibians, alternative sampling may be necessary. Two alternatives are (1) capture–release studies and (2) collection of "sentinel" sympatric amphibian species. Capture–release studies can be used to collect swabs of external tissues, blood, or fecal samples. Swabs appear to be a reliable technique to test for *B. dendrobatidis* (Kriger *et al.* 2006); however for ranaviruses, false-positive and -negative test results are greater than for tail clips and both of these non-lethal techniques have more false results than testing internal organs (D. L. Miller and M. J. Gray, unpublished results). Swabs are typically performed in the oral then cloacal regions, and the swab stored in its packaging container or a microcentrifuge tube. Swabs should be put on ice and frozen similar to tissues. An accepted protocol for swabbing amphibians for *B. dendrobatidis* testing using PCR (discussed in section 26.3.4) has been reported by Brem *et al.* (2007) and can be found at www.amphibianark.org/chytrid.htm. Briefly, the amphibian should be gently but firmly swabbed in a sweeping motion five times at each of the following five locations (for a total of 25 times): rear feet (toe webbing), inner thighs, and ventral abdomen. Occasionally, modifications to this technique are necessary for salamanders (Brem *et al.* 2007). Swabs for *B. dendrobatidis* testing by PCR may be stored in 70% ethanol. Collecting common sympatric species for health assessment can provide insight into the presence of amphibian pathogens at a site, but does not allow for direct health assessment of the species of concern, which may differ in susceptibility.

Shipment of live, freshly dead, or frozen specimens must be via an overnight courier and according to the specific courier guidelines. For fixed specimens, overnight shipment is unnecessary. General guidelines for shipment include triple packaging and labeling each layer of packaging with a waterproof writing utensil. Commonly, the first package layer is a specimen in a Whirl-Pak® bag. The second layer is a larger sealable plastic bag in which multiple specimens are placed. If the first package layer contains liquid (e.g. ethanol), paper towel should be added to the second package to absorb any liquid if a spill occurs. The third package typically is a padded box or shipping cooler. For frozen specimens, adequate ice packs or dry ice should be added around the secondary package. It is vital that the package contains a detailed list of all contents, a description of requested services, and the contact information of the shipper. The tracking information should be provided to the recipient prior to package arrival.

26.3.4 Diagnostics

Several tools are available for diagnosing amphibian diseases but generally require some level of specialized expertise to perform. Commonly used diagnostic tools include necropsy, histology, cytology, bacterial culture, virus isolation, fecal floatation, electron microscopy, molecular modalities, and radiology. Most of these tests can be performed on samples collected from dead or live amphibians. Fresh or frozen tissues can be used for most tests, and are necessary for virus isolation. Frozen tissues are not appropriate for histology or cytology; rather, preserved tissues are used. Although formalin-fixed specimens are preferred for histological examination, ethanol-fixed specimens may also be used. Blood can be used for cell counts to assess immune function and to look for inclusion bodies that can be diagnostic for certain pathogens. Blood also may be tested for the presence of antibody response to various diseases. Examples of laboratories that currently test for amphibian diseases in Australia, New Zealand, Europe, and the USA include Australian Animal Health Laboratories (AAHL), Geelong, Victoria, Australia (www.csiro.au/places/aahl.html), Gribbles Veterinary Pathology, Australia and New Zealand (www.gribblesvets.com/), Exomed, Berlin, Germany (www.exomed.de/), Hohenheim University (R. Marschang), Stuttgart, Germany, Wildlife Epidemiology, Zoological Society of London (ZSL), London, UK, The University of Georgia Veterinary Diagnostic and Investigational Laboratory, Tifton, GA, USA (www.vet.uga.edu/dlab/tifton/index.php), University of Florida (J. Wellehan), College of Veterinary Medicine, Gainesville, FL, USA, and National Wildlife Health Center, Madison, WI, USA (www.nwhc.usgs.gov/).

There are advantages and disadvantages to the various tests available (Table 26.6). Necropsy allows for identification and documentation of external and internal gross changes. Histological and cytological examination allows for identification of changes at the cellular level and is generally necessary to document disease versus infection. Virus isolation is the process of culturing a virus which is necessary to determine the presence of live virus and to perform some molecular tests used in identifying viral species (e.g. sodium dodecyl sulphate/polyacrylamide-gel electrophoresis (SDS/PAGE) and restriction fragment length polymorphism (RFLP)). One caveat is that some viruses are difficult to culture, thus infection cannot be ruled out based solely on negative isolation results. Electron microscopy is used for identifying key features of parasites or other infectious agents (e.g. *B. dendrobatidis*, *Ranavirus*, herpesvirus), documenting intracellular changes or changes to the cellular surface, and confirmation of cultured virus. Electron microscopy can be performed on fresh, fixed, or paraffin-embedded tissues. Radiology allows

Table 26.6 *Advantage and disadvantages of diagnostics tests for amphibian pathogens given the type of sample.*

Sample type	Tests	Pathogen	Advantages	Disadvantages
Live animal	Necropsy, histology, cytology, hematology, virus isolation, bacterial culture, toxicological analysis, parasitology, PCR	Viruses, bacteria, fungi, parasites, toxins	Observe behavior, least chance of contamination, blood collection is possible	Difficulty in transport, stressful to animal
Fresh tissue (including whole dead organisms)	Necropsy if whole animal, histology, cytology, virus isolation, bacterial culture, toxicological analysis, arasitology, PCR	Viruses, bacteria, fungi, parasites, toxins	Can isolate live pathogens	If advanced postmortem autolysis, then of limited value
Frozen tissue	Virus isolation, bacterial culture, PCR	Viruses, bacteria, fungi, parasites, toxins	Can isolate live pathogens	Limited value for histology (freeze artifact)
Swab	Virus isolation, bacterial culture, PCR	Viruses, bacteria, fungi	Non-lethal, may detect shedders	False positives and negatives are possible
Fixed tissue	Histology, PCR	Parasites, bacteria, fungi, viruses,	Can see cellular changes due to disease	Cannot isolate live pathogens

for documentation of bone structure or the presence of foreign bodies, including certain parasites (e.g. *Ribeiroia metacecariae*).

Molecular testing is becoming increasingly popular and affordable for disease diagnostics. It is especially useful for endangered species, as non-lethal sampling can yield accurate results. Specifically, it can be performed on fresh, fixed or paraffin-embedded tissues, swabs, blood, and feces. For testing via PCR, one caveat is that a positive PCR result only confirms the presence of the pathogen whether it is dead or alive. Thus, it is important to perform supportive tests (e.g. virus isolation, histological examination) to differentiate between infection and disease. Either conventional or real-time PCR (qPCR) may be used, depending on the availability of known primer sequences and the purpose of the test. For quantifying viral presence and infection (Yuan *et al.* 2006; Storfer *et al.* 2007), qPCR is most ideal (Brunner *et al.* 2005; Pallister *et al.* 2007). However, if sequencing is desired, which is often necessary to identify the species of a pathogen, conventional PCR is necessary.

There are three standard methods for characterizing amphibian malformations: (1) dissection, (2) radiography, and (3) clearing and staining. Dissection of a carcass is tedious, as it usually requires careful removal of muscles from limbs, head, and axial skeleton. Dermestid beetles (*Dermestes maculatus*) might be used to remove muscles and soft tissues from an amphibian carcass, but reassembly of bones of vertebrae, limbs and digits can be very challenging and time-consuming. Radiography is the preferred diagnostic method for investigating and documenting abnormalities of the skeleton (Meteyer *et al.* 2000). A major limitation of radiography is that cartilage is invisible; hence, detection of abnormalities of cartilage is not possible. Instead, the clearing-and-staining method commonly used in teratological studies of embryos is recommended when both cartilage and bone need to be examined (Kimmel and Trammell 1981; Schotthoefer *et al.* 2003). This method involves "clearing" the skin, muscles, and viscera by immersion in potassium hydroxide. The bones are stained red with Alizarin Red, and cartilage is stained blue using Alcian Blue stain. Clearing and staining is the preferred method to evaluate larval amphibian skeletal abnormalities.

Regardless of the diagnostic tests employed, interpretation of the test results must be done with caution and knowledge of the amphibian pathogen that is being tested (Table 26.6). The type of sample must be considered when targeting a pathogen. For example, infection of *Ranavirus* is best diagnosed from internal organs otherwise environmental contamination (e.g. water or soil) cannot be ruled out. Nonetheless, documentation of ranaviruses from skin surfaces does provide evidence of environmental exposure. In contrast, *B. dendrobatidis* is commonly tested from skin surfaces in adults or mouth parts in larvae, because this pathogen infects only keratinized tissue (Kriger *et al.* 2006; Skerratt *et al.* 2008). However, histology is generally required to distinguish between *B. dendrobatidis* exposure and infection when gross lesions are not observed. In contrast, some pathogens (e.g. alveolates, *Ichthyophonus* spp., *Ribeiroia ondatrae*) may not be identifiable from external swab preparations and often specialized techniques are required (e.g. clearing or radiography for *R. ondatrae*). In addition, one must keep in mind that there is a difference between malformations and deformities. Malformations are those abnormalities that arise during growth and development (organogenesis) in which the organ or structure fails to form normally. A deformity is an abnormality that naturally occurs to a normal organ or structure, such as an amputation or wound. It is often difficult to determine, even in radiographs, whether an abnormality is a malformation or a deformity.

26.4 Biosecurity: preventing disease transmission

Lethal infectious diseases of amphibians may be endemic and emerge in response to stressors, whether anthropogenic or natural (Carey *et al.* 2003). Disease emergence also may occur through geographical transport of pathogens (Jancovich *et al.* 2005; Storfer *et al.* 2007). Ranaviruses, *B. dendrobatidis*, the alveolate organism, and *Ichthyophonus* spp. are well established in many regions of the world; however, it is likely that some amphibian species have never been exposed to these agents. Further, in areas with multiple endemic pathogen strains or species (e.g. ranaviruses), slight variations in genetic coding can increase virulence (Williams *et al.* 2005). Thus, an endemic strain may function as a novel pathogen to amphibian populations outside the region where the pathogen evolved. This may be especially true with amphibian pathogens given the limited mobility of their host. Hence, prevention of the spread of endemic diseases to naïve populations or species remains a high conservation priority. Health examinations of amphibian populations and good biosecurity methods need to be employed because often little is known about the life cycles of infectious diseases, modes of transmission, and the persistence of the pathogen within and outside the amphibian host.

Preventing mechanical transmission of pathogens and contaminants from one location to another by equipment, supplies and people is the purpose of biosecurity. Biosecurity involves three equally important aspects: (1) safety of the humans and animals in the area, (2) decontamination or disinfection of field equipment, and (3) restriction on transporting amphibians among watersheds.

26.4.1 Human and animal safety

Whenever sampling amphibians for disease, the priority must be personal safety and health. For standard monitoring, biologists should wear gloves and waterproof footwear that can be easily disinfected (e.g. rubber boots). If a die-off is observed, it is important to note whether other vertebrates (e.g. birds, fish) are dead or appear morbid. If so, there is a greater chance the animal deaths are due to toxins, which may present a significant human health risk. In cases with a multiple wildlife taxa die-off, field personnel should leave the site immediately without collecting specimens and notify the nearest public health department and wildlife agency. Persons leaving a multiple-taxa mortality site should wash and disinfect boots, waders, nets, and field equipment and change clothes before entering a vehicle and leaving the site (discussed in section 26.4.2).

Few infectious diseases of amphibians are contagious to humans. Potential zoonotic diseases that may be carried by amphibians include certain *Salmonella* spp., *Yersinia* spp., *Chlamydophila* spp. (formerly *Chlamydia*)), and some toxin-producing mycobacteria (e.g. *Mycobacterium liflandii*) that can cause skin ulceration. In addition, Gray *et al.* (2007c) demonstrated that *Rana catesbeiana* metamorphs were suitable hosts for the human pathogen *Escherichia coli* O157:H7. We also demonstrated recently that tadpoles could maintain this pathogen in aquatic mesocosms (M. J. Gray and D. L. Miller, unpublished results). Thus, disposable gloves should be worn whenever handling amphibians, and hands washed thoroughly with soap and warm water after removing gloves. In the field, hands can be soaked in a 2% clorhexidine solution for 1 min or disposable antibacterial wipes used. Avoid exposure of surface water to soaps and disinfectants, as they may negatively affect local flora and fauna. Clothing that becomes stained with feces or skin secretions should be removed as soon as possible and washed in color-safe bleach.

The skin secretions of many amphibians contain potent irritants and toxins. For example, newts (Salamandridae), toads (Bufonidae), and poison-dart frogs (Dendrobatidae) exude toxic skin secretions. Skin secretions of certain newts (e.g. *Taricha*) may cause temporary blindness lasting several hours if the secretions get into the eyes. The parotoid gland secretions of giant toads (*Bufo marinus*), if ingested, can rapidly cause heart malfunction in humans and animals. When handling toads, it is best to avoid touching the parotoid glands. After handling amphibians, avoid touching your eyes or mouth prior to washing hands.

26.4.2 Washing and disinfecting equipment

Cleaning equipment and waders is recommended when leaving any amphibian breeding site, whether it is known that pathogens are present or not (see also www.nwhc.usgs.gov/). Cleaning is a three-step process: (1) washing with a soap or detergent, (2) rinsing thoroughly with clean water, and (3) disinfecting of the objects via a chemical disinfectant. Common soaps or detergents are not disinfectants but are useful in removing sediments and vegetation. Biodegradable soaps should be used in the field and not discarded into surface waters, as many are toxic to amphibians, fish, and invertebrates. Chemical disinfectants need to remain in contact with cleaned and rinsed surfaces for several minutes to kill microorganisms.

Common disinfectants used are chlorhexidine and sodium hypochlorite (bleach). Bleach is often preferred because it is cost effective, easily obtained, and effective against most bacteria and many viruses. The US Fish and Wildlife Service and American Fisheries Society – Fish Health Section (USFWS and AFS-FHS) (2005) recommend 10 min of exposure of a 0.05% bleach solution (i.e. 28.4 g of 6.15% sodium hypochlorite in 3.8 L of clean water) for disinfection of field equipment and surfaces for *B. dendrobatidis*, and, although not conclusive, a 0.5% solution (i.e. 312 g of 6.15% sodium hypochlorite per 3.8 L of water) is recommended to destroy myxosporeans. However, bleach is not very effective at inactivating *Ranavirus*, and requires at least a 3% concentration (Bryan *et al.* 2009). It should be noted that this concentration can be toxic to amphibians. In contrast, chlorhexidine used at a dosage that is safe for amphibians (0.75% for a 1 min exposure) has been shown to inactivate *Ranavirus* (Bryan *et al.* 2009). Further, it is important to keep in mind that the shelf-life of bleach solutions is influenced by exposure to light, air, and organic material, and solutions should be discarded after 5–7 days. After disinfection, equipment may be allowed to air dry or rinsed with fresh, clean water. Alternatively, if carrying large quantities of water is not possible because multiple fields sites are to be visited, surface water from the subsequent site (i.e. where the equipment will be used next) can serve as the rinse water. If mountain systems with stream watersheds are sampled, we recommend that researchers begin sampling at higher elevations and work towards lower sites. If a disease agent is present at higher elevations, it is likely to be at lower elevations due to downstream transmission. Hence, if accidental transmission occurs during travel on fomites, it is less likely to be a novel introduction.

26.4.3 Movement of animals and disease management

Introducing captive-raised or moving wild amphibians into new locations may be necessary because of population declines or extirpations. It is important to understand the initial cause of the die-off to ensure the factor no longer exists. In the case of diseases, environmental testing for the etiologic agent should be done before reintroductions or translocations. For pathogens, existing amphibian species also should be tested to ensure they are not functioning as a reservoir. It hinders conservation efforts to release species with high susceptibility if the pathogen remains at a site. Simultaneously, testing of the source population should be performed prior to reintroduction to avoid introduction of pathogens into the wild. Non-lethal testing as described previously generally can be used. Alternatively, in the case of translocations, lethal testing of common closely related resident species from the donor environment can provide some assurance that the target species is not infected.

Amphibians (dead or alive) from a mortality site should be considered contagious specimens. Morbid animals and carcasses should not be released or discarded at the same or other sites because this may facilitate the spread or persistence of infectious diseases. Dead amphibians that are not used for testing should be placed in double-layered plastic trash bags and disposed by burial or incineration. Removal of carcasses is a good strategy to help thwart the spread of infectious diseases.

While some serious infectious diseases of amphibians (e.g. *B. dendrobatidis*, nematode lungworms (*Rhabdias* spp.)) are readily treated and eliminated from captive populations, some important infectious diseases have no known treatments (e.g. ranaviruses, alveolates) or no practical treatment in the wild. Treatment of any disease varies by the pathogen involved as well as the host. Some pathogens are resistant to many treatments (e.g. antibiotic-resistant bacteria) and some hosts may be sensitive to a particular treatment (e.g. Methylene Blue may be toxic to tadpoles at concentrations over 2 mg/ml). Generally, it is best to contact a veterinarian with experience in amphibians for proper treatment of disease. However, some treatments (i.e. elevated temperature for *B. dendrobatidis* or dermosporidium, sea salt or Methylene Blue for *Saprolegnia*, chlorhexidine for bacteria and *Ranavirus*) may be attempted by the non-veterinarian and treatment guidelines can be found in Wright and Whitaker (2001) and Poole (2008). As a general rule, treatment for disease is only applicable to captive environments; however, it can be a valuable conservation tool for amphibians slated for release.

In the event that animals destined for release test positive for a treatable disease, the animal and any others that may have been exposed should be treated. Following treatment, a minimum of two negative test results with 1 month between tests should be obtained. If the animal does not test negative, the treatment should be repeated. Only animals that test negative should be released into the wild. In addition, if one animal in a group of 10 housed together tests positive for a pathogen, all of the animals should be treated, regardless of individual test results. Current guidelines for treatment and release have been established by the Association of Zoos and Aquariums (Poole 2008). Testing at the appropriate life stage for the host and disease agent is important.

26.5 Conclusions

Amphibians are declining globally and emerging infectious diseases are one of the causes. Natural resource agencies and conservation organizations should consider establishing amphibian disease surveillance programs that monitor populations for at least the two pathogens linked to catastrophic die-offs: *Ranavirus* and *B. dendrobatidis*. Further, the OIE has listed these pathogens as notifiable diseases, mandating that *Ranavirus-* and *B. dendrobatidis*-free status be verified prior to movement of amphibians for commerce. Herein, we have provided guidance on collection, storing, and shipping protocol of amphibians to diagnostic laboratories for disease testing. We encourage readers to use the Internet to locate a wildlife diagnostic laboratory in your area.

Given that pathogens can cause significant mortality that have trickle-down effects on ecosystem processes (Whiles *et al.* 2006), biologists must be prudent to decontaminate field equipment and footwear when moving among amphibian breeding sites. We also recommend that natural resource agencies consider implementing wildlife laws that prevent the use of amphibians as fishing bait. Transmission of *Ranavirus* in western North America has been attributed to the movement and sale of *A. tigrinum* larvae (Storfer *et al.* 2007; Picco and Collins 2008). We also encourage natural resource agencies to develop public educational brochures on the threat of amphibian diseases and the benefits of decontaminating recreational gear when leaving watersheds. Finally, prudent land stewardship undoubtedly reduces the likelihood of disease emergence by decreasing the effect of anthropogenic stressors. We encourage support of existing or development of new conservation programs that help landowners establish undisturbed buffers around amphibian breeding sites.

26.6 References

Baer, C. K. (2006). *Guidelines for Euthanasia of Nondomestic Animals*. American Association of Zoo Veterinarians, Yulee, FL.

Blaustein, A. R. and Wake, D. B. (1995). The puzzle of declining amphibian populations. *Scientific American*, **272**, 52–7.

Boone, M. D. and Bridges, C. M. (2003). Effects of carbaryl on green frog (*Rana clamitans*) tadpoles: timing of exposure versus multiple exposures. *Environmental Toxicology and Chemistry*, **22**, 2695–2702.

Brem, F., Mendelson, III, J. R., and Lips, K. R. (2007). *Field-Sampling Protocol for Batrachochytrium dendrobatidis from Living Amphibians, using Alcohol Preserved Swabs*, version 1.0 (18 July 2007). www.amphibianark.org/pdf/Field%20sampling% 20protocol%20for%20amphibian%20chytrid%20fungi%201.0.pdf. Conservation International, Arlington, VA.

Brunner, J. L., Schock, D. M., Davidson, E. W., and Collins, J. P. (2004). Intraspecific reservoirs: complex life history and the persistence of a lethal ranavirus. *Ecology*, **85**, 560–6.

Brunner, J. L., Richards, K., and Collins, J. P. (2005). Dose and host characteristics influence virulence of ranavirus infections. *Oecologia*, **144**, 399–406.

Brunner, J. L., Schock, D. M., and Collins, J. P. (2007). Transmission dynamics of the amphibian ranavirus *Ambystoma tigrinum* virus. *Diseases of Aquatic Organisms*, **77**, 87–95.

Bryan, L. K., Baldwin, C. A., Gray, M. J., and Miller, D. L. (2009). Efficacy of select disinfectants at inactivating *Ranavirus*. *Diseases of Aquatic Organisms*, **84**, 89–94.

Carey, C., Bradford, D. F., Brunner, J. L., Collins, J. P., Davidson, E. W., Longcore, J. E., Ouellet, M., Pessier, A. P., and Schock, D. M. (2003). Biotic factors in amphibian declines. In G. Linder, S. K. Krest, and D. W. Sparling (eds), *Amphibian Declines: an Integrated Analysis of Multiple StressorEffects*, pp. 153–208. Society of Environmental Toxicology and Chemistry, Pensacola, FL.

Cashins, S. D., Alford, R. A., and Skerratt, L. F. (2008). Lethal effect of latex, nitrile, and vinyl gloves on tadpoles. *Herpetological Review*, **39**, 298–301.

Converse, K. A. and Green, D. E. (2005a). Diseases of tadpoles. In S. K. Majumdar, J. Huffman, F. J. Brenner, and A. I. Panah (eds), *Wildlife Diseases: Landscape Epidemiology, Spatial Distribution, and Utilization of Remote Sensing Technology*, pp. 72–88. Pennsylvania Academy of Science, Easton, PA.

Converse, K. A. and Green, D. E. (2005b). Diseases of salamanders. In S. K. Majumdar, J. Huffman, F. J. Brenner, and A. I. Panah (eds), *Wildlife Diseases: Landscape Epidemiology, Spatial Distribution, and Utilization of Remote Sensing Technology*, pp. 118–30. Pennsylvania Academy of Science, Easton, PA.

Cunningham, A. A., Hyatt, A. D., Russell, P., and Bennett, P. M. (2007). Experimental transmission of a ranavirus disease of common toads (*Bufo bufo*) to common frogs (*Rana temporaria*). *Epidemiology and Infection*, **135**, 1213–16.

Daszak, P., Berger, L., Cunningham, A. A., Hyatt, A. D., Green, D. E., and Speare, R. (1999). Emerging infectious diseases and amphibian population declined. *Emerging Infectious Diseases*, **5**, 735–48.

Davidson, C., Benard, M. F., Shaffer, H. B., Parker, J. M., O'Leary, C., Conlon, J. M., and Rollins-Smith, L. A. (2007). Effects of chytrid and carbaryl exposure on survival, growth and skin peptide defenses in foothill yellow-legged frogs. *Environmental Science and Technology*, **41**, 1771–6.

Davis, A. K., Yabsley, M. J., Keel, M. K., and Maerz, J. C. (2007). Discovery of a novel alveolate pathogen affecting southern leopard frogs in Georgia: description of the disease and host effects. *EcoHealth*, **4**, 310–17.

Forson, D. D. and Storfer, A. (2006). Atrazine increases ranavirus susceptibility in the tiger salamander, *Ambystoma tigrinum*. *Ecological Applications*, **16**, 2325–32.

Gray, M. J., Miller, D. L., Schmutzer, A. C., and Baldwin, C. A. (2007a). Frog virus 3 prevalence in tadpole populations inhabiting cattle-access and non-access wetlands in Tennessee, U.S.A. *Diseases of Aquatic Organisms*, **77**, 97–103.

Gray, M. J., Smith, L. M., Miller, D. L., and Bursey, C. R. (2007b). Influence of agricultural land use on trematode occurrence in southern Great Plains amphibians, U.S.A. *Herpetological Conservation and Biology*, **2**, 23–8.

Gray, M. J., Rajeev, S., Miller, D. L., Schmutzer, A. C., Burton, E. C., Rogers, E. D., and Hickling, G. J. (2007c). Preliminary evidence that American bullfrogs (*Rana catesbeiana*) are suitable hosts for *Escherichia coli* O157:H7. *Applied and Environmental Microbiology*, **73**, 4066–8.

Green, D. E. and Converse, K. A. (2005a). Diseases of amphibian eggs. In S. K. Majumdar, J. Huffman, F. J. Brenner, and A. I. Panah (eds), *Wildlife Diseases: Landscape Epidemiology, Spatial Distribution, and Utilization of Remote Sensing Technology*, pp. 62–71. Pennsylvania Academy of Science, Easton, PA.

Green, D. E. and Converse, K. A. (2005b). Diseases of frogs & toads. In S. K. Majumdar, J. Huffman, F. J. Brenner, and A. I. Panah (eds), *Wildlife Diseases: Landscape Epidemiology, Spatial Distribution, and Utilization of Remote Sensing Technology,* pp. 89–117. Pennsylvania Academy of Science, Easton, PA.

Green, D. E., Converse, K. A., and Schrader, A. K. (2002). Epizootiology of sixty-four amphibian morbidity and mortality events in the USA, 1996–2001. *Annals of the New York Academy of Sciences,* **969**, 323–39.

Greer, A. L. and Collins, J. P. (2007). Sensitivity of a diagnostic test for amphibian ranavirus varies with sampling protocol. *Journal of Wildlife Diseases,* **43**, 525–32.

Griffiths, R. A. and Pavajeau, L. (2008). Captive breeding, reintroduction, and the conservation of amphibians. *Conservation Biology,* **22**, 852–61.

Gutleb, A. G., Bronkhorst, M., van den Berg, J. H. J., and Murk, A. J. (2001). Latex laboratory gloves-an unexpected pitfall in amphibian toxicology assays with tadpoles. *Environmental Toxicology and Pharmacology,* **10**, 119–21.

Harp, E. M. and Petranka, J. W. (2006). Ranavirus in wood frogs (*Rana sylvatica*): potential sources of transmission within and between ponds. *Journal of Wildlife Diseases,* **42**, 307–18.

Jancovich, J. K., Davidson, E. W., Morado, J. F., Jacobs, B. L., and Collins, J. P. (1997). Isolation of a lethal virus from the endangered tiger salamander *Ambystoma tigrinum stebbinsi. Diseases of Aquatic Organisms,* **31**, 161–7.

Jancovich, J. K., Davidson, E. W., Parameswaran, N., Mao, J., Chinchar, V. G., Collins, J. P., Jacobs, B. L., and Storfer, A. (2005). Evidence of emergence of an amphibian iridoviral disease because of human-enhanced spread. *Molecular Ecology,* **14**, 213–24.

Johnson, P. T. J. and Lunde, K. B. (2005). Parasite infection and limb malformation: a growing problem in amphibian conservation. In M. Lannoo (ed.), *Amphibian Declines: the Conservation Status of United States Species,* pp. 124–38, University of California Press, Berkeley, CA.

Kimmel, C. A. and Trammell, C. (1981). A rapid procedure for routine double staining of cartilage and bone in fetal and adult animals. *Stain Technology,* **56**, 271–3.

Kriger, K. M., Hines, H. B., Hyatt, A. D., Boyle, D. G., and Hero, J.-M. (2006). Techniques for detecting chytridiomycosis in wild frogs: comparing histology with real-time Taqman PCR. *Diseases of Aquatic Organisms,* **71**, 141–8.

Lavilla-Pitogo, C. R. and de la Pena, L. D. (2004). *Diseases in Farmed Mud Crabs Scylla spp.: Diagnosis, Prevention, and Control.* SEAFDEC Aquaculture Department, Iloilo, Philippines.

Lips, K. R., Brem, F., Brenes, R., Reeve, J. D., Alford, R. A., Voyles, J., Carey, C., Livo, L., Pessier, A. P., and Collins, J. P. (2006). Emerging infectious disease and the loss of biodiversity in a neotropical amphibian community. *Proceedings of the National Academy of Sciences USA,* **103**, 3165–70.

Longcore, J. E., Pessier, A. P., and Nichols, D. K. (1999). *Batrochochytrium dendrobatidis* gen.et sp. nov., a chytrid pathogenic to amphibians. *Mycologia,* **91**, 219–27.

Majji, S., LaPatra, S., Long, S. M., Sample, R., Bryan, L., Sinning, A., and Chinchar, V. G. (2006). *Rana catesbeiana* virus Z (RCV-Z): a novel pathogenic ranavirus. *Diseases of Aquatic Organisms,* **73**, 1–11.

Mauel, M. J., Miller, D. L., Frazier, K., and Hines, II, M. E. (2002). Bacterial pathogens isolated from cultured bullfrogs (*Rana catesbeiana*). *Journal of Veterinary Diagnostic Investigation*, **14**, 431–3.

Meteyer, C. U., Loeffler, I. K., Fallon, J. F., Converse, K. A., Green, E., Helgen, J. C., Kersten, S., Levey, R., Eaton-Poole, L., and Burkhart, J. G. (2000). Hind limb malformations in free-living northern leopard frogs (*Rana pipiens*) from Maine, Minnesota, and Vermont suggest multiple etiologies. *Teratology*, **62**, 151–71.

Miller, D. L., Bursey, C. R., Gray, M. J., and Smith, L. M. (2004). Metacercariae of *Clinostomum attenuatum* in *Ambystoma tigrinum mavortium*, *Bufo cognatus* and *Spea multiplicata* from west Texas. *Journal of Helminthology*, **78**, 373–6.

Miller, D. L., Rajeev, D., Brookins, M., Cook, J., Whittington, L., and Baldwin, C. A. (2008). Concurrent infection with *Ranavirus*, *Batrachochytrium dendrobatidis*, and *Aeromonas* in a captive anuran colony. *Journal of Zoo and Wildlife Medicine*, **39**, 445–9.

Miller, D. L., Gray, M. J., Rajeev, S., Schmutzer, A. C., Burton, E. C., Merrill, A. and Baldwin, C. (2009). Pathological findings in larval and juvenile anurans inhabiting farm ponds in Tennessee, U. S.A. *Journal of Wildlife Diseases*, **45**, 314–24.

Muths, E., Gallant, A. L., Campbell, E. H. C., Battaglin, W. A., Green, D. E., Staiger, J. S., Walls, S. C., Gunzburger, M. S., and Kearney, R. F. (2006). *The Amphibian Research and Monitoring Initiative (ARMI): 5-year report*. U.S. Geological Survey Scientific Investigations Report 5224. U.S. Geological Survey, Reston, VA.

Ouellet, M., Bonin, J., Rodrigue, J., DesGranges, J. L., and Lair, S. (1997). Hindlimb deformities (ectromelia, ectrodactyly) in free-living anurans from agricultural habitats. *Journal of Wildlife Diseases*, **33**, 95–104.

Pallister, J., Gould, A., Harrison, D., Hyatt, A., Jancovich, J., and Heine, H. (2007). Development of real-time PCR assays for the detection and differentiation of Australian and European ranaviruses. *Journal of Fish Diseases*, **30**, 427–38.

Picco, A. M. and Collins, J. P. (2008). Amphibian commerce as a likely source of pathogen pollution. *Conservation Biology*, **22**, 1582–9.

Poole, V. A. (2008). *Amphibian Husbandry Resource Guide*, edn 1.1. Association of Zoos and Aquarium's Amphibian Taxon Advisory Group, Yulee, FL.

Relyea, R. A. (2005). The lethal impact of Roundup on aquatic and terrestrial amphibians. *Ecological Applications*, **15**, 1118–24.

Relyea, R. A. (2009). A cocktail of contaminants: how mixtures of pesticides at low concentrations affect aquatic communities. *Oecologia*, **159**, 363–76.

Rostand. J. (1958). *Les Anomalies des Amphibiens Anoures*. Sedes, Paris.

Schock, D. M., Bollinger, T. K., Chinchar, V. G., Jancovich, J. K., and Collins, J. P. (2008). Experimental evidence that amphibian ranaviruses are multi-host pathogens. *Copeia*, **2008**, 133–43.

Schotthoefer, A. M., Koehler, A. V., Meteyer, C. U., and Cole, R. A. (2003). Influence of *Ribeiroia ondatrae* (Trematoda: Digenea) infection on limb development and survival of Northern leopard frogs (*Rana pipiens*): Effects of host-stage and parasite exposure level. *Canadian Journal of Zoology*, **81**, 1144–53.

Skerratt, L. F., Berger, L., Speare, R., Cashins, S., McDonald, K. R., Phillott, A. D., Hines, H. B., and Kenyon, N. (2007). Spread of chytridiomycosis has caused the rapid global decline and extinction of frogs. *EcoHealth*, **4**, 125–34.

Skerratt, L. F., Berger, L., Hines, H. B., McDonald, K. R., Mendez, D., and Speare, R. (2008). Survey protocol for detecting chytridiomycosis in all Australian frog populations. *Diseases of Aquatic Organisms* **80**, 85–94.

Storfer, A., Alfaro, M. E., Ridenhour, B. J., Jancovich, J. K., Mech, S. G., Parris, M. J., and Collins, J. P. (2007). Phylogenetic concordance analysis shows an emerging pathogen is novel and endemic. *Ecology Letters*, **10**, 1075–83.

Storrs, S. I. and Semlitsch, R. D. (2008). Variation in somatic and ovarian development: predicting susceptibility of amphibians to estrogenic contaminants. *General and Comparative Endocrinology*, **156**, 524–30.

USFWS and AFS-FHS (US Fish and Wildlife Service and American Fisheries Society – Fish Health Section) (2005). *AFS-FHS Blue Book: suggested procedures for the detection and identification of certain finfish and shellfish pathogens*, 2005 edition. American Fisheries Society – Fish Health Section, Bethesda, MD.

Wake, D. B. and Vredenburg, V. T. (2008). Are we in the midst of the sixth mass extinction? A view from the world of amphibians. *Proceedings of the National Academy of Sciences USA*, **105**, 11466–73.

Walker, S. F., Salas, M. D., Jenkins, D., Garner, T. W.J., Cunningham, A. A., Hyatt, A. D., Bosch, J., and Fisher, M. C. (2007). Environmental detection of *Batrachochytrium dendrobatidis* in a temperate climate. *Diseases of Aquatic Organisms*, **77**, 105–12.

Whiles, M. R., Lips, K. R., Pringle, C. M., Kilham, S. S., Bixby, R. J., Brenes, R., Connelly, S., Colon-Gaud, J. C., Hunte-Brown, M., Huryn, A. D., Montgomery, C., and Peterson, S. (2006). The effects of amphibian population declines on the structure and function of Neotropical stream ecosystems. *Frontiers in Ecology and the Environment*, **4**, 27–34.

Williams, T., Barbosa-Solomieu, V., and Chinchar, V. G. (2005). A decade of advances in iridovirus research. In K. Maramorosch and A. Shatkin (eds), *Advances in Virus Research*, vol. 65, pp. 173–248. Academic Press, New York.

Wright, K. M. and Whitaker, B. R. (2001). *Amphibian Medicine and Captive Husbandry*. Krieger Publishing, Malabar, FL.

Yip, M. J., Porter, J. L., Fyfe, J. A. M., Lavender, C. J., Portaels, F., Rhodes, M., Kator, H., Colorni, A., Jenkin, G. A., and Stinear, T. (2007). Evolution of *Mycobacterium ulcerans* and other mycolactone-producing mycobacteria from a common *Mycobacterium marinum* progenitor. *Journal of Bacteriology*, **189**, 2021–9.

Yuan, J. S., Reed, A., Chen, F., and Stewart, Jr, C. N. (2006). Statistical analysis of real-time PCR data. *BMC Bioinformatics*, **7**, 85–97.

27

Conservation and management

C. Kenneth Dodd, Jr

27.1 Introduction

Considering the large number of amphibian species and the diversity of life histories, potential conservation strategies are likewise diverse and may be complex. Beyond a few guiding principles, such as the need for planning and setting objectives, understanding life-history constraints, and protecting existing habitats and ecosystem function, there are no standardized approaches that will be applicable to all situations. In the brief discussion that follows, I outline some important considerations and various management approaches that have proved successful in particular instances, and I provide references that contain more extended discussions. Specific information on management options for amphibian conservation are available in Beebee (1996), Gent and Gibson (1998), Scoccianti (2001), Kingsbury and Gibson (2002), Semlitsch (2003), Bailey *et al.* (2006), and Mitchell *et al.* (2006).

Amphibian conservation requires an integrated landscape approach to management, rather than solely a species-oriented focus. The reason for this is simple: amphibians do not live in a biotic or physical vacuum in nature. Lindenmayer *et al.* (2007) present a number of conceptual ideas for landscape conservation that are apropos when focusing on amphibians; these apply equally to all management options (Table 27.1). Amphibian conservation options range from the rather simple and inexpensive to the very complex and expensive. When planning, the overriding consideration should be "first, do no harm" to the species, its community, or its habitat. High-technology-based approaches may work no better than simple and inexpensive approaches, and care should be exercised to maximize the benefits from the human resources and funds available. The primary objectives of management should always focus on the species or community of concern, and not on peripheral or extended objectives, such as positive publicity.

Table 27.1 *Conceptual ideas for landscape conservation that are important when planning for amphibian management. Adapted from Lindenmayer et al. (2007) and Gardner et al. (2007).*

Conceptual ideas important for amphibian management
Plan for habitat connectivity, particularly between aquatic and terrestrial habitats
Develop a long-term approach, both in management and in the evaluation of objectives
Manage all aspects of the habitat, not individual components, regarding landscape features, species, or stressors
Manage for change, not a static condition
Manage species and ecosystems using explicit experimental approaches and at multiple scales, rather than resorting to "cookbook" approaches

27.1.1 Statutory protection

The simplest approach to amphibian conservation is to legislate protection for species or habitats. In some cases, this can be extremely effective at preventing habitat destruction or alteration, adverse trade, prevention of take, or similar threats. Legislative approaches work best where there is a direct threat to amphibians or crucial habitats, and where the laws or regulations will be enforced. Measures designed to decrease or eliminate threats also may result from legislative protection, such as mandated mitigation requirements or directives to develop recovery plans. Active management and research must accompany legislative protection; simply putting a species on a list could lead to a false sense of security concerning its protection and, without enforcement, offers no long-term conservation benefits.

27.1.2 Protecting habitats

Coupled with legislative protection is the need for control over habitats where imperilled amphibian species or communities reside. Habitat protection can be achieved through outright purchase by governments or private organizations or, in some cases, by executive order transferring land from one level of controlled use to a more restrictive category (such as by changing land-zoning categories). Habitat purchase allows direct control and incorporation of land into parks, preserves, refuges, or wildlife management units that may allow for multiple use as long as it does not conflict with the conservation objectives outlined in the purchase plan.

In addition to direct purchase, lands may be protected via conservation easements or agreements, tax incentives, or by payment subsidies. Agreements can be flexible to ensure conservation yet allow landowners access and use of their lands. In all cases, however, management plans must ensure that long-term research and monitoring is carried out, and that land use occurs in accordance with initial agreements. All land acquisition and alternative conservation programs require early, careful, fiscal planning in order to ensure long-term continuity of protection, management and assessments of how well objectives are being achieved.

People residing adjacent to protected lands can be incorporated into long-term management programs to provide local or regional economic incentives to protection. Such an approach would be especially desirable where key amphibian habitats are located in proximity to economically disadvantaged human settlements. Providing educational programs, outlets for local arts, and even minimal health care could go a long way to ensure that crucial habitats are protected from vandalism or incursion. This approach works well in both developed and underdeveloped regions, for example, where local residents have been hired as caretakers. Residents thus come to have a stake in the protected area and in the species being protected.

27.2 Managing amphibian populations

I suggest that there are five cardinal rules when developing management programs for amphibians (Dodd *et al.* 2010).

- Understand amphibian life histories. Conservation and management can only proceed with knowledge of amphibian biology, including but not limited to habitat requirements, spatial use of habitats, reproduction, diet, predators, and population biology. Life-history and demographic variables form the biological constraints to management efforts.
- Address causes of decline. Species cannot be managed or recovered unless the causes of decline or need for conservation are understood and corrected. It makes no sense to place imperilled individuals in a continuing position of threat.
- Consider human-based constraints. Finances, personnel, logistics, time, regulations, local sensitivities and collaboration, and administrative policy must be factored into a management program prior to its initiation. The best of intentions will fail unless these human-based constraints are adequate to complete the program in accord with the biological requirements of the species.
- Have achievable biological objectives, plans, and measures of success. These factors should be clearly stated, and management efforts should

be directed at achieving specific goals. Management plans are guidelines, readily changed to accommodate new insights; they should not be altered to meet political objectives.

- Monitor conservation outcomes. Assessing the effectiveness of conservation research and management is essential to determining effectiveness. Monitoring should be considered as fundamental research designed to address specific questions, and must be long-term in nature.

27.3 Wetland breeding sites

In considering management options to conserve wetland-breeding amphibians, Semlitsch (2000a) suggested the following: (1) maintenance or restoration of temporary wetlands with a diverse array of hydroperiods, (2) protection of terrestrial buffer zones of natural vegetation to protect core breeding sites, (3) protection of amphibian communities by invasion of fish predators (primarily in reference to seasonal wetlands), (4) protection of the integrity of ecological connectivity across a landscape, (5) restriction of chemical use on site, but especially in ditches, streams, and wetlands, (6) prohibiting the release of captive or exotic specimens, and (7) identification and resolution of current management practices with those necessary for amphibian conservation.

In general, wetlands of all sizes serve as breeding sites or habitats for adult amphibians, from small tree holes to large lakes. A variety of regulations protect rivers, streams, and lakes in many parts of the world, but it is important to recognize the value of small, seasonal wetlands to amphibians (Semlitsch 2000b); these small isolated wetlands are usually overlooked and regulations rarely protect them. Any management plan on a landscape level must take them into account, and plans for their protection, restoration, or creation are vital to the survival of many amphibian species.

27.3.1 Wetland integrity and hydroperiod

Managing wetlands for amphibians involves the maintenance of breeding and dormancy sites (e.g. lake or pond bottoms, small streams) in conditions favorable for habitation and larval development. Conservation involves preventing physical disturbances, regulating the influx of pollutants, preventing the establishment of fish or non-indigenous predators and vegetation, and controlling vegetative succession where habitats are spatially restricted or limited in distribution.

Wetlands may be surrounded by plastic sheeting to minimize the temporary inflow of siltation and pollutants, such as during road construction, but in cases where disruption to adjacent habitats is permanent, walls or barriers must be built.

Walls can be constructed of pre-cast concrete or other materials (section 27.5.2). Barriers might incorporate passages to allow some animals to cross, or walls could be sloped in such a manner (/‾‾\) as to allow animals to crawl up and across the barrier. This design would block silt or contaminated inflow yet not serve as a barrier to migration.

Wetland ditching is counterproductive to amphibian management, even though some agricultural 'best management practices' advocate it under the guise of 'precision levelling' (Trauth *et al.* 2006). Maintaining hydroperiod (the length of time water is available within a wetland) is critically important. Many amphibians breed in temporary or seasonal wetlands free of fish predators. Periodic drying ensures that fish do not become established. Many processes disrupt the hydroperiod, such as ditching adjacent habitats or pumping underground water to supply municipal needs. In such cases, water can be diverted back to the wetland during the time necessary for amphibian development. Ground water may have different pH and chemical properties than surficial wetlands, however, so it may be necessary to experimentally test developmental rates and success prior to augmenting breeding ponds with ground water (Seigel *et al.* 2006).

Non-indigenous predators (other amphibians, fish, crayfish) are extremely detrimental to amphibians breeding in seasonal wetlands; they should never be intentionally released. Fish may be killed by draining water from temporary ponds during the non-breeding season, or by applying a piscicide such as rotenone. In streams, non-native fish have been removed by using a combination of rotenone and direct netting. However, some piscicides may be detrimental to larval amphibians, so care must be taken when choosing chemical applications. Never apply a chemical unless there is published data showing that it is safe. Otherwise, it will be necessary to conduct experimental evaluation prior to field use.

Invasive vertebrates such as Cuban treefrogs (*Osteopilus septentrionalis*), American bullfrogs (*Rana catesbeiana*), and wild hogs (*Sus* spp.) may be extremely detrimental to amphibians at breeding sites. Hogs and cattle (*Bos* spp.) can be excluded by fencing, but fencing may be counterproductive for bullfrogs inasmuch as native amphibians also need to enter or leave the wetland. Species with extended larval periods can be removed by draining ponds to kill larvae, but adult American bullfrogs must be shot or manually removed. Once removed, high and fine-meshed perimeter fences can be used to prevent immigration, since bullfrogs move over large distances, even in desert conditions. Crayfish also must be manually removed or trapped because of their burrowing ability.

The larvae of Cuban treefrogs have detrimental impacts on native tadpoles (Smith 2005), and adults consume or out-compete native species for retreat sites. PVC pipes can be used to artificially establish retreat sites and, once colonized, individuals can be removed and euthanized. Whether this technique will reduce populations is unknown, so such a removal plan may require extended effort. Although not exotic to their locale, manual removal of water snakes (*Natrix maura*) has proved effective in helping to recover and re-establish populations of *Alytes muletensis* on Mallorca (Román 2003).

Vegetation encroaching upon wetlands may seriously impact them by extracting additional water, changing the chemistry and pH, physically changing the habitat structure of calling and egg-deposition sites, and enclosing the canopy, thus altering thermal regimes. Invasive plants have detrimental effects on larval development, and the secondary compounds they produce may be toxic to some species (Maerz *et al.* 2005a, 2005b; Brown *et al.* 2006). Vegetation may be mechanically or selectively removed or thinned, but care should be taken when using herbicides inasmuch as they or their surfactants may have lethal or sublethal effects on developing larvae (e.g. Hayes 2004).

27.3.2 Wetland creation and restoration

Wetland creation has proven very successful in providing oviposition sites for many pond-breeding amphibians, for both rare and common species (Table 27.2; Anonymous undated; Aplin *et al.* 2000). Newly created breeding ponds are usually placed in areas near former breeding sites, the exact location dependent on existing canopy cover, land use, and vegetation. Created ponds should be located in proximity to non-breeding habitats, and dispersal corridors must be included to accommodate terrestrial foraging activity and winter/dry season retreats. Ponds may be dug by machine or by hand, with subsequent seeding of native vegetation around the shoreline in such a manner to accommodate the breeding habits of targeted species. For example, some species require shallow emergent vegetation for egg deposition, whereas others require arboreal and structured vegetation as calling sites. Zooplankton may be added to provide a food source for tadpoles.

Once dug, many created ponds will require a liner of clay, bentonite, foil, rubber or plastic to allow water retention. The ability to seal ponds may dictate the size of the pond or wetland to be created inasmuch as small wetlands are easier to line than large wetlands. Clay works well in some habitats, but often deteriorates through time, especially if the ponds are seasonal and the bottom is exposed to air for long periods. A "permanent" liner may not allow for seasonal water fluctuation, however, except through evaporation. Thus, many newly created ponds

Table 27.2 *Programs featuring breeding pond creation or restoration as part of amphibian habitat management.*

Location	Species	Type of constructed pond	Reference
New South Wales, Australia	10 frogs	Constructed farm ponds	Hazell et al. (2004)
Prince Edward Island, Canada	6 frogs	Dredged wetlands	Stevens et al. (2002)
3500 ponds, Denmark	7 frogs, 2 salamanders	Dredging existing ponds, dug ponds	Fog (1997)
Samsø, Denmark	Bufo viridis	Dredged ponds, vegetation removal	Amtkjaer (1995)
Lolland, Denmark	Hyla arborea	Dredging existing ponds, dug ponds	Hels and Fog (1995)
France	Pelobates fuscus	Use of bentonite and plastic-liners	Eggert and Guyetant (2002)
Muensterland, Germany	5 frogs, 3 salamanders	Dug and dredged ponds	Schwartze (2002)
Salzburg, Germany	3 frogs, 4 salamanders	Foil-lined permanent ponds	Kyek et al. (2007)
Arribes del Duero, Spain	4 frogs, 4 salamanders	Permanent dug ponds	Alarcos et al. (2003)
Gipuzkoa, Spain	Hyla meridionalis	Permanent dug ponds	Rubio and Etxezarreta (2003)
UK	Rana temporaria	Dug ponds, mostly seasonal	Williams (2005)
Arkansas, USA	Rana sylvatica	Permanent dug wildlife ponds	Cartwright et al. (1998)
Colorado, USA	Bufo boreas	Dug ponds, 1.5–6 m deep, little or no vegetation restoration	Pearl and Bowerman (2006)
Illinois, USA	11 frogs, 6 salamanders	Earthen dams in shallow valley, ponds permanent	Palis (2007)
Minnesota, USA	7 frogs, Ambystoma tigrinum	Filled ditches, reflooded basins	Lehtinen and Galatowitsch (2001)
North Carolina, USA	Rana sylvatica, Ambystoma maculatum	Seasonal and permanent ponds, modified golf course ponds, blocked stream channels	Petranka et al. (2003)
Ohio, USA	7 frogs, 2 salamanders	Complex ponds with shallow and deep sections, revegetation, zooplankton and some amphibians stocked	Weyrauch and Amon (2002)
South Carolina, USA	13 frogs, 4 salamanders	Plastic-lined permanent dug ponds	Pechmann et al. (2001)

are permanent and require vigilance to ensure that fish predators are not inadvertently (through sheet flow) or purposely introduced. Weyrauch and Amon (2002) provided an innovative U-shaped design for created ponds whereby one arm of the U is dug shallow and the other deep. Natural evaporation allows the water to fluctuate in hydroperiod between the arms, and provides amphibians with a choice of water depth.

Once ponds are dug, they are filled naturally by precipitation. Thus, ponds are best created prior to the season when most precipitation occurs, allowing pond chemistry and limnology to stabilize before the breeding season. Water may be added to ensure that at least some amount is available between construction and the onset of rainy periods. Most invertebrates and amphibians will naturally colonize newly created ponds, often very rapidly. However, some researchers have combined pond creation with stocking or relocation in order to speed-up the colonization process or to target certain declining species. As Petranka et al. (2003) noted, perturbations (disease, drought) sometimes hinder efforts to establish amphibians at created ponds, and monitoring needs to continue for approximately 5 years or more to evaluate success. The longer the pond is maintained, the more likely that additional species will colonize it, and that amphibian populations will become established.

Existing breeding ponds may be restored or enhanced, usually by mechanically deepening the pond to remove infill or muck and by removing encroaching vegetation. Trees, such as willows, are cut within the pond basin and along the pond's perimeter. Herbicides must be used with extreme care at amphibian breeding sites, since these or their surfactants may have detrimental effects on amphibians. In fire-dependent habitats, controlled burns through pond basins may help to periodically re-establish breeding ponds by burning off muck and vegetation. Typically, pond restoration is carried out during the non-breeding time of the year when individuals are dispersed. Biebighauser (2002) has prepared a free guidebook to creating and maintaining vernal ponds.

27.3.3 Core habitat and buffer zones

Key amphibian habitats include wetland breeding sites, retreat sites, dispersal corridors, and unique or limited habitats, such as caves, rock faces, steeply sided slopes, or areas where populations may be restricted in distribution, such as on mountain tops or in the spray zone of waterfalls. Each such habitat constitutes a critical area where the integrity of the landscape must be fully protected to ensure population survival. Although Semlitsch and Jensen (2001) advanced the idea of core habitats for wetland-breeding amphibians, the concept may be expanded to include these other unique habitats in terms of management.

Apart from the critical conservation area, zones of protection may extend outward depending on the extent to which amphibians travel away from the site. The critical area requires a buffer to ensure its integrity, and the "core" habitat to be protected becomes the critical area plus its associated buffer zones. For wetland-breeding amphibians, a core area for conservation would include the wetland site, plus both aquatic and terrestrial buffer zones that would ensure a functioning community (see Figure 27.1). For a terrestrial salamander habitat specialist, a core conservation area might include the immediate steep slopes or rock face on which it resides, plus terrestrial buffer zones to allow for complete canopy cover over the slope or rock face, allowing retention of shade and high humidity. Thus, the concept of a "buffer" zone should be expanded to include as much habitat as necessary to maintain functionality, rather than

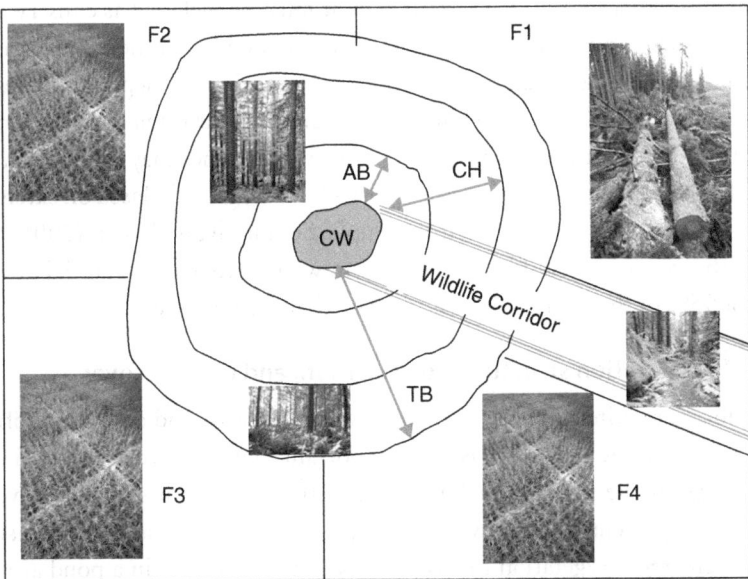

Fig. 27.1 Protection of a pond-breeding amphibian community at a landscape scale in a tree plantation. A core wetland (CW) is surrounded by three areas of protection, the aquatic buffer zone (AB), a core habitat (CH; AB+an area where most individuals disperse), and a terrestrial buffer zone (TB). The CW is linked to other wetlands via wildlife corridors. The surrounding tree farm is managed in rotations, from currently harvested (F1) to various-age stands (F2–F4). The level of human use varies from total protection in CW, to recreational use in CH, to light selective harvesting in TB, to rotational clear- or selective cutting (F1–F4). This configuration provides a mosaic of habitats that allows wetland-breeding amphibian to persist through time, yet permit some human recreational and commercial activity.

be synonymous with a minimum area necessary to protect the critical area from destruction or alteration.

Certainly core areas and associated buffer zones need to be managed essentially free from adverse effects on resident amphibians. However, some human uses may be allowed within a series of outermost concentric buffer zones such that layers of allowed uses surround the core conservation area. For example, a core protection area might extend 100 m from a wetland breeding site (a core area plus aquatic and terrestrial buffer zones) where no forestry would be allowed. However, very selective cutting could be allowed at distances of 100–150 m, with even less restrictions on the types of activity at further than 150 m (see Figure 27.1). Given that amphibians are not evenly distributed around a core site or present at all times of the year, developing an effective core plus buffer zone conservation and management area requires good data on spatial distribution.

A common question is, how much habitat must be included as conservation buffer zones for amphibians? As Semlitsch and Jensen (2001) noted, there is no one-size-fits-all value, and the size of buffer zones will vary depending on species and community. In temperate habitats, amphibians often routinely travel more than 500 m from breeding sites; thus, buffer zones may need to extend much farther than some previous authorities have suggested. Based on extensive life-history and movement data, for example, Semlitsch and Jensen (2001) suggested that a core plus buffer zone of 164 m would provide substantial protection for 95% of the salamander community that they studied.

27.3.4 Vegetation structure, composition, and canopy cover

Amphibians require a variety of vegetative structure around a wetland; often, the structure of the vegetation is more important than the species composition. Knowledge of the breeding habits of local species is the best guide to vegetation management; that is, whether amphibians need elevated calling sites, shallow emergent vegetation for cover, or woody debris within a pond around or on which to deposit eggs. In complex amphibian communities, managers should plan for a diversity of vegetative structure both within wetlands and in associated uplands. Canopy cover is particularly important as it affects thermal regimes. Many species do not breed in enclosed canopy, so tree and vegetation removal will be necessary as wetlands undergo succession.

27.3.5 Water quality

Each species of wetland breeding amphibian has its unique water-quality prerequisites for tadpole development. Part of successfully managing amphibians is knowing these requirements and monitoring them during the course of surveys

(Chapter 7). Acidity is particularly important, as it affects the amount of oxygen available to the developing embryo. As vegetation succession occurs, layers of peat and muck accumulate and may change the pH to non-favorable levels. Such muck can be mechanically removed to raise pH to pre-accumulation levels, or ponds may be limed (with $CaCO_3$ or $Ca(OH)_2$) to increase pH to circumneutral levels (Bellemakers and van Dam 1992; Beattie *et al.* 1993).

27.4 Terrestrial habitats

Terrestrial habitats are critically important for amphibians, although they are often overlooked. Managing amphibians terrestrially involves knowledge of species richness and movement patterns, especially seasonal migratory pathways and orientation during breeding seasons. Species that do not breed in water (e.g. many tropical frogs and plethodontid salamanders) or spend only part of their life cycle on land need moist or humid sites containing cover, surface debris, subsurface burrows, and retreats during cold or dry seasons. Activities which remove natural canopy cover, alter leaf litter, do not mimic natural disturbances (e.g. fires in non fire-evolved ecosystems), or disturb subsurface retreats (e.g. roller-chopping and disking) should be avoided.

27.4.1 Contiguous habitats and edge effects

Large contiguous habitats are ideal for conservation, whether they be rainforest, deciduous woodlands, or prairie grasslands. The more such habitats are fragmented, diminished, or shredded, the less desirable they are for maintaining amphibian biodiversity. Maximum effort should be placed on preserving such habitats. However, nearly all remnants of native habitats are likely to contain some amphibians, depending on location and spatial scale. Since edges expose amphibians to barriers for dispersal, predators, and altered biophysical conditions, landscape configurations designed to minimize edges and maintain interior areas far from edges have the best chances of protecting amphibians through time. A mosaic of breeding and retreat sites must be included within contiguous areas, and corridors of various widths, lengths, and vegetation structure should connect principle amphibian habitats.

27.4.2 Silviculture

One of the most contentious issues in amphibian conservation management is the effect of silvicultural practices on amphibians (deMaynadier and Hunter 1995). However, there are still important principles which should be kept in mind. Certainly, practices which completely destroy wetland breeding sites should be prevented, and buffer zones need to be implemented around streams and wetlands to avoid desiccation and changes in thermal regimes and water chemistry.

The most extreme form of forestry is clear-cutting all trees and vegetation. Many species of terrestrial amphibians, particularly salamanders and ground-dwelling tropical frogs, decline or disappear after clear-cutting, and surviving populations may take many decades to return to pre-cut abundance. If left to succession in the absence of mechanical roller chopping, however, even such severely degraded sites can recover if amphibian source populations are nearby and wetlands with buffers remain intact. However, these sites cannot recover if wetlands are destroyed, large areas are clear-cut, roller-chopping eliminates all subsurface retreats, such as root tunnels, and all surface debris is eliminated manually or by burning. Herbicides and fertilizers also may be detrimental to many species.

One solution to these problems is to manage for a mosaic of habitats. Some areas (including breeding sites and associated buffer zones) would be off limits to cutting, whereas other areas would be cut on a rotational basis, thus providing different aged stands at various locations in the landscape (see Figure 27.1). Patches would have to be sufficiently large to be economical yet retain the diversity of amphibian species and allow for inter-patch migration. If plantation forestry is involved, patches of native vegetation could be retained within the landscape mosaic. A second solution to forest management would be to employ selective cutting that allows for canopy cover, yet does not significantly affect surface and subsurface conditions.

27.4.3 Restoring degraded lands

Natural succession may lead to eventual restoration of degraded amphibian habitats, as long as source populations are within colonizing distance and the habitat has not been critically disturbed (e.g. wetlands, water table, subsurface retreats). However, some habitats will not recover without extensive renovation, such as mine sites, degraded or silted streams, or sites where vegetation has changed to the extent where specialized amphibians can no longer occur. In such cases, extensive and usually expensive vegetation and habitat restoration (including wetland creation, altering water chemistry, and building retreat sites) will be required. Habitat generalists have the best success in repopulating recovered terrestrial sites, but unaided recolonization and recovery, especially by terrestrial species, may take decades. Still, there have been successes. For example, endangered heath sites have been reclaimed in Britain for *Bufo calamita* by removing encroaching terrestrial scrub vegetation and intensively managing the pH of wetlands (Beebee 1996).

27.5 Migratory and dispersal routes

One of the primary goals of amphibian conservation management at the landscape scale is to ensure connectivity between breeding sites, foraging areas, and winter/dry season retreat sites. Amphibians are sometimes capable of long-distance movements, and roads (even unpaved), habitat edges, and open areas have profound effects on movement propensity by some species, but not others.

27.5.1 Corridors between habitat fragments

In forested habitats, the best way to ensure amphibian connectivity is to retain forested corridors between breeding sites and fragmented larger areas. Forested corridors should also be retained along the riparian zones of creeks, streams, and rivers; corridors should extend on both sides of the stream. The size of the corridor will be different in different habitats, but should allow a swath of at least 30–50 m on each side to reduce edge effects, particularly desiccation and exposure to sunlight.

When open areas surround fragmented forests, many amphibians will not cross the ecotonal boundary. However, patches of forest trees or dense shrubs can be planted at regular intervals between patches to serve as cover sites. Small created or restored wetlands interspersed along the corridor also facilitate movement between core fragments. In savannas or grasslands, heaps of rocks, woody debris (log piles), and created wetlands can also be used to facilitate movement between habitat patches. The objective is to create a mosaic of complex habitats along the way to provide amphibians a stepping-stone pattern across unfavorable or hostile habitats (Chan-McLeod and Moy 2007).

27.5.2 Crossing transportation corridors

One of the most formidable barriers to amphibians are the literally thousands of kilometres of roads, highways, rail lines, and other transportation corridors which fragment habitats. Inasmuch as many species migrate between breeding sites and foraging/retreat sites, crossing highways leads to mass mortality which affects population persistence. The effects of roads may extend hundreds of meters away from the actual roadway. As a result, an extensive literature has evolved on methods to allow amphibians and other species to cross.

The most effective way to assist amphibians across roads is to construct a barrier wall to prevent access, coupled with a series of underpasses, culverts, or tunnels to allow transit to the other side (Figure 27.2a–d). The walls serve to prevent trespass, while at the same time funnelling the migrating amphibians toward the underpass or culvert. Walls may be made of pre-cast concrete using

Fig. 27.2 (a) Overview of drift fence/culvert system around an amphibian breeding pond; Arcegno, Canton of Ticino, Switzerland. Photograph by P. Schlup, with permission. (b) Combining a tunnel with water to allow for a high-humidity microclimate during transit underneath a roadway; near Wiedlisbach, Canton of Berne, Switzerland. Photograph by S. Zumbach, with permission. (c) A recently constructed tunnel/barrier wall allowing water flow. Note the seamless connection between the tunnel and barriers; near Belp, Canton of Berne, Switzerland. Photograph by B. Lüscher, with permission. (d) Combined barrier wall and culvert system, Paynes Prairie State Preserve, Alachua County, FL, USA. The lip at the top of the wall discourages trespass by small treefrogs. However, the lack of vegetation maintenance severely reduces the effectiveness of the wall by allowing small vertebrates (frogs, snakes, mammals) to climb up and over the barrier lip. Photograph by C. Kenneth Dodd, Jr, October 2008.

a variety of lip (or overhang) designs to discourage climbing amphibians from crossing over the wall; slick, rather than rugose, surfaces also help to discourage climbing. The size of the culverts has varied among projects, but most provide for light and moisture access (to prevent animals from desiccating and provide an incentive for them to enter) and a moist dirt, mud or water substrate.

Culverts or tunnels may be built well under a raised road bed, or even level with a roadway surface with a grate forming the ceiling when raised road beds are not present; a permanent barrier wall need not be present. In such cases, drift fences constructed of highway cloth or sheet plastic can be used to funnel amphibians toward the underpass. Some amphibians will enter even narrow tunnels and successfully cross a highway, but others may require culverts of more than 1 m diameter. The length of the culvert may have important effects of the propensity of an amphibian to enter and cross successfully. For long culverts, as across more than four lanes, the underpass might have to be constructed in sections, and access to light and moisture become critical. Entrances should allow easy access in both directions, and brush or other woody debris may provide retreat sites and refuges from predators. Excellent discussions showing diagrams and photographs of various types of amphibian passages are provided in Langton (1989), ALASV (1994), Percsy (1995), Glandt *et al.* (2003), Brodziewska (2005), and in the Proceedings of the International Conferences on Ecology and Transportation (www.icoet.net/) (also see Figure 27.2a–d). Overpasses generally do not work as well for amphibians as they do for mammals.

27.6 Intensive manipulation of individuals

27.6.1 Captive breeding

Captive breeding of imperilled amphibians is a high-technology-based option of last resort; many programs are very costly, and few have been successful (Dodd 2005; Griffiths and Pavajeau 2008). Although many amphibians have high fecundity, short generation times, and can be kept in reasonable numbers, other species do not have these characteristics, especially those in decline, or they are very difficult to maintain and breed in captivity. Prior to undertaking captive breeding for conservation or reintroduction, a careful plan must be developed that takes into consideration housing, cleanliness, disease screening, specialized cover, diet and biophysical requirements, and the potential effects of long-term captivity on fitness. For example, Kraaijeveld-Smit *et al.* (2006) showed that predator escape behavior and heterozygosity of *Alytes muletensis* could be maintained for several generations in captivity, but that these traits start to deteriorate over time, thus potentially affecting long-term conservation programs. Conservation is best carried out *in situ* unless the threats become so dire that extinction is likely without intervention and removal to a secure location.

27.6.2 Relocation, repatriation, translocation (RRT)

Moving animals out of harm's way (relocation), into areas currently or previously occupied (repatriation), or even to areas not previously occupied (translocations) has been popular in wildlife management. In some cases, RRT programs have been undertaken without consideration of alternative options or with a firm understanding of the full requirements of RRT projects; they are often expensive and labor-intensive. As a result, many amphibian RRT programs have been unsuccessful, or the ultimate fate of the RRT is still unknown (Dodd and Seigel 1991; Dodd 2005). There have been some successful RRT programs (Román 2003; Rubio and Etxezarreta 2003; Bell *et al.* 2004; Kinne 2004), and such projects are still often advocated.

There are a number of criteria to consider, in addition to funding and staffing, prior to undertaking RRT projects (Dodd and Seigel 1991): (1) the causes of the original decline should be known, and steps must be taken to rectify them, (2) the biological constraints on conservation relating to life history, habitat, and demography should be understood, (3) population genetic and social structure should be evaluated, (4) individuals should be screened for disease, and (5) long-term habitat protection and monitoring should be a part of the RRT project. These criteria need to be considered for both the donor and recipient individuals and populations, where appropriate, especially when a resident population remains at or in the vicinity of a recipient site. To avoid problems of half-way technology, RRT should be considered only after other less manipulative options have been considered.

Various life stages have been used in RRT programs and to augment existing populations. Captive-reared animals may been used as alternative sources for all life stages, particularly juveniles and adults (Bloxam and Tonge 1995) for RRT, rather than removing individuals from existing natural populations. Most projects use larvae or eggs for RRT, as these are the easiest life-history stages to obtain and most abundant for mass movement. Presumably, larvae metamorphosing from a breeding site are more likely to remain in the vicinity than adults or subadults found or reared in one location but released in another. Adult mortality may be high in unfamiliar surroundings, and adults generally are scarcer or less available than eggs or larvae for RRT.

Most successful RRT programs employ a stage-based release protocol through time, rather than planning for a single release. This helps ensure that individuals become established, especially when presented with stochastic environmental conditions. Animals may have to be moved for years prior to successful establishment. Pre-movement activities could include population censuses, assessment and preparation of suitable habitat at recipient sites, disease and genetic screening, predator removal, and education.

27.6.3 Disease and biosecurity

Diseases and parasites, including bacteria (*Aeromonas, Citrobacter*), fungi (*Batrachochytrium dendrobatidis, Saprolegnia*), mesomycetozoa (*Amphibiocystidium, Ichthyophonus*), a newly described alveolate pathogen (Davis *et al.* 2007), and trematodes (*Ribeiroia ondatrae*) have been shown to seriously affect amphibian species and populations (see Chapter 26). In terms of management, biosecurity protocols aim to stop the spread of these diseases throughout a landscape and between continents.

Pathogens may be transported by wildlife, moved when hosts are used as bait, stocked with fish, sold as pets, deliberately released, inadvertently moved on vehicles, boats, and other forms of human transportation, used in medical tests or for educational and research purposes, or transported during catastrophic events, such as floods and tropical storms. Pathogens may move as organisms or propagules, and they can be transported in water, air, or on human and wildlife vectors.

In some instances, regulations can be established which prohibit the deliberate release of nonindigenous species or of individuals held in captivity. Shipments of stocked fish should be screened for amphibians, and both water and fish treated to ensure that diseases are not spread; such treatments are often already mandated when endangered fish species are released. Likewise, amphibians in the pet trade and food industry, particularly the American bullfrog (*R. catesbeiana*), should be screened for amphibian chytrid since they are vectors of this insidious disease. In cases where ponds are known to be infected by amphibian pathogens, public entry may be prohibited and attempts should be made to limit access by wildlife.

27.7 Conclusion

Preventing the decline of amphibians is challenging in the global economy of the twenty-first century, one that requires biologists to use a complex array of innovative and integrated approaches. The greatest threat to amphibians remains the loss or alteration of habitats, so primary emphasis should be placed on maintaining large tracts of undisturbed areas and to link already fragmented habitats. Conservationists must use tested and scientifically based options, and research must address questions in the field, laboratory, and through statistical modelling. Approaches need to focus on the parsimonious (what is most likely to work) rather than on the latest technological advancement, unless they are synonymous. Amphibian conservation is in triage mode; researchers are competing with many others for scarce resources. It is our responsibility to use these resources wisely for amphibians as we navigate through the Earth's sixth great extinction.

27.8 References

Alarcos, G., Ortiz, M. E., Lizana, M., Aragón, A., and Fernández Benéitez, M. J. (2003). La colonización de medios acuáticos por anfibios como herramienta para su conservación: el ejemplo de Arribes del Duero. *Munibe*, **16**, 114–27.

ALASV(ArbeitsgruppeunterLeitungderAbteilungStraßenbaudesVerkehrsministeriums) (1994). Amphibienschutz. Leitfaden für Schutzmaßnahmen an Straßen. *Schrifttenreihe der Straßenbauverwaltung Baden Württemberg, Heft,* **4**.

Amtkjaer, J. (1995). Increasing populations of the green toad (*Bufo viridis*) due to a pond project on the island of Samsø. *Memoranda Societatis pro Fauna et Flora Fennica,* **71**, 77–81.

Anonymous (undated). *Garden Ponds as Amphibian Sanctuaries.* British Herpetological Society, Montrose.

Aplin, K., Paino, A., and Sleep, L. (2000). *Building Frog Friendly Gardens.* Western Australian Museum, Perth.

Bailey, M. A., Holmes, J. N., Buhlmann, K. A., and Mitchell, J. C. (2006). *Habitat management guidelines for amphibians and reptiles of the southeastern United States.* Technical Publication HMG-2. Partners for Amphibian and Reptile Conservation, Montgomery, AL.

Beattie, R. C., Aston, R. J., and Milner, A. G. P. (1993). Embryonic and larval survival of the common frog (*Rana temporaria*) with particular reference to acidic and limed ponds. *Herpetological Journal,* **3**, 43–8.

Beebee, T. J. C. (1996). *Ecology and Conservation of Amphibians.* Chapman & Hall, London.

Bell, B. D., Pledger, S., and Dewhurst, P. L. (2004). The fate of a population of the endemic frog *Leiopelma pakeka* (Anura: Leiopelmatidae) translocated to restored habitat on Maud Island, New Zealand. *New Zealand Journal of Zoology,* **31**, 123–31.

Bellemakers, M. J. S. and van Dam, H. (1992). Improvement of breeding success of the moor frog (*Rana arvalis*) by liming of acid moorland pools and the consequences of liming for water chemistry and diatoms. *Environmental Pollution,* **78**, 165–71.

Biebighauser, T. R. (2002). *A Guide to Creating Vernal Ponds.* U.S. Department of Agriculture, Forest Service, Morehead, KY.

Bloxam, Q. M. C. and Tonge, S. (1995). Amphibians: suitable candidates for breeding-release programmes. *Biodiversity & Conservation,* **4**, 636–44.

Brodziewska, J. (2005). Wildlife tunnels and fauna bridges in Poland: past, present, and future, 1997–2013. *International Conference on Ecology and Transportation 2005 Proceedings,* pp. 448–60.

Brown, C. J., Blossey, B., Maerz, J. C., and Joule, S. J. (2006). Invasive plant and experimental venue affect tadpole performance. *Biological Invasions,* **8**, 327–38.

Cartwright, M. E., Trauth, S. E., and Wilhide, J. D. (1998). Wood frog (*Rana sylvatica*) use of wildlife ponds in northcentral Arkansas. *Journal of the Arkansas Academy of Science,* **52**, 32–4.

Chan-McLeod, A. C. A. and Moy, A. (2007). Evaluating residual tree patches as stepping stones and short-term refugia for red-legged frogs. *Journal of Wildlife Management,* **71**, 1836–44.

Davis, A. K., Yabsley, M. J., Keel, M. K., and Maerz, J. C. (2007). Discovery of a novel alveolate pathogen affecting southern leopard frogs in Georgia: description of the disease and host effects. *EcoHealth,* **4**, 310–17.

deMaynadier, P. G. and Hunter, Jr, M. L. (1995). The relationship between forest management and amphibian ecology: a review of the North American literature. *Environmental Review,* **3**, 230–61.

Dodd, Jr, C. K. (2005). Population manipulations. In M. Lannoo (ed.), *Amphibian Declines. The Conservation Status of United States Species*, pp. 265–70. University of California Press, Berkeley, CA.

Dodd, Jr, C. K. and Seigel, R. A. (1991). Relocation, repatriation, and translocation of amphibians and reptiles: are they conservation strategies that work? *Herpetologica*, **47**, 336–50.

Dodd, Jr, C. K., Loman, J., Cogalniceanu, D., and Puky, M. (2010). Monitoring amphibian populations, in press. In H. H. Heatwole and J. W. Wilkenson (eds), *Amphibian Biology*, vol. 9. *Conservation and Decline of Amphibians*. Surrey Beatty & Sons, Chipping Norton, NSW.

Eggert, C. and Guyetant, R. (2002). Safeguard of a spadefoot toad (*Pelobates fuscus*) population: a French experience. *Atti del terzo Convegno "Salvaguardia Anfibi", Lugano, 23–24 giugno 2000, Cogecstre Ediz. Penne*, **2002**, 47–52.

Fog, K. (1997). A survey of the results of pond projects for rare amphibians in Denmark. *Memoranda pro Societatis Fauna et Flora Fennica*, **73**, 91–100.

Gardner, T. A., Barlow, J., and Peres, C. A. (2007). Paradox, presumption and pitfalls in conservation biology: the importance of habitat change for amphibians and reptiles. *Biological Conservation*, **138**, 166–79.

Gent, T. and Gibson, S. (1998). *Herpetofauna Worker's Manual.* Joint Nature Conservation Committee, Peterborough.

Glandt, D., Schneeweiß, N., Geiger A., and Kronshage, A. (eds.) (2003). Beiträge zum Technischen Amphibienschutz. *Zeitschrift für Feldherpetologie*, Supplement 2.

Griffiths, R. A. and Pavajeau, L. (2008). Captive breeding, reintroduction, and the conservation of amphibians. *Conservation Biology*, **22**, 852–61.

Hayes, T. B. (2004). There is no denying this: defusing the confusion about atrazine. *BioScience*, **54**, 1138–49.

Hazell, D., Hero, J.-M., Lindenmayer, D., and Cunningham, R. (2004). A comparison of constructed and natural habitat for frog conservation in an Australian agricultural landscape. *Biological Conservation*, **119**, 61–71.

Hels, T. and Fog, K. (1995). Does it help to restore ponds? A case of the tree frog (*Hyla arborea*). *Memoranda pro Societatis Fauna et Flora Fennica*, **71**, 93–5.

Kingsbury, B. and Gibson, J. (2002). *Habitat management guidelines for amphibians and reptiles of the Midwest*. Technical Publication HMG-1. Partners for Amphibian and Reptile Conservation, Montgomery, AL.

Kinne, O. (2004). Successful re-introduction of the newts *Triturus cristatus* and *T. vulgaris*. *Endangered Species Research*, **4**, 1–16.

Kraaijeveld-Smit, F. J. L., Griffiths, R. A., Moore, R. D., and Beebee, T. J. C. (2006). Captive breeding and the fitness of reintroduced species: a test of the responses to predators in a threatened amphibian. *Journal of Applied Ecology*, **43**, 360–5.

Kyek, M., Maletzky, A., and Achleitner, S. (2007). Large scale translocation and habitat compensation of amphibian and reptile populations in the course of the redevelopment of a waste disposal site. *Zeitschrift für Feldherpetologie*, **14**, 175–90.

Langton, T. E. S. (ed.) (1989). *Amphibians and Roads. Proceedings of the Toad Tunnel Conference, Rendsburg, Federal Republic of Germany, 7–8 January 1989*. ACO Polymer Products, Shefford.

Lehtinen, R. M. and Galatowitsch, S. M. (2001). Colonization of restored wetlands by amphibians in Minnesota. *American Midland Naturalist*, **145**, 388–96.

Lindenmayer, D., Hobbs, R. J., Montague-Drake, R., Alexandra, J., Bennett, A., Burgman, M., Cale, P., Calhoun, A., Cramer, V., Cullen, P. *et al.* (2007). A checklist for ecological management of landscapes for conservation. *Ecology Letters*, **10**, 1–14.

Maerz, J. C., Blossey, B., and Nuzzo, V. (2005a). Green frogs show reduced foraging success in habitats invaded by Japanese knotweed. *Biodiversity & Conservation*, **14**, 2901–11.

Maerz, J. C., Brown, C. J., Chapin, C. T., and Blossey, B. (2005b). Can secondary compounds of an invasive plant affect larval amphibians? *Functional Ecology*, **19**, 970–5.

Mitchell, J. C., Breisch, A. R., and Buhlmann, K. A. (2006). *Habitat management guidelines for amphibians and reptiles of the northeastern United States*. Technical Publication HMG-3. Partners for Amphibian and Reptile Conservation, Montgomery, AL.

Palis, J. G. (2007). If you build it, they will come: herpetofaunal colonization of constructed wetlands and adjacent terrestrial habitat in the Cache River drainage of southern Illinois. *Transactions of the Illinois State Academy of Science*, **100**, 177–89.

Pearl, C. A. and Bowerman, J. (2006). Observations of rapid colonization of constructed ponds by western toads (*Bufo boreas*) in Oregon, USA. *Western North American Naturalist*, **66**, 397–401.

Pechmann, J. H. K., Estes, R. A., Scott, D. E., and Gibbons, J. W. (2001). Amphibian colonization and use of ponds created for trial mitigation of wetland loss. *Wetlands*, **21**, 93–111.

Percsy, C. (1995). *Les Batraciens sur nos Routes. Ministère de la Région Wallonne*. Brochure technique no. 1. Service de la Conservation de la Nature et des Espaces verts, Wallonne.

Petranka, J. W., Murray, S. S., and Kennedy, C. A. (2003). Responses of amphibians to restoration of a southern Appalachian wetland: perturbations confound post-restoration assessment. *Wetlands*, **23**, 278–90.

Román, A. (2003). El ferreret, la gestión de una especie en estado crítico. *Munibe*, **16**, 90–9.

Rubio, X. and Etxezarreta, J. (2003). Plan de reintroducción y seguimiento de la ranita meridional (*Hyla meridionalis*) en Mendizorrotz (Gipuzkoa, País Vasco) (1998–2003). *Munibe*, **16**, 160–77.

Schwartze, M. (2002). Neuanlage und Verbesserungen von Kleingewässern für den Laubfrosch und andere Amphibien – eine Untersuchung im östlichen Münsterland (NRW). *Zeitschrift für Feldherpetologie*, **9**, 61–73.

Scoccianti, C. (2001). *Amphibia: aspetti di ecologia della conservazione/Amphibia: Aspects of Conservation Ecology*. WWF Italia, Sezione Toscana, Editore Guido Persichino Grafica, Florence.

Seigel, R. A., Dinsmore, A., and Richter, S. C. (2006). Using well water to increase hydroperiod as a management option for pond-breeding amphibians. *Wildlife Society Bulletin*, **34**, 1022–7.

Semlitsch, R. D. (2000a). Principles for management of aquatic-breeding amphibians. *Journal of Wildlife Management*, **64**, 615–31.

Semlitsch, R. D. (2000b). Size does matter: the value of small isolated wetlands. *National Wetlands Newsletter*, **22**, 5–6, 13.

Semlitsch, R. D. (ed.) (2003). *Amphibian Conservation*. Smithsonian Books, Washington DC.

Semlitsch, R. D. and Jensen, J. B. (2001). Core habitat, not buffer zone. *National Wetlands Newsletter*, **23**, 5–6, 11.

Smith, K. G. (2005). Effects of nonindigenous tadpoles on native tadpoles in Florida: evidence of competition. *Biological Conservation*, **123**, 433–41.

Stevens, C. E., Diamond, A. W., and Gabor, T. S. (2002). Anuran call surveys on small wetlands in Prince Edward Island, Canada restored by dredging of sediments. *Wetlands*, **22**, 90–9.

Trauth, J. B., Trauth, S. E., and Johnson, R. L. (2006). Best management practices and drought combine to silence the Illinois Chorus Frog in Arkansas. *Wildlife Society Bulletin*, **34**, 514–18.

Weyrauch, S. L. and Amon, J. P. (2002). Relocation of amphibians to created seasonal ponds in southwestern Ohio. *Ecological Restoration*, **20**, 31–6.

Williams, L. R. (2005). Restoration of ponds in a landscape and changes in common frog (*Rana temporaria*) populations, 1983–2005. *Herpetological Bulletin*, **94**, 22–9.

Index

Figures and tables are indexed in bold. See under

The manufacturer's authorised representative in the EU for product safety is
Oxford University Press España S.A. of el Parque Empresarial San Fernando
de Henares, Avenida de Castilla, 2 – 28830 Madrid (www.oup.es/en).

Printed in the USA/Agawam, MA
December 9, 2024

878519.004